EXERCISES IN ALGEBRA

ALGEBRA, LOGIC AND APPLICATIONS

A series edited by

R. Göbel
Universität Gesamthochschule, Essen, Germany
A. Macintyre
The Mathematical Institute, University of Oxford, UK

EXERCISES IN ALGEBRA

A Collection of Exercises in Algebra,
Linear Algebra and Geometry

Expanded English edition

Edited by

Alexei I. Kostrikin
Moscow State University, Russia

Translated by

V. A. Artamonov

CRC Press
Taylor & Francis Group
Boca Raton London New York

CRC Press is an imprint of the
Taylor & Francis Group, an **informa** business

First published 1996 by Gordon and Breach Science Publishers

Published 2021 by CRC Press
Taylor & Francis Group
6000 Broken Sound Parkway NW, Suite 300
Boca Raton, FL 33487-2742

ISBN 13: 978-2-88449-029-0 (hbk)
ISBN 13: 978-2-88449-030-6 (pbk)

Visit the Taylor & Francis Web site at
http://www.taylorandfrancis.com

and the CRC Press Web site at
http//www.crcpress.com

British Library Cataloguing in Publication Data

Exercises in Algebra: Collection of
Exercises in Algebra, Linear Algebra and
Geometry. – 2 Rev. ed. – (Algebra, Logic &
Applications Series, ISSN 1041–5394; Vol. 6)
   I. Kostrikin, A. I.   II. Artamonov, V. A.
   III. Series
   512.0076

Contents

Preface to the Russian edition

Well presented collections of exercises for courses in higher algebra and linear algebra and geometry already exist (see, for example [5, 6]), and the publication of a new textbook of a similar nature needs to be justified. Changes in the above-mentioned courses at Moscow State University meant that new topics had been introduced, while the treatment of some other traditional topics had been shortened or even dropped altogether. As a result, university teachers had to use a large number of different books, of varying levels, when conducting seminars. In order to improve this situation, the professors of Higher Algebra at Moscow State University decided to prepare a new collection of exercises which would correspond to the modernized three-semester course.

The work assumed a collective nature from the very beginning. The author responsible for each chapter decided on the selection and amount of material, using criteria based on personal experience, whilst always attempting to moderate the quantity involved. In effect this approach has meant a lack of some standard numerical examples, and emphasis on the most noteworthy features. Thus the book contains the type of exercises actually offered in seminars. Almost all the sections contain exercises of a higher level of difficulty, for all of which hints are supplied. These exercises form the smaller part of each section, especially in the sections intended for the first semester. However their role increases in importance towards the end of the courses. The most difficult exercises may be presented and discussed in additional algebra seminars.

Exercises in Algebra was preceded by three rotaprint publications: *Basic Algebra, Linear Algebra and Geometry* and *Additional Chapters in Algebra* and the three parts of this book have similar contents to these. This arrangement of material follows the traditional structure of the lectures at the Department of Mechanics and Mathematics of the Moscow State University, and has been accepted in new curriculum planning at most universities in Russia. Of course the actual content of lectures and the order of exposition of material depends on the individual lecturer. Therefore it must be possible for textbooks and exercises to be used in a sufficiently flexible way. In any case, the authors have consciously allowed for some parallel and repeated material in different sections.

Theoretical comment is reduced to a minimum. However the material is arranged so that independent use of the book becomes increasingly important, especially in the final part. The theoretical basis of Parts One and Three can be found in [1] and that for Part Two in [2]. A significant number of exercises in this book have been taken from the collections of exercises mentioned in the references.

Lists of the definitions and symbols used, which may be helpful to the reader, are given at the end of the book. Any definition not listed will be found in the section 'Theoretical Material'. The latter contains the basic statements which are necessary for the completion of the exercises.

The authors wish to thank V.V. Batyrev, who improved the text. They are especially grateful to the professors of Higher Algebra and Number Theory at St. Petersburg University and to the professors of Algebra and Mathematical Logic at Kiev University, who have carefully reviewed the book and made many concrete suggestions.

The authors are grateful to Professor G.V. Dorofeev, who has given considerable attention to the principles used in ordering the content of the book and in standardizing the symbols, thus removing parallel material, as mentioned above, where it was excessive.

Postgraduate students have helped in checking the solutions and answers to the exercises. Nevertheless it is possible that a number of errors, proportional to the number of authors, remain, and we would be obliged if readers would give us their comments. We hope to be able to take into account diverse points of view and further improve the text.

References

1. Kostrikin, A.I., *Introduction to algebra*. Berlin: Springer-Verlag, 1982.

2. Kostrikin, A.I. and Manin, Y.I., *Linear algebra and geometry*. Reading, UK: Gordon and Breach, 1989.

3. Kurosh, A.G., *A course in higher algebra [Russian]*. Moscow: Nauka, 1971.

4. Skornyakov, L.A., *Elements of algebra [Russian]*. Moscow: Nauka, 1980.

5. Faddeev, D.K. and Sominsky, I.S., *Collection of exercises in higher algebra [Russian]*. Moscow: Nauka, 1974.

6. Proskuryakov, I.V., *Collection of exercises in linear algebra [Russian]*. Moscow: Nauka, 1974.

7. Ikramov, H.D., *Exercises in linear algebra [Russian]*. Moscow: Nauka, 1975.

8. Horn, R. and Johnson, C.R., *Matrix analysis*. London: Cambridge University Press, 1986.

9. Barbeau, E.J., *Polynomials*. New York: Springer-Verlag, 1989.

10. Latyshev, V.N., *Convex polytopes and linear programming [Russian]*. Ulyanovsk: Moscow State University Publications, Ulyanovsk branch, 1992.

Foreword to the English edition

The main purpose behind the publication of *Exercises in Algebra*, as mentioned in the Preface, seems to have been achieved, judging by the comments received and by the experience of the authors — V.A. Artamonov, Y.A. Bahturin, E.S. Golod, V.A. Iskovskih, V.N. Latyshev, A.V. Mikhalev, A.P. Mishina, A.Y. Olshansky, A.A. Panchishkin, I.V. Proskuryakov, A.N. Rudakov, L.A. Skornyakov, A.L. Shmelkin, E.B. Vinberg — at Moscow State University. At the same time, some considerable defects were revealed, namely an insufficient number of computational exercises; a lack of series of typical exercises; inconvenient numbering of exercises and answers, and a too close juxtaposition of exercises of differing levels of difficulty.

The present edition, prepared mainly by V.A. Artamonov, and with the participation of virtually the same authors as the previous one, is aimed at removing these defects. The size of the book has been considerably increased, and not only because standard exercises have been added: special exercises, some of them fairly difficult, have also been included. These exercises, partially extracted from journals and monographs, can satisfy the demands of outstanding students and can help in the choice of topics for future research. They are located at the ends of some sections, after the symbol * * *.

The authors wish to thank E.V. Pankratiev and M.V. Kondratieva for preparing the camera-ready copy for the book.

We hope that the continuous numbering system will considerably simplify the use of the book. We look forward to a positive reaction from readers, and will gratefully consider all suggestions for elimination of any discrepancies which may have slipped in.

The authors are grateful to the Gordon and Breach Publishing Group for their willingness to publish *Exercises in Algebra* in the same series as *Linear Algebra and Geometry*. The education of mathematics students is usually based not only on lectures but also on seminars where students have the opportunity to discuss and solve exercises. This principle seems to apply in both the East and the West. It should be mentioned, as has been noted before, that: 'every textbook is written taking into account the traditions in a given university or, more generally, in the universities of a given country. My algebra textbook is no exception. At the same time, the exchange of ideas in the area of mathematics teaching in different countries is no less important than the exchange of ideas in research'[1].

This book was typeset using the AMS-TEX macro software package.

A.I. Kostrikin

Reference

1.   Kostrikin, A.I., *Introduction to algebra*. Berlin: Springer-Verlag, 1982.

PART ONE
FOUNDATIONS OF ALGEBRA

CHAPTER 1

Sets and maps

1 Operations on subsets. Calculation of the number of elements

101. Let A_i $(i \in I)$, B be subsets of X. Prove that:

a) $\left(\bigcup_{i \in I} A_i \right) \cap B = \bigcup_{i \in I} (A_i \cap B);$

b) $\left(\bigcap_{i \in I} A_i \right) \cup B = \bigcap_{i \in I} (A_i \cup B);$

c) $\overline{\bigcup_{i \in I} A_i} = \bigcap_{i \in I} \overline{A_i};$

d) $\overline{\bigcap_{i \in I} A_i} = \bigcup_{i \in I} \overline{A_i}.$

102. Let X be an arbitrary set, and 2^X be the set of all its subsets. Prove that the operation $\triangle$ of *symmetrical difference*

$$A \triangle B = (A \cap \bar{B}) \cup (\bar{A} \cap B)$$

on the set 2^X has the following properties:

a) $A \triangle B = B \triangle A;$

b) $(A \triangle B) \triangle C = A \triangle (B \triangle C);$

c) $A \triangle \emptyset = A;$

d) for any subset $A \subset X$ there exists a subset $B \subset X$ such that $A \triangle B = \emptyset;$

e) $(A \triangle B) \cap C = (A \cap B) \triangle (A \cap C);$

3

f) $A \triangle B = (A \cup B) \setminus (A \cap B)$;

g) $A \triangle B = (A \setminus B) \cup (B \setminus A)$.

103. For finite sets $A_1, \ldots, A_n$ prove that

$$\left| \bigcup_{i=1}^{n} A_i \right| = \sum_{i=1}^{n} |A_i| - \sum_{1 \le i < j \le n} |A_i \cap A_j| + \ldots$$

$$+ (-1)^{k-1} \sum_{1 \le i_1 < \cdots < i_k \le n} |A_{i_1} \cap \cdots \cap A_{i_k}| + \cdots + (-1)^{n-1} |A_1 \cap \cdots \cap A_n|.$$

104. Prove that for any integer $n > 1$

$$\phi(n) = n \left(1 - \frac{1}{p_1} \right) \left(1 - \frac{1}{p_2} \right) \cdots \left(1 - \frac{1}{p_r} \right)$$

where $p_1, p_2, \ldots, p_r$ are all different prime divisors of n, and $\phi(n)$ is the *Euler function*.

105. What is the maximal number of subsets which can be obtained starting from the given n subsets of a fixed set via the operations of intersection, union and complement?

106. Let A, B, C be subsets of a set. Prove that $A \cap B \subseteq C$ if and only if $A \subseteq \overline{B} \cup C$.

2 Calculation of the number of maps and of the number of subsets. Binomial coefficients

201. Let X be a set of people in a room, Y be a set of chairs in this room. Suppose that

a) each chair is associated with a person who is sitting on it;

b) each person is associated with a chair on which he is sitting.

In what cases do a) and b) define a map $X \to Y$ and $Y \to X$? In which cases are these maps injective, surjective or bijective?

202. Prove that for an infinite set X and a finite subset Y there exists a bijective map $X \setminus Y \to X$.

203. Let $f : X \to Y$ be a map. The map $g : Y \to X$ is *left (right) inverse* for f, if $g \circ f = 1|_X$ ($f \circ g = 1_Y$, respectively). Prove that

a) a map f is injective if and only if it has left inverse;

b) the map f is surjective if and only if it has right inverse.

204. Establish a bijective correspondence between the family of all mappings from a set X into the set $\{0, 1\}$ and a set 2^X (see 102). Calculate $|2^X|$, if $|X| = n$.

205. Let $|X| = m$, $|Y| = n$. Find the number of all

a) maps

b) injective maps

c) bijective maps

d) surjective maps

from the set X into the set Y.

206. Let $|X| = n$. Find the number $\binom{n}{m}$ of all subsets of X of cardinality m.

207. Let $|X| = n$. Find the number of all subsets of X of even cardinalities.

208. Prove the *binomial formula of Newton:*

$$(a + b)^n = \sum_{i=0}^{n} \binom{n}{i} a^i b^{n-i} \qquad (n \in \mathbb{N}).$$

209. Let $|X| = n$ and $m_1 + \cdots + m_k = n$ ($m_i \geq 0$). Find the number $\binom{n}{m_1, \ldots, m_k}$ of ordered partitions of X into k subsets containing respectively $m_1, \ldots, m_k$ elements.

210. Prove that:

a) $$(x_1 + \cdots + x_k)^n = \sum_{\substack{(m_1,\ldots,m_k) \\ m_1+\cdots+m_k=n,\ m_i \geq 0}} \binom{n}{m_1, \ldots, m_k} x_1^{m_1} \ldots x_k^{m_k};$$

b) $$\sum_{\substack{(m_1,\ldots,m_k) \\ m_1+\cdots+m_k=n,\ m_i \geq 0}} \binom{n}{m_1, \ldots, m_k} = k^n.$$

211. Prove that:

a) $$\binom{n}{m} = \binom{n}{n-m};$$

b) $$\sum_{i=0}^{n} \binom{n}{i} = 2^n;$$

c) $$\sum_{i=0}^{n} (-1)^i \binom{n}{i} = 0;$$

d) $\displaystyle\sum_{i=1}^{n} i\binom{n}{i} = n2^{n-1}$;

e) $\displaystyle\sum_{i=1}^{n} (-1)^i i\binom{n}{i} = 0, \ (n > 1)$;

f) $\displaystyle\sum_{i=0}^{m} \binom{p}{i}\binom{q}{m-i} = \binom{p+q}{m}$;

g) $\displaystyle\binom{n}{k-1} + \binom{n}{k} = \binom{n+1}{k}$, $\ 1 \le k \le n$;

h) $\displaystyle\sum_{i=1}^{r} \binom{r+1}{i}(1^i + 2^i + \cdots + n^i) = (n+1)^{r+1} - (n+1)$;

i) $\displaystyle\binom{k}{k} + \binom{k+1}{k} + \cdots + \binom{n+k}{k} = \binom{n+k+1}{k+1}$;

j) $\displaystyle\sum_{i=0}^{n} \frac{p(p+1)\ldots(p+i-1)}{i!} = \frac{(p+1)\ldots(p+n)}{n!}$;

k) $\displaystyle\sum_{i=k}^{n-l} \binom{i}{k}\binom{n-i}{l} = \binom{n+1}{k+l+1}$, $\ $ where $n \geqslant k+l \geq 0$.

212. Prove that $x^m + x^{-m}$ is a polynomial of degree m in $x + x^{-1}$.

213. Find the number of partitions of a number n into an ordered sum of k non-negative integers.

3   Permutations

301. Multiply the permutations in the indicated and in the inverse order:

a) $\begin{pmatrix} 1 & 2 & 3 & 4 & 5 \\ 3 & 4 & 1 & 5 & 2 \end{pmatrix} \times \begin{pmatrix} 1 & 2 & 3 & 4 & 5 \\ 5 & 3 & 1 & 2 & 4 \end{pmatrix}$;

b) $\begin{pmatrix} 1 & 2 & 3 & 4 & 5 & 6 \\ 3 & 6 & 4 & 5 & 2 & 1 \end{pmatrix} \times \begin{pmatrix} 1 & 2 & 3 & 4 & 5 & 6 \\ 2 & 4 & 1 & 5 & 6 & 3 \end{pmatrix}$;

c) $\begin{pmatrix} 1 & 2 & 3 & 4 & 5 \\ 2 & 1 & 3 & 5 & 4 \end{pmatrix} \times \begin{pmatrix} 1 & 2 & 3 & 4 & 5 \\ 4 & 5 & 3 & 2 & 1 \end{pmatrix}$;

d) $\begin{pmatrix} 1 & 2 & 3 & 4 & 5 & 6 \\ 3 & 5 & 1 & 6 & 2 & 4 \end{pmatrix} \times \begin{pmatrix} 1 & 2 & 3 & 4 & 5 & 6 \\ 6 & 3 & 4 & 2 & 1 & 5 \end{pmatrix}.$

302. Decompose the permutation into a product of disjoint cycles:

a) $\begin{pmatrix} 1 & 2 & 3 & 4 & 5 & 6 & 7 \\ 5 & 4 & 1 & 7 & 3 & 6 & 2 \end{pmatrix};$

b) $\begin{pmatrix} 1 & 2 & 3 & 4 & 5 & 6 & 7 \\ 3 & 1 & 6 & 7 & 5 & 2 & 4 \end{pmatrix};$

c) $\begin{pmatrix} 1 & 2 & 3 & 4 & 5 & 6 & 7 \\ 3 & 7 & 6 & 5 & 1 & 2 & 4 \end{pmatrix};$

d) $\begin{pmatrix} 1 & 2 & 3 & 4 & 5 & 6 & 7 \\ 4 & 3 & 6 & 7 & 1 & 5 & 2 \end{pmatrix};$

e) $\begin{pmatrix} 1 & 2 & 3 & 4 & \dots & 2n-1 & 2n \\ 2 & 1 & 4 & 3 & \dots & 2n & 2n-1 \end{pmatrix};$

f) $\begin{pmatrix} 1 & 2 & \dots & n & n+1 & n+2 & \dots & 2n \\ n+1 & n+2 & \dots & 2n & 1 & 2 & \dots & n \end{pmatrix}.$

303. Write down the permutations in the standard form:

a) $(136)(247)(5)$;

b) (1654237);

c) $(135\dots2n-1)(246\dots2n)$.

304. Multiply the permutations:

a) $[(135)(2467)] \cdot [(147)(2356)]$;

b) $[(13)(57)(246)] \cdot [(135)(24)(67)]$.

305. Find the number of inversions in the sequences

a) $2, 3, 5, 4, 1$;

b) $6, 3, 1, 2, 5, 4$;

c) $1, 9, 6, 3, 2, 5, 4, 7, 8$;

d) $7, 5, 6, 4, 1, 3, 2$;

e) $1, 3, 5, 7, \dots, 2n-1, 2, 4, 6, 8, \dots, 2n$;

f) $2, 4, 6, \dots, 2n, 1, 3, 5, \dots, 2n-1$;

g) $k, k+1, \dots, n, 1, 2, \dots, k-1$;

h) $k, k+1, \ldots, n, k-1, k-2, \ldots, 2, 1$.

306. Determine the parity of the permutations:

a) $\begin{pmatrix} 1 & 2 & 3 & 4 & 5 & 6 & 7 \\ 5 & 6 & 4 & 7 & 2 & 1 & 3 \end{pmatrix}$;

b) $\begin{pmatrix} 1 & 2 & 3 & 4 & 5 & 6 & 7 & 8 \\ 3 & 5 & 2 & 1 & 6 & 4 & 8 & 7 \end{pmatrix}$;

c) $\begin{pmatrix} 3 & 5 & 6 & 4 & 2 & 1 & 7 \\ 2 & 4 & 1 & 7 & 6 & 5 & 3 \end{pmatrix}$;

d) $\begin{pmatrix} 2 & 7 & 5 & 4 & 8 & 3 & 6 & 1 \\ 3 & 5 & 8 & 7 & 2 & 6 & 1 & 4 \end{pmatrix}$;

e) $\begin{pmatrix} 1 & 2 & 3 & \cdots & \cdot & \cdot & \cdot & \cdots & n-1 & n \\ 2 & 4 & 6 & \cdots & 1 & 3 & 5 & \cdots & \cdot & \cdot \end{pmatrix}$;

f) $\begin{pmatrix} 1 & 2 & 3 & \cdots & \cdot & \cdot & \cdot & \cdots & n \\ 1 & 3 & 5 & \cdots & 2 & 4 & 6 & \cdots & \cdot \end{pmatrix}$;

g) $\begin{pmatrix} 1 & 2 & 3 & \cdots & n-1 & n \\ n & n-1 & n-2 & \cdots & 2 & 1 \end{pmatrix}$;

h) $\begin{pmatrix} 1 & 2 & 3 & 4 & \cdots & n-1 & n \\ n & 1 & n-1 & 2 & \cdots & \cdots & \cdots \end{pmatrix}$.

307. Determine the parity of the permutations:

a) $(123 \ldots k)$;

b) $(i_1 i_2 \ldots i_k)$;

c) $(1473)(67248)(32)$;

d) $(i_1 i_2)(i_3 i_4)(i_5 i_6) \ldots (i_{2k-1} i_{2k})$;

e) $(i_1 \ldots i_p)(j_1 \ldots j_q)(k_1 \ldots k_r)(l_1 \ldots l_s)$.

308. Let the number of inversions in the second row of the permutation

$$\begin{pmatrix} 1 & 2 & \cdots & n \\ a_1 & a_2 & \cdots & a_n \end{pmatrix}$$

be equal to k. Find the number of inversions in the second row of the permutation

$$\begin{pmatrix} 1 & 2 & \cdots & n \\ a_n & a_{n-1} & \cdots & a_1 \end{pmatrix}.$$

309. Consider the permutations $\begin{pmatrix} 1 & 2 & \cdots & n \\ a_1 & a_2 & \cdots & a_n \end{pmatrix}$ of degree n.

a) In which row $(a_1, \ldots, a_n)$ is the number of inversions the greatest?

b) How many inversions has the index 1, situated in the kth place in the second row?

c) How many inversions has the index n, situated in the kth place in the second row?

310. Interchange the indices q and $q + 1$ in a sequence $a_1, \ldots, a_n$ of indices $1, 2, \ldots, n$, where $1 \le q \le n - 1$. Prove that the number of inversions is changed by ± 1.

311. Suppose that the number of inversions in the second row of the permutation

$$\sigma = \begin{pmatrix} 1 & 2 & \cdots & n \\ a_1 & a_2 & \cdots & a_n \end{pmatrix}$$

is equal to k. Prove that:

a) σ is a product of k adjacent transpositions $(q, q+1)$, where $1 \le q \le n-1$;

b) σ is not a product of less than k adjacent transpositions.

312. Let $\pi, \sigma \in S_n$, and σ be a cycle of length k. Prove that $\pi \sigma \pi^{-1}$ is also a cycle of length k.

313. Find the modification of the decomposition of a permutation into a product of disjoint cycles, after multiplication of the permutation by some transposition. What happens with the *decrement of the permutation*?

314. Prove that any permutation $\sigma \in S_n$ can be represented as a product of the transpositions:

a) $(12), (13), \ldots, (1, n)$;

b) $(12), (23), \ldots, (n-1, n)$.

315. Prove that any permutation $\sigma \in S_n$ is a product of several factors, where each factor is equal either to the cycle (12) or to the cycle $(123 \ldots n)$.

316. Prove that any even permutation is

a) a product of threefold cycles;

b) a product of cycles of the form $(123), (124), \ldots, (12n)$.

317. Let f_{ij} be either a binomial $x_i - x_j$ or a binomial $x_j - x_i$, where i and j are arbitrary integers, $1 \le i < j \le n$, and let $f(x_1, \ldots, x_n)$ be the product of all these binomials. Prove that

$$f(x_{\sigma(1)}, \ldots, x_{\sigma(n)}) = (\mathrm{sgn}\sigma) \cdot f(x_1, \ldots, x_n)$$

for any permutation $\sigma \in S_n$.

*   *   *

318. Let T be a set of transpositions in S_n and Γ a graph with a set of vertices $1, 2, \ldots, n$ and a set of edges T. Prove that

a) any permutation in S_n is a product of transpositions from the set T, if and only if the graph Γ is connected;

b) if $|T| < n - 1$, then there exists a permutation in S_n which is not a product of transpositions from T.

319. Let k be an integer such that $1 \le k \le \binom{n}{2}$. Prove that there exists a permutation $\begin{pmatrix} 1 & 2 & \cdots & n \\ i_1 & i_2 & \cdots & i_n \end{pmatrix} \in S_n$ where k equals the number of inversions in the second row.

320. Find the sum of the numbers of inversions in the second row of all permutations $\begin{pmatrix} 1 & 2 & \cdots & n \\ i_1 & i_2 & \cdots & i_n \end{pmatrix}$.

321. Let $|X| = m$, $|Y| = n$, $\sigma \in S_X$, $\tau \in S_Y$. Put $\xi \in S_{X \times Y}$ where

$$\xi(x, y) = (\sigma(x), \tau(y)) \quad (x \in X, \quad y \in Y).$$

Find:

a) $\mathrm{sgn}\xi$ in terms of $\mathrm{sgn}\sigma$ and $\mathrm{sgn}\tau$;

b) the lengths of disjoint cycles in the decomposition of the permutation ξ, if $k_1, \ldots, k_s$ and $l_1, \ldots, l_t$ are the lengths of disjoint cycles in decompositions of permutations σ and τ (including cycles of length 1). Deduce from b) a new proof of the case a).

322. Let $d = d(\sigma)$ be the decrement of a permutation σ. Prove that

a) $\mathrm{sgn}\sigma = (-1)^d$;

b) the permutation σ is a product of d transpositions;

c) the permutation σ is not a product of less than d transpositions.

323. Let $\sigma \in S_n$. Then $\sigma = \alpha\beta$, where $\alpha, \beta \in S_n$ and $\alpha^2 = \beta^2 = \varepsilon$.

4 Recurrence relations. Induction

401. Let $f(x) = x^2 - ax - b$ be the characteristic polynomial of the recurrent equation

$$u(n) = au(n-1) + bu(n-2) \quad (n \geq n_0 + 2). \tag{$*$}$$

Prove that

a) the function $u(n) = \alpha^n$ satisfies $(*)$ if and only if α is a root of $f(x)$;

b) the function $u(n) = n\alpha^n$ satisfies $(*)$ if and only if α is a double root of $f(x)$;

c) if $f(x)$ has distinct roots α_1 and α_2, then any solution of $(*)$ is of the form

$$u(n) = C_1 \alpha_1^n + C_2 \alpha_2^n,$$

where the constants C_1 and C_2 can be defined uniquely;

d) if $f(x)$ has the double root α, then any solution of $(*)$ is of the form

$$u(n) = C_1 \alpha^n + C_2 n\alpha^n,$$

where the constants C_1 and C_2 can be defined uniquely if $\alpha \neq 0$.

402. Solve the recurrent equations ($n_0 = 0$):

a) $u(n) = 3u(n-1) - 2u(n-2), \quad u(0) = -2, \quad u(1) = 1$;

b) $u(n) = -2u(n-1) - u(n-2), \quad u(0) = -1, \quad u(1) = -1$.

403. Prove that if $a \neq 1$, then

$$1 + 2a + 3a^2 + \cdots + na^{n-1} = \frac{na^{n+1} - (n+1)a^n + 1}{(a-1)^2}.$$

404. Calculate $u(0) + u(1) + \cdots + u(n)$, where $n \geq 2$ and

a) $u(n) = 5u(n-1) - 4u(n-2), \quad u(0) = 0, \quad u(1) = 3$;

b) $u(n) = 2u(n-1) - u(n-2), \quad u(0) = 1, \quad u(1) = -1$;

c) $u(n) = 4u(n-1) - 4u(n-2), \quad u(0) = -2, \quad u(1) = 0$.

405. Let $u(0) = 0$, $u(1) = 1$ and $u(n) = u(n-1) + u(n-2)$, where $n \geq 2$. The integers $u(n)$ are called *Fibonacci numbers*. Find $u(n)$.

406. Let $a \geq -1$. Prove that for any positive integer n the inequality $(1+a)^n \geq 1 + na$ holds.

407. Prove that for any integer $n \geq 2$ the inequality $\dfrac{4^n}{n+1} < \dfrac{(2n)!}{(n!)^2}$ holds.

408. Prove that for any positive integer n:

a) $(n+1)(n+2)\ldots(n+n) = 2^n \cdot 1 \cdot 3 \cdot 5 \cdot \ldots \cdot (2n-1)$;

b) $1\cdot 2 + 2\cdot 3 + \cdots + (n-1)\cdot n = \dfrac{(n-1)n(n+1)}{3}$;

c) $1\cdot 2\cdot 3 + 2\cdot 3\cdot 4 + \cdots + n(n+1)(n+2) = \dfrac{n(n+1)(n+2)(n+3)}{4}$;

d) $\dfrac{1}{4\cdot 5} + \dfrac{1}{5\cdot 6} + \dfrac{1}{6\cdot 7} + \cdots + \dfrac{1}{(n+3)(n+4)} = \dfrac{n}{4(n+4)}$;

e) $\left(1-\dfrac{1}{4}\right)\left(1-\dfrac{1}{9}\right)\ldots\left(1-\dfrac{1}{(n+1)^2}\right) = \dfrac{n+2}{2n+2}$;

f) $1 - \dfrac{1}{2} + \dfrac{1}{3} - \dfrac{1}{4} + \cdots + \dfrac{1}{2n-1} - \dfrac{1}{2n} = \dfrac{1}{n+1} + \dfrac{1}{n+2} + \cdots + \dfrac{1}{2n}$;

g) $\dfrac{1}{1\cdot 2\cdot 3} + \dfrac{1}{2\cdot 3\cdot 4} + \cdots + \dfrac{1}{n(+1)(n+2)} = \dfrac{1}{2}\left(\dfrac{1}{2} - \dfrac{1}{(n+1)(n+2)}\right)$.

409. Prove that for any positive integer n:

a) $n^3 + 5n$ is divisible by 6;

b) $2n^3 + 3n^2 + 7n$ is divisible by 6;

c) $n^5 - n$ is divisible by 30;

d) $2^{2n} - 1$ is divisible by 3;

e) $11^{6n+3} + 1$ is divisible by 148;

f) $n^3 + (n+1)^3 + (n+2)^3$ is divisible by 9;

g) $7^{2n} - 4^{2n}$ is divisible by 33.

410. For any natural number n prove the inequality

$$\frac{n}{2} < 1 + \frac{1}{2} + \frac{1}{3} + \cdots + \frac{1}{2^n - 1} \leq n.$$

* * *

411. Let $u(n)$ be the sequence of Fibonacci integers. Prove that:

a) $u(1) + \cdots + u(n) = u(n+2) - 1$;

b) $u(1)^2 + \cdots + u(n)^2 = u(n)u(n+1)$;

c) $u(n+1)^2 - u(n-1)^2 = u(2n)$;

d) $u(1)^3 + \cdots + u(n)^3 = \dfrac{1}{10}[u(3n+2) + (-1)^{n+1}6u(n-1) + 5]$;

e) $u(m+n) = u(m)u(n-1) + u(m+1)u(n)$;

f) if n divides m, then $u(n)$ divides $u(m)$;

g) $(u(n), u(m)) = u((n, m))$.

412. Let $u_u(t) = 0$, $u_1(t) = 1$ and $u_n(t) = tu_{n-1}(t) - u_{n-2}(t)$. Prove that

a) $u_n(t) = t^{n-1} - \dbinom{n-2}{1}t^{n-3} + \dbinom{n-3}{2}t^{n-5} + \ldots$;

b) if $t = 2\cos\theta$, then $u_n(t) = \dfrac{\sin n\theta}{\sin\theta}$;

c) $u_n(t)^2 - u_k(t)^2 = u_{n-k}(t)u_{n+k}(t)$, where $k = 0, 1, \ldots, n$;

d) $u_{n+1}(t)^2 - u_n(t)^2 = u_{2n+1}(t)$.

413. Let r and $\cos r\pi$ be rational numbers. Prove that $\cos r\pi = 0, \pm\dfrac{1}{2}, \pm 1$.

414. How many parts are contained in a partition of a plane by n lines which are in a general position (i.e. no two of them are parallel and the intersection of any three of them is empty)?

5   Summations

501. Find the sums:

a) $1^2 + 2^2 + \cdots + n^2$;   b) $1^3 + 2^3 + \cdots + n^3$.

502. Prove that the sum $1^k + 2^k + \cdots + n^k$ is a polynomial in n of degree $k+1$.

$$* \quad * \quad *$$

503. Let $N(\sigma) = |\{i \mid \sigma(i) = i\}|$ be the number of fixed elements of a permutation $\sigma \in S_n$ and let

$$\sum_{\sigma \in S_n} (N(\sigma))^s = \gamma(s)n! \quad (1 \le s \le n).$$

Prove that $\gamma(1) = 1$, that $\gamma(s)$ does not depend on n and that

$$\gamma(s+1) = \gamma(s) + \dbinom{s}{1}\gamma(s-1) + \cdots + \dbinom{s}{k}\gamma(s-k) + \cdots + \dbinom{s}{s-1}\gamma(1) + 1.$$

504. Prove that

$$\sum_{d|n} \mu(d) = \begin{cases} 1 & \text{if } n = 1, \\ 0 & \text{if } n > 1, \end{cases}$$

where $\mu(n)$ is the *Möbius function*.

505. Let $f(n)$ and $g(n)$ be two functions $\mathbb{N} \to \mathbb{N}$. Prove that the following conditions are equivalent

a) $g(n) = \sum_{d|n} f(d), \qquad f(n) = \sum_{d|n} \mu(d) g\left(\frac{n}{d}\right),$

b) $g(n) = \prod_{d|n} f(d), \qquad f(n) = \prod_{d|n} g\left(\frac{n}{d}\right)^{\mu(d)}.$

506. Prove that the Euler function $\phi(n)$ and the Möbius function $\mu(n)$ satisfy the relation

$$\sum_{d|n} \frac{\mu(d)}{d} = \frac{\phi(n)}{n}.$$

CHAPTER 2

Arithmetic spaces and linear equations

6 Arithmetic spaces

601. Find the linear combination $3a_1 + 5a_2 - a_3$ of the vectors

$$a_1 = (4, 1, 3, -2), \quad a_2 = (1, 2, -3, 2), \quad a_3 = (16, 9, 1, -3).$$

602. Find the vector x from the equations:

a) $a_1 + 2a_2 + 3a_3 + 4x = 0$,
where $a_1 = (5, -8, -1, 2), a_2 = (2, -1, 4, -3), a_3 = (-3, 2, -5, 4)$;

b) $3(a_1 - x) + 2(a_2 + x) = 5(a_3 + x)$,
where $a_1 = (2, 5, 1, 3), a_2 = (10, 1, 5, 10), a_3 = (4, 1, -1, 1)$.

603. Find out whether the following systems of vectors are linearly independent:

a) $a_1 = (1, 2, 3), a_2 = (3, 6, 7)$;

b) $a_1 = (4, -2, 6), a_2 = (6, -3, 9)$;

c) $a_1 = (2, -3, 1), a_2 = (3, -1, 5), a_3 = (1, -4, 3)$;

d) $a_1 = (5, 4, 3), a_2 = (3, 3, 2), a_3 = (8, 1, 3)$;

e) $a_1 = (4, -5, 2, 6), a_2 = (2, -2, 1, 3), a_3 = (6, -3, 3, 9)$,
$a_4 = (4, -1, 5, 6)$;

f) $a_1 = (1, 0, 0, 2, 5), a_2 = (0, 1, 0, 3, 4), a_3 = (0, 0, 1, 4, 7)$,
$a_4 = (2, -3, 4, 11, 12)$.

604. In a given system of vectors of the same length we choose fixed coordinates (common for all vectors), and we preserve their order. The system of vectors obtained is called the *shortened* system, and the original system is called the *extended* system.

Prove that

a)   the shortened system of a linearly dependent system of vectors is linearly dependent;

b)   the extended system of a linearly independent system of vectors is linearly independent.

605. Prove that if vectors a_1, a_2, a_3 are linearly dependent and a_3 is not a linear combination of a_1 and a_2, then a_1 and a_2 are dependent.

606. Prove that if vectors $a_1, a_2, \ldots, a_k$ are linearly independent and vectors $a_1, a_2, \ldots, a_k, b$ are linearly dependent, then b is a linear combination of $a_1, a_2, \ldots, a_k$.

607. Let an independent system of vectors $a_1, \ldots, a_k$ be given. Find out whether the following systems of vectors are dependent

a) $b_1 = 3a_1 + 2a_2 + a_3 + a_4,$
     $b_2 = 2a_1 + 5a_2 + 3a_3 + 2a_4,$
     $b_3 = 3a_1 + 4a_2 + 2a_3 + 3a_4;$

b) $b_1 = 3a_1 + 4a_2 - 5a_3 - 2a_4 + 4a_5,$
     $b_2 = 8a_1 + 7a_2 - 2a_3 + 5a_4 - 10a_5,$
     $b_3 = 2a_1 - a_2 + 8a_3 - a_4 + 2a_5;$

c) $b_1 = a_1, \quad b_2 = a_1 + a_2, \quad b_3 = a_1 + a_2 + a_3, \quad \ldots,$
     $b_k = a_1 + a_2 + \cdots + a_k;$

d) $b_1 = a_1, \quad b_2 = a_1 + 2a_2, \quad b_3 = a_1 + 2a_2 + 3a_3, \quad \ldots,$
     $b_k = a_1 + 2a_2 + 3a_3 + \cdots + ka_k;$

e) $b_1 = a_1 + a_2, \quad b_2 = a_2 + a_3, \quad b_3 = a_3 + a_4, \quad \ldots,$
     $b_{k-1} = a_{k-1} + a_k \quad b_k = a_k + a_1;$

f) $b_1 = a_1 - a_2, \quad b_2 = a_2 - a_3, \quad b_3 = a_3 - a_4, \quad \ldots,$
     $b_{k-1} = a_{k-1} - a_k \quad b_k = a_k - a_1.$

608. Suppose that we have the vectors

$$a_1 = (0, 1, 0, 2, 0,), \quad a_2 = (7, 4, 1, 8, 3,), \quad a_3 = (0, 3, 0, 4, 0,),$$
$$a_4 = (1, 9, 5, 7, 1,), \quad a_5 = (0, 1, 0, 5, 0).$$

Do there exist coefficients c_{ij} such that the vectors

$$b_i = \sum_{j=1}^{5} c_{ij} a_j \quad (i = 1, 2, 3, 4, 5)$$

are linearly dependent?

609. Find all values of λ such that the vector b is a linear combination of vectors a_1, a_2, a_3, where

a) $a_1 = (2, 3, 5)$, $a_2 = (3, 7, 8)$, $a_3 = (1, -6, 1)$, $b = (7, -2, \lambda)$;

b) $a_1 = (4, 4, 3)$, $a_2 = (7, 2, 1)$, $a_3 = (4, 1, 6)$, $b = (5, 9, \lambda)$;

c) $a_1 = (3, 4, 2)$, $a_2 = (6, 8, 7)$, $a_3 = (15, 20, 11)$, $b = (9, 12, \lambda)$;

d) $a_1 = (3, 2, 5)$, $a_2 = (2, 4, 7)$, $a_3 = (5, 6, \lambda)$, $b = (1, 3, 5)$;

e) $a_1 = (3, 2, 6)$, $a_2 = (5, 1, 3)$, $a_3 = (7, 3, 9)$, $b = (\lambda, 2, 5)$.

610. Find all bases of the systems of vectors:

a) $a_1 = (1, 2, 0, 0)$, $a_2 = (1, 2, 3, 4)$, $a_3 = (3, 6, 0, 0)$;

b) $a_1 = (4, -1, 3, -2)$, $a_2 = (8, -2, 6, -4)$, $a_3 = (3, -1, 4, -2)$, $a_4 = (6, -2, 8, -4)$;

c) $a_1 = (1, 2, 3, 4)$, $a_2 = (2, 3, 4, 5)$, $a_3 = (3, 4, 5, 6)$, $a_4 = (4, 5, 6, 7)$;

d) $a_1 = (2, 1, -3, 1)$, $a_2 = (2, 2, -6, 2)$, $a_3 = (6, 3, -9, 3)$, $a_4 = (1, 1, 1, 1)$;

e) $a_1 = (3, 2, 3)$, $a_2 = (2, 3, 4,)$, $a_3 = (3, 2, 3)$, $a_4 = (4, 3, 4)$, $a_5 = (1, 1, 1)$.

611. When has a system of vectors a unique basis?

612. Find a basis of the systems of vectors, and express, in terms of this basis, the other vectors of the systems:

a) $a_1 = (5, 2, -3, 1)$, $a_2 = (4, 1, -2, 3)$, $a_3 = (1, 1, -1, -2)$, $a_4 = (3, 4, -1, 2)$, $a_5 = (7, -6, -7, 0)$;

b) $a_1 = (2, -1, 3, 5)$, $a_2 = (4, -3, 1, 3)$, $a_3 = (3, -2, 3, 4)$, $a_4 = (4, -1, -15, 17)$,

c) $a_1 = (1, 2, 3, -4)$, $a_2 = (2, 3, -4, 1)$, $a_3 = (2, -5, 8, -3)$, $a_4 = (5, 26, -9, -12)$, $a_5 = (3, -4, 1, 2)$;

d) $a_1 = (2, 3, -4, -1)$, $a_2 = (1, -2, 1, 3)$, $a_3 = (5, -3, -1, 8)$, $a_4 = (3, 8, -9, -5)$;

e) $a_1 = (2, 2, 7, -1)$, $a_2 = (3, -1, 2, 4)$, $a_3 = (1, 1, 3, 1)$;

f) $a_1 = (3, 2, -5, 4)$, $a_2 = (3, -1, 3, -3)$, $a_3 = (3, 5, -13, 11)$;

g) $a_1 = (2, 1)$, $a_2 = (3, 2)$, $a_3 = (1, 1)$, $a_4 = (2, 3)$,

h) $a_1 = (2, 1, -3)$, $a_2 = (3, 1, -5)$, $a_3 = (4, 2, -1)$, $a_4 = (1, 0, -7)$;

i) $a_1 = (2, 3, 5, -4, 1)$, $a_2 = (1, -1, 2, 3, 5)$, $a_3 = (3, 7, 8, -11, -3)$,
    $a_4 = (1, -1, 1, -2, 3)$;

j) $a_1 = (2, -1, 3, 4, -1)$, $a_2 = (1, 2, -3, 1, 2)$, $a_3 = (5, -5, 12, 11, -5)$,
    $a_4 = (1, -3, 6, 3, -3)$;

k) $a_1 = (4, 3, -1, 1, -1)$, $a_2 = (2, 1, -3, 2, -5)$, $a_3 = (1, -3, 0, 1, -2)$,
    $a_4 = (1, 5, 2, -2, 6)$.

613. Let the vectors $a_1, a_2, \ldots, a_k$ be linearly independent. Find all bases of the system of vectors

$$b_1 = a_1 - a_2, \quad b_2 = a_2 - a_3, \quad b_3 = a_3 - a_4, \ldots, b_{k-1} = a_{k-1} - a_k, \quad b_k = a_k - a_1.$$

614. Let a system of vectors

$$a_i = (a_{i1}, a_{i2}, \ldots, a_{in}) \quad (i = 1, 2, \ldots, s; \ s \le n)$$

be given. Prove that if $|a_{jj}| > \sum_{\substack{i=1, \\ i \ne j}}^{s} |a_{ij}|$ for any $j = 1, \ldots, s$, then the system of vectors is linearly independent.

615. Prove that if integer vectors $a_1, a_2, \ldots a_k \in \mathbb{Z}^n$ are linearly dependent over the field $\mathbb{Q}$, then there exist coprime integers $\lambda_1, \lambda_2, \ldots \lambda_k$ such that $\lambda_1 a_1 + \lambda_2 a_2 + \cdots + \lambda_k a_k = 0$.

616. Prove that if a system of integer vectors is linearly independent over a residue field, modulo p for some prime p, then the system of vectors is independent over the field of rationals.

617. Let a system of integer vectors be independent over the field $\mathbb{Q}$. Prove that there exist finitely many primes p such that these vectors are linearly dependent modulo p.

618. For the following systems of integer vectors find all primes p such that these systems are dependent modulo p

a) $a_1 = (0, 1, 1, 1)$, $a_2 = (1, 0, 1, 1)$, $a_3 = (1, 1, 0, 1)$, $a_4 = (1, 1, 1, 0)$;

b) $a_1 = (1, 0, 1, 1)$, $a_2 = (2, 3, 4, 3)$, $a_3 = (1, 3, 1, 1)$.

7 Rank of a matrix

701. Find the rank of the following matrices with the help of bordering minors and elementary row and column operations:

a) $\begin{pmatrix} 8 & 2 & 2 & -1 & 1 \\ 1 & 7 & 4 & -2 & 5 \\ -2 & 4 & 2 & -1 & 3 \end{pmatrix}$;

b) $\begin{pmatrix} 1 & 7 & 7 & 9 \\ 7 & 5 & 1 & -1 \\ 4 & 2 & -1 & -3 \\ -1 & 1 & 3 & 5 \end{pmatrix}$;

c) $\begin{pmatrix} 4 & 1 & 7 & -5 & 1 \\ 0 & -7 & 1 & -3 & -5 \\ 3 & 4 & 5 & -3 & 2 \\ 2 & 5 & 3 & -1 & 3 \end{pmatrix}$;

d) $\begin{pmatrix} 8 & -4 & 5 & 5 & 9 \\ 1 & -3 & -5 & 0 & -7 \\ 7 & -5 & 1 & 4 & 1 \\ 3 & -1 & 3 & 2 & 5 \end{pmatrix}$;

e) $\begin{pmatrix} -6 & 4 & 8 & -1 & 6 \\ -5 & 2 & 4 & 1 & 3 \\ 7 & 2 & 4 & 1 & 3 \\ 2 & 4 & 8 & -7 & 6 \\ 3 & 2 & 4 & -5 & 3 \end{pmatrix}$;

f) $\begin{pmatrix} 77 & 32 & 6 & 5 & 3 \\ 32 & 14 & 3 & 2 & 1 \\ 6 & 3 & 1 & 0 & 0 \\ 5 & 2 & 0 & 1 & 0 \\ 4 & 1 & 0 & 0 & 1 \end{pmatrix}$;

g) $\begin{pmatrix} 1 & 1 & 1 & 1 \\ 4 & 3 & 2 & 1 \\ 1 & 4 & 1 & 1 \\ 5 & 1 & 1 & 1 \\ 1 & 1 & 3 & 1 \\ 1 & 1 & 1 & 2 \end{pmatrix}$;

h) $\begin{pmatrix} 3 & 1 & 1 & 2 & -1 \\ 0 & 2 & -1 & 1 & 2 \\ 4 & 3 & 2 & -1 & 1 \\ 12 & 9 & 8 & -7 & 3 \\ -12 & -5 & -8 & 5 & 1 \end{pmatrix}$;

i) $\begin{pmatrix} 1 & 1 & 0 & 0 & 0 & 0 \\ 0 & 1 & 1 & 0 & 0 & 0 \\ 0 & 0 & 1 & 1 & 0 & 0 \\ 0 & 0 & 0 & 1 & 1 & 0 \\ 0 & 0 & 0 & 0 & 1 & 1 \\ 1 & 0 & 0 & 0 & 0 & 1 \end{pmatrix}$;

j) $\begin{pmatrix} 1 & 1 & 0 & 0 & 0 \\ 0 & 1 & 1 & 0 & 0 \\ 0 & 0 & 1 & 1 & 0 \\ 0 & 0 & 0 & 1 & 1 \\ 1 & 0 & 0 & 0 & 1 \end{pmatrix}$;

k) $\begin{pmatrix} 1 & 1 & 0 & 0 & \dots & 0 & 0 \\ 0 & 1 & 1 & 0 & \dots & 0 & 0 \\ \multicolumn{7}{c}{\dotfill} \\ 0 & 0 & 0 & 0 & \dots & 1 & 1 \\ 1 & 0 & 0 & 0 & \dots & 0 & 1 \end{pmatrix}$.

702. Find the rank of the following matrices for various values of the parameter λ:

a) $\begin{pmatrix} 7-\lambda & -12 & 6 \\ 10 & -19-\lambda & 10 \\ 12 & -24 & 13-\lambda \end{pmatrix}$;

b) $\begin{pmatrix} 1-\lambda & 0 & 0 & 0 \\ 0 & 1-\lambda & 0 & 0 \\ 0 & 0 & 2-\lambda & 3 \\ 0 & 0 & 0 & 3-\lambda \end{pmatrix}$;

c) $\begin{pmatrix} 3 & 4 & 2 & 2 \\ 3 & 17 & 7 & 1 \\ 1 & 10 & 4 & \lambda \\ 4 & 1 & 1 & 3 \end{pmatrix}$;

d) $\begin{pmatrix} 1 & \lambda & -1 & 2 \\ 2 & -1 & \lambda & 5 \\ 1 & 10 & -6 & 1 \end{pmatrix}$;

e) $\begin{pmatrix} 1 & 1 & 2 & 3 \\ 1 & 2-\lambda^2 & 2 & 3 \\ 2 & 3 & 1 & 5 \\ 2 & 3 & 1 & 9-\lambda^2 \end{pmatrix}$;

f) $\begin{pmatrix} -\lambda & 1 & 2 & 3 & 1 \\ 1 & -\lambda & 3 & 2 & 1 \\ 2 & 3 & -\lambda & 1 & 1 \\ 3 & 2 & 1 & -\lambda & 1 \end{pmatrix}$;

g) $\begin{pmatrix} \lambda & 1 & 2 & \ldots & n-1 & 1 \\ 1 & \lambda & 2 & \ldots & n-1 & 1 \\ 1 & 2 & \lambda & \ldots & n-1 & 1 \\ \multicolumn{6}{c}{\cdots\cdots\cdots\cdots\cdots\cdots} \\ 1 & 2 & 3 & \ldots & \lambda & 1 \\ 1 & 2 & 3 & \ldots & n & 1 \end{pmatrix}$;

h) $\begin{pmatrix} 1 & \lambda & \lambda^2 & \ldots & \lambda^n \\ 2 & 1 & \lambda & \ldots & \lambda^{n-1} \\ 2 & 2 & 1 & \ldots & \lambda^{n-2} \\ \multicolumn{5}{c}{\cdots\cdots\cdots\cdots\cdots} \\ 2 & 2 & 2 & \ldots & 1 \end{pmatrix}$.

703. Prove that if the rank of a matrix A is not changed after adjoining any column of a matrix B with the same number of rows, then it is not changed after joining all columns of B.

704. Prove that the rank of a product of matrices does not exceed the rank of each factor.

705. Prove that the rank of the matrix $(A|B)$, which is obtained by joining a matrix B to a matrix A, does not exceed the sum of the ranks of A and B.

706. Prove that the rank of a sum of matrices does not exceed the sum of their ranks.

707. Prove that every matrix of rank r can be presented as a sum of r matrices of rank 1, but it cannot be presented as a sum of a fewer number of them.

708. Prove that if the rank of a matrix is equal to r, then the minor situated at the intersection of any r linearly independent rows and linearly independent columns is not equal to 0.

709. Let A be a square matrix of size $n > 1$ and r be its rank. Find the rank of the adjoint matrix $\hat{A} = (A_{ij})$, where A_{ij} is the cofactor of the element a_{ji} in A.

710. Let A and B be matrices with real entries and with the same number of rows. Prove that

$$r \begin{pmatrix} A & B \\ 2A & -5B \end{pmatrix} = r(A) + r(B).$$

711. Let A and B be square matrices of a fixed size. Prove that

$$r \begin{pmatrix} A & AB \\ B & B+B^2 \end{pmatrix} = r(A) + r(B).$$

712. Prove that every matrix of rank 1 has a decomposition

$$\begin{pmatrix} b_1 c_1 & b_1 c_2 & \cdots & b_1 c_n \\ b_2 c_1 & b_2 c_2 & \cdots & b_2 c_n \\ \cdots\cdots\cdots\cdots\cdots\cdots\cdots \\ b_m c_1 & b_m c_2 & \cdots & b_m c_n \end{pmatrix} = {}^t B \cdot C,$$

where $B = (b_1, b_2, \ldots, b_m)$, $C = (c_1, c_2, \ldots, c_n)$.

713. Let A_1, A_2, ..., A_k be matrices with the same number of rows, and $C = (c_{ij})$ be a nonsingular matrix of size k.

Prove that the rank of the matrix

$$\begin{pmatrix} c_{11} A_1 & c_{12} A_2 & \cdots & c_{1k} A_k \\ \cdots\cdots\cdots\cdots\cdots\cdots\cdots\cdots \\ c_{k1} A_1 & c_{k2} A_2 & \cdots & c_{kk} A_k \end{pmatrix}$$

is equal to the sum of the ranks of A_1, A_2, ..., A_k.

714. Prove that the rectangular matrix

$$\begin{pmatrix} A & B \\ C & D \end{pmatrix},$$

where A is a nonsingular matrix of size n, has the rank n if and only if $D = CA^{-1}B$; in this case

$$\begin{pmatrix} A & B \\ C & D \end{pmatrix} = \begin{pmatrix} A \\ C \end{pmatrix} (E_n \quad A^{-1}B).$$

* * *

715. Prove that by elementary row operations of type II, every nonsingular matrix can be reduced to the form

$$\begin{pmatrix} 1 & 0 & \cdots & 0 & 0 \\ 0 & 1 & \cdots & 0 & 0 \\ \cdots\cdots\cdots\cdots\cdots\cdots \\ 0 & 0 & \cdots & 1 & 0 \\ 0 & 0 & \cdots & 0 & d \end{pmatrix}.$$

716. Prove that any matrix with the determinant 1 is a product of *elementary matrices* $E + \lambda E_{ij}$ $(i \neq j)$.

717. Prove that if the rows (columns) of a matrix A are linearly dependent, then A can be reduced, by elementary row and column operations of type II, to the form

$$\begin{pmatrix} E_r & 0 \\ 0 & 0 \end{pmatrix},$$

where E_r is the identity matrix of size r.

718. Let A and B be matrices of sizes $m \times n$, $n \times t$ and of ranks $r(A), r(B)$, respectively. Prove that the rank of the matrix AB is not less than $r(A) + r(B) - n$.

719. Prove that by elementary row operations any matrix can be reduced to a matrix of the form

$$\begin{pmatrix}
0 & \cdots & 0 & 1 & * & \cdots & * & 0 & * & \cdots & * & 0 & * & \cdots & * & 0 & * & \cdots & * \\
0 & \cdots & 0 & 0 & 0 & \cdots & 0 & 1 & * & \cdots & * & 0 & * & \cdots & * & 0 & * & \cdots & * \\
0 & \cdots & 0 & 0 & 0 & \cdots & 0 & 0 & 0 & \cdots & 0 & 1 & * & \cdots & * & 0 & * & \cdots & * \\
\cdots\cdots\cdots\cdots\cdots\cdots\cdots\cdots\cdots\cdots\cdots\cdots & \vdots & \vdots & \vdots & \ddots & \vdots \\
\cdots\cdots\cdots\cdots\cdots\cdots\cdots\cdots\cdots\cdots\cdots\cdots & * & 0 & * & \cdots & * \\
0 & \cdots & 0 & 0 & 0 & \cdots & 0 & 0 & 0 & \cdots & 0 & 0 & 0 & \cdots & 0 & 1 & * & \cdots & * \\
0 & \cdots & 0 & 0 & 0 & \cdots & 0 & 0 & 0 & \cdots & 0 & 0 & 0 & \cdots & 0 & 0 & 0 & \cdots & 0 \\
\cdots\cdots\cdots\cdots\cdots\cdots\cdots\cdots\cdots\cdots\cdots\cdots\cdots\cdots\cdots\cdots\cdots\cdots \\
0 & \cdots & 0 & 0 & 0 & \cdots & 0 & 0 & 0 & \cdots & 0 & 0 & 0 & \cdots & 0 & 0 & 0 & \cdots & 0
\end{pmatrix}$$

and that this form is unique.

8   Systems of linear equations

801. Find the general and a particular solution of these systems of linear equations, applying the method of Gauss:

a)
$$5x_1 + 3x_2 + 5x_3 + 12x_4 = 10,$$
$$2x_1 + 2x_2 + 3x_3 + 5x_4 = 4,$$
$$x_1 + 7x_2 + 9x_3 + 4x_4 = 2;$$

b)
$$-9x_1 + 6x_2 + 7x_3 + 10x_4 = 3,$$
$$-6x_1 + 4x_2 + 2x_3 + 3x_4 = 2,$$
$$-3x_1 + 2x_2 - 11x_3 - 15x_4 = 1;$$

c)
$$-9x_1 + 10x_2 + 3x_3 + 7x_4 = 7,$$
$$-4x_1 + 7x_2 + x_3 + 3x_4 = 5,$$
$$7x_1 + 5x_2 - 4x_3 - 6x_4 = 3;$$

d)
$$12x_1 + 9x_2 + 3x_3 + 10x_4 = 13,$$
$$4x_1 + 3x_2 + x_3 + 2x_4 = 3,$$
$$8x_1 + 6x_2 + 2x_3 + 5x_4 = 7;$$

e)
$$-6x_1 + 9x_2 + 3x_3 + 2x_4 = 4,$$
$$-2x_1 + 3x_2 + 5x_3 + 4x_4 = 2,$$
$$-4x_1 + 6x_2 + 4x_3 + 3x_4 = 3;$$

f)
$$8x_1 + 6x_2 + 5x_3 + 2x_4 = 21,$$
$$3x_1 + 3x_2 + 2x_3 + x_4 = 10,$$
$$4x_1 + 2x_2 + 3x_3 + x_4 = 8,$$
$$3x_1 + 3x_2 + x_3 + x_4 = 15,$$
$$7x_1 + 4x_2 + 5x_3 + 2x_4 = 18;$$

g)
$$2x_1 + 5x_2 - 8x_3 = 8,$$
$$4x_1 + 3x_2 - 9x_3 = 9,$$
$$2x_1 + 3x_2 - 5x_3 = 7,$$
$$x_1 + 8x_2 - 7x_3 = 12;$$

h)
$$6x_1 + 4x_2 + 5x_3 + 2x_4 + 3x_5 = 1,$$
$$3x_1 + 2x_2 - 2x_3 + x_4 \qquad = -7,$$
$$9x_1 + 6x_2 + x_3 + 3x_4 + 2x_5 = 2,$$
$$3x_1 + 2x_2 + 4x_3 + x_4 + 2x_5 = 3.$$

802. Analyse the system of equations and find the general solution corresponding to the values of the parameter λ:

a)
$$8x_1 + 6x_2 + 3x_3 + 2x_4 = 5,$$
$$-12x_1 - 3x_2 - 3x_3 + 3x_4 = -6,$$
$$4x_1 + 5x_2 + 2x_3 + 3x_4 = 3,$$
$$\lambda x_1 + 4x_2 + x_3 + 4x_4 = 2;$$

b)
$$-6x_1 + 8x_2 - 5x_3 - x_4 = 9,$$
$$-2x_1 + 4x_2 + 7x_3 + 3x_4 = 1,$$
$$-3x_1 + 5x_2 + 4x_3 + 2x_4 = 3,$$
$$-3x_1 + 7x_2 + 17x_3 + 7x_4 = \lambda;$$

c)
$$2x_1 + 5x_2 + x_3 + 3x_4 = 2,$$
$$4x_1 + 6x_2 + 3x_3 + 5x_4 = 4,$$
$$4x_1 + 14x_2 + x_3 + 7x_4 = 4,$$
$$2x_1 - 3x_2 + 3x_3 + \lambda x_4 = 7;$$

d)
$$2x_1 - x_2 + 3x_3 + 4x_4 = 5,$$
$$4x_1 - 2x_2 + 5x_3 + 6x_4 = 7,$$
$$6x_1 - 3x_2 + 7x_3 + 8x_4 = 9,$$
$$\lambda x_1 - 4x_2 + 9x_3 + 10x_4 = 11;$$

e)
$$2x_1 + 3x_2 + x_3 + 2x_4 = 3,$$
$$4x_1 + 6x_2 + 3x_3 + 4x_4 = 5,$$
$$6x_1 + 9x_2 + 5x_3 + 6x_4 = 7,$$
$$8x_1 + 12x_2 + 7x_3 + \lambda x_4 = 9;$$

f)
$$\lambda x_1 + x_2 + x_3 = 1,$$
$$x_1 + \lambda x_2 + x_3 = 1,$$
$$x_1 + x_2 + \lambda x_3 = 1;$$

g)
$$\lambda x_1 + x_2 + x_3 + x_4 = 1,$$
$$x_1 + \lambda x_2 + x_3 + x_4 = 1,$$
$$x_1 + x_2 + \lambda x_3 + x_4 = 1,$$
$$x_1 + x_2 + x_3 + \lambda x_4 = 1;$$

h)
$$(1 + \lambda)x_1 + x_2 + x_3 = 1,$$
$$x_1 + (1 + \lambda)x_2 + x_3 = \lambda,$$
$$x_1 + x_2 + (1 + \lambda)x_3 = \lambda^2;$$

i)
$$(1 + \lambda)x_1 + x_2 + x_3 = \lambda^2 + 3\lambda,$$
$$x_1 + (1 + \lambda)x_2 + x_3 = \lambda^3 + 3\lambda^2,$$
$$x_1 + x_2 + (1 + \lambda)x_3 = \lambda^4 + 3\lambda^3.$$

803. Find all vectors in a space $\mathbb{R}^n$, whose image is equal to the vector $b \in \mathbb{R}^m$ under the linear map $\mathbb{R}^n \to \mathbb{R}^m$ given by the matrix A:

a) $\quad A = \begin{pmatrix} 3 & 2 & -5 \\ 3 & 4 & -9 \\ 5 & 2 & -8 \\ 8 & 1 & -7 \end{pmatrix}, \qquad\qquad b = \begin{pmatrix} 7 \\ 9 \\ 8 \\ 12 \end{pmatrix};$

b) $\quad A = \begin{pmatrix} 1 & -3 & -3 & -14 \\ 2 & -6 & -3 & -1 \\ 3 & -9 & -5 & -6 \end{pmatrix}, \qquad b = \begin{pmatrix} 8 \\ -5 \\ -4 \end{pmatrix};$

c) $\quad A = \begin{pmatrix} 1 & -2 & -1 & -2 & -3 \\ 3 & -6 & -2 & -4 & -5 \\ 3 & -6 & -4 & -8 & -13 \\ 2 & -4 & -1 & -1 & -2 \end{pmatrix}, \qquad b = \begin{pmatrix} -2 \\ -3 \\ -9 \\ -1 \end{pmatrix};$

d) $\quad A = \begin{pmatrix} 1 & 9 & 4 & -5 \\ 3 & 2 & 2 & 5 \\ 2 & 3 & 2 & 2 \\ 1 & 7 & 6 & -1 \\ 2 & 2 & 3 & 4 \end{pmatrix}, \qquad b = \begin{pmatrix} 1 \\ 3 \\ 2 \\ 7 \\ 5 \end{pmatrix};$

e) $\quad A = \begin{pmatrix} 1 & 1 & 3 & -2 & 3 \\ 2 & 2 & 4 & -1 & 3 \\ 3 & 3 & 5 & -2 & 3 \\ 2 & 2 & 8 & -3 & 9 \end{pmatrix}, \qquad b = \begin{pmatrix} 1 \\ 2 \\ 1 \\ 2 \end{pmatrix};$

f) $\quad A = \begin{pmatrix} 3 & -6 & -1 & 4 \\ 1 & -2 & -3 & 7 \\ 2 & -4 & -14 & 31 \end{pmatrix}, \qquad b = \begin{pmatrix} -7 \\ -5 \\ -10 \end{pmatrix};$

g) $\quad A = \begin{pmatrix} 2 & 1 & 3 & -2 & 1 \\ 6 & 3 & 5 & -4 & 3 \\ 2 & 1 & 7 & -4 & 1 \\ 4 & 2 & 2 & -3 & 3 \end{pmatrix}, \qquad b = \begin{pmatrix} 4 \\ 5 \\ 11 \\ 6 \end{pmatrix};$

h) $\quad A = \begin{pmatrix} 8 & 6 & 5 & 2 \\ 3 & 3 & 2 & 1 \\ 4 & 2 & 3 & 1 \\ 3 & 5 & 1 & 1 \\ 7 & 4 & 5 & 2 \end{pmatrix}, \qquad b = \begin{pmatrix} 21 \\ 10 \\ 8 \\ 15 \\ 18 \end{pmatrix}.$

804. Find the general solution, and a fundamental system of solutions, of the systems of linear equations:

a)
$$x_1 + x_2 - 2x_3 + 2x_4 = 0,$$
$$3x_1 + 5x_2 + 6x_3 - 4x_4 = 0,$$
$$4x_1 + 5x_2 - 2x_3 + 3x_4 = 0,$$
$$3x_1 + 8x_2 + 24x_3 - 19x_4 = 0;$$

b)
$$x_1 - x_3 = 0,$$
$$x_2 - x_4 = 0,$$
$$- x_1 + x_3 - x_5 = 0,$$
$$- x_2 + x_4 - x_6 = 0,$$
$$- x_3 + x_5 = 0,$$
$$- x_4 + x_6 = 0;$$

c)
$$x_1 - x_3 + x_5 = 0,$$
$$x_2 - x_4 + x_6 = 0,$$
$$x_1 - x_2 + x_5 - x_6 = 0,$$
$$x_2 - x_3 + x_6 = 0,$$
$$x_1 - x_4 + x_5 = 0;$$

d)
$$x_1 + x_2 = 0,$$
$$x_1 + x_2 + x_3 = 0,$$
$$x_2 + x_3 + x_4 = 0,$$
$$\dots\dots\dots\dots\dots\dots$$
$$x_{n-2} + x_{n-1} + x_n = 0,$$
$$x_{n-1} + x_n = 0.$$

805. Find a basis of the kernel of the linear mappings given by the matrices:

a) $\begin{pmatrix} 3 & 5 & -4 & 2 \\ 2 & 4 & -6 & 3 \\ 11 & 17 & -8 & 4 \end{pmatrix}$;

b) $\begin{pmatrix} 3 & 5 & 3 & 2 & 1 \\ 5 & 7 & 6 & 4 & 3 \\ 7 & 9 & 9 & 6 & 5 \\ 4 & 8 & 3 & 2 & 0 \end{pmatrix}$;

c) $\begin{pmatrix} 6 & 9 & 2 \\ -4 & 1 & 1 \\ 5 & 7 & 4 \\ 2 & 5 & 3 \end{pmatrix}$;

d) $\begin{pmatrix} 5 & 7 & 6 & -2 & 2 \\ 8 & 9 & 9 & -3 & 4 \\ 7 & 1 & 6 & -2 & 6 \\ 4 & -1 & 3 & -1 & 4 \end{pmatrix}$;

$$
\text{e)} \begin{pmatrix} 5 & 6 & -2 & 7 & 4 \\ 2 & 3 & -1 & 4 & 2 \\ 7 & 9 & -3 & 5 & 6 \\ 5 & 9 & -3 & 1 & 6 \end{pmatrix}; \qquad \text{f)} \begin{pmatrix} 3 & 4 & 1 & 2 & 3 \\ 5 & 7 & 1 & 3 & 4 \\ 4 & 5 & 2 & 1 & 5 \\ 7 & 10 & 1 & 6 & 5 \end{pmatrix}.
$$

806. Solve the systems of equation by applying Cramer's rule:

a)
$$
\begin{aligned}
2x_1 - \ x_2 &= 1, \\
x_1 + 16x_2 &= 17;
\end{aligned}
$$

b)
$$
\begin{aligned}
2x_1 + 5x_2 &= 1, \\
3x_1 + 7x_2 &= 2;
\end{aligned}
$$

c)
$$
\begin{aligned}
x_1 \cos \alpha + x_2 \sin \alpha &= \cos \beta, \\
-x_1 \sin \alpha + x_2 \cos \alpha &= \sin \beta;
\end{aligned}
$$

d)
$$
\begin{aligned}
2x_1 + \ x_2 + \ x_3 &= 3, \\
x_1 + 2x_2 + \ x_3 &= 0, \\
x_1 + \ x_2 + 2x_3 &= 0;
\end{aligned}
$$

e)
$$
\begin{aligned}
x_1 + x_2 + x_3 &= 6, \\
-x_1 + x_2 + x_3 &= 0, \\
x_1 - x_2 + x_3 &= 2;
\end{aligned}
$$

f)
$$
\begin{aligned}
2x_1 + 3x_2 + 5x_3 &= 10, \\
3x_1 + 7x_2 + 4x_3 &= 3, \\
x_1 + 2x_2 + 2x_3 &= 3.
\end{aligned}
$$

807. Find a polynomial $f(x)$ of degree 2 with real coefficients such that $f(1) = 8$, $f(-1) = 2$, $f(2) = 14$.

808. Find a polynomial $f(x)$ of degree 3 such that $f(-2) = 1$, $f(-1) = 3$, $f(1) = 13$, $f(2) = 33$.

809. Find a polynomial $f(x)$ of degree 4 such that $f(-3) = -77$, $f(-2) = -13$, $f(-1) = 1$, $f(1) = -1$, $f(2) = -17$.

810. Solve the systems of congruences:

a)
$$
\begin{aligned}
2x + \ y - \ z &\equiv 1, \\
x + 2y + \ z &\equiv 2, \quad (\text{mod } 5); \\
x + \ y - \ z &\equiv -1,
\end{aligned}
$$

b)
$$
\begin{aligned}
3x + 2y + 5z &\equiv 1, \\
2x + 5y + 3z &\equiv 1, \quad (\text{mod } 17). \\
5x + 3y + 2z &\equiv 4.
\end{aligned}
$$

811. Prove that if the determinant of a square matrix (a_{ij}) of size n with integer entries is coprime with an integer m, then the system of congruences

$$
(a_{i1}x_1 + a_{i2} + \cdots + a_{in}) \equiv b_i \quad (\text{mod } m) \quad (i = 1, 2, \ldots, n)
$$

has a unique solution modulo m.

* * *

812. Let A be an integer matrix and d be the least absolute value of its entries. Prove that if, under *integral elementary row and column operations* of A the number d does not decrease, then d divides all entries of A.

813. Prove that, with the help of elementary row and column operations over the ring of integers $\mathbb{Z}$, any integer matrix can be transformed to a matrix $\begin{pmatrix} A & 0 \\ 0 & 0 \end{pmatrix}$, where $A = \mathrm{diag}\{d_1, \ldots, d_r\}$ and $d_i | d_{i+1}$ $(i = 1, 2, \ldots, r-1)$.

814. Prove that if the square integer system of linear equations is definite modulo any prime p, then it is definite over the ring of integers.

815. Find out whether the square integer system of linear equations, compatible modulo any prime p, is compatible over the ring of integers.

816. Prove that the following systems of equations have unique solutions modulo almost all primes (with finitely many exceptions). Solve these systems modulo the exceptional primes.

a)
$$\begin{aligned} x_1 + 2x_2 + 2x_3 &= 2, \\ 2x_1 + x_2 - 2x_3 &= 1, \\ 2x_1 - 2x_2 + x_3 &= 1; \end{aligned}$$

b)
$$\begin{aligned} x_1 + x_2 + x_3 &= 1, \\ x_1 + x_2 + x_4 &= 1, \\ x_1 + x_3 + x_4 &= 1, \\ x_2 + x_3 + x_4 &= 1; \end{aligned}$$

c)
$$\begin{aligned} x_1 + x_2 + x_3 + x_4 &= 1, \\ x_1 + x_2 - x_3 - x_4 &= 1, \\ x_1 + x_2 + x_3 - x_4 &= 1, \\ x_1 - x_2 - x_3 + x_4 &= 0. \end{aligned}$$

817. Prove that any system of linear equations with real coefficients can be transformed by elementary row operations of type II to a row-echelon system.

818. Prove that the greatest common divisor of minors of fixed size k of an integer matrix is not changed under integral elementary row and column operations.

819. Prove that if the integer matrix, with the help of integral elementary row and column operations, is reduced to $\begin{pmatrix} A & 0 \\ 0 & 0 \end{pmatrix}$, where $A = \mathrm{diag}\{d_1, d_2, \ldots, d_r\}$, $d_i \neq 0$ and $d_i | d_{i+1}$, then integers $d_1, d_2, \ldots, d_r$ are uniquely defined (up to their signs).

820. Two sets of variables are *integrally equivalent*, if they are connected by the relation $\begin{pmatrix} y_1 \\ \vdots \\ y_n \end{pmatrix} = U \begin{pmatrix} x_1 \\ \vdots \\ x_n \end{pmatrix}$, where U is an integer matrix whose determinant is equal to ± 1. Prove that the system of equations

$$\sum_{j=1}^{n} a_{ij} x_j = b_i \quad (i = 1, 2, \ldots, m),$$

where a_{ij}, b_i are integers, is integrally equivalent to a system of equations of the form
$$d_i y_i = c_i \quad (i = 1, 2, \ldots, n),$$
and the set of variables $(y_1, \ldots, y_n)$ is integrally equivalent to the set $(x_1, \ldots, x_n)$.

821. Prove that an integer system of equations has an integer solution if and only if the greatest common divisors of all minors of size k of the matrix of the system and of the extended one coincide for any integer k.

822. Prove that an integer system of equations has an integer solution if and only if it has a solution modulo any prime p.

823. Justify the following practical way of determining all integer solutions of a system of equations
$$\sum_{j=1}^{n} a_{ij} x_j = b_i \quad (i = 1, 2, \ldots, m)$$

with integer coefficients.

Take the matrix $\begin{pmatrix} A & b \\ E_n & 0 \end{pmatrix}$ of size $(n+m) \times (n+1)$. Applying only integral elementary operations to the first m rows and n columns, reduce this matrix to the

form $\begin{pmatrix} D & C \\ U & 0 \end{pmatrix}$, where $C = \begin{pmatrix} c_1 \\ \vdots \\ c_m \end{pmatrix}$, $\quad |\det U| = 1$,

$$D = \begin{pmatrix} d_1 & 0 & \cdots & 0 & 0 & \cdots & 0 \\ 0 & d_2 & \cdots & 0 & 0 & \cdots & 0 \\ \cdots & \cdots & \cdots & \cdots & \cdots & \cdots & \cdots \\ 0 & 0 & \cdots & d_r & 0 & \cdots & 0 \\ 0 & 0 & \cdots & 0 & 0 & \cdots & 0 \\ \cdots & \cdots & \cdots & \cdots & \cdots & \cdots & \cdots \\ 0 & 0 & \cdots & 0 & 0 & \cdots & 0 \end{pmatrix} \quad d_i | d_{i+1}, \quad d_1 \neq 0, \ldots, d_r \neq 0.$$

The original system is compatible if $d_i | c_i$ for $i = 1, \ldots, r$, $c_k = 0$ for $k > r$ and the general solution is given by the formula

$$\begin{pmatrix} x_1 \\ \vdots \\ x_n \end{pmatrix} = U \begin{pmatrix} c_1/d_1 \\ \vdots \\ c_r/d_r \\ y_{r+1} \\ \vdots \\ y_n \end{pmatrix},$$

where $y_{r+1}, y_{r+2}, \ldots, y_n$ are arbitrary integers.

824. Find all integer solutions of the following systems of equations:

a) $2x_1 + 3x_2 + 4x_3 = 5$;

b) $2x_1 + 3x_2 - 11x_3 - 15x_4 = 1$,
            $4x_1 - 6x_2 + 2x_3 + 3x_4 = 2$,
            $2x_1 - 3x_2 + 5x_3 + 7x_4 = 1$.

825. Let A and B be matrices of the same size and suppose that homogeneous systems of linear equations with matrices A and B are equivalent. Prove that B can be obtained from A by elementary row operations.

826. Let a system of linear complex equations $AX = b$ with a square nonsingular matrix A be given. Assume that the sum of the absolute values of the elements of each row in the matrix $E + A$ is less than 1. Let X_0 be an arbitrary column. Define inductively $X_{m+1} = (A + E)X_m - b$. Then the sequence X_m converges to a solution of the system $AX = b$.

CHAPTER 3

Determinants

9 Determinants of sizes two and three

901. Calculate the determinants:

a) $\begin{vmatrix} 3 & 5 \\ 5 & 3 \end{vmatrix}$;

b) $\begin{vmatrix} ab & ac \\ bd & cd \end{vmatrix}$;

c) $\begin{vmatrix} \cos\alpha & -\sin\alpha \\ \sin\alpha & \cos\alpha \end{vmatrix}$;

d) $\begin{vmatrix} \sin\alpha & \sin\beta \\ \cos\alpha & \cos\beta \end{vmatrix}$;

e) $\begin{vmatrix} \log_b a & 1 \\ 1 & \log_a b \end{vmatrix}$;

f) $\begin{vmatrix} \cos\alpha + i\sin\alpha & 1 \\ 1 & \cos\alpha - i\sin\alpha \end{vmatrix}$;

g) $\begin{vmatrix} a+bi & c+di \\ -c+di & a-bi \end{vmatrix}$.

902. Calculate the determinants:

a) $\begin{vmatrix} 1 & 2 & 3 \\ 5 & 1 & 4 \\ 3 & 2 & 5 \end{vmatrix}$;

b) $\begin{vmatrix} -1 & 5 & 4 \\ 3 & -2 & 0 \\ -1 & 3 & 6 \end{vmatrix}$;

c) $\begin{vmatrix} 0 & 2 & 2 \\ 2 & 0 & 2 \\ 2 & 2 & 0 \end{vmatrix}$;

d) $\begin{vmatrix} 1 & 2 & 3 \\ 4 & 5 & 6 \\ 7 & 8 & 9 \end{vmatrix}$;

e) $\begin{vmatrix} a & b & c \\ b & c & a \\ c & a & b \end{vmatrix}$;

f) $\begin{vmatrix} 0 & a & 0 \\ b & c & d \\ 0 & e & 0 \end{vmatrix}$;

g) $\begin{vmatrix} \sin\alpha & \cos\alpha & 1 \\ \sin\beta & \cos\beta & 1 \\ \sin\gamma & \cos\gamma & 1 \end{vmatrix}$;

h) $\begin{vmatrix} 1 & 0 & 1+i \\ 0 & 1 & i \\ 1-i & -i & 1 \end{vmatrix}$;

31

i) $\begin{vmatrix} 1 & \varepsilon & \varepsilon^2 \\ \varepsilon^2 & 1 & \varepsilon \\ \varepsilon & \varepsilon^2 & 1 \end{vmatrix}$ $\left(\varepsilon = -\dfrac{1}{2} + i\dfrac{\sqrt{3}}{2} \right)$;

j) $\begin{vmatrix} 1 & 1 & 1 \\ 1 & \varepsilon & \varepsilon^2 \\ 1 & \varepsilon^2 & \varepsilon \end{vmatrix}$ $\left(\varepsilon = \cos\dfrac{4}{3}\pi + i\sin\dfrac{4}{3}\pi \right)$.

10 Expanding a determinant. Inductive definition

1001. Find out what products occur in the expansion of the determinants of appropriate sizes and find out the signs of these products:

a) $a_{13}a_{22}a_{31}a_{46}a_{55}a_{64}$;

b) $a_{31}a_{13}a_{52}a_{45}a_{24}$;

c) $a_{34}a_{21}a_{46}a_{17}a_{73}a_{54}a_{62}$.

1002. Find indices i, j, k such that the product $a_{51}a_{i6}a_{1j}a_{35}a_{44}a_{6k}$ occurs in the expansion of a determinant of size six, with a minus sign.

1003. In the expansion of the determinant

$$\begin{vmatrix} x & 1 & 2 & 3 \\ x & x & 1 & 2 \\ 1 & 2 & x & 3 \\ x & 1 & 2 & 2x \end{vmatrix}$$

find all the products containing x^4 and x^3.

1004. Making use of the definition, calculate the following determinants:

a) $\begin{vmatrix} a_{11} & 0 & 0 & \ldots & 0 \\ a_{21} & a_{22} & 0 & \ldots & 0 \\ a_{31} & a_{32} & a_{33} & \ldots & 0 \\ \hdotsfor{5} \\ a_{n1} & a_{n2} & a_{n3} & \ldots & a_{nn} \end{vmatrix}$; b) $\begin{vmatrix} 0 & \ldots & 0 & 0 & a_{1n} \\ 0 & \ldots & 0 & a_{2,n-1} & a_{2n} \\ \hdotsfor{5} \\ a_{n1} & \ldots & a_{n,n-2} & a_{n,n-1} & a_{nn} \end{vmatrix}$;

c) $\begin{vmatrix} a & 3 & 0 & 5 \\ 0 & b & 0 & 2 \\ 1 & 2 & c & 3 \\ 0 & 0 & 0 & d \end{vmatrix}$; d) $\begin{vmatrix} 1 & 0 & 2 & a \\ 2 & 0 & b & 0 \\ 3 & c & 4 & 5 \\ d & 0 & 0 & 0 \end{vmatrix}$;

e)
$$\begin{vmatrix} a_{11} & a_{12} & a_{13} & a_{14} & a_{15} \\ a_{21} & a_{22} & a_{23} & a_{24} & a_{25} \\ a_{31} & a_{32} & 0 & 0 & 0 \\ a_{41} & a_{42} & 0 & 0 & 0 \\ a_{51} & a_{52} & 0 & 0 & 0 \end{vmatrix}.$$

1005. Find the expansion of the determinant as a polynomial in t,

$$\begin{vmatrix} -t & 0 & 0 & \ldots & 0 & a_1 \\ a_2 & -t & 0 & \ldots & 0 & 0 \\ \multicolumn{6}{c}{\ldots\ldots\ldots\ldots\ldots\ldots} \\ 0 & 0 & 0 & \ldots & -t & 0 \\ 0 & 0 & 0 & \ldots & a_n & -t \end{vmatrix}.$$

1006. Calculate the determinant of the matrix in which all entries of the principal diagonal are equal to 1 and entries of the jth column are equal to $a_1, a_2, \ldots, a_{j-1}, a_{j+1}, \ldots, a_n$, and all the other entries of the matrix are equal to 0.

1007. Let

$$\sigma = \begin{pmatrix} 1 & 2 & \ldots & n \\ i_1 & i_2 & \ldots & i_n \end{pmatrix} \in S_n,$$

and A be a square matrix of size n with entries a_{rs}, where $a_{rs} = 1$ for $s = i_r$ and $a_{rs} = 0$ otherwise. Prove that the determinant of A is equal to the sign of the permutation σ.

11 Basic properties of a determinant

1101. What will happen with a determinant of size n if

a) all its entries are multiplied by -1;

b) each entry a_{ik} is multiplied by c^{i-k} ($c \neq 0$);

c) each entry is replaced by the symmetric one with respect to the secondary diagonal;

d) each entry is replaced by the symmetric one with respect to the 'center' of the determinant;

e) we rotate the determinant counter-clockwise by $90°$ around the 'center'?

1102. What will happen with a determinant of size n if

a)   the first column is moved to the last column and the other columns are moved to the left while retaining their order;

b)   its rows are written down in inverse order?

1103. What will happen to a determinant if

a)   we add to each column, starting with the second, the previous column;

b)   we add to each column, starting with the second, all previous columns;

c)   we subtract from each row, except from the last, the next row and from the last row we subtract the former first row;

d)   we add to each column, starting with the second, the previous column, and to the first column we add the former last one?

1104. Prove that the determinant of a skew-symmetric matrix of odd size is equal to 0.

1105. Integers 20604, 53227, 25755, 20927 and 289 are divisible by 17. Prove that the determinant

$$\begin{vmatrix} 2 & 0 & 6 & 0 & 4 \\ 5 & 3 & 2 & 2 & 7 \\ 2 & 5 & 7 & 5 & 5 \\ 2 & 0 & 9 & 2 & 7 \\ 0 & 0 & 2 & 8 & 9 \end{vmatrix}$$

is divisible by 17.

1106. Calculate the determinant without using its expansion:

$$\begin{vmatrix} x & y & z & 1 \\ y & z & x & 1 \\ z & x & y & 1 \\ \dfrac{x+z}{2} & \dfrac{x+y}{2} & \dfrac{y+z}{2} & 1 \end{vmatrix}.$$

1107. What is the value of the determinant in which the sum of rows with even indices is equal to the sum of all rows with odd indices?

1108. Prove that any determinant is equal to the half-sum of two determinants, one of which is obtained from the given one by addition of a number b to all entries of the ith row, while the other one is obtained by similar addition of the number $-b$.

1109. Prove that if all entries of a determinant of size n are differentiable functions in one variable, then the derivative of this determinant is equal to the sum of n determinants D_i, where all rows of D_i, except the ith one, are the same as in D, and the ith row consists of derivatives of entries of the ith row of D.

1110. Calculate the determinants:

a)
$$\begin{vmatrix} a_1 + x & x & \cdots & x \\ a_1 & a_2 + x & \cdots & x \\ \cdots\cdots\cdots\cdots\cdots\cdots\cdots & & & \\ x & x & \cdots & a_n + x \end{vmatrix};$$

b)
$$\begin{vmatrix} a_1 + x & a_2 & \cdots & a_n \\ a_1 & a_2 + x & \cdots & a_n \\ \cdots\cdots\cdots\cdots\cdots\cdots\cdots & & & \\ a_1 & a_2 & \cdots & a_n + x \end{vmatrix};$$

c)
$$\begin{vmatrix} 1 + x_1 y_1 & 1 + x_1 y_2 & \cdots & 1 + x_1 y_n \\ 1 + x_2 y_1 & 1 + x_2 y_2 & \cdots & 1 + x_2 y_n \\ \cdots\cdots\cdots\cdots\cdots\cdots\cdots\cdots\cdots & & & \\ 1 + x_n y_1 & 1 + x_n y_2 & \cdots & 1 + x_n y_n \end{vmatrix};$$

d)
$$\begin{vmatrix} f_1(a_1) & f_1(a_2) & \cdots & f_1(a_n) \\ f_2(a_1) & f_2(a_2) & \cdots & f_2(a_n) \\ \cdots\cdots\cdots\cdots\cdots\cdots\cdots & & & \\ f_n(a_1) & f_n(a_2) & \cdots & f_n(a_n) \end{vmatrix},$$

where $f_i(x)$ are polynomials of degree, at most, $n - 2$ $(i = 1, 2, \ldots, n)$;

e)
$$\begin{vmatrix} 1 + a_1 + b_1 & a_1 + b_1 & \cdots & a_1 + b_n \\ a_2 + b_1 & 1 + a_2 + b_2 & \cdots & a_2 + b_n \\ \cdots\cdots\cdots\cdots\cdots\cdots\cdots\cdots\cdots\cdots & & & \\ a_n + b_1 & a_n + b_1 & \cdots & 1 + a_n + b_n \end{vmatrix}.$$

12 Expanding a determinant according to the elements of a row or a column

1201. Calculate the determinant by expanding it according to the elements of the third row

$$\begin{vmatrix} 2 & -3 & 4 & 1 \\ 4 & -2 & 3 & 2 \\ a & b & c & d \\ 3 & -1 & 4 & 3 \end{vmatrix}.$$

1202. Calculate the determinant by expanding it according to the second row

$$\begin{vmatrix} 5 & a & 2 & -1 \\ 4 & b & 4 & -3 \\ 2 & c & 3 & -2 \\ 4 & d & 5 & -4 \end{vmatrix}.$$

1203. Calculate the determinants:

a)
$$\begin{vmatrix} x & y & 0 & \ldots & 0 & 0 \\ 0 & x & y & \ldots & 0 & 0 \\ 0 & 0 & x & \ldots & 0 & 0 \\ \multicolumn{6}{c}{\ldots\ldots\ldots\ldots\ldots} \\ 0 & 0 & 0 & \ldots & x & y \\ y & 0 & 0 & \ldots & 0 & x \end{vmatrix};$$

b)
$$\begin{vmatrix} a_0 & a_1 & a_2 & \ldots & a_{n-1} & a_n \\ -y_1 & x_1 & 0 & \ldots & 0 & 0 \\ 0 & -y_2 & x_2 & \ldots & 0 & 0 \\ \multicolumn{6}{c}{\ldots\ldots\ldots\ldots\ldots} \\ 0 & 0 & 0 & \ldots & x_{n-1} & 0 \\ 0 & 0 & 0 & \ldots & -y_n & x_n \end{vmatrix};$$

c)
$$\begin{vmatrix} a_0 & -1 & 0 & 0 & \ldots & 0 & 0 \\ a_1 & x & -1 & 0 & \ldots & 0 & 0 \\ a_2 & 0 & x & -1 & \ldots & 0 & 0 \\ \multicolumn{7}{c}{\ldots\ldots\ldots\ldots\ldots} \\ a_{n-1} & 0 & 0 & 0 & \ldots & x & -1 \\ a_n & 0 & 0 & 0 & \ldots & 0 & x \end{vmatrix};$$

d)
$$\begin{vmatrix} n!a_0 & (n-1)!a_1 & (n-2)!a_2 & \ldots & a_n \\ -n & x & 0 & \ldots & a_n \\ 0 & -(n-1) & x & \ldots & 0 \\ \multicolumn{5}{c}{\ldots\ldots\ldots\ldots\ldots} \\ 0 & 0 & 0 & \ldots & x \end{vmatrix};$$

e)
$$\begin{vmatrix} 1 & 2 & 3 & \ldots & n-1 & n \\ -1 & x & 0 & \ldots & 0 & 0 \\ 0 & -1 & x & \ldots & 0 & 0 \\ \multicolumn{6}{c}{\ldots\ldots\ldots\ldots\ldots} \\ 0 & 0 & 0 & \ldots & x & 0 \\ 0 & 0 & 0 & \ldots & -1 & x \end{vmatrix};$$

f)
$$\begin{vmatrix} n & -1 & 0 & 0 & \ldots & 0 & 0 \\ n-1 & x & -1 & 0 & \ldots & 0 & 0 \\ n-2 & 0 & x & -1 & \ldots & 0 & 0 \\ \multicolumn{7}{c}{\dotfill} \\ 2 & 0 & 0 & 0 & \ldots & x & -1 \\ 1 & 0 & 0 & 0 & \ldots & 0 & x \end{vmatrix} ;$$

g)
$$\begin{vmatrix} 1 & 0 & 0 & 0 & \ldots & 0 & 1 \\ 1 & a_1 & 0 & 0 & \ldots & 0 & 0 \\ 1 & 1 & a_2 & 0 & \ldots & 0 & 0 \\ 1 & 0 & 1 & a_3 & \ldots & 0 & 0 \\ \multicolumn{7}{c}{\dotfill} \\ 1 & 0 & 0 & 0 & \ldots & 1 & a_n \end{vmatrix} ;$$

h)
$$\begin{vmatrix} a_1 & 0 & \ldots & 0 & b_1 \\ 0 & a_2 & \ldots & b_2 & 0 \\ \multicolumn{5}{c}{\dotfill} \\ 0 & b_{2n-1} & \ldots & a_{2n-1} & 0 \\ b_{2n} & 0 & \ldots & 0 & a_{2n} \end{vmatrix} ;$$

i)
$$\begin{vmatrix} a_0 & 1 & 1 & 1 & \ldots & 1 \\ 1 & a_1 & 0 & 0 & \ldots & 0 \\ 1 & 0 & a_2 & 0 & \ldots & 0 \\ \multicolumn{6}{c}{\dotfill} \\ 1 & 0 & 0 & 0 & \ldots & a_n \end{vmatrix} .$$

1204. Prove that the $(n+1)$th member $u(n+1)$ of the Fibonacci sequence (see Exercise 405) is equal to the determinant

$$\begin{vmatrix} 1 & 1 & 0 & 0 & \ldots & 0 & 0 \\ -1 & 1 & 1 & 0 & \ldots & 0 & 0 \\ 0 & -1 & 1 & 1 & \ldots & 0 & 0 \\ \multicolumn{7}{c}{\dotfill} \\ 0 & 0 & 0 & 0 & \ldots & -1 & 1 \end{vmatrix}$$

of size n.

13 Calculating a determinant with the help of elementary operations

1301. Calculate the determinants:

a)
$$\begin{vmatrix} 1 & 2 & 3 & 4 \\ -3 & 2 & -5 & 13 \\ 1 & -2 & 10 & 4 \\ -2 & 9 & -8 & 25 \end{vmatrix} ;$$

b)
$$\begin{vmatrix} 1 & -1 & 1 & -2 \\ 1 & 3 & -1 & 3 \\ -1 & -1 & 4 & 3 \\ -3 & 0 & -8 & -13 \end{vmatrix} ;$$

c) $\begin{vmatrix} 7 & 6 & 9 & 4 & -4 \\ 1 & 0 & -2 & 6 & 6 \\ 7 & 8 & 9 & -1 & -6 \\ 1 & -1 & -2 & 4 & 5 \\ -7 & 0 & -9 & 2 & -2 \end{vmatrix};$

d) $\begin{vmatrix} 4 & 4 & -1 & 0 & -1 & 8 \\ 2 & 3 & 7 & 5 & 2 & 3 \\ 3 & 2 & 5 & 7 & 3 & 2 \\ 1 & 2 & 2 & 1 & 1 & 2 \\ 1 & 7 & 6 & 6 & 5 & 7 \\ 2 & 1 & 1 & 2 & 2 & 1 \end{vmatrix};$

e) $\begin{vmatrix} 1 & 5 & 3 & 5 & -4 \\ 3 & 1 & 2 & 9 & 8 \\ -1 & 7 & -3 & 8 & -9 \\ 3 & 4 & 2 & 4 & 7 \\ 1 & 8 & 3 & 3 & 5 \end{vmatrix};$

f) $\begin{vmatrix} -5 & -7 & -2 & 2 & -2 & 16 \\ 0 & 0 & 4 & 0 & -5 & 0 \\ 2 & 0 & -2 & 0 & 2 & 0 \\ 6 & 4 & 6 & -1 & 15 & -5 \\ 5 & -4 & 10 & 1 & 14 & 6 \\ 3 & 0 & -2 & 0 & 3 & 0 \end{vmatrix};$

g) $\begin{vmatrix} 1001 & 1002 & 1003 & 1004 \\ 1002 & 1003 & 1001 & 1002 \\ 1001 & 1001 & 1001 & 999 \\ 1001 & 1000 & 998 & 999 \end{vmatrix};$

h) $\begin{vmatrix} 27 & 44 & 40 & 55 \\ 20 & 64 & 21 & 40 \\ 13 & -20 & -13 & 24 \\ 46 & 45 & -55 & 84 \end{vmatrix};$

i) $\begin{vmatrix} 30 & 20 & 15 & 12 \\ 20 & 15 & 12 & 15 \\ 15 & 12 & 15 & 20 \\ 12 & 15 & 20 & 30 \end{vmatrix};$

j) $\begin{vmatrix} \frac{1}{2} & \frac{1}{3} & \frac{1}{2} & 1 \\ \frac{1}{3} & \frac{1}{2} & 1 & \frac{1}{2} \\ \frac{1}{2} & 1 & \frac{1}{2} & \frac{1}{3} \\ 1 & \frac{1}{2} & \frac{1}{3} & \frac{1}{2} \end{vmatrix};$

k) $\begin{vmatrix} 1 & 10 & 100 & 1000 & 10000 & 100000 \\ 0.1 & 2 & 30 & 400 & 5000 & 60000 \\ 0 & 0.1 & 3 & 60 & 1000 & 15000 \\ 0 & 0 & 0.1 & 4 & 100 & 2000 \\ 0 & 0 & 0 & 0.1 & 5 & 150 \\ 0 & 0 & 0 & 0 & 0.1 & 6 \end{vmatrix};$

l) $\begin{vmatrix} 4 & -2 & 0 & 5 \\ 3 & 2 & -2 & 1 \\ -2 & 1 & 3 & -1 \\ 2 & 3 & -6 & -3 \end{vmatrix};$

m) $\begin{vmatrix} 4 & 3 & 3 & 5 \\ 3 & 4 & 3 & 2 \\ 3 & 2 & 5 & 4 \\ 2 & 4 & 2 & 3 \end{vmatrix};$

n) $\begin{vmatrix} 3 & 2 & 4 & 5 \\ 4 & -3 & 2 & -4 \\ 5 & -2 & -3 & -7 \\ -3 & 4 & 2 & 9 \end{vmatrix};$

o) $\begin{vmatrix} 14 & 13 & 3 & -13 \\ -7 & -4 & 2 & 10 \\ 21 & 23 & 0 & -23 \\ 7 & 12 & -2 & -6 \end{vmatrix};$

p) $\begin{vmatrix} 6 & 3 & 8 & -4 \\ 5 & 6 & 4 & 2 \\ 0 & 3 & 4 & 2 \\ 4 & 1 & -4 & 6 \end{vmatrix};$

q) $\begin{vmatrix} 2 & 4 & 6 & -5 \\ 1 & 6 & 5 & 4 \\ -3 & 2 & 4 & 6 \\ 4 & 5 & 2 & 3 \end{vmatrix}.$

1302. Calculate the following determinants by reducing them to the triangular form:

a) $\begin{vmatrix} 1 & 2 & 3 & \ldots & n \\ -1 & 0 & 3 & \ldots & n \\ -1 & -2 & 0 & \ldots & n \\ \hdotsfor{5} \\ -1 & -2 & -3 & \ldots & 0 \end{vmatrix}$;
b) $\begin{vmatrix} 1 & n & n & \ldots & n \\ n & 2 & n & \ldots & n \\ n & n & 3 & \ldots & n \\ \hdotsfor{5} \\ n & n & n & \ldots & n \end{vmatrix}$;

c) $\begin{vmatrix} 1 & \ldots & 1 & 1 & 1 \\ a_1 & \ldots & a_1 & a_1 - b_1 & a_1 \\ a_2 & \ldots & a_2 - b_2 & a_2 & a_2 \\ \hdotsfor{5} \\ a_n - b_n & \ldots & a_n & a_n & a_n \end{vmatrix}$;

d) $\begin{vmatrix} x_1 & a_{12} & a_{13} & \ldots & a_{1n} \\ x_1 & x_2 & a_{23} & \ldots & a_{2n} \\ x_1 & x_2 & x_3 & \ldots & a_{3n} \\ \hdotsfor{5} \\ x_1 & x_2 & x_3 & \ldots & x_3 \end{vmatrix}$;
e) $\begin{vmatrix} 1 & 2 & 3 & \ldots & n-2 & n-1 & n \\ 2 & 3 & 4 & \ldots & n-1 & n & n \\ 3 & 4 & 5 & \ldots & n & n & n \\ \hdotsfor{7} \\ n & n & n & \ldots & n & n & n \end{vmatrix}$;

f) $\begin{vmatrix} 1 & x & x^2 & x^3 & \ldots & x^n \\ a_{11} & 1 & x & x^2 & \ldots & x^{n-1} \\ a_{21} & a_{22} & 1 & x & \ldots & x^{n-2} \\ \hdotsfor{6} \\ a_{n1} & a_{n2} & a_{n3} & a_{n4} & \ldots & 1 \end{vmatrix}$;

g) $\begin{vmatrix} 1 & 1 & \ldots & 1 & -n \\ 1 & 1 & \ldots & -n & 1 \\ \hdotsfor{5} \\ 1 & -n & \ldots & 1 & 1 \\ -n & 1 & \ldots & 1 & 1 \end{vmatrix}$;
h) $\begin{vmatrix} a & b & \ldots & b & b \\ b & a & \ldots & b & b \\ \hdotsfor{5} \\ b & b & \ldots & a & b \\ b & b & \ldots & b & a \end{vmatrix}$;

i) $\begin{vmatrix} 1 & a_1 & a_2 & \ldots & a_n \\ 1 & a_1 + b_1 & a_2 & \ldots & a_n \\ 1 & a_1 & a_2 + b_2 & \ldots & a_n \\ \hdotsfor{5} \\ 1 & a_1 & a_2 & \ldots & a_n + b_n \end{vmatrix}$.

1303. Calculate the determinant

$$\begin{vmatrix} a & a+h & a+2h & \ldots & a+(n-2)h & a+(n-1)h \\ a+(n-1)h & a & a+1h & \ldots & a+(n-3)h & a+(n-2)h \\ \hdotsfor{6} \\ a+h & a+2h & a+3h & \ldots & a+(n-1)h & a \end{vmatrix}.$$

14 Calculating special determinants

1401. Calculate the following determinants applying recurrence relations (see Exercise 401):

a) $\begin{vmatrix} 2 & 1 & 0 & \ldots & 0 \\ 1 & 2 & 0 & \ldots & 0 \\ 0 & 1 & 2 & \ldots & 0 \\ \multicolumn{5}{c}{\dotfill} \\ 0 & 0 & 0 & \ldots & 2 \end{vmatrix}$;

b) $\begin{vmatrix} 3 & 2 & 0 & \ldots & 0 \\ 1 & 3 & 2 & \ldots & 0 \\ 0 & 1 & 3 & \ldots & 0 \\ \multicolumn{5}{c}{\dotfill} \\ 0 & 0 & 0 & \ldots & 3 \end{vmatrix}$;

c) $\begin{vmatrix} 5 & 6 & 0 & 0 & 0 & \ldots & 0 & 0 \\ 4 & 5 & 2 & 0 & 0 & \ldots & 0 & 0 \\ 0 & 1 & 3 & 2 & 0 & \ldots & 0 & 0 \\ 0 & 0 & 1 & 3 & 2 & \ldots & 0 & 0 \\ \multicolumn{8}{c}{\dotfill} \\ 0 & 0 & 0 & 0 & 0 & \ldots & 3 & 2 \\ 0 & 0 & 0 & 0 & 0 & \ldots & 1 & 3 \end{vmatrix}$;

d) $\begin{vmatrix} 1 & 2 & 0 & 0 & 0 & \ldots & 0 & 0 \\ 3 & 4 & 3 & 0 & 0 & \ldots & 0 & 0 \\ 0 & 2 & 5 & 3 & 0 & \ldots & 0 & 0 \\ 0 & 0 & 2 & 5 & 3 & \ldots & 0 & 0 \\ \multicolumn{8}{c}{\dotfill} \\ 0 & 0 & 0 & 0 & 0 & \ldots & 5 & 3 \\ 0 & 0 & 0 & 0 & 0 & \ldots & 2 & 5 \end{vmatrix}$;

e) $\begin{vmatrix} 3 & 2 & 0 & 0 & \ldots & 0 & 0 & 0 \\ 1 & 3 & 1 & 0 & \ldots & 0 & 0 & 0 \\ 0 & 2 & 3 & 2 & \ldots & 0 & 0 & 0 \\ 0 & 0 & 1 & 3 & \ldots & 0 & 0 & 0 \\ \multicolumn{8}{c}{\dotfill} \\ 0 & 0 & 0 & 0 & \ldots & 1 & 3 & 1 \\ 0 & 0 & 0 & 0 & \ldots & 0 & 2 & 3 \end{vmatrix}$;

f) $\begin{vmatrix} \alpha+\beta & \alpha\beta & 0 & 0 & \ldots & 0 \\ 1 & \alpha+\beta & \alpha\beta & 0 & \ldots & 0 \\ 0 & 1 & \alpha+\beta & \alpha\beta & \ldots & 0 \\ \multicolumn{6}{c}{\dotfill} \\ 0 & 0 & 0 & 0 & \ldots & \alpha+\beta \end{vmatrix}$;

g)
$$\begin{vmatrix} 1 & 1 & 1 & \ldots & 1 \\ 1 & 2 & 2^2 & \ldots & 2^n \\ 1 & 3 & 3^2 & \ldots & 3^n \\ \hdotsfor{5} \\ 1 & n+1 & (n+1)^2 & \ldots & (n+1)^n \end{vmatrix};$$

h)
$$\begin{vmatrix} a^n & (a-1)^n & \ldots & (a-n)^n \\ a^{n-1} & (a-1)^{n-1} & \ldots & (a-n)^{n-1} \\ \hdotsfor{4} \\ a & a-1 & \ldots & a-n \\ 1 & 1 & \ldots & 1 \end{vmatrix};$$

i)
$$\begin{vmatrix} 1 & \ldots & 1 \\ x_1+1 & \ldots & x_n+1 \\ x_1^2+x_1 & \ldots & x_n^2+x_n \\ \hdotsfor{3} \\ x_1^{n-1}+x_1^{n-2} & \ldots & x_n^{n-1}+x_n^{n-2} \end{vmatrix};$$

j)
$$\begin{vmatrix} a_1^n & a_1^{n-1}b_1 & a_1^{n-2}b_1^2 & \ldots & b_1^n \\ a_2^n & a_2^{n-1}b_2 & a_2^{n-2}b_2^2 & \ldots & b_2^n \\ \hdotsfor{5} \\ a_{n+1}^n & a_{n+1}^{n-1}b_{n+1} & a_{n+1}^{n-2}b_{n+1}^2 & \ldots & b_{n+1}^n \end{vmatrix};$$

k)
$$\begin{vmatrix} 1 & x_1 & x_1^2 & \ldots & x_1^{s-1} & x_1^{s+1} & \ldots & x_1^n \\ 1 & x_2 & x_2^2 & \ldots & x_2^{s-1} & x_2^{s+1} & \ldots & x_2^n \\ \hdotsfor{8} \\ 1 & x_n & x_n^2 & \ldots & x_n^{s-1} & x_n^{s+1} & \ldots & x_n^n \end{vmatrix};$$

l)
$$\begin{vmatrix} 1+x_1 & 1+x_1^2 & \ldots & 1+x_1^n \\ 1+x_2 & 1+x_2^2 & \ldots & 1+x_2^n \\ \hdotsfor{4} \\ 1+x_n & 1+x_n^2 & \ldots & 1+x_n^n \end{vmatrix};$$

m)
$$\begin{vmatrix} 0 & 1 & 1 & \ldots & 1 & 1 \\ 1 & 0 & x & \ldots & x & x \\ 1 & x & 0 & \ldots & x & x \\ \hdotsfor{6} \\ 1 & x & x & \ldots & 0 & x \\ 1 & x & x & \ldots & x & 0 \end{vmatrix};$$

n)
$$\begin{vmatrix} a & x & x & \ldots & x \\ y & a & x & \ldots & x \\ y & y & a & \ldots & x \\ \hdotsfor{5} \\ y & y & y & \ldots & a \end{vmatrix}.$$

15   Determinant of a product of matrices

1501. Calculate the determinant

$$\begin{vmatrix} a & b & c & d \\ -b & a & d & -c \\ -c & -d & a & b \\ -d & c & -b & a \end{vmatrix}$$

by squaring it.

1502. Calculate the following determinants factorizing them as a product of determinants:

a)
$$\begin{vmatrix} \cos(\alpha_1 - \beta_1) & \cos(\alpha_1 - \beta_2) & \cdots & \cos(\alpha_1 - \beta_n) \\ \cos(\alpha_2 - \beta_1) & \cos(\alpha_2 - \beta_2) & \cdots & \cos(\alpha_2 - \beta_n) \\ \vdots & \vdots & \ddots & \vdots \\ \cos(\alpha_n - \beta_1) & \cos(\alpha_n - \beta_2) & \cdots & \cos(\alpha_n - \beta_n) \end{vmatrix};$$

b)
$$\begin{vmatrix} \dfrac{1 - a_1^n b_1^n}{1 - a_1 b_1} & \cdots & \dfrac{1 - a_1^n b_n^n}{1 - a_1 b_n} \\ \vdots & \ddots & \vdots \\ \dfrac{1 - a_n^n b_1^n}{1 - a_n b_1} & \cdots & \dfrac{1 - a_n^n b_n^n}{1 - a_n b_n} \end{vmatrix};$$

c)
$$\begin{vmatrix} (a_0 + b_0)^n & \cdots & (a_0 + b_n)^n \\ \vdots & \ddots & \vdots \\ (a_n + b_0)^n & \cdots & (a_n + b_n)^n \end{vmatrix};$$

d)
$$\begin{vmatrix} s_0 & s_1 & s_2 & \cdots & s_{n-1} \\ s_1 & s_2 & s_3 & \cdots & s_n \\ s_2 & s_3 & s_4 & \cdots & s_{n+1} \\ \vdots & \vdots & \vdots & \ddots & \vdots \\ s_{n-1} & s_n & s_{n+1} & \cdots & s_{2n-2} \end{vmatrix},$$

where $s_k = x_1^k + x_2^k + \cdots + x_n^k$.

1503. Prove that the determinant of the circulant matrix

$$\begin{vmatrix} a_1 & a_2 & a_3 & \cdots & a_n \\ a_n & a_1 & a_2 & \cdots & a_{n-1} \\ a_{n-1} & a_n & a_1 & \cdots & a_{n-2} \\ \cdots\cdots\cdots\cdots\cdots\cdots\cdots\cdots \\ a_2 & a_3 & a_4 & \cdots & a_1 \end{vmatrix}$$

is equal to $f(\varepsilon_1) f(\varepsilon_2) \ldots f(\varepsilon_n)$, where $f(x) = a_1 + a_2 x + \cdots + a_n x^{n-1}$, and ε_1, $\varepsilon_2 \ldots, \varepsilon_n$ are all roots of 1 of order n.

1504. Calculate the determinants:

a)
$$\begin{vmatrix} a & b & c & d \\ d & a & b & c \\ c & d & a & b \\ b & c & d & a \end{vmatrix};$$

b)
$$\begin{vmatrix} 1 & \alpha & \alpha^2 & \cdots & \alpha^{n-1} \\ \alpha^{n-1} & 1 & \alpha & \cdots & \alpha^{n-2} \\ \alpha^{n-2} & \alpha^{n-1} & 1 & \cdots & \alpha^{n-3} \\ \cdots & \cdots & \cdots & \cdots & \cdots \\ \alpha & \alpha^2 & \alpha^3 & \cdots & 1 \end{vmatrix}.$$

16 Additional exercises

1601. Find the maximal value of a determinant of size three, whose entries are

a) integers 0 or 1;

b) integers 1 or -1.

1602. Prove that a determinant of size n is equal to zero if entries at the intersection of some k of its rows and of some l of its columns are equal to zero, and $k + l > n$.

1603. Let D be a determinant of size $n > 1$, and D_1 and D_2 be determinants obtained from D by replacing each element a_{ij} by its cofactor A_{ij} for D_1 and by the minor M_{ij} for D_2. Prove that $D_1 = D_2$.

1604. The *adjoint* matrix $\tilde{A}$ of a square matrix A of size n is the matrix in which the cofactor A_{ji} is situated at ij. Prove that

a) $|\tilde{A}| = |A|^{n-1}$;

b) $\tilde{\tilde{A}} = |A|^{n-2}A$ if $n > 2$, and $\tilde{\tilde{A}} = A$ if $n = 2$.

1605. *Binet–Cauchy formula.* Let $A = (a_{ij})$, $B = (b_{ij})$ be matrices of size $m \times n$, and $A_{i_1,\dots,i_m}$ and $B_{i_1,\dots,i_m}$ be minors of size m in matrices A and B, respectively, composed by columns with indices $i_1, \dots, i_m$, and

$$c_{ij} = \sum_{k=1}^{n} a_{ik}b_{jk}, \quad C = (c_{ij}), \quad (i = 1, \dots, m; \quad j = 1, \dots, m).$$

Prove that

$$\det C = \sum_{1 \le i_1 < i_2 < \cdots < i_m \le n} A_{i_1,\dots,i_m} B_{i_1,\dots,i_m},$$

if $m \le n$, and $\det C = 0$, if $m > n$.

1606. Let A and B be matrices of sizes $p \times n$ and $n \times k$ respectively, and let

$$A\begin{pmatrix} i_1 & \cdots & i_m \\ j_1 & \cdots & j_m \end{pmatrix}, \quad B\begin{pmatrix} i_1 & \cdots & i_m \\ j_1 & \cdots & j_m \end{pmatrix}$$

be minors of matrices A and B, situated at the intersection of rows with indices $i_1, \ldots, i_m$ and of columns with indices $j_1, \ldots, j_m$. Let $C = AB$. Prove that

$$C\begin{pmatrix} i_1 & \cdots & i_m \\ j_1 & \cdots & j_m \end{pmatrix} = \sum_{1 \le k_1 < k_2 < \cdots < k_m \le n} A\begin{pmatrix} i_1 & \cdots & i_m \\ k_1 & \cdots & k_m \end{pmatrix} B\begin{pmatrix} k_1 & \cdots & k_m \\ j_1 & \cdots & j_m \end{pmatrix}$$

if $m \le n$, and $C\begin{pmatrix} i_1 & \cdots & i_m \\ j_1 & \cdots & j_m \end{pmatrix} = 0$, if $m > n$.

1607. Prove that the sum of the principal minors of size k in a matrix $A \cdot {}^t A$ is equal to the sum of the squares of all minors of size k in A.

1608. Let

$$D = \begin{vmatrix} a_{11} & \cdots & a_{1n} \\ \cdots\cdots\cdots\cdots \\ a_{n1} & \cdots & a_{nn} \end{vmatrix}.$$

Prove that

$$\begin{vmatrix} a_{11} & \cdots & a_{1n} & x_1 \\ \cdots\cdots\cdots\cdots\cdots\cdots \\ a_{n1} & \cdots & a_{nn} & x_n \\ x_1 & \cdots & x_n & z \end{vmatrix} = Dz - \sum_{i,j=1}^{n} A_{ij} x_i x_j.$$

1609. Prove that the sum of the cofactors of the elements of a row in a determinant is not changed, if we add the same number to all elements of the matrix.

1610. Prove that if all the entries in a row (column) of a determinant are equal to 1, then the sum of the cofactors of all entries in the determinant is equal to the determinant.

1611. Let

$$A = \begin{vmatrix} a_{11} & \cdots & a_{1n} \\ \cdots\cdots\cdots\cdots \\ a_{n1} & \cdots & a_{nn} \end{vmatrix}, \quad B = \begin{vmatrix} b_{11} & \cdots & b_{1k} \\ \cdots\cdots\cdots\cdots \\ b_{k1} & \cdots & b_{kk} \end{vmatrix},$$

$$D = \begin{vmatrix} a_{11}b_{11} & \cdots & a_{1n}b_{11} & a_{11}b_{12} & \cdots & a_{1n}b_{12} & \cdots & a_{11}b_{1k} & \cdots & a_{1n}b_{1k} \\ \cdots\cdots\cdots\cdots\cdots\cdots\cdots\cdots\cdots\cdots\cdots\cdots\cdots\cdots\cdots\cdots \\ a_{n1}b_{11} & \cdots & a_{nn}b_{11} & a_{n1}b_{12} & \cdots & a_{nn}b_{12} & \cdots & a_{n1}b_{1k} & \cdots & a_{nn}b_{1k} \\ \cdots\cdots\cdots\cdots\cdots\cdots\cdots\cdots\cdots\cdots\cdots\cdots\cdots\cdots\cdots\cdots \\ a_{11}b_{k1} & \cdots & a_{1n}b_{k1} & a_{11}b_{k2} & \cdots & a_{1n}b_{k2} & \cdots & a_{11}b_{kk} & \cdots & a_{1n}b_{kk} \\ \cdots\cdots\cdots\cdots\cdots\cdots\cdots\cdots\cdots\cdots\cdots\cdots\cdots\cdots\cdots\cdots \\ a_{n1}b_{k1} & \cdots & a_{nn}b_{k1} & a_{n1}b_{k2} & \cdots & a_{nn}b_{k2} & \cdots & a_{n1}b_{kk} & \cdots & a_{nn}b_{kk} \end{vmatrix},$$

(*D* being the determinant of the matrix of size *nk* which is the Kroneker product of *A* and *B*).

Prove that $D = A^k B^n$.

1612. The *continuant* is the determinant

$$(a_1 a_2 \dots a_n) = \begin{vmatrix} a_1 & 1 & 0 & 0 & \dots & 0 & 0 \\ -1 & a_2 & 1 & 0 & \dots & 0 & 0 \\ 0 & -1 & a_3 & 1 & \dots & 0 & 0 \\ \dots & \dots & \dots & \dots & & & \\ 0 & 0 & 0 & 0 & \dots & -1 & a_n \end{vmatrix}.$$

a) Find the expression of $(a_1 a_2 \dots a_n)$ as a polynomial in $a_1, \dots, a_n$.

b) Write down the expansion of the continuant according to the elements of the first *k* rows.

c) Establish the following relation of the continuant with continuous fractions

$$\frac{(a_1 a_2 \dots a_n)}{(a_2 a_3 \dots a_n)} = a_1 + \cfrac{1}{a_2 + \cfrac{1}{a_3 + \dots + \cfrac{1}{a_n}}}.$$

1613. Prove that if A, B, C, D are square matrices of size *n* and $C \cdot {}^t D = D \cdot {}^t C$, then

$$\begin{vmatrix} A & B \\ C & D \end{vmatrix} = |A \cdot {}^t D - B \cdot {}^t C|.$$

1614. Prove that if A, B, C, D are square matrices of size *n*, where either *C* or *D* is a nondegenerate matrix, then $CD = DC$ implies

$$\begin{vmatrix} A & B \\ C & D \end{vmatrix} = |AD - BC|.$$

1615. Calculate the determinant $\begin{vmatrix} cE & A \\ A & cE \end{vmatrix}$, where

$$A = \begin{vmatrix} a & 1 & 0 & 0 & \dots & 0 \\ 1 & a & 1 & 0 & \dots & 0 \\ 0 & 1 & a & 1 & \dots & 0 \\ \dots & \dots & \dots & \dots & & \\ 0 & 0 & 0 & 0 & \dots & a \end{vmatrix}.$$

1616. Prove that the determinant which is obtained by deleting the kth column from the matrix

$$(a_{ij}) = \left(\binom{j-1}{i-1} \right) \quad (i = 1, \ldots, n+1; \; j = 1, \ldots, n+2),$$

is equal to $\binom{n+1}{k-1}$.

1617. Prove that

$$\begin{vmatrix} \dfrac{1}{2!} & \dfrac{1}{3!} & \dfrac{1}{4!} & \cdots & \dfrac{1}{(2k+2)!} \\[2mm] 1 & \dfrac{1}{2!} & \dfrac{1}{3!} & \cdots & \dfrac{1}{(2k+1)!} \\[2mm] 0 & 1 & \dfrac{1}{2!} & \cdots & \dfrac{1}{(2k)!} \\[2mm] \vdots & \vdots & \vdots & \ddots & \vdots \\[2mm] 0 & 0 & 0 & \cdots & \dfrac{1}{2!} \end{vmatrix} = 0 \quad (k \in \mathbb{N}).$$

1618. *Euler's identity.* Multiply matrices

$$\begin{pmatrix} x_1 & x_2 & x_3 & x_4 \\ x_2 & -x_1 & -x_4 & x_3 \\ x_3 & x_4 & -x_1 & -x_2 \\ x_4 & -x_3 & x_2 & -x_1 \end{pmatrix} \quad \text{and} \quad \begin{pmatrix} y_1 & y_2 & y_3 & y_4 \\ y_2 & -y_1 & -y_4 & y_3 \\ y_3 & y_4 & -y_1 & -y_2 \\ y_4 & -y_3 & y_2 & -y_1 \end{pmatrix},$$

and prove that

$$
\begin{aligned}
(x_1^2 + x_2^2 + x_3^2 + x_4^2)(y_1^2 + y_2^2 + y_3^2 + y_4^2) =& (x_1 y_1 + x_2 y_2 + x_3 y_3 + x_4 y_4)^2 \\
&+ (x_1 y_2 - x_2 y_1 - x_3 y_4 + x_4 y_3)^2 \\
&+ (x_1 y_3 + x_2 y_4 - x_3 y_1 - x_4 y_2)^2 \\
&+ (x_1 y_4 - x_2 y_3 + x_3 y_2 - x_4 y_1)^2.
\end{aligned}
$$

1619. Calculate the determinant of the matrix (a_{ij}) of size n, where

a) a_{ij} is equal to 1, if i divides j, and is equal to 0 otherwise;

b) a_{ij} is equal to the number of common divisors of indices i and j.

1620. Prove that the determinant of the matrix (d_{ij}) of size n, where d_{ij} is the greatest common divisor of numbers i and j, is equal to $\phi(1)\phi(2)\ldots\phi(n)$.

1621. Let $x_1 \ldots x_n$, $y_1 \ldots y_n$ be some numbers, and $x_i y_j \neq 1$ for all $i, j = 1, \ldots, n$, and let $\Delta(x_1 \ldots x_n)$, $\Delta(y_1 \ldots y_n)$ be Vandermonde determinants. Prove that

$$\Delta(x_1, \ldots, x_n)\Delta(y_1, \ldots, y_n) = \det\left(\frac{1}{1 - x_i y_j}\right)_{i,j=1,\ldots,n} \cdot \prod_{i,j=1}^{n}(1 - x_i y_j).$$

CHAPTER 4

Matrices

17 Operations on matrices

1701. Multiply the matrices:

a) $\begin{pmatrix} 1 & n \\ 0 & 1 \end{pmatrix} \cdot \begin{pmatrix} 1 & m \\ 0 & 1 \end{pmatrix}$;

b) $\begin{pmatrix} \cos\alpha & \sin\alpha \\ \sin\alpha & \cos\alpha \end{pmatrix} \cdot \begin{pmatrix} \cos\beta & \sin\beta \\ \sin\beta & \cos\beta \end{pmatrix}$;

c) $\begin{pmatrix} 3 & -4 & 5 \\ 2 & -3 & 1 \\ 3 & -5 & -1 \end{pmatrix} \cdot \begin{pmatrix} 3 & 29 \\ 2 & 18 \\ 0 & -3 \end{pmatrix}$;

d) $\begin{pmatrix} 1 & 5 & 3 \\ 2 & -3 & 1 \end{pmatrix} \cdot \begin{pmatrix} 2 & -3 & 5 \\ -1 & 4 & -2 \\ 3 & -1 & 1 \end{pmatrix}$;

e) $\begin{pmatrix} 1 & 2 & 1 \\ 3 & 1 & 3 \\ 1 & 2 & 1 \end{pmatrix} \cdot \begin{pmatrix} 1 & 3 & 1 \\ 2 & 1 & 2 \\ 1 & 3 & 1 \end{pmatrix}$;

f) $\begin{pmatrix} 1 & -1 & 3 \\ -1 & 1 & -3 \\ 2 & -2 & 6 \end{pmatrix} \cdot \begin{pmatrix} 1 & 5 & 2 \\ 0 & 3 & -1 \\ 2 & 1 & -1 \end{pmatrix}$;

g) $\begin{pmatrix} 1 & 2 & 0 & 0 \\ 2 & 1 & 0 & 0 \\ 0 & 0 & 1 & 3 \\ 0 & 0 & 3 & 1 \end{pmatrix} \cdot \begin{pmatrix} 1 & 1 & 0 & 0 \\ 1 & 1 & 0 & 0 \\ 0 & 0 & 1 & -1 \\ 0 & 0 & -1 & 1 \end{pmatrix}$;

h)
$$\begin{pmatrix} 1 & 1 & 0 & 0 \\ 1 & 2 & 0 & 0 \\ 0 & 0 & 3 & 1 \\ 0 & 0 & 1 & 1 \end{pmatrix} \cdot \begin{pmatrix} 1 & 1 & 0 & 0 \\ 1 & 3 & 0 & 0 \\ 0 & 0 & 1 & 2 \end{pmatrix}.$$

1702. Carry out the operations:

a)
$$\begin{pmatrix} 3 & 0 & 2 & 0 \\ 0 & 1 & 2 & 1 \\ 2 & 3 & 0 & 0 \end{pmatrix} \cdot \begin{pmatrix} 1 & -2 & 2 \\ 2 & -1 & 1 \\ -1 & 1 & -2 \\ 2 & 2 & -1 \end{pmatrix} + \begin{pmatrix} -2 & 0 & -3 \\ 0 & 6 & -3 \\ 5 & -2 & 8 \end{pmatrix};$$

b)
$$\begin{pmatrix} 3 & 0 & 2 \\ 0 & 1 & 3 \\ 2 & 2 & 0 \\ 0 & 1 & 0 \end{pmatrix} \cdot \begin{pmatrix} 1 & 2 & -1 & 2 \\ -2 & -1 & 1 & 2 \\ 2 & 1 & 1 & 2 \end{pmatrix} + \begin{pmatrix} 0 & -4 & 6 & 1 \\ 2 & 2 & -5 & -2 \\ 2 & -2 & 6 & 4 \\ 1 & 3 & 0 & 1 \end{pmatrix}.$$

1703. Calculate:

a)
$$\begin{pmatrix} 1 & 2 & 2 \\ 2 & 1 & -2 \\ 2 & -2 & 1 \end{pmatrix}^2;$$

b)
$$\begin{pmatrix} 0 & 1 & 0 & 0 \\ 0 & 0 & 1 & 0 \\ 0 & 0 & 0 & 1 \\ 0 & 0 & 0 & 0 \end{pmatrix}^2;$$

c)
$$\begin{pmatrix} 1 & 1 & 1 & 1 \\ 1 & 1 & -1 & -1 \\ 1 & -1 & 1 & -1 \\ 1 & -1 & -1 & 1 \end{pmatrix}^2;$$

d)
$$\begin{pmatrix} 0 & 1 & 0 & 0 \\ 0 & 0 & 2 & 0 \\ 0 & 0 & 0 & 3 \\ 0 & 0 & 0 & 0 \end{pmatrix}^2.$$

1704. Calculate:

a)
$$\begin{pmatrix} \cos \alpha & \sin \alpha \\ -\sin \alpha & \cos \alpha \end{pmatrix}^n;$$

b)
$$\begin{pmatrix} \lambda & 1 \\ 0 & \lambda \end{pmatrix}^n;$$

c)
$$\begin{pmatrix} 2 & 1 \\ 5 & 3 \end{pmatrix} \cdot \begin{pmatrix} 1 & 0 \\ 1 & 1 \end{pmatrix} \cdot \begin{pmatrix} 3 & -1 \\ -5 & 2 \end{pmatrix}^n.$$

1705. Calculate the value of the polynomial $f(x)$ at the matrix A:

a) $f(x) = x^3 - 2x^2 + 1;$ $A = \begin{pmatrix} 2 & 1 & 0 \\ 0 & 2 & 0 \\ 1 & 1 & 1 \end{pmatrix};$

b) $f(x) = x^3 - 3x + 2;$ $A = \begin{pmatrix} 2 & 1 & 1 \\ 1 & 2 & 1 \\ 1 & 1 & 2 \end{pmatrix}.$

1706. Prove that if the matrices A and B commute, then

$$(A + B)^n = \sum_{i=0}^{n} \binom{n}{i} A^i B^{n-i}.$$

Find an example of two matrices A, B, for which this formula is not valid.

1707. Consider the square matrix

$$H = \begin{pmatrix} 0 & 1 & 0 & \cdots & 0 \\ 0 & 0 & 1 & \cdots & 0 \\ \cdots\cdots\cdots\cdots\cdots \\ 0 & 0 & 0 & \cdots & 1 \\ 0 & 0 & 0 & \cdots & 0 \end{pmatrix}.$$

Calculate all powers of H.

1708. Consider the square matrix

$$J = \begin{pmatrix} \lambda & 1 & 0 & \cdots & 0 \\ 0 & \lambda & 1 & \cdots & 0 \\ \cdots\cdots\cdots\cdots\cdots \\ 0 & 0 & \cdots & \lambda & 1 \\ 0 & 0 & 0 & \cdots & \lambda \end{pmatrix}$$

of size n. Prove that if $f(x)$ is a polynomial, then

$$f(J) = \begin{pmatrix} f(\lambda) & \dfrac{f'(\lambda)}{1!} & \dfrac{f''(\lambda)}{2!} & \cdots & \dfrac{f^{n-2}(\lambda)}{(n-2)!} & \dfrac{f^{(n-1)}(\lambda)}{(n-1)!} \\[2mm] 0 & f(\lambda) & \dfrac{f'(\lambda)}{1!} & \cdots & \dfrac{f^{(n-3)}(\lambda)}{(n-3)!} & \dfrac{f^{(n-2)}(\lambda)}{(n-2)!} \\[2mm] \vdots & \vdots & \vdots & \ddots & \vdots & \vdots \\[2mm] 0 & 0 & 0 & \cdots & f(\lambda) & \dfrac{f'(\lambda)}{1!} \\[2mm] 0 & 0 & 0 & \cdots & 0 & f(\lambda) \end{pmatrix}.$$

1709. Let C, A be square matrices of the same size and $f(x)$ be a polynomial. Prove that $f(CAC^{-1}) = Cf(A)C^{-1}$.

1710. Calculate e^A, where

a) $\quad A = \begin{pmatrix} 2 & 1 \\ -4 & -2 \end{pmatrix}$;

b) $\quad A = \begin{pmatrix} 0 & 1 & 2 \\ 0 & 0 & 6 \\ 0 & 0 & 0 \end{pmatrix}$.

1711. Calculate ln A, where

a) $\quad A = \begin{pmatrix} 3 & 1 \\ -4 & -1 \end{pmatrix}$;

b) $\quad A = \begin{pmatrix} 1 & 1 & 0 & \dots & 0 \\ 0 & 1 & 1 & \dots & 0 \\ & & \dots\dots\dots & & \\ 0 & 0 & 0 & \dots & 1 \end{pmatrix}$.

1712. Let $A = (a_{ij})$ be a matrix of size $m \times n$. Prove that $A = \sum_{i,j} a_{ij} E_{ij}$, where E_{ij} are the *matrix units*.

1713. Prove that $E_{ij} E_{pq} = \delta_{ip} E_{iq}$.

1714. Let A be an arbitrary matrix. Calculate $E_{ij} A$.

1715. Let A be an arbitrary matrix. Calculate $A E_{ij}$.

1716. Let A be a square matrix such that $E_{i1} A = A E_{i1}$ for any matrix unit E_{i1}. Prove $A = \lambda E$ for some scalar λ.

1717. Let A be a square matrix, and $E_{ii} A = A E_{ii}$ for every index i. Prove that A is diagonal.

1718. Suppose that the square matrix A commutes with all nonsingular matrices. Prove that $A = \lambda E$ for some scalar λ.

1719. Find all matrices A of size n such that $\operatorname{tr} AX = 0$ for any matrix X of size n.

1720. Prove that the trace of the product of two matrices does not depend on the order of the factors.

1721. Prove that if C is a nonsingular matrix then $\operatorname{tr} CAC^{-1} = \operatorname{tr} A$ for any matrix A of the same size.

1722. For what λ does the equation $[X, Y] = \lambda E$ have a solution, where $[X, Y]$ is the *commutator* of matrices X and Y?

1723. Prove that for any square matrices A, B, C

a) $[A, BC] = [A, B]C + B[A, C]$;

b) $[[A, B], C] + [[B, C], A] + [[C, A], B] = 0$.

1724. Prove that for any matrices of size 2, $[[A, B]^2, C] = 0$.

1725. Let $A, B, \dots, D_1$ be square matrices of the same size. Find the expression of the product of matrices

$$\begin{pmatrix} A & B \\ C & D \end{pmatrix} \cdot \begin{pmatrix} A_1 & B_1 \\ C_1 & D_1 \end{pmatrix}$$

in terms of the given matrices.

$$* \quad * \quad *$$

1726. Let A be a triangular real matrix commuting with ^tA. Prove that A is diagonal.

1727. Let $A = (a_{ij}) \in \mathbf{M}_n(\mathbb{R})$ be a symmetric nonsingular matrix such that $a_{ij} = 0$ if $|i - j| \geq k$ for some fixed index $k < n$. Assume that $A = {}^tB \cdot B$ where $B = (b_{ij})$ is upper triangular. Prove that $b_{ij} = 0$ for $j - i \geq k$.

1728. Prove that any matrix with zero trace is a sum of commutators of matrices with zero traces.

1729. For the matrix

$$\begin{pmatrix} 0 & 1 & 0 & \ldots & 0 \\ 0 & 0 & 1 & \ldots & 0 \\ & & \cdots\cdots\cdots & & \\ 0 & 0 & 0 & \ldots & 1 \\ 0 & 0 & 0 & \ldots & 0 \end{pmatrix}$$

find matrices A and B such that

$$[A, X] = X, \quad [A, B] = -B, \quad [X, B] = A.$$

18 Matrix equations. Inverse matrix

1801. Solve the systems of matrix equations:

a) $\qquad X + Y = \begin{pmatrix} 1 & 1 \\ 0 & 1 \end{pmatrix}, \qquad 2X + 3Y = \begin{pmatrix} 1 & 1 \\ 0 & 1 \end{pmatrix};$

b) $\qquad 2X - Y = \begin{pmatrix} 0 & 1 \\ -1 & 0 \end{pmatrix}, \qquad -4X + 2Y = \begin{pmatrix} 0 & -2 \\ 2 & 0 \end{pmatrix}.$

1802. Prove that the square matrix X of size 2 is a root of the polynomial

$$X^2 - (\operatorname{tr} X)X + \det X = 0.$$

1803. Solve the matrix equations:

a) $\qquad \begin{pmatrix} 1 & 3 \\ 1 & 2 \end{pmatrix} X = \begin{pmatrix} 1 & 1 \\ 1 & 1 \end{pmatrix};$

b) $\qquad X \begin{pmatrix} -1 & 1 \\ 3 & -4 \end{pmatrix} = \begin{pmatrix} -2 & -1 \\ 3 & 4 \end{pmatrix};$

c) $\begin{pmatrix} 2 & -1 \\ 4 & -2 \end{pmatrix} X = \begin{pmatrix} 1 & 3 \\ 2 & 6 \end{pmatrix};$

d) $X \begin{pmatrix} 2 & -1 \\ 4 & -2 \end{pmatrix} = \begin{pmatrix} 1 & 3 \\ 6 & 2 \end{pmatrix};$

e) $\begin{pmatrix} 3 & 1 \\ 2 & 1 \end{pmatrix} X \begin{pmatrix} 1 & 3 \\ 1 & 2 \end{pmatrix} = \begin{pmatrix} 3 & 3 \\ 2 & 2 \end{pmatrix};$

f) $\begin{pmatrix} 1 & 2 & -3 \\ 3 & 2 & -4 \\ 2 & -1 & 0 \end{pmatrix} X = \begin{pmatrix} 1 & -3 & 0 \\ 10 & 2 & 7 \\ 10 & 7 & 8 \end{pmatrix};$

g) $X \begin{pmatrix} 5 & 3 & 1 \\ 1 & -3 & -2 \\ -5 & 2 & 1 \end{pmatrix} = \begin{pmatrix} -8 & 3 & 0 \\ -5 & 9 & 0 \\ -2 & 15 & 0 \end{pmatrix};$

h) $\begin{pmatrix} 1 & 1 & \dots & 1 \\ 0 & 1 & \dots & 1 \\ \cdots\cdots\cdots \\ 0 & 0 & \dots & 1 \end{pmatrix} X = \begin{pmatrix} 1 & 2 & 3 & \dots & n \\ 0 & 1 & 2 & \dots & n-1 \\ \cdots\cdots\cdots\cdots\cdots \\ 0 & 0 & 0 & \dots & 1 \end{pmatrix};$

i) $X \begin{pmatrix} 1 & 1 & -1 \\ 2 & 1 & 0 \\ 1 & -1 & 1 \end{pmatrix} = \begin{pmatrix} 1 & -1 & 3 \\ 4 & 3 & 2 \\ 1 & -2 & 5 \end{pmatrix};$

j) $\begin{pmatrix} 1 & 2 & 1 \\ 2 & 1 & 2 \\ 1 & 2 & 1 \end{pmatrix} X = \begin{pmatrix} 2 & 1 & 0 \\ 1 & 1 & 2 \\ -1 & 2 & 1 \end{pmatrix};$

k) $\begin{pmatrix} 2 & 1 & 0 \\ 1 & 2 & 0 \\ 0 & 0 & 1 \end{pmatrix} X \begin{pmatrix} 0 & 0 & 1 \\ 0 & 1 & 0 \\ 1 & 0 & 0 \end{pmatrix} = \begin{pmatrix} 0 & 1 & 0 \\ 1 & 0 & 0 \\ 0 & 0 & 0 \end{pmatrix};$

l) $X \begin{pmatrix} 1 & 1 & 1 \\ 1 & 2 & 3 \\ 1 & 4 & 9 \end{pmatrix} = \begin{pmatrix} 1 & 2 & 3 \\ 2 & 4 & 6 \\ 3 & 6 & 9 \end{pmatrix};$

m) $\begin{pmatrix} 1 & 2 & 3 \\ 2 & 3 & 1 \\ 3 & 1 & 2 \end{pmatrix} X = \begin{pmatrix} 0 & 0 & 1 \\ 1 & 0 & 0 \\ 0 & 1 & 0 \end{pmatrix};$

n) $\begin{pmatrix} 1 & 1 & 0 \\ 2 & 1 & 2 \\ 0 & 1 & 1 \end{pmatrix} X = \begin{pmatrix} 5 & -1 & 2 \\ -6 & 6 & 6 \\ -2 & 1 & 7 \end{pmatrix}.$

1804. Let A, B be matrices of sizes $m \times n$ and $m \times k$, respectively. Prove that the matrix equation $AX = B$, where X is a matrix of size $n \times k$, has a solution if and only if the rank of A coincides with rank of the augmented matrix $(A|B)$.

1805. Let A be a square matrix. Prove that the matrix equation $AX = B$ has a unique solution if and only if A is nonsingular.

1806. Let A be a matrix of size $n \times m$ where $m \neq n$. Prove that for any natural number k there exists a matrix B of size $n \times k$ such that either the matrix equation $AX = B$ has no solutions or the solution of $AX = B$ is not unique.

1807. Prove that the system of equations

$$\sum_{j=1}^{n} a_{ij} X_j = B_i \quad (i = 1, 2, \ldots, n),$$

where X_j and B_i are matrices of size $p \times q$ has a unique solution if and only if $\det(a_{ij}) \neq 0$.

1808. With the help of the adjoint matrix find the inverse of the matrix:

a) $\begin{pmatrix} 1 & 3 \\ 0 & 1 \end{pmatrix}$;

b) $\begin{pmatrix} 1 & 0 \\ 3 & 2 \end{pmatrix}$;

c) $\begin{pmatrix} 1 & 2 \\ 3 & 5 \end{pmatrix}$;

d) $\begin{pmatrix} 1 & 3 \\ 2 & 7 \end{pmatrix}$;

e) $\begin{pmatrix} 5 & 0 & 0 \\ 0 & 3 & 0 \\ 0 & 0 & -2 \end{pmatrix}$;

f) $\begin{pmatrix} 1 & 0 & 0 \\ 0 & 1 & 0 \\ 3 & 0 & 1 \end{pmatrix}$;

g) $\begin{pmatrix} 6 & 0 & 0 \\ 0 & 1 & 2 \\ 0 & 3 & 5 \end{pmatrix}$;

h) $\begin{pmatrix} 1 & 3 & 0 \\ 2 & 7 & 0 \\ 0 & 0 & 7 \end{pmatrix}$;

i) $\begin{pmatrix} 1 & 1 & 0 \\ 0 & 1 & 0 \\ 0 & 3 & 3 \end{pmatrix}$;

j) $\begin{pmatrix} 2 & 0 & 0 \\ 3 & 1 & 1 \\ 0 & 0 & 2 \end{pmatrix}$;

k) $\begin{pmatrix} \cos \alpha & -\sin \alpha \\ \sin \alpha & \cos \alpha \end{pmatrix}$.

1809. Using elementary row operations, find the inverse of the matrix:

a) $\begin{pmatrix} 1 & 0 & 0 & 0 \\ 0 & 0 & 1 & 0 \\ 0 & 0 & 0 & 1 \\ 0 & 1 & 0 & 0 \end{pmatrix}$;

b) $\begin{pmatrix} 0 & 0 & 1 & 0 \\ 1 & 0 & 0 & 0 \\ 0 & 0 & 0 & 1 \\ 0 & 1 & 0 & 0 \end{pmatrix}$;

c) $\begin{pmatrix} 2 & 0 & 0 & 0 \\ 0 & 0 & 0 & 1 \\ 0 & 2 & 0 & 0 \\ 0 & 0 & 1 & 0 \end{pmatrix}$;

d) $\begin{pmatrix} 0 & 0 & 0 & -1 \\ 0 & 0 & 2 & 0 \\ 1 & 0 & 0 & 0 \\ 0 & 3 & 0 & 0 \end{pmatrix}$;

e) $\begin{pmatrix} 1 & 1 & \cdots & 1 \\ 0 & 1 & \cdots & 1 \\ \vdots & & & \vdots \\ 0 & 0 & \cdots & 1 \end{pmatrix}$;

f) $\begin{pmatrix} 1 & 0 & 0 & \cdots & 0 & 0 \\ 1 & 1 & 0 & \cdots & 0 & 0 \\ 0 & 1 & 1 & \cdots & 0 & 0 \\ \vdots & & & & & \vdots \\ 0 & 0 & 0 & \cdots & 1 & 1 \end{pmatrix}$;

g) $\begin{pmatrix} 2 & 5 & 7 \\ 6 & 3 & 4 \\ 5 & -2 & -3 \end{pmatrix}$;

h) $\begin{pmatrix} 3 & -4 & 5 \\ 2 & -3 & 1 \\ 3 & -5 & -1 \end{pmatrix}$;

i) $\begin{pmatrix} 2 & 7 & 3 \\ 3 & 9 & 4 \\ 1 & 5 & 3 \end{pmatrix}$;

j) $\begin{pmatrix} 1 & 2 & 2 \\ 2 & 1 & -2 \\ 2 & -2 & 1 \end{pmatrix}$;

k) $\begin{pmatrix} 1 & 2 & 3 & 4 \\ 2 & 3 & 1 & 2 \\ 1 & 1 & 1 & -1 \\ 1 & 0 & -2 & -6 \end{pmatrix}$.

1810. Find the inverse of the square matrix:

a) $\begin{pmatrix} A & 0 \\ B & C \end{pmatrix}$;

b) $\begin{pmatrix} A & B \\ 0 & C \end{pmatrix}$.

where A, C are nonsingular matrices.

1811. Find the inverse of the matrix:

a) $\begin{pmatrix} 1 & 2 & 0 & 0 \\ 2 & 3 & 0 & 0 \\ 1 & -1 & 1 & 3 \\ 0 & 1 & 0 & 2 \end{pmatrix}$;

b) $\begin{pmatrix} 2 & 3 & 1 & 2 \\ 1 & 1 & 2 & 0 \\ 0 & 0 & 1 & -1 \\ 0 & 0 & 1 & -2 \end{pmatrix}$.

1812. Let A, B, C, D be nonsingular matrices. Prove

$$\begin{pmatrix} A & B \\ C & D \end{pmatrix}^{-1} = \begin{pmatrix} (A - BD^{-1}C)^{-1} & (C - DB^{-1}A)^{-1} \\ (B - AC^{-1}D)^{-1} & (D - CA^{-1}B)^{-1} \end{pmatrix}.$$

1813. What is the value of the determinant of

a)   an orthogonal matrix;

b)   an unitary matrix?

1814. What is the value of the determinant of an integer matrix A if its inverse A^{-1} is also an integer matrix?

1815. Let A be a square matrix of size n whose entries are polynomials in a variable t, and suppose that $\det A$ is a nonzero polynomial. Prove that there exists a unique matrix B whose entries are polynomials in t such that $AB = BA = (\det A)E$. Find B if

a) $A = \begin{pmatrix} 1-t & 1+t \\ 1+t^2 & t^3 \end{pmatrix}$;

b) $A = \begin{pmatrix} t & 1 & t \\ -1 & 1 & 1 \\ -t & 1 & t \end{pmatrix}$.

1816. Prove that in the matrix ring over a field

a)   an invertible matrix is not a zero divisor;

b)   any matrix is either invertible or a left and right zero divisor.

1817. Prove that if a matrix $E + AB$ is invertible then the matrix $E + BA$ is invertible too.

1818. Let A and B be matrices of sizes $n \times m$ and $m \times n$, respectively, and suppose that AB and BA are the identity matrices of sizes n and m. Prove that $m = n$.

1819. Let A be a matrix of size $m \times n$ of rank m. Prove that there exists a matrix X of size $n \times m$ such that AX is the identity matrix of size m.

1820. What will happen with the matrix A^{-1} if, in A

a)   we interchange the ith and the jth rows;

b)   we add the jth row multiplied by c to the ith one;

c)   we multiply the ith row by a number $c \neq 0$;

d)   we apply operations a) – c) to columns?

1821. Prove that $(AB)^{-1} = B^{-1}A^{-1}$.

$$* \quad * \quad *$$

1822. Let $\hat{X}$ be the adjoint matrix (see 1604) of a square matrix X. Prove that $\widehat{(AB)} = \hat{B}\hat{A}, \ \widehat{(A^{-1})} = (\hat{A})^{-1}, \ \widehat{({}^t A)} = {}^t(\hat{A})$.

1823. Let B and C be rows of length n such that $C^t B \neq -1$ and let E be the identity matrix of size n. Prove that the matrix $E + {}^t BC$ is invertible.

1824. Let B, C be rows of length n such that $C^t B = -1$ and let E be the identity matrix of size n. Prove that the rank of $E + {}^t BC$ is equal to $n - 1$.

19   Special matrices

1901. Prove that $E_{ii} - E_{jj} = [E_{ij}, E_{ji}]$ if $i \neq j$.

1902. Factorize the matrix into a product of elementary matrices:

a) $\begin{pmatrix} 1 & 2 \\ 4 & 5 \end{pmatrix}$;

b) $\begin{pmatrix} 0 & 1 & 1 \\ 1 & 0 & 1 \\ 1 & 1 & 0 \end{pmatrix}$.

1903. Using the properties of elementary matrices, multiply the following matrices:

a)
$$\begin{pmatrix} 1 & 2 & 3 & 4 \\ 1 & 3 & 5 & 7 \\ 1 & 2 & 4 & 8 \\ 1 & 1 & 1 & 1 \end{pmatrix} \cdot \begin{pmatrix} 1 & 0 & 0 & 0 \\ 0 & 2 & 0 & 0 \\ 0 & 0 & 3 & 0 \\ 0 & 0 & 0 & 4 \end{pmatrix};$$

b)
$$\begin{pmatrix} 1 & 0 & 0 & 0 \\ 0 & 2 & 0 & 0 \\ 0 & 0 & 3 & 0 \\ 0 & 0 & 0 & 4 \end{pmatrix} \cdot \begin{pmatrix} 1 & 2 & 3 & 4 \\ 1 & 3 & 5 & 7 \\ 1 & 2 & 4 & 8 \\ 1 & 1 & 1 & 1 \end{pmatrix};$$

c)
$$\begin{pmatrix} 1 & 2 & 3 & 4 \\ 1 & 3 & 5 & 7 \\ 1 & 2 & 4 & 8 \\ 1 & 1 & 1 & 1 \end{pmatrix} \cdot \begin{pmatrix} 1 & 0 & 0 & 0 \\ 0 & 1 & 0 & 0 \\ 2 & 0 & 1 & 0 \\ -3 & 0 & 0 & 1 \end{pmatrix};$$

d)
$$\begin{pmatrix} 1 & 0 & 0 & 0 \\ 1 & 1 & 0 & 0 \\ 2 & 0 & 1 & 0 \\ -3 & 0 & 0 & 1 \end{pmatrix} \cdot \begin{pmatrix} 1 & 2 & 3 & 4 \\ 1 & 3 & 5 & 7 \\ 1 & 2 & 4 & 8 \\ 1 & 1 & 1 & 1 \end{pmatrix}.$$

1904. Prove the following properties of the transposes:

a) $^t(A + B) = {}^tA + {}^tB$;

b) $^t(\lambda A) = \lambda\, {}^tA$;

c) $^t(AB) = {}^tB \cdot {}^tA$;

d) $({}^tA)^{-1} = {}^t(A^{-1})$;

e) $^t({}^tA) = A$.

1905. Prove that any matrix has a unique representation as the sum of a *symmetric* matrix and a *skew-symmetric* one.

1906. Prove that

a) if matrices A and B are *orthogonal* then matrices A^{-1} and AB are orthogonal;

b) if complex matrices A and B are *unitary* then matrices A^{-1} and AB are also unitary.

1907. Prove that

a) the product of two symmetric or skew-symmetric matrices is symmetric if and only if these matrices commute;

b) the product of a symmetric matrix and a skew-symmetric one is skew-symmetric if and only if these matrices commute.

1908. Under what condition is the product of two *Hermitian* or *skew-Hermitian* matrices Hermitian?

1909. Prove that for any complex square matrix X there exists a matrix Y such that $XYX = X$, $YXY = Y$, and matrices XY and YX are Hermitian.

1910. Prove that the inverse of a (skew-)symmetric matrix is (skew-)symmetric.

1911. Prove that if both of the matrices A and B are symmetric or skew-symmetric then the commutator $[A, B]$ is skew-symmetric.

1912. Is it the case that any skew-symmetric matrix is a sum of commutators of skew-symmetric matrices?

1913. Find all symmetric and skew-symmetric orthogonal matrices of size 2.

1914. Find all *lower nil-triangular* matrices commuting with all lower nil-triangular matrices of the same size.

1915. Prove that the sum of two commuting *nilpotent* matrices is a nilpotent matrix. Is this statement valid for non-commuting matrices?

1916. Prove that if matrices A, B and $[A, B]$ are nilpotent and matrices A, B commute with $[A, B]$, then $A + B$ is nilpotent.

1917. Prove that if the matrix A of size 2 is nilpotent then $A^2 = 0$.

1918. Prove that any lower nil-triangular matrix is nilpotent.

1919. Prove that if the matrix A is nilpotent then the matrices $E - A$ and $E + A$ are invertible.

1920. Prove that if the matrix A is nilpotent and the constant term of the polynomial $f(t)$ is not equal to zero then the matrix $f(A)$ is invertible.

1921. Solve the equation $AX + X + A = 0$ where A is a nilpotent matrix.

1922. Prove that a nilpotent matrix of size 2 has zero trace.

1923. Prove that the product of two commuting *periodic* matrices is a periodic matrix. Is this statement valid for non-commuting matrices?

1924. Prove that the matrix CAC^{-1} is nilpotent (periodic) if and only if the matrix A is nilpotent (periodic).

1925. Let σ be a permutation of the set $\{1, 2, \ldots, n\}$ and $A_\sigma = (\delta_{i\sigma(j)})$, where δ_{ij} denotes the *Kronecker delta*. Prove that

a) the matrix A_σ is periodic;

b) for any permutations σ and τ, we have $A_{\sigma\tau} = A_\sigma A_\tau$;

c) A_σ can be factorized into a product of at most $n - 1$ elementary matrices.

1926. Prove that the product of upper-triangular matrices is upper-triangular.

1927. Prove that the inverse of a unitriangular matrix is again unitriangular.

CHAPTER 5

Complex numbers

20 Complex numbers in algebraic form

2001. Calculate the expressions:

a) $(2+i)(3-i)+(2+3i)(3+4i)$;

b) $(2+i)(3+7i)-(1+2i)(5+3i)$;

c) $(4+i)(5+3i)-(3+i)(3-i)$;

d) $\dfrac{(5+i)(7-6i)}{3+i}$;

e) $\dfrac{(5+i)(3+5i)}{2i}$;

f) $\dfrac{(1+3i)(8-i)}{(2+i)^2}$;

g) $\dfrac{(2+i)(4+i)}{1+i}$;

h) $\dfrac{(3-i)(1-4i)}{2-i}$;

i) $(2+i)^3+(2-i)^3$;

j) $(3+i)^3-(3-i)^3$;

k) $\dfrac{(1+i)^5}{(1-i)^3}$;

l) $\left(-\dfrac{1}{2}\pm\dfrac{\sqrt{3}}{2}i\right)^3$.

2002. Calculate $i^{77}, i^{98}, i^{-57}, i^n$, where n is an integer.

2003. Prove the equalities:

a) $(1+i)^{8n}=2^{4n}, \quad (n\in\mathbb{Z})$;

b) $(1+i)^{4n}=(-1)2^{2n}, \quad (n\in\mathbb{Z})$.

2004. Solve the systems of equations:

a) $(1+i)z_1+(1-i)z_2=1+i$,
$(1-i)z_1+(1+i)z_2=1+3i$;

b) $iz_1+(1+i)z_2=2+2i$,
$2iz_1+(3+2i)z_2=5+3i$;

c) $(1 - i)z_1 - 3iz_2 = -i,$
     $2z_1 - (3 + 3i)z_2 = 3 - i;$

d) $2z_1 - (2 + i)z_2 = -i,$
     $(4 - 2i)z_1 - 5z_2 = -1 - 2i;$

e) $x + iy - 2z = 10,$
     $x - y + 2iz = 20,$
     $ix + 3iy - (1 + i)z = 30.$

2005. Find real numbers x and y which satisfy the equations:

a) $(2 + i)x + (1 + 2i)y = 1 - 4i;$ b) $(3 + 2i)x + (1 + 3i)y = 4 - 9i.$

2006. Prove that

   a)   a complex number z is real if and only if $\bar{z} = z;$

   b)   a complex number z is purely imaginary if and only if $\bar{z} = -z.$

2007. Prove that

   a)   the product of two complex numbers is real if and only if one of them is conjugate to the other multiplied by a real factor;

   b)   a sum and a product of two complex numbers are real if and only if these numbers are either real or conjugate.

2008. Find all complex numbers conjugate to their own

   a)   square;

   b)   cube.

2009. Prove that if the number z is obtained from the given complex numbers $z_1, z_2, \ldots, z_n$ with the help of finitely many operations of addition, subtraction, multiplication and division, then the number $\bar{z}$ is obtained from $\bar{z}_1, \bar{z}_2, \ldots, \bar{z}_n$ with the help of the same operations.

2010. Prove that the determinant

$$\begin{vmatrix} z_1 & \bar{z}_1 & a \\ z_2 & \bar{z}_2 & b \\ z_3 & \bar{z}_3 & c \end{vmatrix},$$

is a purely imaginary number, provided z_1, z_2, z_3 are complex numbers and a, b, c are real numbers.

2011. Solve the equations:

a) $z^2 = i;$ b) $z^2 = 3 - 4i;$

c) $z^2 = 5 - 12i$;

d) $z^2 - (1+i)z + 6 + 3i = 0$;

e) $z^2 - 5z + 4 + 10i = 0$;

f) $z^2 + (2i - 7)z + 13 - i = 0$.

21 Complex numbers in trigonometric form

2101. Find the trigonometric form of the numbers:

a) 5;

b) i;

c) -2;

d) $-3i$;

e) $1+i$;

f) $1-i$;

g) $1+i\sqrt{3}$;

h) $-1+i\sqrt{3}$;

i) $-1-i\sqrt{3}$;

j) $1-i\sqrt{3}$;

k) $\sqrt{3}+i$;

l) $-\sqrt{3}+i$;

m) $-\sqrt{3}-i$;

n) $\sqrt{3}-i$;

o) $1+i\dfrac{\sqrt{3}}{3}$;

p) $2+\sqrt{3}+i$;

q) $1-(2+\sqrt{3})i$;

r) $\cos\alpha - i\sin\alpha$;

s) $\sin\alpha + i\cos\alpha$;

t) $\dfrac{1+i\tan\alpha}{1-i\tan\alpha}$;

u) $1+\cos\varphi + i\sin\varphi$, $\varphi \in [-\pi, \pi]$;

v) $\dfrac{\cos\phi + i\sin\varphi}{\cos\psi + i\sin\psi}$.

2102. Calculate the expressions:

a) $(1+i)^{1000}$;

b) $(1+i\sqrt{3})^{150}$;

c) $(\sqrt{3}+i)^{30}$;

d) $\left(1+\dfrac{\sqrt{3}}{2}+\dfrac{i}{2}\right)^{24}$;

e) $(2-\sqrt{2}+i)^{12}$;

f) $\left(\dfrac{1-i\sqrt{3}}{1+i}\right)^{12}$;

g) $\left(\dfrac{\sqrt{3}+i}{1-i}\right)^{30}$;

h) $\dfrac{(-1+i\sqrt{3})^{15}}{(1-i)^{20}} + \dfrac{(-1-i\sqrt{3})^{15}}{(1+i)^{20}}$.

2103. Solve the equations:

a) $|z| + z = 8 + 4i$;

b) $|z| - z = 8 + 12i$.

2104. Prove the following properties of absolute values of complex numbers:

a) $|z_1 \pm z_2| \le |z_1| + |z_2|$;

b) $||z_1| - |z_2|| \le |z_1 \pm z_2|$;

c) $|z_1 + z_2| = |z_1| + |z_2|$ if and only if the vectors z_1 and z_2 have the same direction;

d) $|z_1 + z_2| = ||z_1| - |z_2||$ if and only if the vectors z_1 and z_2 have the opposite direction.

2105. Prove that

a) if $|z| < 1$ then $|z^2 - z + i| < 3$;

b) if $|z| \le 2$ then $1 \le |z^2 - 5| \le 9$;

c) if $|z| < \dfrac{1}{2}$ then $|(1 + i)z^3 + iz| < \dfrac{3}{4}$.

2106. Prove the inequality

$$|z_1 - z_2| \le ||z_1| - |z_2|| + \min\{|z_1|, |z_2|\} \cdot |\arg z_1 - \arg z_2|.$$

In what case does this inequality turn into an equality?

2107. Prove that

$$|z_1| + |z_2| = \left|\frac{z_1 + z_2}{2} - \sqrt{z_1 z_2}\right| + \left|\frac{z_1 + z_2}{2} + \sqrt{z_1 z_2}\right|.$$

2108. Prove the formula of de Moivre:

$$[r(\cos \varphi + i \sin \varphi)]^n = r^n(\cos n\varphi + i \sin n\varphi)$$

for integers $n, r \ne 0$.

2109. Calculate for $n \in \mathbb{Z}$ the expressions:

a) $(1 + i)^n$;

b) $\left(\dfrac{1 - i\sqrt{3}}{2}\right)^n$;

c) $\left(\dfrac{1 - i \tan \alpha}{1 + i \tan \alpha}\right)^n$;

d) $(1 + \cos \varphi + i \sin \varphi)^n$.

2110. Prove that if $z + z^{-1} = 2 \cos \varphi$, then $z^n + z^{-n} = 2 \cos n\varphi$, where $n \in \mathbb{Z}$.

2111. Represent the following functions as polynomials in $\sin x$ and $\cos x$:

a) $\sin 4x$;

b) $\cos 4x$;

c) $\sin 5x$;

d) $\cos 5x$.

2112. Prove the equalities:

a) $\cos nx = \sum_{k=0}^{[n/2]} (-1)^k \binom{n}{2k} \cos^{n-2k} x \cdot \sin^{2k} x$;

b) $\sin nx = \sum_{k=0}^{[(n-1)/2]} (-1)^k \binom{n}{2k+1} \cos^{n-2k-1} x \cdot \sin^{2k+1} x$.

2113. Express the following functions as linear combinations of $\sin kx$ and $\cos kx$, $k \in \mathbb{Z}$:

a) $\sin^4 x$;

b) $\cos^4 x$;

c) $\sin^5 x$;

d) $\cos^5 x$.

2114. Prove the equalities:

a) $\cos^{2m} x = \dfrac{1}{2^{2m-1}} \left[\sum_{k=0}^{m-1} \binom{2m}{k} \cos(2m - 2k)x + \dfrac{1}{2}\binom{2m}{m} \right]$;

b) $\cos^{2m+1} x = \dfrac{1}{2^{2m}} \sum_{k=0}^{m} \binom{2m+1}{k} \cos(2m + 1 - 2k)x$;

c) $\sin^{2m} x = \dfrac{(-1)^m}{2^{2m-1}} \left[\sum_{k=0}^{m-1} (-1)^k \binom{2m}{k} \cos(2m - 2k)x + \dfrac{(-1)^m}{2}\binom{2m}{m} \right]$;

d) $\sin^{2m+1} x = \dfrac{(-1)}{2^{2m}} \sum_{k=0}^{m} (-1)^k \binom{2m+1}{k} \sin(2m + 1 - 2k)x$.

22   Roots of complex numbers. Cyclotomic polynomials

2201. Prove that if a complex number z is a root of order n of a real number a, then the conjugate number $\bar{z}$ is a root of order n of a too.

2202. Prove that if

$$\sqrt[n]{z} = \{z_1, z_2, \ldots, z_n\},$$

then

$$\sqrt[n]{\bar{z}} = \{\bar{z}_1, \bar{z}_2, \dots, \bar{z}_n\}.$$

2203. Which sets $\sqrt[n]{z}$ contain a real number?

2204.[1] Let z and w be complex numbers. Prove the equalities:

a) $\sqrt[n]{z^n w} = z \sqrt[n]{w}$;

b) $\sqrt[n]{-z^n w} = -z \sqrt[n]{w}$;

c) $\sqrt[n]{zw} = u \sqrt[n]{w}$, where u is one of the values of $\sqrt[n]{z}$.

2205. Prove that the union of sets $\sqrt[n]{z}$ and $\sqrt[n]{-z}$ is equal to the set $\sqrt[2n]{z^2}$.

2206. Is the equality $\sqrt[ns]{z^s} = \sqrt[n]{z}$ ($s > 1$) valid?

2207. Calculate

a) $\sqrt[6]{i}$;

b) $\sqrt[10]{512(1 - i\sqrt{3})}$;

c) $\sqrt[6]{2\sqrt{2}(1 - i)}$;

d) $\sqrt[3]{1}$;

e) $\sqrt[4]{1}$;

f) $\sqrt[6]{i}$;

g) $\sqrt[3]{i}$;

h) $\sqrt[4]{-4}$;

i) $\sqrt[6]{64}$;

j) $\sqrt[8]{16}$;

k) $\sqrt[6]{-27}$;

l) $\sqrt[4]{8\sqrt{3}i - 8}$;

m) $\sqrt[4]{-72(1 - i\sqrt{3})}$;

n) $\sqrt[3]{1 + i}$;

o) $\sqrt[3]{2 - 2i}$;

p) $\sqrt[4]{-\dfrac{18}{1 + i\sqrt{3}}}$;

q) $\sqrt[4]{\dfrac{7 - 2i}{1 + i\sqrt{2}} + \dfrac{4 + 14i}{\sqrt{2} + 2i} - (8 - 2i)}$;

r) $\sqrt[3]{\dfrac{1 - 5i}{1 + i} - 5\dfrac{1 + 2i}{2 - i} + 2}$;

s) $\sqrt[4]{\dfrac{-2 + 2\sqrt{3}i}{2 + i\sqrt{5}} - 5\dfrac{\sqrt{3} + i}{2\sqrt{5} + 5i}}$.

---

[1] By definition, the set zA is equal to $\{za \mid a \in A\}$.

2208. Find, in two different ways, the roots of 1 of order 5 and express in radicals:

a) $\cos \dfrac{2\pi}{5}$;

b) $\sin \dfrac{2\pi}{5}$;

c) $\cos \dfrac{4\pi}{5}$;

d) $\sin \dfrac{4\pi}{5}$.

2209. Solve the equations:

a) $$(z+1)^n + (z-1)^n = 0;$$

b) $$(z+1)^n - (z-1)^n = 0;$$

c) $$(z+i)^n + (z-i)^n = 0.$$

2210. Express in radicals the roots of 1 of orders 2, 3, 4, 6, 8, 12.

2211. Find the product of all roots of 1 of order n.

2212. Let $\varepsilon_k = \cos \dfrac{2\pi k}{n} + i \sin \dfrac{2\pi k}{n}$ $(0 \le k < n)$. Prove that

a) $\sqrt[n]{1} = \{\varepsilon_0, \varepsilon_1, \ldots, \varepsilon_{n-1}\}$;

b) $\varepsilon_k = \varepsilon_1^k$ $(0 \le k < n)$;

c) $\varepsilon_k \varepsilon_l = \begin{cases} \varepsilon_{k+l}, & \text{if } k+l < n, \quad (0 \le k < n, \\ \varepsilon_{k+l-n}, & \text{if } k+l \ge n \quad 0 \le l < n); \end{cases}$

d) the set $\mathbf{U}_n$ of order n roots of 1 is a cyclic group of order n with respect to multiplication;

e) every cyclic group of order n is isomorphic to the group $\mathbf{U}_n$.

2213. Prove that

a) if the integers r and s are coprime and $\alpha^r = \alpha^s = 1$, then $\alpha = 1$;

b) if d is the greatest common divisor of the integers r and s, then $\mathbf{U}_r \cap \mathbf{U}_s = \mathbf{U}_d$;

c) if integers r and s are coprime, then any root of 1 of order rs has a unique presentation as the product of a root of order r and a root of order s.

2214. Prove that the following statements are equivalent:

a) ε is a primitive root of 1 of order n;

b) the order of ε in the group $\mathbf{U}_n$ is equal to n;

c) ε is a generator of the group $\mathbf{U}_n$.

2215. Prove that if ε is a primitive root of order n of 1 then $\bar{\varepsilon}$ is also a primitive root of order n of 1.

2216. Prove that if integers r and s are coprime, then ε is a primitive root of order rs of 1 if and only if ε is the product of a primitive root of order r and a primitive root of order s.

2217.

a) Let z be a primitive root of order n of 1. Calculate $1+2z+3z^2+\cdots+nz^{n-1}$.

b) Let z be a primitive root of order $2n$ of 1. Calculate $1 + z + \cdots + z^{n-1}$.

c) Let z be a root of 1 and $z^n \pm z^m \pm 1 = 0$. Find n and m.

2218. Prove that

a) the number of primitive roots of order n of 1 is equal to $\phi(n)$ (see Exercise 104);

b) if integers m and n are coprime then $\phi(mn) = \phi(m)\phi(n)$.

2219. Prove that if z is a primitive root of odd order n of 1 then $-z$ is a primitive root of order $2n$.

$$* \quad * \quad *$$

2220. Denote by $\sigma(n)$ the sum of all primitive roots of order n of 1. Prove that

a) $\sigma(1) = 1$;

b) if $n > 1$, then $\sum_{d|n} \sigma(d) = 0$;

c) $\sigma(p) = -1$, if p is prime;

d) $\sigma(p^k) = 0$, if p is prime, $k > 1$;

e) $\sigma(rs) = \sigma(r) \cdot \sigma(s)$, if integers r and s are coprime;

f) the function $\sigma(n)$ coincides with the Möbius function $\mu(n)$.

2221. Let d be the (positive) greatest common divisor of the integer s and the natural number n and let ε_i be a primitive root of order n of 1 ($i = 1, 2, \ldots, \phi(n)$). Prove the equality

$$\sum_{i=1}^{\phi(n)} \varepsilon_i^s = \frac{\phi(n)}{\phi\left(\frac{n}{d}\right)} \mu\left(\frac{n}{d}\right).$$

2222. Is the number $\dfrac{2+i}{2-i}$ a root of some order of 1?

2223. Find the cyclotomic polynomials $\Phi_n(x)$ when n is equal to:

a) 1;　b) 2;　c) 3;　d) 4;　e) 6;　f) 12;　g) p, where p is prime;　h) p^k, where p is prime, $k > 1$.

2224. Prove the following properties of cyclotomic polynomials:

a) $\displaystyle\prod_{d|n} \Phi_d(x) = x^n - 1$;

b) $\Phi_{2n}(x) = \Phi_n(-x)$ (n is an odd integer greater than 1);

c) $\Phi_n(x) = \displaystyle\prod_{d|n}(x^d - 1)^{\mu\left(\frac{n}{d}\right)}$;

d) if k is divisible by any prime divisor of n, then

$$\Phi_n(x) = \Phi_k\left(x^{\frac{n}{k}}\right);$$

e) if n is divisible by a prime number p and it is not divisible by p^2, then

$$\Phi_n(x) = \Phi_{\frac{n}{p}}(x^p)\left(\Phi_{\frac{n}{p}}(x)\right)^{-1}.$$

2225. Find the cyclotomic polynomials for $n = 10, 14, 15, 30, 36, 100, 216, 288, 1000$.

2226. Prove that in any cyclotomic polynomial

a) all coefficients are integers;

b) the leading coefficient is equal to 1;

c) the constant term is equal to -1 if $n = 1$ and is equal to 1 if $n > 1$.

2227. Find the sum of all coefficients in the cyclotomic polynomial $\Phi_n(x)$.

23　Calculation of sums and products with the help of complex numbers

2301. Calculate the sums:

a) $1 - \dbinom{n}{2} + \dbinom{n}{4} - \dbinom{n}{6} + \ldots$;

b) $\binom{n}{1} - \binom{n}{3} + \binom{n}{5} - \binom{n}{7} + \cdots$;

c) $1 + \binom{n}{4} + \binom{n}{8} + \cdots$;

d) $\binom{n}{1} + \binom{n}{5} + \binom{n}{9} + \cdots$.

2302. Prove the equalities:

a) $\cos x + \cos 2x + \cdots + \cos nx = \dfrac{\sin \dfrac{nx}{2} \cos \dfrac{(n+1)x}{2}}{\sin \dfrac{x}{2}}$ $(x \neq 2k\pi, k \in \mathbb{Z})$;

b) $\sin x + \sin 2x + \cdots + \sin nx = \dfrac{\sin \dfrac{nx}{2} \sin \dfrac{(n+1)x}{2}}{\sin \dfrac{x}{2}}$ $(x \neq 2k\pi, k \in \mathbb{Z})$;

c) $\cos \dfrac{\pi}{n} + \cos \dfrac{3\pi}{n} + \cos \dfrac{5\pi}{n} + \cdots + \cos \dfrac{(2n-1)\pi}{n} = 0$;

d) $\sin \dfrac{\pi}{n} + \sin \dfrac{3\pi}{n} + \sin \dfrac{5\pi}{n} + \cdots + \sin \dfrac{(2n-1)\pi}{n} = 0$;

e) $\dfrac{1}{n} \sum\limits_{k=0}^{n-1} (x + \varepsilon_k y)^n = x^n + y^n$ $(\varepsilon_0, \varepsilon_1, \ldots, \varepsilon_{n-1}$ are roots of 1 of order n);

f) $x^{2n+1} - 1 = (x - 1) \prod\limits_{k=1}^{n} \left(x^2 - 2x \cos \dfrac{\pi k}{2n+1} + 1 \right)$;

g) $x^{2n} - 1 = (x^2 - 1) \prod\limits_{k=1}^{n-1} \left(x^2 - 2x \cos \dfrac{\pi k}{n} + 1 \right)$;

h) $\prod\limits_{k=1}^{n-1} \sin \dfrac{\pi k}{2n} = \dfrac{\sqrt{n}}{2^{n-1}}$;

i) $\prod\limits_{k=1}^{n} \sin \dfrac{\pi k}{2n+1} = \dfrac{\sqrt{2n+1}}{2^n}$.

*   *   *

2303. Solve the equation

$$\cos\varphi + \binom{n}{1}\cos(\varphi+\alpha)x + \binom{n}{2}\cos(\varphi+2\alpha)x^2 + \cdots + \binom{n}{n}\cos(\varphi+n\alpha)x^n = 0.$$

2304. Prove that

a) $\quad 1 + \binom{n}{3} + \binom{n}{6} + \cdots = \dfrac{1}{3}\left(2^n + 2\cos\dfrac{\pi n}{3}\right);$

b) $\quad \binom{n}{1} + \binom{n}{4} + \binom{n}{7} + \cdots = \dfrac{1}{3}\left(2^n + 2\cos\dfrac{(n-2)\pi}{4}\right);$

c) $\quad \binom{n}{2} + \binom{n}{5} + \binom{n}{8} + \cdots = \dfrac{1}{3}\left(2^n + 2\cos\dfrac{(n-4)\pi}{3}\right);$

d) $\quad 2\cos mx = (2\cos x)^m - \dfrac{m}{1}(2\cos x)^{m-2} + \dfrac{m(m-3)}{1\cdot 2}(2\cos x)^{m-4} +$

$\qquad \cdots + +(-1)^k \dfrac{m(m-k-1)\dots(m-2k+1)}{k!}(2\cos x)^{m-2k} + \dots.$

2305. Find the sums

a) $\quad \cos x + \binom{n}{1}\cos 2x + \cdots + \binom{n}{n}\cos(n+1)x;$

b) $\quad \sin x + \binom{n}{1}\sin 2x + \cdots + \binom{n}{n}\sin(n+1)x;$

c) $\quad \sin^2 x + \sin^2 3x + \cdots + \sin^2(2n-1)x;$

d) $\quad \cos x + 2\cos 2x + 3\cos 3x + \cdots + n\cos nx;$

e) $\quad \sin x + 2\sin 2x + 3\sin 3x + \cdots + n\sin nx.$

2306. Prove

a) $\quad \cos^2 x + \cos^2 2x + \cdots + \cos^2 nx = \dfrac{n}{2} + \dfrac{\cos(n+1)x\sin nx}{2\sin x};$

b) $\quad \sin^2 x + \sin^2 2x + \cdots + \sin^2 nx = \dfrac{n}{2} - \dfrac{\cos(n+1)x\sin nx}{2\sin x}.$

2307. Prove that for an odd natural number m

$$\frac{\sin mx}{\sin x} = (-4)^{(m-1)/2}\prod_{1\le j\le (m-1)/2}\left(\sin^2 x - \sin^2\frac{2\pi j}{m}\right).$$

24 Complex numbers and geometry of a plane

2401. Show in a plane the points which correspond to the numbers $5,\ -2,$ $-3i,\ \pm 1 \pm i\sqrt{3}$.

2402. Find complex numbers corresponding to

a) the vertices of a square with the center at the origin with sides of length 1 which are parallel to the axes of coordinates;

b) the vertices of the regular triangle with the center at the origin, with one side parallel to an axis of coordinates, one vertex situated on the negative real half-axis and with the radius of the circumscribed circle equal to 1;

c) the vertices of the regular hexagon with the center at the point $2 + i\sqrt{3}$, one side parallel to the abscissa and with the radius of the circumscribed circle equal to 2;

d) the vertices of the regular n-gon, with the center at the origin, such that 1 is one of its vertices.

2403. Explain the geometric meaning of the expression $|z_1 - z_2|$, where z_1 and z_2 are given complex numbers.

2404. Indicate the geometric meaning of the number $\arg \dfrac{z_1 - z_2}{z_2 - z_3}$, where $z_1, z_2,$ z_3 are distinct complex numbers.

2405. Where on a plane are the points corresponding to

a) the complex numbers z_1, z_2, z_3, such that

$$z_1 + z_2 + z_3 = 0, \quad |z_1| = |z_2| = |z_3| \neq 0;$$

b) the complex numbers z_1, z_2, z_3, z_4, such that

$$z_1 + z_2 + z_3 + z_4 = 0, \quad |z_1| = |z_2| = |z_3| = |z_4| \neq 0.$$

2406. Show in a plane the set of points corresponding to the complex numbers z, which satisfy the conditions:

a) $|z| = 1;$ b) $\arg z = \dfrac{\pi}{3};$ c) $|z| \leq 2;$ d) $|z - 1 - i| < 1;$

e) $|z + 3 + 4i| \leq 5;$ f) $2 < |z| < 3;$ g) $1 \leq |z - 2i| < 2;$

h) $|\arg z| < \pi/6;$

i) $\alpha < \arg(z - z_0) < \beta$, where $-\pi < \alpha < \beta \leq \pi$ and z_0 is a given complex number;

j) $|\Re z| \le 1$;   k) $-1 < \Re i z < 0$;   l) $|\Im z| = 1$;   m) $|\Re z + \Im z| < 1$;

n) $|z-1| + |z+1| = 3$;   o) $|z+2| - |z-2| = 3$;   p) $|z-2| = \Re z + 2$.

2407. Prove the identity

$$|z + w|^2 + |z - w|^2 = 2|z|^2 + 2|w|^2$$

and indicate its geometric meaning.

2408. Let the complex numbers z_1, z_2, z_3 correspond to the vertices A_1, A_2, A_3 of a parallelogram. Find the number corresponding to the vertex A_4 which is opposite to A_2.

2409. Find the complex numbers corresponding to the opposite vertices of a square, if two of its adjacent vertices correspond to the numbers z and w.

2410. Find the complex numbers corresponding to the vertices of a regular n-gon if two of its adjacent vertices correspond to numbers z_0 and z_1.

2411. Show in a plane the set of points corresponding to complex numbers $z = \dfrac{1 + ti}{1 - ti}$, where $t \in \mathbb{R}$.

2412. Prove that

a) points on a plane, corresponding to the complex numbers z_1, z_2, z_3, are situated on one line if and only if there exist real numbers $\lambda_1, \lambda_2, \lambda_3$, not all of them equal to zero, such that

$$\lambda_1 z_1 + \lambda_2 z_2 + \lambda_3 z_3 = 0, \quad \lambda_1 + \lambda_2 + \lambda_3 = 0;$$

b) points on a plane, corresponding to distinct complex numbers z_1, z_2, z_3, are situated in one line if and only if the number $\dfrac{z_1 - z_3}{z_2 - z_3}$ is a real one;

c) points on a plane, corresponding to distinct complex numbers z_1, z_2, z_3, z_4, which are not situated in one line, belong to a circle if and only if their double ratio $\dfrac{z_1 - z_3}{z_2 - z_3} : \dfrac{z_1 - z_4}{z_2 - z_4}$ is a real number.

2413. Show in a plane the set of points corresponding to the complex numbers z satisfying the equality $\left|\dfrac{z - z_1}{z - z_2}\right| = \lambda$, where $z_1, z_2 \in \mathbb{C}$ and λ is a positive real number.

2414. Find min $|3 + 2i - z|$ if $|z| \le 1$.

2415. Find max $|1 + 4i - z|$ if $|z - 10i + 2| \le 1$.

2416. A *lemniscate*. Show in the plane the set of points corresponding to complex numbers z satisfying the equality $|z^2 - 1| = \lambda$. For $\lambda = 1$ write down the equation of the obtained curve in polar coordinates.

2417. The *extended complex plane* is the complex plane, complemented by 'the point at infinity' ∞. Prove that if (z_1, z_2, z_3) and (w_1, w_2, w_3) are two triples of distinct points of the extended complex plane, then there exists a linear-fractional transformation

$$w = \frac{az + b}{cz + d} \quad (a, b, c, d \in \mathbb{C}, \quad ad - bc \neq 0),$$

which maps the first triple to the second one.

2418. Prove that if, in each of the two quadruples (z_1, z_2, z_3, z_4), (w_1, w_2, w_3, w_4) of the points of the extended complex plane all entries are distinct, then there exists a linear-fractional transformation which maps one of these quadruples into the other if and only if their double ratios

$$\frac{z_1 - z_3}{z_2 - z_3} : \frac{z_1 - z_4}{z_2 - z_4} = \frac{w_1 - w_3}{w_2 - w_3} : \frac{w_1 - w_4}{w_2 - w_4}$$

coincide.

2419. Prove that a linear-fractional transformation of the extended complex plane maps lines and circles to lines and circles.

2420. Prove that the linear-fractional transformation $w = \dfrac{az + b}{cz + d}$, $(ad - bc = 1)$, is mapping the real axis to itself if and only if the matrix $\begin{pmatrix} a & b \\ c & d \end{pmatrix}$ is proportional to a real matrix.

2421. Explain the geometric meaning of the linear-fractional transformation $w = \dfrac{1}{z}$.

2422. Explain the geometric meaning of the transformation of the complex plane given by the formula $w = z^n$ $(n \geq 2)$.

$$* \quad * \quad *$$

2423. Prove that the *Zhukovsky function*, $w = \dfrac{1}{2}\left(z + \dfrac{1}{z}\right)$, maps the hyperbola (ellipses) with the centres at $-1, 1$ into ellipses with the same centers.

2424. Prove that any linear-fractional transformation, which maps the open upper half-plane onto the interior of the unit disc with the center at the origin is of the form

$$w = a\frac{z - b}{z - \bar{b}}, \quad \text{where} \quad |a| = 1, \quad \Im b > 0.$$

2425. Prove that any linear-fractional transformation, which maps the unit disc with the center at the origin into itself, is of the form

$$w = a\frac{z - b}{1 - z\bar{b}}, \quad \text{where} \quad |a| = 1, \quad b < 1.$$

2426. For what complex numbers a does the function $z \to z + az^2$ map the disc $|z| \le 1$ bijectively into itself?

CHAPTER 6

Polynomials

25 Division with a remainder. Euclidean algorithm

2501. Divide the polynomial $f(x)$, with a remainder, by the polynomial $g(x)$:

a) $f(x) = 2x^4 - 3x^2 + 4x^2 - 5x + 6,$ $g(x) = x^2 - 3x + 1;$

b) $f(x) = x^3 - 3x^2 - x - 1,$ $g(x) = 3x^2 - 2x + 1.$

2502. Find the greatest common divisor of the polynomials:

a) $x^4 + x^3 - 3x^2 - 4x - 1$ and $x^3 + x^2 - x - 1;$

b) $x^6 + 2x^4 - 4x^3 - 3x^2 + 8x - 5$ and $x^5 + x^2 - x + 1;$

c) $x^5 + 3x^2 - 2x + 2$ and $x^6 + x^5 + x^4 - 3x^2 + 2x - 6;$

d) $x^4 + x^3 - 4x + 5$ and $2x^3 - x^2 - 2x + 2;$

e) $x^5 + x^4 - x^3 - 2x - 1$ and $3x^4 + 2x^3 + x^2 + 2x - 2;$

f) $x^6 - 7x^4 + 8x^3 - 7x + 7$ and $3x^5 - 7x^3 + 3x^2 - 7;$

g) $x^5 - 2x^4 + x^3 + 7x^2 - 12x + 10$ and $3x^4 - 6x^3 + 5x^2 + 2x - 2;$

h) $x^5 + 3x^4 - 12x^3 - 52x^2 - 52x - 12$ and $x^4 + 3x^3 - 6x^2 - 22x - 12;$

i) $x^5 + x^4 - x^3 - 3x^2 - 3x - 1$ and $x^4 - 2x^3 - x^2 - 2x + 1;$

j) $x^4 - 4x^3 + 1$ and $x^3 - 3x^2 + 1;$

k) $x^4 - 10x^2 + 1$ and $x^4 - 4\sqrt{2}x^3 + 6x^2 + 4\sqrt{2}x + 1.$

2503. Find the greatest common divisor of the polynomials $f(x)$ and $g(x)$ and its linear expression in terms of $f(x)$ and $g(x)$:

a) $f(x) = x^4 + 2x^3 - x^2 - 4x - 2$, $\quad g(x) = x^4 + x^3 - x^2 - 2x - 2$;

b) $f(x) = 3x^3 - 2x^2 + x + 2$, $\qquad\qquad g(x) = x^2 - x + 1$.

2504. Let $d(x)$ be the greatest common divisor of $f(x)$ and $g(x)$. Prove that

a) there exist polynomials $u(x)$, $v(x)$ such that $\deg u(x) < \deg g(x) - \deg d(x)$ and $d(x) = f(x)u(x + g(x)v(x)$;

b) in the case a) we have also $\deg v(x) < \deg f(x) - \deg d(x)$;

c) the polynomials $u(x)$, $v(x)$ in a) are uniquely defined.

2505. Using the method of indeterminate coefficients, find polynomials $u(x)$ and $v(x)$ such that $f(x)u(x) + g(x)v(x) = 1$:

a) $\qquad f(x) = x^4 - 4x^3 + 1$, $\qquad\qquad g(x) = x^3 - 3x^2 + 1$;

b) $\qquad f(x) = x^3$, $\qquad\qquad\qquad\qquad g(x) = (1 - x)^2$;

c) $\qquad f(x) = x^4$, $\qquad\qquad\qquad\qquad g(x) = (1 - x)^4$.

2506. Find polynomials $u(x)$ and $v(x)$ such that

$$x^m u(x) + (1 - x)^n v(x) = 1.$$

2507. Find the greatest common divisor and its expression in terms of f and g over the field $\mathbf{F}_2$:

a) $\qquad f = x^5 + x^4 + 1$, $\qquad\qquad g = x^4 + x^2 + 1$;

b) $\qquad f = x^5 + x^3 + x + 1$, $\qquad\qquad g = x^4 + 1$;

c) $\qquad f = x^5 + x + 1$, $\qquad\qquad\quad g = x^4 + x^3 + 1$;

d) $\qquad f = x^5 + x^3 + x$, $\qquad\qquad\quad g = x^4 + x + 1$.

2508. After factoring out multiple irreducible factors, factorize the given polynomials into irreducible factors:

a) $\qquad x^6 - 15x^4 + 8x^3 + 51x^2 - 72x + 27$;

b) $\qquad x^5 - 6x^4 + 16x^3 - 24x^2 + 20x - 8;$

c) $\qquad x^5 - 10x^3 - 20x^2 - 15x - 4;$

d) $\qquad x^6 - 6x^4 - 4x^3 + 9x^2 + 12x + 4;$

e) $\qquad x^6 - 2x^5 - x^4 - 2x^3 + 5x^2 + 4x + 4;$

f) $\qquad x^7 - 3x^6 + 5x^5 - 7x^4 + 7x^3 - 5x^2 + 3x - 1;$

g) $\qquad x^8 + 2x^7 + 5x^6 + 6x^5 + 8x^4 + 6x^3 + 5x^2 + 2x + 1.$

2509. Let K be a field and $f \in K[[x]]$, $g \in K[x]\backslash K$. Do there exist elements $r \in K[x]$, $h \in K[[x]]$, such that $f = hg + r$ and either $r = 0$, or $\deg r < \deg g$?

26   Simple and multiple roots over fields of characteristic zero

2601. Divide the polynomial $f(x)$, with remainder, by $x - x_0$ and calculate the value $f(x_0)$:

a) $f(x) = x^4 - 2x^3 + 4x^2 - 6x + 8,$ $x_0 = 1;$

b) $f(x) = 2x^5 - 5x^3 - 8x,$ $x_0 = -3;$

c) $f(x) = 3x^5 + x^4 - 19x^2 - 13x - 10,$ $x_0 = 2;$

d) $f(x) = x^4 - 3x^3 - 10x^2 + 2x + 5,$ $x_0 = -2;$

e) $f(x) = x^5,$ $x_0 = 1;$

f) $f(x) = x^4 + 2x^3 - 3x^2 - 4x + 1,$ $x_0 = -1;$

g) $f(x) = x^4 - 8x^3 + 24x^2 - 50x + 90,$ $x_0 = 2;$

h) $f(x) = x^4 + 2ix^3 - (1 + i)x^2 - 3x + 7 + i,$ $x_0 = -i;$

i) $f(x) = x^4 + (3 - 8i)x^3 - (21 + 18i)x^2 - (33 - 20i)x + 7 + 18i,$
   $x_0 = -1 + 2i.$

2602. Expand the polynomial $f(x)$ in powers of $x - x_0$ and find the values of its derivatives at the point x_0:

a) $\qquad f(x) = x^5 - 4x^3 + 6x^2 - 8x + 10,$ $\qquad x_0 = 2;$

b) $\qquad f(x) = x^4 - 3ix^3 - 4x^2 + 5ix - 1,$ $\qquad x_0 = 1 + 2i;$

c) $\qquad f(x) = x^4 + 4x^3 + 6x^2 + 10x + 20,$ $\qquad x_0 = -2.$

2603. Determine the multiplicity of the root x_0 of the polynomial $f(x)$:

a) $\qquad f(x) = x^5 - 5x^4 + 7x^3 - 2x^2 + 4x - 8, \qquad\qquad x_0 = 2;$

b) $\qquad f(x) = x^5 + 7x^4 + 16x^3 + 8x^2 - 16x - 16, \qquad x_0 = -2;$

c) $\qquad f(x) = 3x^5 + 2x^4 + x^3 - 10x - 8, \qquad\qquad x_0 = -1;$

d) $\qquad f(x) = x^5 - 6x^4 + 2x^3 + 36x^2 - 27x - 54, \qquad x_0 = 3.$

2604. For what value of a does the polynomial $x^5 - ax^2 - ax + 1$ have -1 as a root, with a multiplicity of at least two?

2605. For what values of a and b is the polynomial $ax^{n+1} + bx^n + 1$ divisible by $(x - 1)^2$?

2606. For what values of a and b does the polynomial $x^5 + ax^3 + b$ have a double nonzero root?

2607. Prove that the polynomials:

a) $x^{2n} - nx^{n+1} + nx^{n-1} - 1;$

b) $x^{2n+1} - (2n + 1)x^{n+1} + (2n + 1)x^n - 1;$

c) $(n - 2m)x^n - nx^{n-m} + nx^m - (n - 2m),$

have 1 as a triple root.

2608. Prove that the polynomial

$$1 + \frac{x}{1!} + \frac{x^2}{2!} + \cdots + \frac{x^n}{n!}$$

has no multiple roots.

2609. Prove that the polynomial

$$a_1 x^{n_1} + a_2 x^{n_2} + \cdots + a_k x^{n_k} \qquad (n_1 < n_2 < \cdots < n_k)$$

has no nonzero roots of multiplicity greater than $k - 1$.

* * *

2610. Determine the multiplicity of the root a of the polynomial

$$\frac{x - a}{2}[f'(x) + f'(a)] - f(x) + f(a)$$

where $f(x)$ is a polynomial.

2611. Prove that over a field of characteristic zero, a polynomial $f(x)$ is divisible by its derivative if and only if $f(x) = a_0(x - x_0)^n$.

2612. Prove that if a polynomial $f(x)$ of degree n has no multiple roots, then $[f'(x)]^2 - f(x)f''(x)$ has no roots of multiplicity greater than $n - 1$.

2613. Consider the recurrence equation

$$u(n + k) = a_0 u(n) + a_1 u(n + 1) + \cdots + a_{k-1} u(n + k - 1), \quad k \neq 0, \quad a_0 \neq 0$$

and put $f(x) = x^k - a_{k-1}x^{k-1} - \cdots - a_0$. Prove that

a) a function $u(n) = n^r a^n, r \geq 0, a \neq 0$, is a solution of this equation if and only if a is a root of $f(x)$ of multiplicity not less than $r + 1$;

b) if $a_1, \ldots, a_m$ are all roots of $f(x)$ of multiplicities $s_1, \ldots, s_m$, then an arbitrary solution $u(n)$ of the recurrence equation is of the form

$$u(n) = \sum_{i=1}^{m} g_i(n)a_i^n,$$

where $g_i(x)$ is a polynomial of degree at most $s_i - 1, i = 1, \ldots, m$.

2614. Let $f(x) = a_0 + a_1 x + \cdots + a_k x^k$. Prove that a nonzero number z is a root of multiplicity at least $r + 1$ if and only if

$$a_0 + a_1 z + a_2 z^2 + \cdots + a_m z^m + \cdots + a_k z^k = 0$$
$$a_1 z + 2a_2 z^2 + \cdots + ma_m z^m + \cdots + ka_k z^k = 0$$
$$a_1 z + 2^2 a_2 z^2 + \cdots + m^2 a_m z^m + \cdots + k^2 a_k z^k = 0$$
$$\cdots\cdots\cdots\cdots\cdots\cdots\cdots\cdots\cdots$$
$$a_1 z + 2^r a_2 z^2 + \cdots + m^r a_m z^m + \cdots + k^r a_k z^k = 0.$$

27 Prime decomposition over $\mathbb{R}$ and $\mathbb{C}$

2701. Factorize the polynomials into linear factors over the field of complex numbers:

a) $x^3 - 6x^2 + 11x - 6$;   b) $x^4 + 4$;   c) $x^6 + 27$;

d) $x^{2n} + x^n + 1$;   e) $\cos(n \arccos x)$;   f) $\sin((2n + 1) \arcsin x)$.

2702. Factorize the polynomials into linear and quadratic factors over the field of real numbers:

a) $x^6 + 27$;

b) $x^4 + 4x^3 + 4x^2 + 1$;

c) $x^4 - ax^2 + 1$, $|a| < 2$;

d) $x^{2^n} + x^n + 1$;

e) $x^6 - x^3 + 1$;

f) $x^{12} + x^8 + x^4 + 1$.

2703. Construct the polynomial with complex coefficients of the least degree having

a)   the double root 1 and the simple roots 2,3 and $1 + i$;

b)   the double root i and the simple root $-1 - i$.

2704. Construct the polynomial with real coefficients of the least degree having

a)   the double root 1 and the simple roots 2, 3 and $1 + i$;

b)   the double root i and the simple root $-1 - i$.

2705. Prove that the polynomial $x^{3m} + x^{3n+1} + x^{3p+2}$ is divisible by $x^2 + x + 1$.

2706. For what values of m, n and p is the polynomial $x^{3m} - x^{3n+1} + x^{3p+2}$ divisible by $x^2 - x + 1$?

2707. For what value of m is the polynomial $(x + 1)^m - x^m - 1$ divisible by $(x^2 + x + 1)^2$?

2708. Find the greatest common divisor of the polynomials:

a) $(x - 1)^3(x + 2)^2(x - 3)(x + 4)$ and $(x - 1)^2(x + 2)(x + 5)$;

b) $(x - 1)(x^2 - 1)(x^3 - 1)(x^4 - 1)$ and $(x + 1)(x^2 + 1)(x^3 + 1)(x^4 + 1)$;

c) $x^m - 1$ and $x^n - 1$;

d) $x^m + 1$ and $x^n + 1$.

2709. Prove that if $f(x^n)$ is divisible by $x - 1$, then $f(x^n)$ is divisible by $x^n - 1$.

2710. Prove that if $a \neq 0$ and $f(x^n)$ is divisible by $(x - a)^k$, then $f(x^n)$ is divisible by $(x^n - a^n)^k$.

2711. Let $F(x) = f_1(x^3) + xf_2(x^3)$ be divisible by $x^2 + x + 1$. Then $f_1(x)$ and $f_2(x)$ are divisible by $x - 1$.

*   *   *

2712. Let the values of the polynomial $f(x)$ be non-negative for all $x \in \mathbb{R}$. Prove that $f(x) = f_1(x)^2 + f_2(x)^2$ for some $f_1(x), f_2(x) \in \mathbb{R}[x]$.

2713. Let f, g be coprime complex polynomials. Then $\max(\deg f, \deg g)$ is less than the number of distinct roots of the polynomial $fg(f + g)$.

2714. Let f, g, h be pairwise coprime complex polynomials, and $f^n + g^n = h^n$. Prove that $n \leq 2$.

28 Polynomials over the field of rationals and over finite fields

2801. Prove that if an irreducible fraction $\dfrac{p}{q}$ is a root of a polynomial $f(x) = a_0 x^n + a_1 x^{n-1} + \cdots + a_{n-1}x + a_n$ with integer coefficients, then

a) $p|a_n$;   b) $q|a_0$;   c) $(p - mq)|f(m)$ for all $m \in \mathbb{Z}$.

2802. Find all rational roots of the polynomials:

a) $x^3 - 6x^2 + 15x - 14$;

b) $x^4 - 2x^3 - 8x^2 + 13x - 24$;

c) $6x^4 + 19x^3 - 7x^2 - 26x + 12$;

d) $24x^4 - 42x^3 - 77x^2 + 56x + 60$;

e) $24x^5 + 10x^4 - x^3 - 19x^2 - 5x + 6$;

f) $10x^4 - 13x^3 + 15x^2 - 18x - 24$;

g) $4x^4 - 7x^2 - 5x - 1$;

h) $2x^3 + 3x^2 + 6x - 4$.

2803. Prove that a polynomial with integer coefficients $f(x)$ has no integer roots if $f(0)$, $f(1)$ are odd integers.

* * *

2804. Let a polynomial $f(x)$ with integer coefficients have the values ± 1 at two integer points x_1, x_2. Prove that $f(x)$ has no rational roots if $|x_1 - x_2| > 2$. If $|x_1 - x_2| \leq 2$ then the only possible rational root is equal to $\frac{1}{2}(x_1 + x_2)$.

2805. Prove that a polynomial which is irreducible over the field of rationals has no multiple complex roots.

2806. A polynomial with integer coefficients is *primitive*, if its coefficients are relatively prime. Prove that the product of primitive polynomials is also primitive.

2807. Prove that if a polynomial with integer coefficients is reducible over the field of rationals then it can be factorized into the product of two polynomials with integer coefficients of lower degrees.

2808. The *Eisenstein criterion of irreducibility*. Let $f(x)$ be a polynomial with integer coefficients and p be a prime number, such that

   a)   the leading coefficient of $f(x)$ is not divisible by p;

   b)   all other coefficients of $f(x)$ are divisible by p;

   c)   the constant term of $f(x)$ is not divisible by p^2.

Prove that the polynomial $f(x)$ is irreducible over the field of rationals.

2809. Prove the irreducibility over the field of rationals of the following polynomials:

   a) $x^4 - 8x^3 + 12x^2 - 6x + 2$;

   b) $x^5 - 12x^3 + 36x - 12$;

   c) $x^{105} - 9$;

   d) $\Phi_p(x) = x^{p-1} + x^{p-2} + \cdots + x + 1$ (p being a prime);

   e) $(x-a_1)(x-a_2)\ldots(x-a_n) - 1$, where $a_1, a_2, \ldots, a_n$ are distinct integers;

   f) $(x - a_1)^2 \ldots (x - a_n)^2 + 1$, where $a_1, \ldots, a_n$ are distinct integers.

2810. Prove that the polynomial $x^n - x - 1$, $n \geq 2$ is irreducible over $\mathbb{Q}$.

2811. Prove that the polynomial $x^n + x + 1$ is irreducible over $\mathbb{Q}$, if $n \not\equiv 2$ (mod 3). Prove that if $n \equiv 2$ (mod 3) then the polynomial $x^n + x + 1$ is divisible over $\mathbb{Z}$ by $x^2 + x + 1$.

2812. Let $f(x) = x^n \pm x^m \pm 1$. Prove that either $f(x)$ is irreducible over $\mathbb{Q}$, or some complex root of 1 is a root of $f(x)$.

2813. Let $f(x) = x^n \pm x^m \pm x^q \pm 1$. Prove that either $f(x)$ is irreducible over $\mathbb{Q}$, or some complex root of 1 is a root of $f(x)$.

2814. Prove that any polynomial with integer coefficients of a positive degree has a root in fields $\mathbb{Z}_p$ for infinitely many primes p.

2815. Prove that if $\mathbf{F}_q$ is the field with q elements, then $x^q - x = \prod_{a \in \mathbf{F}_q}(x - a)$.

2816. Let F be a finite field. Prove that for any mapping $h : F^n \to F$ there exists a polynomial f in the ring $F[x_1, \ldots, x_n]$ such that $f(a_1, \ldots, a_n) = h(a_1, \ldots, a_n)$ for all $a_1, \ldots, a_n \in F$.

2817. Let $f(x)$ be the polynomial from Exercises 2810 or 2811 and have q roots which are complex roots of 1. Prove that there is a factorization $f(x) = g(x)h(x)$ in $\mathbb{Q}[x]$, where all roots of $g(x)$ are roots of 1 and $h(x)$ is irreducible over $\mathbb{Q}$.

2818. Prove that the polynomial $f(x) = x^n + ax \pm 1$, $a \in \mathbb{Z}$, is irreducible over $\mathbb{Q}$, provided $|a| \geq 3$.

2819. Prove that if the polynomial $f(x) = x^n \pm 2x \pm 1$ is reducible over $\mathbb{Q}$, then $f(x) = g(x)(x \pm 1)$, where $g(x)$ is irreducible over $\mathbb{Q}$.

2820. Prove that the polynomial $f(x) = x^n + gx^p + r \in \mathbb{Z}[x]$, $1 \leq p < n$, is irreducible over $\mathbb{Q}$ if $|g| > 1 + |r|^{n-1}$ and $|r|$ is not a dth power for any nontrivial divisor d of the integer n.

2821. The polynomial $f(x) = x^n + a_{n-1}x^{n-1} + \cdots + a_0 \in \mathbb{Z}[x]$ is irreducible over $\mathbb{Q}$ if $|a_{n-1}| > 1 + |a_0| + \cdots + |a_{n-2}|$.

2822. Find:

a)   all irreducible polynomials of degree ≤ 4 over the field $\mathbb{Z}_p$;

b)   all irreducible polynomials of degree 2 over the field $\mathbb{Z}_3$;

c)   the number of irreducible polynomials of degree 5 over the field $\mathbb{Z}_2$;

d)   the number of irreducible monic polynomials of degree 3 and 4 over the field $\mathbb{Z}_3$.

2823. Find the number of irreducible monic polynomials of degree 2 and 3 over the field with 9 elements.

2824. Prove that the polynomial $\Phi_d(x)$ where d divides $p-1$, can be factorized into linear factors over $\mathbb{Z}_p$.

2825. Let $f(x) \in \mathbb{Z}_p[x]$ be irreducible. Prove that the polynomials $f(x)$, $f(x+1), \ldots, f(x+p-1)$ are either distinct or coincide.

2826. Prove that if $a \in \mathbb{Z}_p^*$, then the polynomial $x^p - x - a$ is irreducible over $\mathbb{Z}_p$.

2827. Let b be a nonzero element of $\mathbb{Z}_p$. Prove that $x^p - x - b$ is irreducible over $\mathbb{F}_{p^n}$ if and only if the integer n is not divisible by p.

2828. Prove that if $a \neq 1$ then the polynomial $x^q - ax - b$ has a root in $\mathbb{F}_q$.

2829. Prove that $x^{2n} + x^n + 1$ is irreducible over $\mathbb{Z}_2$ if and only if $n = 3^k$ for some integer $k \geq 0$.

2830. Prove that $x^{4n} + x^n + 1$ is irreducible over $\mathbb{Z}_2$ if and only if $n = 3^k 5^m$ for some integers $k, m \geq 0$.

2831. Find all integers a for which all roots of the polynomial $x^4 - 14x^3 + 61x^2 + 84x + a$ are integers.

2832. Let I_m be the number of distinct monic irreducible polynomials of degree m over the finite field with q elements. Prove that in the power series ring $\mathbb{Q}[[z]]$

$$\frac{1}{1-qz} = \prod_{m=1}^{\infty} \left(\frac{1}{1-z^m}\right)^{I_m}.$$

2833. Under the assumption of Exercise 2832 prove that q^k is equal to the sum $m I_m$ for all divisors m of the integer k.

2834. Let I_m be as defined in Exercise 2832. Prove that

$$I_m = \frac{1}{m} \sum_{d\mid m} \mu(d) q^{mk-1}.$$

29   Rational fractions

2901. Represent the rational fraction as a sum of partial fractions over the field of complex numbers:

a) $\dfrac{x^2}{(x-1)(x+2)(x+3)}$;

b) $\dfrac{1}{x^4+4}$;

c) $\dfrac{x}{(x^2-1)^2}$;

d) $\dfrac{5x^2+6x-28}{(x-1)^3(x+1)^2(x-2)}$;

e) $\dfrac{1}{(x-1)(x-2)(x-3)(x-4)}$;

f) $\dfrac{3+x}{(x-1)(x^2+1)}$;

g) $\dfrac{x^2}{(x^4-1)}$;

h) $\dfrac{1}{x^3-1}$;

i) $\dfrac{n!}{x(x-1)\ldots(x-n)}$;

j) $\dfrac{1}{(x^2-1)^2}$;

k) $\dfrac{1}{(x^n-1)^2}$.

2902. Represent the rational fraction as a sum of partial fractions over the field of real numbers:

a) $\dfrac{x^2}{x^4-16}$;

b) $\dfrac{1}{x^2+4}$;

c) $\dfrac{x}{(x+1)(x^2+1)^2}$;

d) $\dfrac{1}{(x^4-1)^2}$;

e) $\dfrac{1}{\cos(n\arccos x)}$;

f) $\dfrac{1}{f(x)}$, where the polynomial $f(x)$ of degree n has n distinct real roots;

g) $\dfrac{1}{x^3 - 1}$; h) $\dfrac{x^2}{x^6 + 27}$; i) $\dfrac{2x - 1}{x(x + 1)^2(x^2 + x + 1)^2}$;

j) $\dfrac{1}{(x^4 - 1)^2}$; k) $\dfrac{x^{2m}}{x^{2n} + 1}$, $m < n$.

2903. Represent $\dfrac{1}{x^p - x}$ as a sum of partial fractions over $\mathbf{Z}_p$.

2904. Prove that for any nonzero polynomials f, g

$$\frac{(fg)'}{fg} = \frac{f'}{f} + \frac{g'}{g}.$$

2905. Let $f = (x - a_1) \ldots (x - a_n)$. Prove that

$$\frac{f'}{f} = \frac{1}{x - a_1} + \cdots + \frac{1}{x - a_n}.$$

30 Interpolation

3001. Find the polynomial of the least degree with the given table of values:

a)
x	-1	0	1	2	3
$f(x)$	6	5	0	3	2
;

b)
x	1	2	3	4	6
$f(x)$	5	6	1	-4	10

3002. Prove that a polynomial of degree $< n$, with integer values at n successive integer points, has integer values at all integer points.

Does the polynomial have integer coefficients?

3003. Prove that any function $f : F \to F$ on a finite field F with q elements is uniquely represented as a polynomial of degree $< q$.

3004. Prove that the polynomial of degree $< n$, with values $y_1, \ldots, y_n$, at points $x_1, \ldots, x_n$ is equal to

$$g(x) \sum_{i=1}^{n} \frac{y_i}{(x - x_i)g'(x_i)},$$

where $g(x) = (x - x_1) \ldots (x - x_n)$.

3005. The polynomial $f(x)$ of degree at most $n - 1$ has the values $y_1, \ldots, y_n$ for the roots of order n of 1. Find $f(0)$.

3006. Prove that the points $x_1, \ldots, x_n \in \mathbb{C}$ are vertices of the regular n-gon with the center at the point x_0, if and only if for any polynomial $f(x)$ of degree $< n$, the equality

$$f(x_0) = \frac{1}{n}[f(x_1) + \cdots + f(x_n)]$$

is fulfilled.

3007. Let all roots $x_1, \ldots, x_n$ of the polynomial $f(x)$ be distinct.

a)   Prove that for any non-negative integer $s \leq n - 2$

$$\sum_{i=1}^{n} \frac{x_i^s}{f'(x_i)} = 0.$$

b)   Calculate the sum

$$\sum_{i=1}^{n} \frac{x_i^{n-1}}{f'(x_i)}.$$

3008. Find the polynomial of degree $2n$ whose remainder when divided by $x(x - 2) \ldots (x - 2n)$ is equal to -1.

3009. Construct the polynomial $f(x)$ over $\mathbb{Z}_p$, of the least possible degree, such that $f(k) = k^{-1}$ for $k = 1, 2, \ldots, p - 1$.

3010. Construct the polynomial $f(x)$ over $\mathbb{Z}_7$, of the least possible degree, satisfying the conditions $f(0) = 1$, $f(1) = 0$ and $f(k) = k$ for $k = 2, 3, 4, 5, 6$.

3011. Let $\mathbf{F}_q$ be a field with $q > 2$ elements, and c be a generator of the cyclic group $\mathbf{F}_q^*$. Prove that the group of permutations S_q, acting on $\mathbf{F}_q$, is generated by the mappings $f(x) = x + 1$, $h(x) = cx$, $g(x) = x^{q-2}$.

3012. Given the assumptions of Exercise 3011, prove that the alternating group $\mathbf{A}_q$ is generated by the polynomials $c^2 x$, $x + 1$, $(x^{q-2} + 1)^{q-2}$.

3013. Let $k_0, \ldots, k_n$ be natural numbers and x_i, b_{ij} be elements of a field F of zero characteristic, where $i = 0, \ldots, n$, $j = 0, \ldots, k_i - 1$. It is assumed that the elements $x_0, \ldots, x_n$ are distinct. Prove there exists the unique polynomial $f(x) \in F[x]$, of degree at most $k_0 + \cdots + k_n - 1$, such that $f^j(x_i) = b_{ij}$ for all i, j.

3014. Let k_i, F, x_i, b_{ij} be as in Exercise 3013. Put

$$f(x) = \sum_{i=0}^{n} G_i(x) \sum_{k=0}^{k_i-1} \sum_{l=0}^{k} \frac{b_{il}}{l!} \frac{d^l}{dx^l} \left(\frac{1}{G_i(x)} \right) \bigg|_{x=x_i} (x - x_i)^k,$$

where $G_i(x) = \prod_{j \neq i}(x - x_j)^{k_j}$. Prove that $f(x)$ is a polynomial of degree at most $k_0 + \cdots + k_n - 1$ and $f^{(j)}(x_i) = b_{ij}$ for all i, j.

31 Symmetric polynomials. Vieta formulas

3101. Construct the monic polynomial of degree 4 having

a) the roots $1, 2, -3, -4$;

b) the triple root -1 and the simple root i;

c) the roots $2, -1, 1 + i$ and $-i$;

d) the double root 3 and the simple roots $-2, -4$.

3102. Find the sum of squares and the product of all complex roots of the polynomial:

a) $3x^3 + 2x^2 - 1$;

b) $x^4 - x^2 - x - 1$.

3103. Find the sum of the numbers, which are inverse to the complex roots of the polynomials:

a) $3x^3 + 2x^2 - 1$;

b) $x^4 - x^2 - x - 1$.

3104. Find the values of all elementary symmetric polynomials at the complex roots of degree n of 1.

3105. Determine λ such that one of the roots of the polynomial $x^3 - 7x + \lambda$ is equal to the other one multiplied by 2.

3106. The sum of two roots of the polynomial $2x^3 - x^2 - 7x + \lambda$ is equal to 1. Find λ.

3107. Determine the relation between p and q under which the roots x_1, x_2, x_3 of the polynomial $x^3 + px + q$ satisfy the condition $x_3 = \dfrac{1}{x_2} + \dfrac{1}{x_1}$.

3108. *Wilson's criterion.* Prove that $(p - 1)! \equiv -1 \pmod{p}$ if and only if p is prime.

3109. Express in terms of elementary symmetric polynomials:

a) $x_1^2 x_2 + x_1 x_2^2 + x_1^2 x_3 + x_1 x_3^2 + x_2^2 x_3 + x_2 x_3^2$;

b) $x_1^4 + x_2^4 + x_3^4 - 2x_1^2 x_2^2 - 2x_1^2 x_3^2 - 2x_2^2 x_3^2$;

c) $(x_1x_2 + x_3x_4)(x_1x_3 + x_2x_4)(x_1x_4 + x_2x_3)$;

d) $(x_1 + x_2 - x_3 - x_4)(x_1 - x_2 + x_3 - x_4)(x_1 - x_2 - x_3 + x_4)$;

e) $(x_1 + x_2 + 1)(x_1 + x_2 + 1)(x_2 + x_3 + 1)(x_2 + x_3 + 1)$;

f) $(x_1x_2 + x_3)(x_1x_3 + x_2)(x_2x_3 + x_1)$;

g) $(2x_1 - x_2 - x_3)(2x_2 - x_1 - x_3)(2x_3 - x_1 - x_2)$;

h) $(x_1 + x_2)(x_1 + x_3)(x_1 + x_4)(x_2 + x_3)(x_2 + x_4)(x_3 + x_4)$;

i) $x_1^5 x_2^2 + x_1^2 x_2^5 + x_1^5 x_3^2 + x_1^2 x_3^5 + x_2^5 x_3^2 + x_2^2 x_3^5$;

j) $(x_1 - 1)(x_2 - 1)(x_3 - 1)$;

k) $x_1^2 + \ldots$;

l) $x_1^3 + \ldots$;

m) $x_1^2 x_2 x_3 + \ldots$;

n) $x_1^2 x_2^2 + \ldots$;

o) $x_1^3 x_2 x_3 + \ldots$;

p) $x_1^3 x_2^2 + \ldots$.

3110. Find the value of the symmetric polynomial F at the roots of the polynomial $f(x)$:

a) $F = x_1^3(x_2 + x_3) + x_2^3(x_1 + x_3) + x_3^3(x_1 + x_2)$,
$f(x) = x^3 - x^2 - 4x + 1$;

b) $F = x_1^3(x_2x_3 + x_2x_4 + x_3x_4) + x_2^3(x_1x_3 + x_1x_4 + x_3x_4) + x_3^3(x_1x_2 + x_1x_4 + x_2x_4) + x_4^3(x_1x_2 + x_1x_3 + x_2x_3)$, $f(x) = x^4 + x^3 - 2x^2 - 3x + 1$;

c) $F = (x_1 - x_2)^2(x_1 - x_3)^2(x_2 - x_3)^2$, $f(x) = x^3 + a_1x^2 + a_2x + a_2$;

d) $F = x_1^4 x_2 + \ldots$, $f(x) = 3x^3 - 5x^2 + 1$;

e) $F = x_1^3 x_2^3 + \ldots$, $f(x) = 3x^4 - 2x^3 + 2x^2 + x - 1$;

f) $F = (x_1^2 + x_1x_2 + x_2^2)(x_2^2 + x_2x_3 + x_3^2)(x_1^2 + x_1x_3 + x_3^2)$,
$f(x) = 5x^3 - 6x^2 - 7x - 8$.

3111. Let $x_1, \ldots, x_n$ be the roots of the polynomial $x^n + a_{n-1}x^{n-1} + \cdots + a_0$. Prove that any symmetric polynomial in $x_2, x_3, \ldots, x_n$ can be represented as a polynomial in x_1.

3112. Let σ_{ki} be an elementary symmetric polynomial of degree k in the variables $x_1, \ldots, x_{i-1}, x_{i+1}, \ldots, x_n$. Prove that

$$\sigma_{ki} = \sigma_k - x_i\sigma_{k-1} + \cdots + (-1)^{k-1}x_i^{k-1}\sigma_1 + (-1)^k x_i^k.$$

(It is assumed that $\sigma_m = 0$ if $m > n$ and $\sigma_{mi} = 0$ if $m \geq n$.)

3113. Consider the polynomial $\lambda_t = (1 + x_1t)\ldots(1 + x_nt)$ in the variables $x_1, \ldots, x_n, t$. Prove that $\lambda_t = 1 + \sigma_1 t + \sigma_2 t^2 + \cdots + \sigma_n t^n$.

3114. Let λ_t be as in Exercise 3113, and $s_k = x_1^k + \cdots + x_n^k$. Prove that

$$\frac{d}{dt}(\ln \lambda_t) = \sum_{k \geq 0}(-1)^k s_k t^{k-1}.$$

3115. Prove the *Newton's formula*

$$s_k - \sigma_1 s_{k-1} + \sigma_2 s_{k-2} + \cdots + (-1)^{k-1}\sigma_{k-1}s_1 + (1)^k \sigma_k = 0.$$

(It is assumed that $\sigma_k = 0$ if $k > n$.)

3116. Prove that

$$s_k = \begin{vmatrix} \sigma_1 & 1 & 0 & \ldots & 0 & 0 \\ 2\sigma_2 & \sigma_1 & 1 & \ldots & 0 & 0 \\ \cdots\cdots\cdots\cdots\cdots\cdots\cdots\cdots\cdots\cdots\cdots\cdots \\ (k-1)\sigma_{k-1} & \sigma_{k-2} & \sigma_{k-3} & \ldots & \sigma & 1 \\ k\sigma_k & \sigma_{k-1} & \sigma_{k-2} & \ldots & \sigma_2 & \sigma_1 \end{vmatrix}.$$

3117. Prove that

$$\sigma_k = \frac{1}{k!}\begin{vmatrix} s_1 & 1 & 0 & \ldots & 0 & 0 \\ s_2 & s_1 & 2 & \ldots & 0 & 0 \\ \cdots\cdots\cdots\cdots\cdots\cdots\cdots\cdots\cdots\cdots\cdots\cdots \\ s_{k-1} & s_{k-2} & s_{k-3} & \ldots & s_1 & k-1 \\ s_k & s_{k-1} & s_{k-2} & \ldots & s_2 & s_1 \end{vmatrix}.$$

3118. Find the values of s_m at the roots of $\Phi_n(x)$.

3119. Find the values of $s_1, \ldots, s_n$ at the roots of the polynomial

$$x^n + \frac{x^{n-1}}{1!} + \frac{x^{n-2}}{2!} + \cdots + \frac{1}{n!}.$$

3120. Calculate the values of the symmetric polynomials s_k at the complex roots of 1 of degree k.

3121. Solve the systems of equations over the field of complex numbers:

a)
$$
\begin{aligned}
x_1 + x_2 + x_3 &= 0, \\
x_1^2 + x_2^2 + x_3^2 &= 0, \\
x_1^3 + x_2^3 + x_3^3 &= 24;
\end{aligned}
$$

b)
$$
\begin{aligned}
x_1^2 + x_2^2 + x_3^2 &= 6, \\
x_1^3 + x_2^3 - x_1 x_2 x &- 3 = -4, \\
x_1 x_2 + x_1 x_3 + x_2 x_4 &= -3.
\end{aligned}
$$

3122. Prove that the value of any symmetric polynomial with integer coefficients in n variables at the roots of 1 of order n is an integer.

3123. Let ζ be a primitive complex root of order k of 1. Prove that for any complex number a

$$
(x - a)(x\zeta - a)\dots(x\zeta^{k-1} - a) = (-1)^{k+1}(x^k - a^k).
$$

3124. Let ζ be a primitive complex root of order k of 1 and $f(x)$ be a polynomial with complex coefficients. Prove that

a) $f(x)f(x\zeta)\dots f(x\zeta^{k-1}) = h(x^k)$, where $h(x)$ is a polynomial;

b) the roots $h(x)$ are precisely the kth powers of roots of the polynomial $f(x)$.

3125. Find the polynomial of degree 3 whose roots are:

a) cubes of the complex roots of the polynomial $x^3 - x - 1$;

b) fourth powers of the complex roots of the polynomial $2x^3 - x^2 + 2$.

3126. Find the polynomial of degree 4 whose roots are:

a) squares of the complex roots of the polynomial $x^4 + 2x^3 - x + 3$;

b) cubes of the complex roots of the polynomial $x^4 - x - 1$.

$$* \quad * \quad *$$

3127.

a) Let $f(x_1 \dots x_n)$ be an antisymmetric polynomial in $x_1, \dots, x_n$. Prove that $f(x_1 \dots x_n) = \Delta(x_1 \dots x_n)g(x_1 \dots x_n)$, where $\Delta(x_1 \dots x_n)$ is the Vandermonde determinant and $g(x_1 \dots x_n)$ is a symmetric polynomial.

b) Let $h(x_1 \dots x_n)$ be a symmetric polynomial and $h(x_1, x_1, x_3, \dots, x_n) = 0$. Prove $h(x_1 \dots x_n) = \Delta(x_1 \dots x_n)^2 u(x_1 \dots x_n)$ where $u(x_1 \dots x_n)$ is a symmetric polynomial.

3128. Let

$$h_k = \sum_{1 \leq i_1 \leq \cdots \leq i_k \leq n} x_{i_1} \ldots x_{i_k}.$$

and λ_t be as in Exercise 3113. Prove that

a) $\lambda_t^{-1} = \sum_{k \geq 0}(-1)^k h_k t^k$;

b) $\sigma_k - h_1\sigma_{k-1} + \cdots + (-1)^{k-1}h_{k-1}\sigma_1 + (-1)^k h_k = 0, \quad k \geq 1$;

c) each symmetric polynomial is a polynomial in $h_1, \ldots, h_n$.

3129. A *partition* of a number n is a set λ of non-negative integers $\lambda = (\lambda_1, \ldots, \lambda_n)$, where $\lambda_1 + \cdots + \lambda_n = n$ and $\lambda_1 \geq \lambda_2 \geq \cdots \geq \lambda_n \geq 0$. Let $p(n)$ be the number of partitions of the number n. Prove that

$$\prod_{m \geq 0}(1 - t^m)^{-1} = \sum_{n \geq 0} p(n)t^n.$$

3130. Let $\alpha = (\alpha_1, \ldots, \alpha_n)$, $\alpha_1 > \alpha_2 > \cdots > \alpha_n \geq 0$ be a set of natural numbers. Put

$$a_\alpha(x_1 \ldots x_n) = \sum_{G \in S_n}(\text{sgn}\sigma)x_{\sigma 1}^{\alpha_1} \ldots x_{\sigma n}^{\alpha_n}.$$

Prove that

a) $a_\alpha(x_1 \ldots x_n) = \det \begin{pmatrix} x_1^{\alpha_1} & \cdots & x_n^{\alpha_1} \\ \cdots\cdots\cdots\cdots\cdots \\ x_1^{\alpha_n} & \cdots & x_n^{\alpha_n} \end{pmatrix}$;

b) if $\delta = (n-1, n-2, \ldots, 1, 0)$, then $a_\delta(x_1 \ldots x_n)$ is the Vandermonde determinant in $x_n, \ldots, x_1$.

3131. Let $\lambda = (\lambda_1, \ldots, \lambda_n)$ be a partition of some natural number k. Put $\alpha_i = \lambda_i + n - i$ for all i. Let

$$S_\lambda(x_1 \ldots x_n) = \frac{a_\alpha}{a_\delta}.$$

Prove that

a) $S_\lambda(x_1 \ldots x_n)$ is a symmetric polynomial with integer coefficients;

b) $S_\lambda(x_1 \ldots x_n)$ for all $\lambda = (\lambda_1 \ldots \lambda_n)$ form the basis of the linear space of symmetric polynomials in $x_1, \ldots, x_n$;

c) if $\lambda = (1, \ldots, 1)$, then $S_\lambda(x_1 \ldots x_n) = \sigma_n$;

d) if $\lambda = (n, 0 \ldots, 0)$, then $S_\lambda(x_1 \ldots x_n) = h_n$ (see Exercise 3128).

3132. Prove that

a) $\displaystyle\prod_{i,j=1}^{n}(1-x_iy_j)^{-1} = \sum_{\lambda} S_\lambda(x_1, x_2, \dots)S_\lambda(y_1, y_2, \dots);$

b) $\displaystyle\prod_{i,j=1}^{n}(1+x_iy_j) = \sum_{\lambda} S_\lambda(x_1, x_2, \dots)S_{\lambda'}(y_1, y_2, \dots),$

where the summation is taken over all partitions $\lambda = (\lambda_1 \dots \lambda_n)$. λ' is the conjugate partition, i.e. λ'_i is the number of all j such that $\lambda_j \geq i$.

3133. Prove that

$$\sum_{\tau \in S_n} \sigma_k(x_{\tau(1)}y_1, \dots, x_{\tau(n)}y_n) = \sigma_k(x_1 \dots x_n)\sigma_k(y_1 \dots y_n).$$

3134. Let F be the field of fractions of the ring of symmetric polynomials with integer coefficients in $x_1 \dots x_n$. Prove that F coincides with the subfield of $\mathbb{Q}(x_1, \dots, x_n)$, consisting of all symmetric rational fractions.

32   Resultant and discriminant

3201. Calculate the resultant of the polynomials

a) $x^3 - 3x^2 + 2x + 1$ and $2x^2 - x - 1$;

b) $2x^3 - 3x^2 + 2x + 1$ and $x^2 + x + 3$;

c) $2x^3 - 3x^2 - x + 2$ and $x^4 - 2x^2 - 3x + 4$;

d) $3x^3 + 2x^2 + x + 1$ and $2x^3 + x^2 - x - 1$;

e) $2x^4 - x^3 + 3$ and $3x^3 - x^2 + 4$.

3202. Find all values λ at which the polynomials:

a) $x^3 - \lambda x + 2$ and $x^2 + \lambda x + 2$;

b) $x^3 + \lambda x^2 - 9$ and $x^3 + \lambda x - 3$;

c) $x^3 - 2\lambda x + \lambda^3 x$ and $x^2 + \lambda^2 - 2$

have a common root.

3203. Eliminate x from the systems of equations:

a) $x^2 - xy + y^2 = 3,$ b) $x^3 - xy - y^3 + y = 0,$
      $x^2y + xy^2 = 6;$ $x^2 + x - y^2 = 1;$

c) $y^2 - 7xy + 4x^2 + 13x - 2y - 3 = 0,$
   $y^2 - 14xy + 9x^2 + 28x - 4y - 5 = 0;$

d) $y^2 + x^2 - y - 3x = 0,$
   $y^2 - 6xy - x^2 + 11y + 7x - 12 = 0;$

e) $5y^2 - 6xy + 5x^2 - 16 = 0,$
   $y^2 - xy + 2x^2 - y - x - 4 = 0.$

3204. Prove that $R(f, g_1g_2) = R(f, g_1)R(f, g_2)$.

3205. Find the resultant of the polynomials Φ_n and $x^m - 1$.

3206. Find the resultant of the polynomials Φ_n and Φ_m.

3207. Calculate the discriminant of the polynomials:

a) $ax^2 + bx + c;$

b) $x^3 + px + q;$

c) $x^3 + a_1x^2 + a_2x + a_3;$

d) $2x^4 - x^3 - 4x^2 + x + 1;$

e) $x^4 - x^3 - 3x^2 + x + 1.$

3208. Find all values λ at which the polynomials:

a) $x^3 - 3x + \lambda;$

b) $x^4 - 4x + \lambda;$

c) $x^3 - 8x^2 + (13 - \lambda)x - (6 + 2\lambda);$ d) $x^4 - 4x^3 + (2 - \lambda)x + 2x - 2,$

have a multiple root.

3209. Prove that

$$D[(x - a)f(x)] = D[f(x)] \cdot f(a)^2.$$

* * *

3210. Calculate the discriminant of the polynomial

$$x^{n-1} + x^{n-2} + \cdots + 1.$$

3211. Calculate the discriminant of $\Phi_n(x)$.

3212. Calculate the discriminant of the polynomial

$$1 + \frac{x}{1!} + \frac{x^2}{2!} + \cdots + \frac{x^n}{n!}.$$

3213. Let f and g be irreducible polynomials. Prove that

$$D(fg) = D(f)D(g)[R(f, g)]^2.$$

3214. Let j, k be natural numbers and $d = (j, k)$. Prove that
$$R(x^j - a^j, x^k - b^k) = (-1)^j (b^{jkd^{-1}} - a^{jkd^{-1}})^d.$$

3215. Let $n > k > 0$ and $d = (n, k)$. Prove that

$$D(x^n + ax^k + b) = (-1)^{n(n-1)2^{-1}} b^{k-1} \times$$
$$\times \left[n^{nd^{-1}} b^{(n-k)d^{-1}} - (-1)^{nd^{-1}} (n-k)^{(n-k)d^{-1}} k^{kd^{-1}} a^{nd^{-1}} \right].$$

3216. Calculate the discriminant of the polynomial $x^n + a$.

3217. Calculate the discriminant

a) of the Hermite polynomial $P_n(x) = (-1)^n e^{\frac{x^2}{2}} \dfrac{d^n}{dx^n} \left(e^{-\frac{x^2}{2}} \right)$;

b) of the Laguerre polynomial $P_n(x) = (-1)^n e^x \dfrac{d^n}{dx^n} (x^n e^{-x})$;

c) of the Chebyshev polynomial $2 \cos \left(n \arccos \dfrac{x}{2} \right)$.

33 Isolation of roots

3301. Write down the Sturm series and isolate the roots of the polynomials

a) $x^3 - 3x - 1$;

b) $x^3 + x^2 - 2x - 1$;

c) $x^3 - 7x + 7$;

d) $x^3 - x + 5$;

e) $x^3 + 3x - 5$;

f) $x^4 - 12x^2 - 16x - 4$;

g) $x^4 - x - 1$;

h) $2x^4 - 8x^3 + 8x^2 - 1$;

i) $x^4 + x^2 - 1$;

j) $x^4 + 4x^3 - 12x + 9$.

3302. Write down the Sturm series for the real polynomial $x^5 - 5ax^3 + 5a^2x + 2b$. Find the number of real roots of the polynomial according to the sign of the number $a^5 - b^2$.

3303. Write down the Sturm series for the real polynomial $x^n + px + q$. Find the number of real roots of the polynomial according to the parity and the sign of the number $d = -(n-1)^{n-1} p^n - n^n q^{n-1}$.

3304. Write down the Sturm series and find the number of real roots of the polynomial

$$E_n(x) = 1 + \frac{x}{1!} + \frac{x^2}{2!} + \cdots + \frac{x^n}{n!}.$$

3305. Prove that the polynomial $t^3 - 3t + r$ cannot have more than one real root in the segment $[0, 1]$.

3306. Assume that all roots of a polynomial $f(x) \in \mathbb{R}[x]$ are real and

$$f(x) = a(x - a_1)^{k_1} \ldots (x - a_m)^{k_m}, \quad a \neq 0,$$

that where $a_1 < a_2 < \cdots < a_m$. Prove that

a) $f'(x) = na(x - a_1)^{k_1-1} \ldots (x - a_m)^{k_m-1}(x - b_1) \ldots (x - b_{m-1})$
   where $a_1 < b_1 < a_2 < b_2 < \cdots < a_{m-1} < b_{m-1} < a_m$;

b) if the integer k does not exceed the degree of $f(x)$, then the multiple roots of the kth derivative $f^{(k)}(x)$ are precisely the numbers a_i, $k_i \geq k$;

c) if $f(x) = c_n x^n + c_{n-1} x^{n-1} + \cdots + c_0$, where $c_n \neq 0$ and $c_k = c_{k+1} = 0$ for some $k = 0, \ldots, n - 2$, then $c_0 = c_1 = \cdots = c_k = c_{k+1} = 0$.

3307. Let $g(x) = b_n x^n + \cdots + b_0$ be a real polynomial, $b_n, b_0 \neq 0$ and $b_k = b_{k+1}$ for some $k = 1, \ldots, n - 2$. Then $g(x)$ has at least one nonreal root.

3308. Prove that the real polynomial

$$a_n x^n + a_{n-1} x^{n-1} + \cdots + a_3 x^3 + x^2 + x + 1, \quad a_n \neq 0,$$

has at least one nonreal root.

3309. Prove that all complex roots z of the polynomial $nx^n - x^{n-1} - \cdots - 1$ satisfy the condition $|z| \leq 1$.

3310. Prove that all positive roots of the polynomial

$$f(x) = x(x + 1)(x + 2) \ldots (x + n) - 1$$

are less than $\frac{1}{n!}$.

3311. Prove that the polynomial $x^4 - 5x^3 - 4x^2 - 7x + 4$ has no negative roots.

3312. How many roots of the polynomial $x^6 + 6x + 10$ lie in each quadrant of the complex plane?

3313. Let $n_1 < \cdots < n_k$ be natural numbers. Prove that the polynomial

$$1 + x^{n_1} + \cdots + x^{n_k}$$

has no complex roots z such that $|z| < \dfrac{\sqrt{5} - 1}{2}$.

3314. Prove that the absolute values of all complex roots of the real polynomial $x^{n+1} - ax^n + ax - 1$ are equal to 1.

3315. Let k be a natural number and $|a_i| < k$ for $i = 1, \ldots, n$. Then for any root z of the polynomial $a_n x^n + \cdots + a_1 x + 1$, one has $|z| \geq \dfrac{1}{k+1}$.

3316. If all roots of a polynomial $f(x) \in \mathbb{C}[x]$ are located in the upper half-plane, then all roots of $f'(x)$ belong in the same half-plane.

3317. Let D be a convex domain in the complex plane which contains all the roots of a polynomial $f(x) \in \mathbb{C}[x]$. Then all roots of $f'(x)$ belong to D.

3318. Let $f_0, f_1, \ldots, f_n$ be a sequence of real polynomials with positive leading coefficients such that

(1)   the degree of f_k is equal to k, $k = 0, \ldots, n$ and

(2) $f_k = a_k f_{k-1} - c_k f_{k-2}$, where a_k, c_k are real polynomials, and $c_k(r) > 0$ for all $r \in \mathbb{R}$ at $k \geq 2$.

Prove that

a)   the roots of all polynomials f_k are real;

b)   a root of f_{k-1} is located between two roots of f_k.

3319. Determine the number of real roots

a)   of the Hermite polynomial $(-1)^n e^{\frac{x^2}{2}} \dfrac{d^n}{dx^n} e^{-\frac{x^2}{2}}$;

b)   of the Laguerre polynomial $(-1)^n e^x \dfrac{d^n}{dx^n} (x^n e^{-x})$.

3320. Determine all polynomials with coefficients ± 1 having only real roots.

PART TWO
LINEAR ALGEBRA AND GEOMETRY

CHAPTER 7

Vector spaces

In this chapter the coordinates of vectors are recorded in a row. The basis of the space, consisting of vectors $e_1, e_2, \ldots, e_n$, is recorded as the row $(e_1, e_2, \ldots, e_n)$, while for matrix recording, the coordinates of basic vectors are placed in a column.

The *matrix of a change of an 'old' basis by a 'new' one* $(e_1', e_2', \ldots, e_n')$ is the matrix $T = (t_{ij})$ whose columns are the coordinates of the new basic vectors in the old basis.

Thus,

$$(e_1', e_2', \ldots, e_n') = (e_1, e_2, \ldots, e_n)T,$$

and the coordinates of a vector x in the old and new bases are connected by the equality $x_i = \sum_{j=1}^{n} t_{ij} x_j'$, or, in the matrix record

$$\begin{pmatrix} x_1 \\ x_2 \\ \vdots \\ x_n \end{pmatrix} = T \begin{pmatrix} x_1' \\ x_2' \\ \vdots \\ x_n' \end{pmatrix}.$$

34 Concept of a vector space. Bases

3401. Let x, y be vectors, α, β be scalars. Prove that

a) $\alpha x = 0$ if and only if $\alpha = 0$ or $x = 0$;

b) $\alpha x + \beta y = \beta x + \alpha y$ if and only if $\alpha = \beta$ or $x = y$.

3402. For which values of λ

a) does the linear independence of a system of vectors $\{a_1, a_2\}$ imply the linear independence of the system $\{\lambda a_1 + a_2, a_1 + \lambda a_2\}$;

b)   does the linear independence of a system $\{a_1, \ldots, a_n\}$ imply the linear independence of the system $\{a_1 + a_2, a_2 + a_3, \ldots, a_{n-1} + a_n, a_n + \lambda a_1\}$?

3403. Prove the linear independence of the systems of functions:

a) $\sin x, \cos x$;

b) $1, \sin x, \cos x$;

c) $\sin x, \sin 2x, \ldots, \sin nx$;

d) $1, \cos x, \cos 2x, \ldots, \cos nx$;

e) $1, \cos x, \sin x, \cos 2x, \sin 2x, \ldots, \cos nx, \sin nx$;

f) $1, \sin x, \sin^2 x, \ldots, \sin^n x$;

g) $1, \cos x, \cos^2 x, \ldots, \cos^n x$.

3404. Prove the linear independence of the systems of functions:

a) $e^{\alpha_1 x}, \ldots, e^{\alpha_n x}$;

b) $x^{\alpha_1}, \ldots, x^{\alpha_n}$;

c) $(1 - \alpha_1 x)^{-1}, \ldots, (1 - \alpha_n x)^{-1}$,

where $\alpha_1, \ldots, \alpha_n$ are pairwise distinct real numbers.

3405. Prove that in the space of functions of one real variable, vectors $f_1, \ldots, f_n$ are linearly independent if and only if there exist numbers $a_1, \ldots, a_n$ such that $\det\left(f_i(a_j)\right) \neq 0$.

3406.

a)   Let there be defined, in a vector space V over the field $\mathbb{C}$, a new multiplication of vectors by complex numbers by the rule $\alpha \circ x = \bar{\alpha} x$. Prove that V with respect to the operations $+$ and $\circ$ is a vector space. Find its dimension.

b)   Let $\mathbb{C}^n$ be the abelian group of all rows $(a_1, \ldots, a_n)$ of length n, $a_i \in \mathbb{C}$. If $b \in \mathbb{C}$ we put $b \circ (a_1, \ldots, a_n) = (\bar{b}a_1, \ldots, \bar{b}a_n)$. Is $\mathbb{C}^n$ a vector space with respect to the operations $+$ and $\circ$?

3407. Prove that

a)   the group $\mathbb{Z}$ is not isomorphic to the additive group of any vector space;

b)   the group $\mathbb{Z}_n$ is isomorphic to the additive group of a vector space over some field if and only if n is a prime number;

c)   a commutative group A is a vector space over the field $\mathbb{Z}_p$ if and only if $px = 0$ for any $x \in A$;

d) a commutative group A can be turned into a vector space over $\mathbb{Q}$, if and only if it has no elements of finite order (except zero) and, for any natural number n and any $a \in A$, the equation $nx = a$ has a solution in the group A.

3408. Let F be a field and E be its subfield.

a) Prove that F is a vector space over E.

b) If F is finite then $|F| = |E|^n$, where n is the dimension of F as a vector space over E.

c) If F is finite then $|F| = p^m$, where p is the characteristic of F.

d) Find the basis and dimension of $\mathbb{C}$ over $\mathbb{R}$.

e) Let $m_1, \ldots, m_1$ be distinct square-free natural numbers. Prove that the numbers $1, \sqrt{m_1}, \ldots, \sqrt{m_n}$ are linearly independent in $\mathbb{R}$ over $\mathbb{Q}$.

f) Let $r_1, \ldots, r_n$ be distinct rational numbers in the interval $(0, 1)$. Prove that in the space $\mathbb{R}$ over $\mathbb{Q}$ the numbers $2^{r_1}, \ldots, 2^{r_n}$ are independent.

g) Let α be a complex root of an irreducible polynomial over $\mathbb{Q}$, $p \in \mathbb{Q}[x]$. Find the dimension over $\mathbb{Q}$ of the space $\mathbb{Q}[\alpha]$, consisting of all numbers of the form $f(\alpha)$, $f \in \mathbb{Q}[x]$.

3409. Let M be a set consisting of n elements. On the set of its subsets 2^M let there be defined the operations of addition and multiplication by elements of the field $\mathbb{Z}_2$ as in Exercise 102.

$$1X = X, \quad 0X = \emptyset.$$

a) Prove that with respect to these operations the set 2^M is a vector space over the field $\mathbb{Z}_2$, and find its basis and dimension.

b) Let $X_1, \ldots, X_k$ be subsets of M, neither of which is contained in the union of the others. Prove that $\{X_1, \ldots, X_k\}$ is an independent system.

3410. Let the vectors $e_1, \ldots, e_n$ and x be given, in some basis by coordinates:

a) $e_1 = (1, 1, 1)$, $e_2 = (1, 1, 2)$, $e_3 = (1, 2, 3)$, $x = (6, 9, 14)$;

b) $e_1 = (2, 1, -3)$, $e_2 = (3, 2, -5)$, $e_3 = (1, -1, 1)$, $x = (6, 2, -7)$;

c) $e_1 = (1, 2, -1, -2)$, $e_2 = (2, 3, 0, -1)$, $e_3 = (1, 2, 1, 4)$, $e_4 = (1, 3, -1, 0)$, $x = (7, 14, -1, 2)$.

Prove that $(e_1, \ldots, e_n)$ is also a basis of the space and find the coordinates of x in this basis.

3411. Prove that each of the two given systems of vectors S and S' is a basis. Find the matrix of the change of the base S to S'.

a) $S = ((1, 2, 1),\ (2, 3, 3),\ (3, 8, 2))$,
    $S' = ((3, 5, 8),\ (5, 14, 13),\ (1, 9, 2))$;

b) $S = ((1, 1, 1, 1),\ (1, 2, 1, 1),\ (1, 1, 2, 1,),\ (1, 3, 2, 3))$,
    $S' = ((1, 0, 3, 3),\ (-2, -3, -5, -4),\ (2, 2, 5, 4),\ (-2, -3, -4, -4))$.

3412. Prove that in the space $\mathbb{R}[x]_n$ of polynomials of degree $\leq n$ with real coefficients the systems

$$\{1, x, \ldots, x^n\} \text{ and } \{1, x - a, (x - a)^2, \ldots, (x - a)^n\} \quad (a \in \mathbb{R})$$

are bases. Find the coordinates of the polynomial $f(x) = a_0 + a_1 x + \cdots + a_n x^n$ in these bases and the matrix of change from the first basis to the second one.

3413. What happens with the matrix of the change from one basis to another if

a) we interchange two vectors of the first base;

b) we interchange two vectors of the second base;

c) we write the vectors of both bases in inverse order?

3414. Prove that the following systems of vectors are linearly independent and complete them to a basis of the space of rows

a) $a_1 = (2, 2, 7, -1)$, $a_2 = (3, -1, 2, 4)$, $a_3 = (1, 1, 3, 1)$;

b) $a_1 = (2, 3, -4, -1)$, $a_2 = (1, -2, 1, 3)$;

c) $a_1 = (4, 3, -1, 1, 1)$, $a_2 = (2, 1, -3, 2, -5)$,
    $a_3 = (1, -3, 0, 1, -2)$, $a_4 = (1, 5, 2, -2, 6)$;

d) $a_1 = (2, 3, 5, -4, 1)$, $a_2 = (1, -1, 2, 3, 5)$.

35   Subspaces

3501. Find out whether the following sets of vectors form a subspace of appropriate vector spaces:

a) vectors of the plane with the origin O whose ends belong to one of two given lines which are intersecting at the point O;

b) vectors of the plane with the origin O whose ends belong to a given line;

c) vectors of the plane with the origin O whose ends do not belong to given line;

d) vectors of the coordinate plane whose ends belong to the first quadrant;

e) vectors of the space $\mathbb{R}^n$ with integer coordinates;

f) vectors of an arithmetic space F^n, where F is a field, which are solutions of a given system of linear equations;

g) vectors of a linear space which are linear combinations of the given vectors $a_1, \ldots, a_k$;

h) bounded sequences of complex numbers;

i) convergent sequences of real numbers;

j) sequences of real numbers with the fixed limit a;

k) sequences $u(n)$ of elements of a field F satisfying the recurrence equation

$$u(n + k) = f(n) + a_0 u(n) + a_1 u(n + 1) + \cdots + a_{k-1} u(n + k - 1),$$

where $(f(n))$ is a fixed sequence of elements of F, k is a fixed natural number, and $a_i \in F$;

l) polynomials of even degree with coefficients in a field F;

m) polynomials with coefficients in a field F which do not contain even powers of the variable x;

n) elements of the space 2^M (see Exercise 409) of even cardinalities;

o) elements of 2^M of odd cardinalities.

3502. Prove that the following sets of vectors in a space F^n, where F is a field, form subspaces. Find their bases and dimensions:

a) vectors in which the first and last coordinate coincide;

b) vectors in which the coordinates with even indices are equal to 0;

c) vectors in which the coordinates with even indices are equal;

d) vectors of the form $(\alpha, \beta, \alpha, \beta, \ldots)$;

e) vectors which are solutions of a homogeneous system of equations.

3503. Find out if the following sets of matrices of size n over a field F form subspaces of the spaces of matrices $\mathbf{M}_n(F)$. Find their bases and dimensions:

a) all matrices;

b) symmetric matrices;

c) skew-symmetric matrices;

d) nonsingular matrices;

e) singular matrices;

f) matrices with zero trace;

g) matrices commuting with the given set of matrices;

h) matrices X satisfying the equations $A_i X + X B_i = 0$, where $\{A_i, B_i\}$ is the given set of matrices.

3504. Let $\mathbb{R}^S$ be a space of all functions defined on a set S and taking the real values. Find out if the following sets of functions $f(x) \in \mathbb{R}^S$ form a subspace:

a) functions with the value a at the given point $s \in S$;

b) functions with the value a at all points of some fixed subset $T \subseteq S$;

c) functions vanishing at some point of the set S;

d) functions with a limit a at $x \to \infty$ (if $S = \mathbb{R}$);

e) functions with finitely many points of discontinuity (if $S = \mathbb{R}$).

3505. Let K^∞ be a space of infinite sequences of elements of a field K. Find out if the following sets of sequences form subspaces of K^∞:

a) almost zero sequences;

b) sequences in which only finitely many members are equal to zero;

c) sequences in which all elements are distinct from 1.

3506. Prove that in the spaces $\mathbb{R}^\infty$ and $\mathbb{C}^\infty$ the following sets form subspaces:

a) *Cauchy sequences*, that is: for any $\varepsilon > 0$ there exists a number $N \in \mathbb{N}$ such that $|x_n - x_k| < \varepsilon$ for all $n, k > N$;

b) sequences satisfying the *Hilbert condition*: the series $\sum_{i=1}^{\infty} |x_i|^2$ converges;

c) sequences of a polynomial growth, i.e. $|x_n| \le Cn^k$, where C, k are natural numbers depending on the sequence;

d) sequences of an exponential growth, i.e. $|x_n| \le Ce^n$, where C, n are positive real numbers depending on a sequence.

3507. Find out if the following sets of polynomials form subspaces of the spaces $\mathbb{R}[x]_n$ (see Exercise 3412). Find their bases and dimensions:

a) polynomials with a given root $\alpha \in \mathbb{R}$;

b) polynomials with a given root $\alpha \in \mathbb{C} \setminus \mathbb{R}$;

c) polynomials with given roots $\alpha_1, \ldots, \alpha_k \in \mathbb{R}$;

d) polynomials with a given simple root $\alpha \in \mathbb{R}$.

3508. Prove that if a subspace of the vector space $\mathbb{R}[x]_n$ (see Exercise 3412) contains at least one polynomial of degree k for any $k = 0, 1, \ldots, m$ and does not contain polynomials of degree $> m$ then it coincides with $\mathbb{R}[x]_m$.

3509. Let $\mathbb{R}[x_1, \ldots, x_m]$ be a space of polynomials in variables $x_1, \ldots, x_m$. Find:

a) the dimension of the subspace of all homogeneous polynomials of degree k;

b) the dimension of the subspace of symmetric polynomials which are homogeneous polynomials of degree k.

3510. Let V be an n-dimensional vector space over a field F with q elements. Find:

a) the number of vectors in V;

b) the number of bases of V;

c) the number of nonsingular matrices of size n over F;

d) the number of singular matrices of size n over F;

e) the number of k-dimensional subspaces of V;

f) the number of solutions of an equation $AX = 0$, where A is a rectangular matrix of rank r, X is a column of variables of height n.

3511. Find a basis and the dimension of the linear span of the following system of vectors:

a) $a_1 = (1, 0, 0, -1)$, $a_2 = (2, 1, 1, 0)$, $a_3 = (1, 1, 1, 1)$, $a_4 = (1, 2, 3, 4)$, $a_5 = (0, 1, 2, 3)$;

b) $a_1 = (1, 1, 1, 1, 0)$, $a_2 = (1, 1, -1, -1, -1)$, $a_3 = (2, 2, 0, 0, -1)$, $a_4 = (1, 1, 5, 5, 2)$, $a_5 = (1, -1, -1, 0, 0)$.

3512. Let L_1 and L_2 be subspaces of a finite-dimensional vector space V. Prove that

a) if $L_1 \subseteq L_2$, then $\dim L_1 \leq \dim L_2$ and equality occurs only if $L_1 = L_2$;

b) if $\dim(L_1 + L_2) = 1 + \dim(L_1 \cap L_2)$ then the sum $L_1 + L_2$ is equal to one of these subspaces and the intersection $L_1 \cap L_2$ is equal to the other one;

c) if $\dim L_1 + \dim L_2 > \dim V$ then $L_1 \cap L_2 \neq 0$.

3513. Let U, V, W be subspaces of a vector space.

a) Is it possible to assert that $U \cap (V + W) = (U \cap V) + (U \cap W)$?

b) Prove that the previous equality is valid if $V \subseteq U$.

c) Prove that $(U+W) \cap (W+V) \cap (V+U) = [(W+V) \cap U] + [(V+U) \cap W]$.

d) Prove that $\dim[(U + V) \cap W] + \dim(U \cap V) = \dim[(V + W) \cap U] + \dim(V \cap W)$.

e) Prove that $(U \cap V) + (V \cap W) + (W \cap U) \subseteq (U+V) \cap (V+W) \cap (W+U)$ and that the difference of the dimensions of these subspaces is an even number.

3514. Find the dimensions of the sums and intersections of the linear spans of the systems of vectors of the space $\mathbb{R}^4$:

a) $S = \langle (1, 2, 0, 1), (1, 1, 1, 0) \rangle$,
    $T = \langle (1, 0, 1, 0), (1, 3, 0, 1) \rangle$;

b) $S = \langle (1, 1, 1, 1), (1, -1, 1, -1), (1, 3, 1, 3) \rangle$,
    $T = \langle (1, 2, 0, 2), (1, 2, 1, 2), (3, 1, 3, 1) \rangle$;

c) $S = \langle (2, -1, 0, -2), (3, -2, 1, 0), (1, -1, 1, -1) \rangle$,
    $T = \langle (3, -1, -1, 0), (0, -1, 2, 3), (5, -2, -1, 0) \rangle$.

3515. Find bases of the sum and of the intersection of the linear spans $\langle a_1, a_2, a_3 \rangle$ and $\langle b_1, b_2, b_3 \rangle$:

a)
$a_1 = (1, 2, 1)$, $b_1 = (1, 2, 2)$,
$a_2 = (1, 1, -1)$, $b_2 = (2, 3, -1)$,
$a_3 = (1, 3, 3)$, $b_3 = (1, 1, -3)$;

b)
$a_1 = (-1, 6, 4, 7, -2)$, $b_1 = (1, 1, 2, 1, -1)$,
$a_2 = (-2, 3, 0, 5, -2)$, $b_2 = (0, -2, 0, -1, -5)$,
$a_3 = (-3, 6, 5, 6, -5)$, $b_3 = (2, 0, 2, 1, -3)$;

c) $a_1 = (1, 1, 0, 0, -1),$ $b_1 = (1, 0, 1, 0, 1),$
           $a_2 = (0, 1, 1, 0, 1),$ $b_2 = (0, 2, 1, 1, 0),$
           $a_3 = (0, 0, 1, 1, 1),$ $b_3 = (1, 2, 1, 2, -1);$

d) $a_1 = (1, 2, 1, 0),$ $b_1 = (2, -1, 0, -1),$
           $a_2 = (-1, 1, 1, 1),$ $b_2 = (1, -1, 3, 7);$

e) $a_1 = (1, 2, -1, -2),$ $b_1 = (2, 5, -6, -5),$
           $a_2 = (3, 1, 1, 1),$ $b_2 = (-1, 2, -7, -3).$
           $a_3 = (-1, 0, 1, -1),$

3516. Find the system of linear equations defining the systems of vectors:

a) $\langle(1, -1, 1, 0), (1, 1, 0, 1), (2, 0, 1, 1)\rangle;$

b) $\langle(1, -1, 1, -1, 1), (1, 1, 0, 0, 3), (3, 1, 1, -1, 7)\rangle.$

3517. Let $L_1, \ldots, L_k$ be subspaces of a vector space. Prove that

a)   the sum of these subspaces is direct if and only if at least one of its vectors
     has a unique representation

$$x_1 + \cdots + x_k \quad (x_i \in L_i; \quad i = 1, \ldots, k);$$

b)   the condition $L_i \cap L_j = 0$ for any distinct i and j from 1 to k is not
     sufficient for the sum of these subspaces to be direct.

3518. Let the subspaces $U, V \subseteq \mathbb{R}^n$ be given by the equations

$$x_1 + x_2 + \cdots + x_n = 0, \quad x_1 = x_2 = \cdots = x_n.$$

Prove that $\mathbb{R}^n = U \oplus V$ and find the projections of basic vectors into U in parallel
with V and into V in parallel with U.

3519. Put in the space $\mathbb{R}^4$

$$U = \langle(1, 1, 1, 1), (-1, -2, 0, 1)\rangle, \qquad V = \langle(-1, -1, 1, -1), (2, 2, 0, 1)\rangle.$$

Prove that $\mathbb{R}^4 = U \oplus V$, and find the projection of the vector $(4, 2, 4, 4)$ into the
subspace U in parallel with V.

3520. Prove that for any subspace $U \subseteq \mathbb{R}^n$ there exists a subspace V such that
$\mathbb{R}^n = U \oplus V.$

3521. Prove that the space of matrices $\mathbf{M}_n(\mathbb{R})$ is the direct sum of the subspace of symmetric and the subspace of skew-symmetric matrices. Find the projections of the matrix

$$\begin{pmatrix} 1 & 1 & \cdots & 1 \\ 0 & 1 & \cdots & 1 \\ \vdots & \vdots & \ddots & \vdots \\ 0 & 0 & \cdots & 1 \end{pmatrix}$$

into each of these subspaces in parallel with the other one.

3522. Let U be the subspace of skew-symmetric matrices, V be a subspace of upper-triangular matrices in $\mathbf{M}_n(\mathbb{R})$.

a) Prove that $U \oplus V = \mathbf{M}_n(\mathbb{R})$.

b) Find the projections of the matrices E_{ij} into U and into V.

3523. Let U be the subspace of symmetric matrices, V be the subspace of upper-triangular matrices in $\mathbf{M}_n(\mathbb{R})$.

a) Prove that $U \oplus V = \mathbf{M}_n(\mathbb{R})$.

b) Find the projection of the matrix E_{ij} into U and into V.

3524. Let F be a field with q elements, U be a subspace of dimension m of a space V of dimension n over F. Find the number of subspaces W of V such that $V = U \oplus V$.

3525. Let V be a linear space over an infinite field F and $V_1, \ldots, V_k$ be subspaces of V such that $V = V_1 \cup \cdots \cup V_k$. Prove that $V = V_i$ for some $i = 1, \ldots, k$.

3526. Let V be a linear space over a field F, and U, W be subspaces of V such that $U \cup W = V$. Prove that either $V = U$ or $V = W$.

3527. Find an example of a space V over a finite field such that $V = U_1 \cup U_2 \cup U_3$, where U_1, U_2, U_3 are proper subspaces of V.

36   Linear functions and mappings

3601. Let $V_0 \xrightarrow{A_1} V_1 \xrightarrow{A_2} \ldots \xrightarrow{A_m} V_m$ be a sequence of linear mappings of vector spaces. Prove that

$$\sum_{i=1}^{m} \dim \operatorname{Ker} A_i - \sum_{i=1}^{m} \dim(V_i / \operatorname{Im} A_i) = \dim V_0 - \dim V_m.$$

3602. Let F be a field with q elements. Find

a)   the number of linear mappings from F^n to F^k;

b)   the number of linear injective mappings from F^n to F^k;

c)   the number of linear surjective mappings from F^n to F^k.

3603. Let the linear mapping $A : V \to W$ be given in the base (e_1, e_2, e_3) of a space V and in the base (f_1, f_2) of a space W by the matrix $\begin{pmatrix} 0 & 1 & 2 \\ 3 & 4 & 5 \end{pmatrix}$. Find the matrix of the mapping A in the bases $(e_1, e_1 + e_2, e_1 + e_2 + e_3)$ and $(f_1, f_1 + f_2)$.

3604. Let $L = K[x]_1$ (see Exercise 3412), (K is a field). Find the matrix of the linear mapping $A : f(x) \mapsto f(S)$ from the space L into the space $M = M_2(K)$. Here $S = \begin{pmatrix} a & b \\ c & d \end{pmatrix}$ is a fixed matrix, and there are chosen the basis $(1, x)$ in L, and the basis of matrix units in M.

3605. Let $A, B : V \to W$ be linear mappings such that $\dim(\operatorname{Im} A) \leq \dim(\operatorname{Im} B)$. Prove that there exist operators C, D on V and W such that $A = DBC$ and C (or D) can be chosen as nonsingular.

3606. Let $A, B : V \to W$ be linear mappings. Prove that the following conditions are equivalent:

a) $\operatorname{Ker} A \leq \operatorname{Ker} B$;

b) $B = CA$ for some operator C on W.

3607. Let $A, B : V \to W$ be linear mappings. Prove that the following conditions are equivalent:

a) $\operatorname{Im} A \subseteq \operatorname{Im} B$;

b) $A = BD$ for some operator D on V.

3608. Let $A : V \to W$ be linear mappings. Prove that there exists a linear mapping $B : W \to V$ such that $A = ABA$, $B = BAB$.

3609. Let $V = \mathbb{R}[x]_n$ and the mappings α^a ($a \in \mathbb{R}$), β^i, γ^i from the spaces V into $\mathbb{R}$ be given by the rules

$$\alpha^a(f) = f(a), \quad \beta^i(f) = f^{(i)}(0), \quad \gamma^i(f) = \int_0^i f(x)\,dx.$$

Prove that the systems:

a) $\alpha^1, \ldots, \alpha^n$,   b) $\beta^0, \beta^1, \ldots, \beta^n$,   c) $\gamma^0, \gamma^1, \ldots, \gamma^n$,

are bases of the dual space V^*.

3610.

  a) Prove that for any basis of the dual space V^* of V there exists a unique basis of V for which the given basis is dual.

  b) Find such a basis in Exercise 3609a.

  c) Find such a basis in Exercise 3609b.

3611. Prove that for any nonzero linear function f on n-dimensional space V there exists a base $(e_1, \ldots, e_n)$ such that

$$f(x_1e_1 + \cdots + x_ne_n) = x_1$$

for any coordinates $x_1, \ldots, x_n$.

3612. Prove that any k-dimensional subspace of n-dimensional space is the intersection of kernels of some $n - k$ linear functions.

3613. Let f be a nonzero linear function on a vector space V (not necessarily finite-dimensional), and $U = \operatorname{Ker} f$. Prove that

  a) U is a *maximal* subspace of V, i.e. it is not contained in any other subspace different from V;

  b) $V = U \oplus \langle a \rangle$ for any $a \notin U$.

3614. Prove that if two linear functions on a vector space have the same kernels then they differ only by a scalar coefficient.

3615. Prove that n linear functions on n-dimensional space are linearly independent if and only if the intersection of their kernels is the zero subspace.

3616. Prove that the vectors $e_1, \ldots, e_k$ of a finite-dimensional space V are linearly independent if and only if there exist linear functions $f^1, \cdots, f^k \in V^*$ such that $\det \left(f^i(e_j) \right) \neq 0$.

3617. For any subset U of a finite-dimensional space V and for any subset W of the dual space V^* we put

$$U^0 = \{ f \in V^* | f(x) = 0 \text{ for any } x \in U \},$$
$$W^0 = \{ x \in V | f(x) = 0 \text{ for any function } f \in V^* \}.$$

Prove that

  a) U^0 is a subspace of V^* and if U is a subspace then $\dim U + \dim U^0 = \dim V$;

  b) if U_1 and U_2 are subspaces of V then $U_1^0 = U_2^0$ if and only if $U_1 = U_2$;

c)   for any subspace U of the space V

$$(U^0)^0 = U, \quad (U_1 + U_2)^0 = U_1^0 \cap U_2^0, \quad (U_1 \cap U_2)^0 = U_1^0 + U_2^0.$$

3618. Prove that the space of the polynomials $\mathbb{Q}[x]$ is not isomorphic to its dual one.

3619. Let l_1, l_2 be two linear functions on a linear space V and $l_1(x)l_2(x) = 0$ for all $x \in V$. Prove that one of the functions is zero.

3620. Let $l_1, \ldots, l_k$ be linear functions on a linear vector space V over an infinite field. If $l_1(x) \ldots l_k(x) = 0$ for all $x \in V$ then one of the functions is zero.

$$* \quad * \quad *$$

3621. Let K be a finite field with q^n elements, F be a subfield of K with q elements, and $L(x) = x + x^q + \cdots + x^{q^{n-1}}$, $x \in K$. Prove that:

a) L is a linear operator on K considered as a vector space over F;

b)   the kernel of L consists of all elements $a - a^q$, where $a \in K$;

c) F is contained in the kernel of L if and only if the characteristic of K divides n.

CHAPTER 8

Bilinear and quadratic functions

37 General bilinear and sesquilinear functions

Throughout this section it is assumed that the characteristic of the main field is different from two.

3701. What functions of two arguments are bilinear functions on appropriate spaces:

a) $f(x, y) = {}^t x \cdot y$ $(x, y \in F^n$ are columns, F is a field);

b) $f(A, B) = \text{tr}(AB)$ $(A, B \in M_n(F), F$ is a field);

c) $f(A, B) = \text{tr}(AB - BA)$;

d) $f(A, B) = \det(AB)$;

e) $f(A, B) = \text{tr}(A + B)$;

f) $f(A, B) = \text{tr}(A \cdot {}^t B)$;

g) $f(A, B) = \text{tr}({}^t A \cdot B)$;

h) $f(A, B)$ is the (i, j)-entry of a matrix AB;

i) $f(u, v) = \Re(uv); (u, v \in \mathbb{C}, \mathbb{C}$ is considered as a vector space over $\mathbb{R})$;

j) $f(u, v) = \Re(u\bar{v})$;

k) $f(u, v) = |uv|$;

l) $f(u, v) = \Im(u\bar{v})$;

m) $f(u, v) = \displaystyle\int_a^b uv\,dt$ $(u, v$ are continuous functions of t in the interval $[a, b]$);

115

n) $f(u, v) = \int_a^b uv' \, dt$ (u, v are differentiable functions on the segment $[a, b]$ and $u(a) = u(b) = v(a) = v(b) = 0$);

o) $f(u, v) = \int_a^b (u + v)^2 dt$;

p) $f(u, v) = (uv)(\alpha)$ ($u, v \in F[x]$, $\alpha \in F$);

q) $f(u, v) = \frac{d}{dt}(uv)(a)$;

r) $f(u, v) = |u + v|^2 - |u|^2 - |v|^2$ ($u, v \in \mathbb{R}^3$);

s) $f(u, v) = \varepsilon(u \times v)$ ($\times$ is the vector product, $\varepsilon(x)$ is the sum of coordinates of a vector x in the given basis).

3702. Choose a basis and find matrices of appropriate bilinear functions on finite-dimensional spaces from Exercise 3701.

3703. Let F be a field and $F(x)$ be the field of rational functions in the variable x. Prove that the mappings $x \mapsto \varepsilon x^\tau$ induce automorphisms of order two of $F(x)$, where $\varepsilon, \tau = \pm 1$ and $(\varepsilon, \tau) \neq (1, 1)$.

3704. Let p, q be distinct primes. Prove that the real numbers of the form $a + b\sqrt{p} + c\sqrt{q} + d\sqrt{pq}$, where $a, b, c, d \in \mathbb{Q}$ form a subfield $\mathbb{Q}(\sqrt{p}, \sqrt{q})$ of $\mathbb{R}$. Check if the mapping

$$x \to \bar{x} = a - b\sqrt{p} + c\sqrt{q} - d\sqrt{pq}$$

is an automorphism of order two of this field.

3705. Let F be a field with an automorphism $x \to \bar{x}$ of order two. What functions of two arguments are sesquilinear functions on appropriate vector spaces:

a) $f(x, y) = {}^t x \cdot \bar{y}$ ($x, y \in F^n$ are columns);

b) $f(x, y) = \text{tr}(A\bar{B})$ ($A, B \in \mathbf{M}_n(F)$);

c) $f(A, B) = \det(A\bar{B})$;

d) $f(A, B) = \text{tr}(A^t \bar{B})$;

e) $f(A, B) = \text{tr}(A \cdot {}^t B)$;

f) $f(A, B)$ is the (i, j)-entry of the matrix $A\bar{B}$;

g) $f(A, B)$ is the (i, j)-entry of the matrix $A^t \bar{B}$;

h) $f(u, v) = \frac{d}{dt}(u\bar{v})(a)$, ($u, v \in F[x]$, $a \in F$).

3706. Find the matrix of a bilinear function f in a new basis, given the matrix of f in the old basis and the matrix of the change of basis:

a)
$$\begin{pmatrix} 1 & 2 & 3 \\ 4 & 5 & 6 \\ 7 & 8 & 9 \end{pmatrix},$$

$e_1' = e_1 - e_2,$
$e_2' = e_1 + e_3,$
$e_3' = e_1 + e_2 + e_3;$

b)
$$\begin{pmatrix} 0 & 2 & 1 \\ -2 & 2 & 0 \\ -1 & 0 & 3 \end{pmatrix},$$

$e_1' = e_1 + 2e_2 - e_3,$
$e_2' = e_2 - e_3,$
$e_3' = -e_1 + e_2 - 3e_3.$

3707. A sesquilinear function f on two-dimensional complex space with a basis e_1, e_2 is given by a matrix B. Find the matrix B' of the function f with the basis e_1', e_2', where

a) $B = \begin{pmatrix} i+1 & -1 \\ 0 & -i \end{pmatrix},$ $\quad e_1' = e_1 + ie_2, \quad e_2' = +ie_1 + e_2;$

b) $B = \begin{pmatrix} 2 & -1+i \\ -i & 0 \end{pmatrix},$ $\quad e_1' = 2e_1 - ie_2, \quad e_2' = ie_1 + e_2.$

3708. Let a bilinear function f be given in some basis by a matrix F. Find $f(x, y)$ if

a) $F = \begin{pmatrix} 1 & -1 & 1 \\ -2 & -1 & 3 \\ 0 & 4 & 5 \end{pmatrix},$ $\quad \begin{aligned} x &= (1, 0, 3), \\ y &= (-1, 2, -4); \end{aligned}$

b) $F = \begin{pmatrix} i & 1+i & 0 \\ -1+i & 0 & 2-i \\ 2+i & 3-i & -1 \end{pmatrix},$ $\quad \begin{aligned} x &= (1+i, 1-i, 1), \\ y &= (-2+i, -i, 3+2i). \end{aligned}$

3709. Find the value $f(x, y)$ of a sesquilinear function given by a matrix B in some basis of complex space if

a) $B = \begin{pmatrix} 5 & 2 \\ -1 & i \end{pmatrix},$ $\quad x = (i, -2), \quad y = (1-i, 3+i).$

b) $B = \begin{pmatrix} -i & 1-2i \\ -4 & 2+3i \end{pmatrix},$ $\quad x = (2, i+3), \quad y = (-i, 6-2i).$

3710. Let g be a bilinear function with a matrix G in some basis of a space V, and let A be a linear operator on V with a matrix A. Find in this basis the matrix of the bilinear function $f(u, v) = g(u, A(v))$ if

a)
$$G = \begin{pmatrix} 1 & -1 & 0 \\ 2 & 0 & -2 \\ 3 & 4 & 5 \end{pmatrix}, \qquad A = \begin{pmatrix} -1 & 1 & 1 \\ -3 & -4 & 2 \\ 1 & -2 & -3 \end{pmatrix};$$

b)
$$G = \begin{pmatrix} 0 & 1 & 2 \\ 4 & 0 & 3 \\ 5 & 6 & 0 \end{pmatrix}, \qquad A = \begin{pmatrix} 1 & -4 & 3 \\ 4 & -1 & -2 \\ -3 & 2 & 1 \end{pmatrix}.$$

3711. Let g be a sesquilinear function with a matrix G in some basis of a linear space V, A be a linear operator on V with a matrix A. Find the matrix of the sesquilinear function $f(u, v) = g(u, A(v))$ in this basis if

a)
$$G = \begin{pmatrix} -5 + i & 2 \\ -i & 4 + i \end{pmatrix}, \qquad A = \begin{pmatrix} i & 0 \\ 2 & -i \end{pmatrix};$$

b)
$$G = \begin{pmatrix} 4 - i & 2 - i \\ 0 & 1 + i \end{pmatrix}, \qquad A = \begin{pmatrix} -i + 1 & 0 \\ 4 & 3 + i \end{pmatrix}.$$

3712. Find the left and right kernels of a bilinear function f which is given in a basis (e_1, e_2, e_3) by a matrix

a)
$$\begin{pmatrix} 2 & -3 & 1 \\ 3 & -5 & 5 \\ 5 & -8 & 6 \end{pmatrix};$$
b)
$$\begin{pmatrix} 4 & 3 & 2 \\ 1 & 3 & 5 \\ 3 & 6 & 9 \end{pmatrix}.$$

3713. Let $F = \mathbb{Q}(\sqrt{3})$ and $\bar{x} = a - b\sqrt{3}$ if $x = a + b\sqrt{3}$, where $a, b \in \mathbb{Q}$. Find, in a two-dimensional vector space over F, the left and the right kernels of the sesquilinear function which is given in the basis (e_1, e_2) by the matrix

a)
$$\begin{pmatrix} 1 + \sqrt{3}, & -1 - \sqrt{3} \\ 1 & 2 - \sqrt{3} \end{pmatrix};$$
b)
$$\begin{pmatrix} -3, & -2\sqrt{3} \\ 2\sqrt{3}, & 4 \end{pmatrix}.$$

3714. Find the left and the right kernels of the bilinear function $f(x, y) = (x, A(y))$, where A is the linear operator with a matrix A in an orthonormal basis (e_1, e_2, e_3) of Euclidean space:

a) $A = \begin{pmatrix} 5 & -6 & 1 \\ 3 & -5 & -2 \\ 2 & -1 & 3 \end{pmatrix};$ b) $A = \begin{pmatrix} 2 & -1 & 3 \\ 3 & -2 & 2 \\ 5 & -4 & 0 \end{pmatrix}.$

3715. Let $F, \bar{x}$ be as in Exercise 3713. Find the left and the right kernels of the sesquilinear function $f(x, y) = g(x, \mathcal{A}(y))$, where $\mathcal{A}$ is the linear operator with a matrix A, and g is a sesquilinear function defined by the identity matrix on two-dimensional space with a basis (e_1, e_2):

a) $\quad A = \begin{pmatrix} 1 + \sqrt{3}, & 1 \\ 0 & 0 \end{pmatrix}$;
   b) $\quad A = \begin{pmatrix} 0, & 0 \\ 2\sqrt{3} & 0 \end{pmatrix}$.

3716. Let f be a bilinear function with a matrix F on a vector space V, and U be a subspace of V. Find the left and the right orthocomplements of U with respect to f (i.e. the maximal subspaces U_1 and U_2 such that $f(U_1, U) = f(U, U_2) = 0$) if

a) $\quad F = \begin{pmatrix} 4 & 1 & 3 \\ 3 & 3 & 6 \\ 2 & 5 & 9 \end{pmatrix}, \qquad U = \langle (1, -1, 0), (-2, 3, 1) \rangle$;

b) $\quad F = \begin{pmatrix} 6 & -8 & 5 \\ 5 & -5 & 3 \\ 1 & -3 & 2 \end{pmatrix}, \qquad U = \langle (2, 0, -3), (3, 1, -5) \rangle$.

3717. Let $F, \bar{x}$ be as in Exercise 3713 and f be a sesquilinear function with a matrix G on a vector space V, and U be a subspace of V. Find the left and the right orthogonal complements of U with respect to f if

a) $\quad G = \begin{pmatrix} 1 + \sqrt{3} & 2 & -\sqrt{3} \\ 0 & 1 & 2 \\ 1 - \sqrt{3} & 2 & 0 \end{pmatrix}, \qquad U = \langle (1, 0, \sqrt{3}), (0, 2, 1) \rangle$;

b) $\quad G = \begin{pmatrix} 1 & -\sqrt{3} & 2 \\ 2 & 0 & 1 \\ 1 & 1 - \sqrt{3} & \sqrt{3} \end{pmatrix}, \qquad U = \langle (1, -\sqrt{3}, 2) \rangle$.

3718. Let F be a finite field with q^2 elements, and $\bar{x} = x^q$ for all $x \in F$. Assume that K is a finite field containing F and $n = \dim_F K$. Prove that

a) $x \to \bar{x}$ is an automorphism of the second order of F;

b) a function
$$f(x, y) = xy^q + x^{q^2} y^{q^3} + \cdots + x^{q^{2n-2}} y^{q^{2n-1}},$$
is a sesquilinear function on K as a vector space over F;

c) a function $f(x, y)$ from b) is nonsingular;

d) find the left and the right orthogonal complements of F in K with respect to $f(x, y)$.

3719. Let $F = \mathbb{Q}[i]$, $K = F[\sqrt{2}]$. Consider K as a vector space over F. Prove that

a) $\dim_F K = 2$;

b) a function $f(z_1 + z_2\sqrt{2}, t_1 + t_2\sqrt{2}) = z_1\bar{t}_1 + 2z_2\bar{t}_2$ where the bar denotes the complex conjugation, is a sesquilinear function on K as a vector space over F.

c) Find the matrix of f in the basis $e_1 = 1$, $e_2 = \sqrt{2}$.

d) Prove that the function f is nonsingular.

e) Find the left and the right orthogonal complements of F in K with respect to f.

3720. Let $F = \mathbb{C}(x)$ be the field of rational functions with an automorphism such that $x \to \varepsilon x^\tau$, where ε, $\tau = \pm 1$, $(\varepsilon, \tau) \neq (1, 1)$.

a) Prove that the polynomial $y^4 - x \in F[y]$ is irreducible over F.

b) Prove that the function

$$f(u, v) = u(x, y)v(ix^\tau, y) + u(x, iy)v(\varepsilon x^\tau, iy) + u(x, -y)v(\varepsilon x^\tau, -y)$$
$$+ u(x, -iy)v(\varepsilon x^\tau, -iy)$$

on the field $K = F[y]/(y^4 - x)$ as a vector space over F, is sesquilinear.

c) Find the matrix of $f(u, v)$ in the basis $1, y, y^2, y^3$ of the space K over F.

d) Prove that the function f is nonsingular.

e) Find the left and the right orthocomplements of the linear span $\langle 1, y \rangle$ in K.

f) Find a basis $u_0, \dots, u_3$ of K such that $f(u_i, y^j) = \delta_{ij}$, where $i, j = 0, \dots, 3$.

3721. Under what elementary transformations of a basis does the matrix of a bilinear function change as the matrix of a linear operator:

a) $(e_1, \dots, e_i, \dots, e_n) \to (e_1, \dots, \lambda e_i, \dots, e_n)$;

b) $(e_1, \dots, e_i, \dots, e_n) \to (e_1, \dots, e_i + \lambda e_j, \dots, e_n)$ $(j \neq i)$;

c) $(e_1, \dots, e_i, \dots, e_j, \dots, e_n) \to (e_1, \dots, e_j, \dots, e_i, \dots, e_n)$?

3722. Find a relation between the matrices A, B, G of linear operators $\mathcal{A}$, $\mathcal{B}$ and the bilinear (sesquilinear) function g in some basis of a space and the matrix F of a bilinear (sesquilinear) function

$$f(x, y) = g(\mathcal{A}(x), \mathcal{B}(y)).$$

3723. Prove that any bilinear (sesquilinear) function f of rank 1 can be factorized as a product of two linear functions $p(x)q(y)$ $(p(x)q(\bar{y})$, respectively). To what simplest form can the matrix of the function f be transformed by elementary changes of basis?

3724. Let $\mathbf{e} = (e_1, \dots, e_n)$, $\mathbf{e}' = (e'_1, \dots, e'_n)$ be two bases of a space V, and let C be the matrix of a change of $\mathbf{e}$ to $\mathbf{e}'$ and f be the bilinear (sesquilinear) function on V with matrices F and F' in these bases. Find a relation between F and F'.

3725. Prove that the bilinear and sesquilinear functions $\operatorname{tr}(AB)$, $\operatorname{tr}(A^t B)$, $\operatorname{tr}(A\bar{B})$, $\operatorname{tr}(A^t \bar{B})$ on the space $M_n(K)$ are nonsingular.

3726. Prove that the dimensions of the left and the right kernels of a bilinear (sesquilinear) function coincide, although these kernels can be not coincident.

3727. Let f be a nonsingular bilinear (sesquilinear) function on a space V. Prove that for any linear function p there is a unique vector $v \in V$ such that $p(x) = f(x, v)$ for any $x \in V$ and the mapping $p \mapsto v$ is an isomorphism of the spaces V^* and V.

3728. Let F be the matrix of a nonsingular bilinear function f on a real space of dimension n.

a)   Prove that for odd n the matrix $-F$ is not a matrix of f in any basis of V.

b)   Is the statement a) valid for even n?

c)   Is the statement a) valid for even n and for a diagonal matrix F?

3729. For a non-zero bilinear (sesquilinear) function f on a space V, let there exist a number ε such that for any $x, y \in V$

$$f(y, x) = \varepsilon f(x, y)$$

$(\overline{f(y, x)} = \varepsilon f(x, y)$, respectively). Prove that

a) ε is equal either to 1 or to -1;

b)   if U_1 and U_2 are totally isotropic with respect to f subspaces of the same dimension and $U_1 \cap U_2^{\perp} = 0$, then the restriction of f to their sum $U_1 + U_2$ is a nonsingular function;

c)   if W_1 and W_2 are maximal totally isotropic with respect to f subspaces and $W_1 \cap W_2 = 0$, then $\dim W_1 = \dim W_2$;

d) if nonsingular bilinear functions f_1 and f_2 satisfy the previous condition (with the same ε), and with respect to each of them the space V is a direct sum of two totally isotropic subspaces, then the functions f_1 and f_2 are equivalent.

3730. Let f be a bilinear (sesquilinear) function on a space V and for any $x, y \in V$ an equality $f(x, y) = 0$ implies $f(y, x) = 0$. Prove that

a) if f is a bilinear function, then f is either symmetric or skew-symmetric;

b) if f is a sesquilinear function, then f is either Hermitian or skew-Hermitian.

3731. Let f_1, f_2 be bilinear (sesquilinear) functions on a space V with a basis $(e_1, \ldots, e_n)$. Define the bilinear function f on a space W with the basis

$$(a_{11}, a_{12}, \ldots, a_{1n}, a_{21}, \ldots, a_{2n}, \ldots, a_{n1}, \ldots, a_{nn})$$

by setting

$$f(a_{ij}, a_{kl}) = f_1(e_i, e_k) f_2(e_j, e_l).$$

a) Find the matrix of f in the given basis.

b) Prove that if V is a direct sum of subspaces totally isotropic with respect to f_1, then W is a direct sum of subspaces totally isotropic with respect to f.

3732. Find out, without calculations, if the bilinear functions are equivalent:

a) $f_1(x, y) = 2x_1 y_2 - 3x_1 y_3 + x_2 y_3 - 2x_2 y_1 - x_3 y_2 + 3x_3 y_1$,
$f_2(x, y) = x_1 y_2 - x_2 y_1 + 2x_2 y_2 + 3x_1 y_3 - 3x_3 y_1$;

b) $f_1(x, y) = x_1 y_1 + i x_1 y_2$,
$f_2(x, y) = 2x_1 y_1 + (1 + i)x_1 y_2 + (1 - i)x_2 y_1 - i x_2 y_2$.

3733. Reduce these skew-symmetric bilinear functions to the canonical form:

a) $x_1 y_2 - x_1 y_3 - x_2 y_1 + 2x_2 y_3 + x_3 y_1 - 2x_3 y_2$;

b) $2x_1 y_2 + x_1 y_3 - 2x_2 y_1 + 3x_2 y_3 - x_3 y_1 - 3x_3 y_2$;

c) $x_1 y_2 + x_4 y_3 - x_2 y_1 + 2x_2 y_3 - 2x_3 y_2 + 3x_3 y_4 - x_1 y_4 - 3x_4 y_3$;

d) $x_1 y_2 + x_1 y_3 + x_1 y_4 - x_2 y_1 - x_2 y_3 + x_3 y_2 + x_3 y_4 - x_4 y_1 - x_4 y_3$.

3734. Reduce these skew-Hermitian functions on a complex space to the canonical form:

a) $x_1\bar{y}_2 - ix_1\bar{y}_3 - ix_2\bar{y}_3 - \bar{x}_1y_2 + i\bar{x}_1y_3 + i\bar{x}_2y_3 + ix_1\bar{y}_1 - i\bar{x}_1y_1;$

b) $(1+i)x_1\bar{y}_2 + 2x_1\bar{y}_3 + ix_1\bar{y}_4 - (1-i)x_2\bar{y}_3 - (1-i)\bar{x}_1y_3 + i\bar{x}_1y_4 + (1+i)\bar{x}_2y_3 + 2ix_2\bar{y}_2 - 2\bar{x}_1y_3.$

3735. Prove that the function $h(f,g) = \int_0^1 fg'\,dx$ on the space of polynomials of degree ≤ 5, vanishing at points 0 and 1, is skew-symmetric. Find its canonical basis.

3736. Prove that the determinant of an integer skew-symmetric matrix is the square of an integer.

3737. Let f be a skew-symmetric bilinear function on a space V, W be a subspace of V, and $W^\perp$ its orthocomplement with respect to f. Prove that $\dim W - \dim(W \cap W^\perp)$ is an even integer.

3738. Prove that for any complex nonsingular skew-Hermitian matrix A, that there exists a nonsingular complex matrix C such that $CA^t\bar{C}$ is a diagonal matrix and that the entries of the principal diagonal are pure imaginary complex numbers.

3739. Let f be a skew-symmetric bilinear function on a space V, V' be its kernel, and W be the maximal totally isotropic subspace. Prove that

$$\dim W = \frac{\dim V + \dim V'}{2}.$$

3740. Let f be a nonsingular skew-symmetric bilinear function on n-dimensional space V. Let $G = (g_{ij})$ be a skew-symmetric matrix of size n. Prove that there exist vectors $v_1, \ldots, v_n \in V$, such that $g_{ij} = f(v_i, v_j)$.

3741. Let $f(x,y)$ be a Hermitian function on a complex space, $q(x) = f(x,x)$. Prove that

$$4f(x,y) = q(x+y) - q(x-y) + iq(x+iy) - iq(x-iy).$$

3742. Prove that the real and imaginary parts of a Hermitian function on a complex vector space V are symmetric and skew-symmetric functions on V, respectively, considered as $2n$-dimensional vector space.

3743. Prove that if f is a positive definite Hermitian form on a complex space then

$$f(x,y)\overline{f(x,y)} \leq f(x,x) \cdot f(y,y).$$

3744. Let $\mathcal{A}$ be a linear operator, and f be a positive definite Hermitian function on a complex vector space V. Prove that if $f(\mathcal{A}(x), x) = 0$ for all $x \in V$, then $\mathcal{A}$ is the zero operator.

Is this statement valid for symmetric bilinear functions on a real space V?

3745. For what values of n can a nonsingular bilinear function on n-dimensional vector space have:

a)   a totally isotropic subspace of dimension $n - 1$;

b)   a totally isotropic subspace of dimension $n - 2$?

Deduce the formula for the greatest possible dimension of a totally isotropic subspace.

3746. Let $A = (a_{ij}) \in \mathbf{M}_n(\mathbb{R})$ be a symmetric matrix and

$$L(f) = \sum_{i,j=1}^{n} a_{ij} \frac{\partial^2 f}{\partial x_i \partial x_j}$$

be a differential operator on $\mathbb{R}[x_1, \ldots, x_n]$. Prove that

a)   if

$$C \begin{pmatrix} x_1 \\ \vdots \\ x_n \end{pmatrix} = \begin{pmatrix} y_1 \\ \vdots \\ y_n \end{pmatrix}, \quad C \in \mathbf{GL}_n(\mathbb{R}),$$

is a change of variables then

$$L(f) = \sum_{i,j=1}^{n} b_{ij} \frac{\partial^2 f}{\partial y_i \partial y_j},$$

where $(b_{ij}) = C A^t C$;

b)   there exists a nonsingular linear change of the variables in $\mathbb{R}[x_1, \ldots, x_n]$, such that

$$L(f) = \frac{\partial^2 f}{\partial y_1^2} + \cdots + \frac{\partial^2 f}{\partial y_k^2} - \frac{\partial^2 f}{\partial y_{k+1}^2} - \cdots - \frac{\partial^2 f}{\partial y_s^2},$$

with respect to the new variables $y_1, \ldots, y_k$, where $0 \le k \le s \le n$.

38   Symmetric bilinear, Hermitian and quadratic functions

In this section Hermitian functions are considered on complex spaces.

3801. What bilinear functions in Exercise 3701 are symmetric?

3802.

a)   What sesquilinear functions in Exercise 3705 are Hermitian?

b)   Is the function $f(x, y)$ from Exercise 3718b Hermitian?

c)   Is the function $f(u, v)$ from Exercise 3720b Hermitian?

3803. Find out, without calculations, if the bilinear functions are equivalent

$$f_1(x, y) = x_1 y_1 + 2x_1 y_2 + 3x_1 y_3 + 4x_2 y_1 + 5x_2 y_2 + 6x_2 y_3 + 7x_3 y_1$$
$$+ 8x_3 y_2 + 10x_3 y_3,$$
$$f_2(x, y) = 2x_1 y_1 - x_1 y_3 + x_2 y_2 - x_3 y_1 + 5x_3 y_3.$$

3804. Find out, without calculations, for what bilinear functions f there exists a basis in which the matrix of f is diagonal:

a) $-x_1 y_1 - 2x_1 y_2 - 2x_2 y_1 - 3x_2 y_2 + x_3 y_1 - 4x_3 y_3$;

b) $-x_1 y_2 - x_2 y_1 + 3x_2 y_2 + 5x_2 y_3 + 5x_3 y_2 - x_3 y_3$.

3805. Prove that orthocomplements to spaces with respect to a nonsingular symmetric (Hermitian) function satisfy the properties:

a) $(U^\perp)^\perp = U$;

b) $(U_1 + U_2)^\perp = U_1^\perp \cap U_2^\perp$;

c) $(U_1 \cap U_2)^\perp = U_1^\perp + U_2^\perp$.

3806. Find the orthocomplement of the linear span $\langle f_1, f_2 \rangle$ with respect to the bilinear function with the matrix F if

a) $F = \begin{pmatrix} 1 & -1 & -2 \\ -1 & 0 & -3 \\ 2 & -3 & 7 \end{pmatrix}$, $f_1 = (1, 2, 3)$, $f_2 = (4, 5, 6)$;

b) $F = \begin{pmatrix} -1 & 2 & 5 \\ 2 & 2 & 8 \\ 5 & 8 & 29 \end{pmatrix}$, $f_1 = (-3, -15, 21)$, $f_2 = (2, 10, -14)$.

3807. Find the orthocomplement to the linear span $\langle e_1, e_2 \rangle$ with respect to the Hermitian function with the matrix G if

a) $G = \begin{pmatrix} 1 & i & 1-i \\ -i & 0 & -2 \\ 1+i & -2 & -2 \end{pmatrix}$, $\begin{aligned} e_1 &= (i, 1, -1), \\ e_2 &= (1 - 2i, -i, 3); \end{aligned}$

b) $\quad G = \begin{pmatrix} 0 & -2+i & -i \\ -2-i & 2 & -1+i \\ i & -1-i & -1 \end{pmatrix}, \quad \begin{array}{l} e_1 = (-i+1, 2, 0), \\ e_2 = (-1+3i, -3i, 2). \end{array}$

3808. Applying the Jacobi method find the canonical form of the symmetric bilinear functions:

a) $2x_1y_1 - x_1y_2 + x_1y_3 - x_2y_1 + x_3y_1 + 3x_3y_3;$

b) $2x_1y_2 + 3x_1y_3 + 2x_2y_1 - x_2y_3 + 3x_3y_1 - x_3y_2 + x_3y_3.$

3809. Applying the Jacobi method find whether the bilinear functions with the matrices

$$\begin{pmatrix} 1 & 2 & 3 \\ 2 & 0 & -1 \\ 3 & -1 & 3 \end{pmatrix}, \quad \begin{pmatrix} 1 & 3 & 0 \\ 3 & 1 & 1 \\ 0 & 1 & 5 \end{pmatrix}$$

are equivalent

a) over the field of real numbers;

b) over the field of rationals.

3810. What symmetric bilinear functions in Exercises 3701 and 3705 are positive definite?

3811. For what values of λ are the following quadratic functions positive definite:

a) $5x_1^2 + x_2^2 + \lambda x_3^2 + 4x_1x_2 - 2x_1x_3 - 2x_2x_3;$

b) $2x_1^2 + x_2^2 + 3x_3^2 + 2\lambda x_1x_2 + 2x_1x_3;$

c) $x_1^2 + x_2^2 + 5x_3^2 + 2\lambda x_1x_2 - 2x_1x_3 + 4x_2x_3;$

d) $x_1^2 + 4x_2^2 + x_3^2 + 2\lambda x_1x_2 + 10x_1x_3 + 6x_2x_3;$

e) $x_1\bar{x}_1 + ix_1\bar{x}_2 - ix_2\bar{x}_1 + \lambda x_2\bar{x}_2;$

f) $2x_1\bar{x}_1 - (1-i)x_1\bar{x}_3 - (1+i)\bar{x}_1x_3 + 2\lambda x_2\bar{x}_3 + 2\bar{\lambda}\bar{x}_2x_3 + x_2\bar{x}_2 + 5x_3\bar{x}_3?$

3812. Prove that for any positive definite symmetric bilinear (Hermitian) function f, the inequality

$$\sqrt{f(x+y, x+y)} \le \sqrt{f(x, x)} + \sqrt{f(y, y)},$$

holds and equality occurs if and only if $\alpha x = \beta y$, where α, β are non-negative real numbers which are not equal to zero simultaneously.

3813. Prove, without application of the Sylvester criterion, that for positive definiteness of the quadratic function $\sum_{i,j=1}^{n} a_{ij}x_i\bar{x}_j$ the condition $a_{ii} > 0$ ($i = 1, \ldots, n$) is necessary, but not sufficient.

3814. Under what values of λ are the following quadratic functions negative definite:

a) $-x_1^2 + \lambda x_2^2 - x_3^2 + 4x_1x_2 + 8x_2x_3$;

b) $\lambda x_1^2 - 2x_2^2 - 3x_3^2 + 2x_1x_2 - 2x_1x_3 + 2x_2x_3$;

c) $\lambda x_1\bar{x}_1 + 3x_2\bar{x}_2 - ix_1\bar{x}_2 + i\bar{x}_1x_2$;

d) $4x_1\bar{x}_1 - 2x_2\bar{x}_2 - (\lambda + i)x_1\bar{x}_2 - (\bar{\lambda} - i)\bar{x}_1x_2$?

3815. Find the symmetric bilinear (Hermitian) functions associated with the quadratic functions:

a) $x_1^2 + 2x_1x_2 + 2x_2^2 - 6x_1x_3 + 4x_2x_3 - x_3^2$;

b) $x_1x_2 + x_1x_3 + x_2x_3$;

c) $x_1\bar{x}_1 + ix_1\bar{x}_2 - i\bar{x}_1x_2 + 2x_2\bar{x}_2$;

d) $(5 - i)x_1\bar{x}_2 + (6 + i)\bar{x}_1x_2 + x_2\bar{x}_2$.

3816. Find the symmetric bilinear function associated with the quadratic function $q(x) = f(x, x)$, where

a) $f(x, y) = 2x_1y_1 - 3x_1y_2 - 4x_1y_3 + x_2y_1 - 5x_2y_3 + x_3y_3$;

b) $f(x, y) = -x_1y_2 + x_2y_1 - 2x_2y_2 + 3x_2y_3 - x_3y_1 + 2x_3y_3$.

3817. Are the quadratic functions

a) $x_1^2 - 2x_1x_2 + 2x_3^2 + 4x_2x_3 + 5x_3^2$ and $x_1^2 - 4x_1x_2 + 2x_1x_3 + 4x_2^2 + x_3^2$;

b) $2x_1^2 + 9x_2^2 + 3x_3^2 + 8x_1x_2 - 4x_1x_3 - 10x_2x_3$ and
$2x_1^2 + 3x_2^2 + 6x_3^2 - 4x_1x_2 - 4x_1x_3 + 8x_2x_3$;

equivalent over the field of real numbers?

3818. Find the normal form of quadratic functions applying Lagrange's method

a) $x_1^2 + x_2^2 + 3x_3^2 + 4x_1x_2 + 2x_1x_3 + 2x_2x_3$;

b) $x_1^2 + 2x_2^2 + x_3^2 + 2x_1x_2 + 4x_1x_3 + 2x_2x_3$;

c) $x_1^2 - 3x_3^2 - 2x_1x_2 + 2x_1x_3 - 6x_2x_3$;

d) $x_1x_2 + x_1x_3 + x_1x_4 + x_2x_3 + x_2x_4 + x_3x_4$;

e) $x_1\bar{x}_1 - ix_1\bar{x}_2 + i\bar{x}_1x_2 + 2x_2\bar{x}_2$;

f) $(1-i)x_1\bar{x}_2 + (1+i)\bar{x}_1x_2 + (1-2i)x_1\bar{x}_3 + (1+2i)\bar{x}_1x_3 + x_2\bar{x}_3 + \bar{x}_2x_3$;

g) $\displaystyle\sum_{i,j=1}^{n} a_i a_j x_i x_j$, where not all numbers $a_1, \ldots, a_n$ are equal to 0;

h) $\displaystyle\sum_{i=1}^{n} x_i^2 + \sum_{1\le i<j\le n} x_i x_j$;

i) $\displaystyle\sum_{1\le i<j\le n} x_i x_j$;

j) $\displaystyle\sum_{i=1}^{n-1} x_i x_{i+1}$;

k) $\displaystyle\sum_{i=1}^{n} (x_i - s)^2, s = (x_1 + \cdots + x_n)/n$;

l) $\displaystyle\sum_{1\le i<j\le n} |i-j| x_i x_j$.

3819. Are the quadratic functions

a) $x_1^2 - 2x_1x_2 + 2x_1x_3 - 2x_1x_4 + x_2^2 + 2x_2x_3 - 4x_2x_4 + x_3^2 - 2x_4^2$ and
$x_1^2 + x_1x_2 + x_3x_4$;

b) $x_1^2 + 4x_2^2 + x_3^2 + 4x_1x_2 - 2x_1x_3$ and
$x_1^2 + 2x_2^2 - x_3^2 + 4x_1x_2 - 2x_1x_3 - 4x_2x_3$;

equivalent over the field of complex numbers?

3820. Let q be a mapping from a real vector space V into the field $\mathbb{R}$ such that there exist quadratic functions a, b and a bilinear function c for which

$$q(\lambda x + \mu y) = \lambda^2 a(x) + \lambda\mu c(x, y) + \mu^2 b(y)$$

for any $\lambda, \mu \in \mathbb{R}$ and $x, y \in V$. Prove that q is a quadratic function.

3821. Let $f_1, \ldots, f_{r+s}$ be linear functions. Prove that the positive index of inertia of the function

$$q(x) = |f_1(x)|^2 + \cdots + |f_r(x)|^2 - |f_{r+1}(x)|^2 - \cdots - |f_{r+s}(x)|^2$$

does not exceed r and the negative index does not exceed s.

3822. Find the positive and the negative indices of inertia

a)   of the quadratic function $q(x) = \text{tr} x^2$ on the space $\mathbf{M}_n(\mathbb{R})$;

b)   of the quadratic function $q(x) = \text{tr} x \bar{x}$ on the space $\mathbf{M}_n(\mathbb{C})$.

3823. Let f be a nonsingular symmetric bilinear (Hermitian) function on a space of dimension ≥ 3. Prove that if the function f is non-zero on a two-dimensional subspace U, then there exists a three-dimensional subspace $W \supseteq U$, on which the restriction of f is nonsingular.

3824. Let f be a nonsingular symmetric bilinear (Hermitian) function whose negative index of inertia is equal to 1, and $f(v, v) < 0$ for some vector v. Prove that the restriction of f to any subspace, containing v, is nonsingular.

3825. Let f be a nonsingular symmetric bilinear (Hermitian) function on a space of dimension ≥ 3. Prove that any isotropic vector belongs to an intersection of two-dimensional subspaces such that on each of them the restriction of f is nonsingular.

3826. Prove that the dimension of a maximal isotropic subspace with respect to a nonsingular symmetric bilinear (Hermitian) function is equal to the least of its positive and negative indices of inertia.

3827. Find the positive and negative indices of inertia of a nonsingular quadratic function on a $2n$-dimensional vector space, having n-dimensional totally isotropic subspace.

3828. Let a nonsingular quadratic function q on a $2n$-dimensional space V vanish on a n-dimensional subspace U. Prove that

a)   there exists a n-dimensional subspace U' such that

$$V = U \oplus U', \quad q(U') = 0;$$

b)   the function q has in some basis the form

$$x_1 x_2 + x_3 x_4 + \cdots + x_{2n-1} x_{2n}.$$

3829. Prove that if some principal minor of size r of a symmetric matrix is non-zero and all its bordering principal minors of sizes $r + 1$ and $r + 2$ are equal to zero, then the rank of this matrix is equal to r.

3830. Prove that a real symmetric (complex Hermitian) matrix A can be factorized as $A = {}^t C \cdot C$ ($A = {}^t C \cdot \bar{C}$, respectively), where C is a square matrix, if and only if all principal minors of A are non-negative.

3831. Find the dimension of the space of symmetric bilinear functions in n variables.

3832. Associate with each (nondirected) graph Γ, with vertices $v_1, \ldots, v_n$, the quadratic function $q_\Gamma(x) = \sum_{i,j=1}^n a_{ij} x_i x_j$ by setting

$$
a_{ij} = \begin{cases} 2, & \text{if } i = j, \\ -1, & \text{if } v_i \text{ and } v_j \text{ are connected by an edge,} \\ 0, & \text{if } v_i \text{ and } v_j \text{ are not connected by an edge.} \end{cases}
$$

Consider graphs

nsiaer grapns

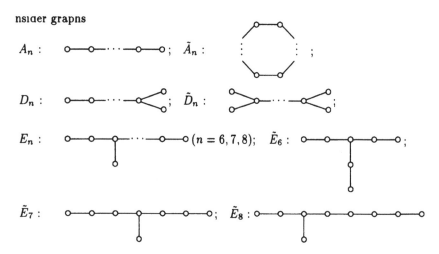

(The number of vertices in the graph Γ_n is equal to n and in the graph $\tilde{\Gamma}_n$ is equal to $n + 1$.)

Prove that the function q_Γ is positive definite for the graphs Γ_n and $q_\Gamma(x) \geq 0$ for all x is positive semidefinite for the graphs $\tilde{\Gamma}_n$:

3833. Let q be a nonsingular quadratic function on a space V over an arbitrary field F. Prove that if there exists a nonzero vector $x \in V$ for which $q(x) = 0$ then the mapping $q : V \to F$ is surjective.

3834. Let $f(x, y)$ be a Hermitian non-negative definite function and $f(z, z) = 0$ for some $z \neq 0$. Prove that $f(z, t) = 0$ for all t.

3835. Let $f(x, y)$ be a positive definite Hermitian function and $f(x, x) = f(y, y) = 1$ for some $x, y \neq 0$. Prove that $|f(x, y)| \leq 1$.

CHAPTER 9

Linear operators

39 Definition of a linear operator. The image, the kernel, the matrix of a linear operator

3901. Which mappings are linear operators on appropriate vector spaces:

a) $x \mapsto a$ (a is a fixed vector);

b) $x \mapsto x + a$ (a is a fixed vector);

c) $x \mapsto \alpha x$ (α is a fixed scalar);

d) $x \mapsto (x, a)b$ (V is an Euclidean space, a, b are fixed vectors);

e) $x \mapsto (a, x)x$ (V is an Euclidean space, a is a fixed vector);

f) $f(x) \mapsto f(ax + b)$ ($f \in \mathbb{R}[x]_n$; a, b are fixed numbers);

g) $f(x) \mapsto f(x + 1) - f(x)$ ($f \in \mathbb{R}[x]_n$);

h) $f(x) \mapsto f^{(k)}(x)$ ($f \in \mathbb{R}[x]_n$);

i) $(x_1, x_2, x_3) \mapsto (x_1 + 2, x_2 + 5, x_3)$;

j) $(x_1, x_2, x_3) \mapsto (x_1 + 3x_3, x_2^3, x_1 + x_3)$;

k) $(x_1, x_2, x_3) \mapsto (x_1, x_2, x_1 + x_2 + x_3)$?

3902. Prove that a linear operator maps a linearly dependent system of vectors to a linearly dependent system.

3903. Prove that on n-dimensional space for any linearly independent system of vectors $a_1, \ldots, a_n$ and an arbitrary system of vectors $b_1, \ldots, b_n$, there exists a unique linear operator, which maps a_i to b_i ($i = 1, \ldots, n$).

3904. Prove that any linear operator on one-dimensional vector space is of the form $x \mapsto \alpha x$, where α is a scalar.

3905. Find the images and the kernels of the linear operators from Exercise 3901.

3906. Prove that the operator of differentiation

a)   is singular on the space of polynomials of degree $\leq n$;

b)   is nonsingular on the space of functions with the bases $(\cos t, \sin t)$.

3907. Prove that any subspace of a vector space is:

a)   the kernel of some linear operator;

b)   the image of some linear operator.

3908. Prove that two linear operators of rank 1 having the same kernels and images are commuting.

3909. Let $\mathcal{A}$ be a F-linear operator on a subspace L of a space V different from V. Prove that there exist infinitely many linear operators on V, whose restriction to L coincides with $\mathcal{A}$, provided the field F is infinite.

3910. Let $\mathcal{A}$ be a linear operator on a space V, and L be a subspace of V. Prove that

a)   the image $\mathcal{A}(L)$ and the preimage $\mathcal{A}^{-1}(L)$ are subspaces of V;

b)   if $\mathcal{A}$ is nonsingular and V is finite-dimensional then

$$\dim \mathcal{A}(L) = \dim \mathcal{A}^{-1}(L) = \dim L.$$

3911. Let $\mathcal{A}$ be a linear operator on a space V, L be a subspace of V, and $L \cap \mathrm{Ker}\mathcal{A} = 0$. Prove that any linearly independent system of vectors in L maps by $\mathcal{A}$ to a linearly independent system.

3912. Prove, for linear operators $\mathcal{A}, \mathcal{B}, \mathcal{C}$, the Frobenious inequality

$$\mathrm{rk}\mathcal{B}\mathcal{A} + \mathrm{rk}\mathcal{A}\mathcal{C} \leq \mathrm{rk}\mathcal{A} + \mathrm{rk}\mathcal{B}\mathcal{A}\mathcal{C}.$$

3913. A linear operator $\mathcal{A}$ is a *pseudoreflection*, if $\mathrm{rk}(\mathcal{A} - \mathcal{E}) = 1$. Prove that any linear operator on a n-dimensional space is a product of at most n pseudoreflections.

3914. Prove that the set of operators $\mathcal{X}$, such that $\mathcal{A}\mathcal{X} = 0$ for a linear operator $\mathcal{A}$ on a n-dimensional space, is a vector space. Find its dimension.

3915. Find the matrix of the operator:

a) $(x_1, x_2, x_3) \mapsto (x_1, x_1 + 2x_2, x_2 + 3x_3)$ on the space $\mathbb{R}^3$ with a basis of unit vectors;

b) of the rotation of the plane through an angle α with an arbitrary ortho-normal basis;

c) of the rotation of the three-dimensional space through an angle $2\pi/3$ around the line which is given in a rectangular system of coordinates by the equations
$x_1 = x_2 = x_3$ with the basis of unit vectors of the coordinate axes;

d) of the projection of the three-dimensional space with the basis (e_1, e_2, e_3) to the axis of the vector e_2 in parallel with the coordinate plane of the vectors e_1 and e_3;

e) $x \mapsto (x, a)a$ on an Euclidean space with the orthonormal basis (e_1, e_2, e_3) if $a = e_1 - 2e_3$;

f) $X \mapsto \begin{pmatrix} a & b \\ c & d \end{pmatrix} \cdot X$ on the space $\mathbf{M}_2(\mathbb{R})$ with a basis of matrix units;

g) $X \mapsto X \cdot \begin{pmatrix} a & b \\ c & d \end{pmatrix}$ on the space $\mathbf{M}_2(\mathbb{R})$ with a basis of matrix units;

h) $X \mapsto {}^tX$ on the space $\mathbf{M}_2(\mathbb{R})$ with a basis of matrix units;

i) $X \mapsto AXB$ (A, B are fixed matrices in the space $\mathbf{M}_2(\mathbb{R})$ with a basis of matrix units;

j) $X \mapsto AX + XB$ (A, B are fixed matrices) on the space $\mathbf{M}_2(\mathbb{R})$ with a basis of matrix units;

k) of the differentiation on the space $\mathbb{R}[x]_n$ with the basis $(1, x, \ldots, x^n)$;

l) of the differentiation on the space $\mathbb{R}[x]_n$ with the basis $(x^n, x^{n-1}, \ldots, 1)$;

m) of the differentiation on the space $\mathbb{R}[x]_n$ with the basis

$$\left(1, x - 1, \frac{(x-1)^2}{2}, \ldots, \frac{(x-1)^n)}{n!}\right).$$

3916. Prove that the space $\mathbb{R}^3$ has a unique linear operator which maps the vectors $(1, 1, 1)$, $(0, 1, 0)$, $(1, 0, 2)$ to the vectors $(1, 1, 1)$, $(0, 1, 0)$, $(1, 0, 1)$, respectively. Find its matrix with a basis of unit vectors.

3917. Let a vector space V be a direct sum of subspaces L_1 and L_2 with bases $(a_1, \ldots, a_k)$ and $(a_{k+1}, \ldots, a_n)$, respectively. Prove that the projection onto L_1 in parallel with L_2 is a linear operator and find its matrix with the basis $(a_1, \ldots, a_n)$.

3918. Find the general form of matrices of linear operators on the n-dimensional space with a basis $(a_1, \ldots, a_k, a_{k+1}, \ldots, a_n)$, which map the given independent vectors $a_1, \ldots, a_k$ $(k < n)$ to the given vectors $b_1, \ldots, b_k$.

3919. Let a linear operator on a space V have, with a basis $(e_1, \ldots, e_4)$, the matrix

$$\begin{pmatrix} 0 & 1 & 2 & 3 \\ 5 & 4 & 0 & -1 \\ 3 & 2 & 0 & 3 \\ 6 & 1 & -1 & 7 \end{pmatrix}.$$

Find the matrix of this operator with the bases:

a) (e_2, e_1, e_3, e_4);

b) $(e_1, \ e_1 + e_2, \ e_1 + e_2 + e_3, \ e_1 + e_2 + e_3 + e_4)$.

3920. Let a linear operator on the space $\mathbb{R}[x]_2$ have, with the basis $(1, x, x^2)$, the matrix

$$\begin{pmatrix} 0 & 0 & 1 \\ 0 & 1 & 0 \\ 1 & 0 & 0 \end{pmatrix}.$$

Find its matrix with the basis

$$(3x^2 + 2x + 1, \ x^2 + 3x + 2, \ 2x^2 + x + 3).$$

3921. Let a linear operator on the space $\mathbb{R}^3$ have, with the basis

$$((8, -6, 7), \ (-16, 7, -13), \ (9, -3, 7)),$$

the matrix

$$\begin{pmatrix} 1 & -18 & 15 \\ -1 & -22 & 20 \\ 1 & -25 & 22 \end{pmatrix}.$$

Find its matrix with the basis

$$((1, -2, 1), \ (3, -1, 2), \ (2, 1, 2)).$$

3922. Let a linear operator $\mathcal{A}$ on a n-dimensional vector space V map linearly independent vectors $a_1, \ldots, a_n$ to vectors $b_1, \ldots, b_n$. Prove that the matrix of $\mathcal{A}$ with some basis $\mathbf{e} = (e_1, \ldots, e_n)$ is equal to BA^{-1}, where the columns of matrices A and B consist of coordinates of the given vectors the bases $\mathbf{e}$.

3923. Find the general form of the matrix of a linear operator $\mathcal{A}$ with a basis whose first k vectors form:

a) a basis of the kernel of $\mathcal{A}$;

b) a basis of the image of $\mathcal{A}$.

3924. Prove that if $f(t) = f_1(t)f_2(t)$ is a factorization of a polynomial $f(t)$ into coprime factors and a linear operator A satisfies the condition $f(A) = 0$, then the matrix of A with some basis has the form $\begin{pmatrix} A_1 & 0 \\ 0 & A_2 \end{pmatrix}$, where $f_1(A_1) = 0$, $f_2(A_2) = 0$.

40 Eigenvectors, invariant subspaces, root subspaces

4001. Find the eigenvectors and eigenvalues

a) of the operator of differentiation on the space $\mathbb{R}[x]_n$;

b) of the operator $X \mapsto {}^t X$ on the space $M_n(\mathbb{R})$;

c) of the operator $X \dfrac{d}{dx}$ on the space $\mathbb{R}[X]_n$;

d) of the operator $\dfrac{1}{X} \displaystyle\int_0^X f(t)dt$ on the space $\mathbb{R}[X]_n$;

e) of the operator $f \mapsto \dfrac{d^n f}{dx^n}$ on the linear span

$$\langle 1, \cos x, \sin x, \ldots, \cos mx, \sin mx \rangle;$$

f) of the operator $f \to \int_0^x f(t)dt$ on the linear span

$$\langle 1, \cos x, \sin x, \ldots, \cos mx, \sin mx \rangle.$$

4002. Prove that the linear operator $f \mapsto f(ax + b)$ on the space $\mathbb{R}[x]_n$ has the set of eigenvalues $1, a, \ldots, a^n$.

4003. Prove that an eigenvector of a linear operator A with an eigenvalue λ is an eigenvector of an operator $f(A)$, where $f(t)$ is a polynomial, with an eigenvalue $f(\lambda)$.

4004. Prove that if an operator A is nonsingular then operators A and A^{-1} have the same eigenvectors.

4005. Prove that all nonzero vectors of a space are eigenvectors of a linear operator A if and only if A is a homothety $x \mapsto \alpha x$, where α is some fixed scalar.

4006. Prove that if a linear operator A on a n-dimensional space has n distinct eigenvalues then any linear operator commuting with A has a basis consisting of its eigenvectors.

4007. Prove that the subspace $V_\lambda(\mathcal{A})$, consisting of all eigenvectors of an operator $\mathcal{A}$ with an eigenvalue λ and a zero vector, is invariant under any linear operator $\mathcal{B}$ commuting with $\mathcal{A}$.

4008. Prove that for any (possibly infinite) set of commuting linear operators on a finite-dimensional complex space

a)   there exists a common eigenvector;

b)   there exists a basis in which the matrices of all these operators are upper-triangular.

4009. Prove that if an operator $\mathcal{A}^2$ has an eigenvalue λ^2 then one of the numbers λ and $-\lambda$ is an eigenvalue of $\mathcal{A}$.

4010. Prove that

a)   coefficients $c_1, \ldots, c_n$ of the polynomial

$$|A - \lambda E| = (-\lambda)^n + c_1(-\lambda)^{n-1} + \cdots + c_n$$

are sums of principal minors of corresponding sizes of the matrix A;

b)   the sum and the product of eigenvalues of the matrix A are equal to its trace and its determinant, respectively.

4011. Prove that any polynomial of degree n with the leading coefficient $(-1)^n$ is a characteristic polynomial of some matrix of size n.

4012. Prove that if A and B are square matrices of the same size then matrices AB and BA have the same characteristic polynomials.

4013. Find eigenvalues of a matrix ${}^t A \cdot A$, where A is a row $(a_1, \ldots, a_n)$.

4014. Prove that all eigenvalues of a matrix are different from zero if and only if this matrix is nonsingular.

4015. Find the eigenvalues and eigenvectors of the linear operators given in some bases by the matrices:

a) $\begin{pmatrix} 2 & -1 & 2 \\ 5 & -3 & 3 \\ -1 & 0 & -2 \end{pmatrix}$; b) $\begin{pmatrix} 0 & 1 & 0 \\ -4 & 4 & 0 \\ -2 & 1 & 2 \end{pmatrix}$; c) $\begin{pmatrix} 4 & -5 & 2 \\ 5 & -7 & 3 \\ 6 & -9 & 4 \end{pmatrix}$;

d) $\begin{pmatrix} 7 & -12 & 6 \\ 10 & -19 & 10 \\ 12 & -24 & 13 \end{pmatrix}$; e) $\begin{pmatrix} 4 & -5 & 7 \\ 1 & -4 & 9 \\ -4 & 0 & 5 \end{pmatrix}$; f) $\begin{pmatrix} 3 & -1 & 0 & 0 \\ 1 & 1 & 0 & 0 \\ 3 & 0 & 5 & -3 \\ 4 & -1 & 3 & -1 \end{pmatrix}$.

4016. Find out if the following matrices can be reduced to diagonal ones by a change of basis over the field $\mathbb{R}$ and over the field $\mathbb{C}$:

a) $\begin{pmatrix} -1 & 3 & -1 \\ -3 & 5 & -1 \\ -3 & 3 & 1 \end{pmatrix};$ b) $\begin{pmatrix} 4 & 7 & -5 \\ -4 & 5 & 0 \\ 1 & 9 & -4 \end{pmatrix};$ c) $\begin{pmatrix} 4 & 2 & -5 \\ 6 & 4 & -9 \\ 5 & 3 & -7 \end{pmatrix};$

d) $\begin{pmatrix} 1 & 1 & 1 & 1 \\ 1 & 1 & -1 & -1 \\ 1 & -1 & 1 & -1 \\ 1 & -1 & -1 & 1 \end{pmatrix}.$

Find this basis and the corresponding form of the matrix.

4017. Given a matrix with the entries $\alpha_1, \ldots, \alpha_n$ on the secondary diagonal, and all other entries equal to zero, under what conditions is the matrix similar to a diagonal matrix?

4018. Given a matrix A of size n, whose entries on the secondary diagonal are equal to 1, and all other of whose entries are equal to zero, find a matrix T such that $B = T^{-1}AT$ is a diagonal matrix. Calculate the matrix B.

4019. Prove that the number of linearly independent eigenvectors of a linear operator $\mathcal{A}$, with an eigenvalue λ, is less or equal to the multiplicity of λ as a root of the characteristic polynomial of $\mathcal{A}$.

4020. Let $\lambda_1, \ldots, \lambda_n$ be eigenvalues of a linear operator $\mathcal{A}$ on an n-dimensional complex space. Find the eigenvalues of $\mathcal{A}$ as an operator on the corresponding $2n$-dimensional real space.

4021. Let $\lambda_1, \ldots, \lambda_n$ be roots of the characteristic polynomial of a matrix A. Find the eigenvalues:

a) of the linear operator $X \mapsto AX^t A$ on the space $\mathbf{M}_n(\mathbb{R})$;

b) of the linear operator $X \mapsto AXA^{-1}$ on the space $\mathbf{M}_n(\mathbb{R})$ (the matrix A is nonsingular).

4022. Find all invariant subspaces of the operator of differentiation on the space $\mathbb{R}[x]_n$.

4023. Prove that a linear span of any system of eigenvectors of a linear operator $\mathcal{A}$ is invariant under $\mathcal{A}$.

4024. Prove that

a) the kernel and the image of a linear operator $\mathcal{A}$ are invariant under $\mathcal{A}$;

b) any subspace containing the image of an operator $\mathcal{A}$ is invariant under $\mathcal{A}$;

c) if a subspace L is invariant under $\mathcal{A}$ then $\mathcal{A}(L)$ and $\mathcal{A}^{-1}(L)$ are invariant under $\mathcal{A}$;

d)   if a linear operator $\mathcal{A}$ is nonsingular then any subspace invariant under $\mathcal{A}$ is invariant under $\mathcal{A}^{-1}$.

4025. Prove that any linear operator on n-dimensional complex space has an invariant subspace of dimension $n - 1$.

4026. Let a linear operator on a vector space over a field K have the matrix

$$\begin{pmatrix} a_1 & 1 & 0 & \ldots & 0 \\ a_2 & 0 & 1 & \ldots & 0 \\ \multicolumn{5}{c}{\ldots\ldots\ldots\ldots\ldots\ldots} \\ a_{n-1} & 0 & 0 & \ldots & 1 \\ a_n & 0 & 0 & \ldots & 0 \end{pmatrix}$$

with some basis, where the polynomial $x^n - a_1 x^{n-1} - \cdots - a_{n-1}x - a_n$ is irreducible over K. Prove that the operator has no nontrivial invariant subspaces.

4027. Let a linear operator $\mathcal{A}$ on an n-dimensional space have, with some basis, a diagonal matrix with distinct entries on the diagonal. Find all subspaces invariant under $\mathcal{A}$.

4028. Find all invariant subspaces for a linear operator having, with some basis, the matrix consisting of a Jordan box.

4029. Find in three-dimensional vector space all subspaces invariant under the linear operator with the matrix

$$\begin{pmatrix} 4 & -2 & 2 \\ 2 & 0 & 2 \\ -1 & 1 & 1 \end{pmatrix}.$$

4030. Find in three-dimensional vector space all subspaces invariant under two linear operators given by the matrices

$$\begin{pmatrix} 5 & -1 & -1 \\ -1 & 5 & -1 \\ -1 & -1 & 5 \end{pmatrix} \quad \text{and} \quad \begin{pmatrix} -6 & 2 & 3 \\ 2 & -3 & 6 \\ 3 & 6 & 2 \end{pmatrix}.$$

4031. Find in $\mathbb{R}[X]_n$ and $\mathbb{C}[X]_n$ all subspaces invariant under operators

a) $\mathcal{A}(f) = X \dfrac{df}{dX};$ b) $\mathcal{A}(f) = \dfrac{1}{X} \displaystyle\int_0^X f(t)\,dt.$

4032. Find in the linear span of functions

$$\langle \cos x, \sin x, \ldots, \cos nx, \sin nx \rangle$$

all subspaces invariant under the operator

a) $A(f) = \dfrac{df}{dx}$;

b) $A(f) = \displaystyle\int_0^x f(t)\,dt.$

4033. Let A, B be linear operators on a finite-dimensional vector space V over the field $\mathbb{C}$ such that $A^2 = B^2 = \mathcal{E}$. Prove that there exists either a one-dimensional or two-dimensional subspace of V which is invariant under A and B.

4034. Prove that a complex vector space with only one line invariant under a linear operator A is indecomposable into a direct sum of nonzero subspaces invariant under A.

4035. Find the eigenvalues and the root subspaces of the linear operator given in some bases by the matrix:

a) $\begin{pmatrix} 4 & -5 & 2 \\ 5 & -7 & 3 \\ 6 & -9 & 4 \end{pmatrix}$;

b) $\begin{pmatrix} 1 & -3 & 4 \\ 4 & -7 & 8 \\ 6 & -7 & 7 \end{pmatrix}$;

c) $\begin{pmatrix} 2 & 6 & -15 \\ 1 & 1 & -5 \\ 1 & 2 & -6 \end{pmatrix}$;

d) $\begin{pmatrix} 0 & -2 & 3 & 2 \\ 1 & 1 & -1 & -1 \\ 0 & 0 & 2 & 0 \\ 1 & -1 & 0 & 1 \end{pmatrix}.$

4036. Prove that a linear operator on a complex vector space has a diagonal matrix in some basis if and only if all its root vectors are eigenvectors.

4037. Prove that if a linear operator on a complex vector space has a diagonal matrix in some basis then its restriction to any invariant subspace L also has a diagonal matrix in some basis of L.

4038. Prove that any root subspace of a linear operator A is invariant under any linear operator B commuting with A.

4039. Prove that if a matrix of a linear operator A is reduced to the Jordan form then any invariant subspace L is a direct sum of intersections of L with the root subspaces of A.

* * *

4040. Let $A \in \mathbf{M}_n(\mathbb{C})$. Consider an operator L_A on the space $\mathbf{M}_{n \times m}(\mathbb{C})$ where $L_A(X) = AX$. Find the eigenvalues of L_A. Find the root subspaces of L_A where A is an upper-triangular matrix.

4041. Let $A \in \mathbf{M}_n(\mathbb{C})$, $B \in \mathbf{M}_m(\mathbb{C})$ and A, B have no common eigenvalues. Prove that

a) if X is a matrix of size $n \times m$ and $AX - XB = 0$ then $X = 0$;

b) the equation $AX - XB = C$ where X, C are matrices of size $n \times m$ has a unique solution.

4042. Let $\mathcal{A}$ be a linear operator on a finite-dimensional complex vector space V. Prove that there exists a basis of V in which the matrix of $\mathcal{A}$ is upper-triangular.

4043. Let $\mathcal{A}$ be a linear operator on a finite-dimensional real space V. Prove that there exists a basis of V in which the matrix of $\mathcal{A}$ has the block-upper-triangular form

$$\begin{pmatrix} \boxed{A_1} & & & * \\ & \boxed{A_2} & & \\ & & \ddots & \\ 0 & & & \boxed{A_n} \end{pmatrix},$$

where the square blocks $A_1, \ldots, A_n$ have size two at most.

4044. Let $\mathcal{A}, \mathcal{B}$ be linear operators on a finite-dimensional complex vector space and let the rank of the operator $\mathcal{AB} - \mathcal{BA}$ not exceed 1. Prove that there exists a common eigenvector for $\mathcal{A}$ and $\mathcal{B}$.

41 Jordan canonical form and its applications. Minimal polynomial

4101. Find the Jordan form of the matrix:

a) $\begin{pmatrix} 1 & -3 & 4 \\ 4 & -7 & 8 \\ 6 & -7 & 7 \end{pmatrix}$; b) $\begin{pmatrix} 4 & -5 & 7 \\ 1 & -4 & 9 \\ -4 & 0 & 5 \end{pmatrix}$; c) $\begin{pmatrix} 4 & 6 & 0 \\ -3 & -5 & 0 \\ -3 & -6 & 1 \end{pmatrix}$;

d) $\begin{pmatrix} 3 & 0 & 8 \\ 3 & -1 & 6 \\ -2 & 0 & -5 \end{pmatrix}$; e) $\begin{pmatrix} -2 & 8 & 6 \\ -4 & 10 & 6 \\ 4 & -8 & -4 \end{pmatrix}$; f) $\begin{pmatrix} 1 & -3 & 0 & 3 \\ -2 & -6 & 0 & 13 \\ 0 & -3 & 1 & 3 \\ -1 & -4 & 0 & 8 \end{pmatrix}$;

g) $\begin{pmatrix} 3 & -1 & 1 & -7 \\ 9 & -3 & -7 & -1 \\ 0 & 0 & 4 & -8 \\ 0 & 0 & 2 & -4 \end{pmatrix}$; h) $\begin{pmatrix} 1 & -1 & 0 & 0 & \ldots & 0 & 0 \\ 0 & 1 & -1 & 0 & \ldots & 0 & 0 \\ 0 & 0 & 1 & -1 & \ldots & 0 & 0 \\ \hdotsfor{7} \\ 0 & 0 & 0 & 0 & \ldots & 1 & -1 \\ 0 & 0 & 0 & 0 & \ldots & 0 & 1 \end{pmatrix}$;

i) $\begin{pmatrix} 1 & 1 & 1 & \cdots & 1 \\ 0 & 1 & 1 & \cdots & 1 \\ 0 & 0 & 1 & \cdots & 1 \\ \cdots\cdots\cdots\cdots\cdots \\ 0 & 0 & 0 & \cdots & 1 \end{pmatrix}$;

j) $\begin{pmatrix} n & n-1 & n-2 & \cdots & 1 \\ 0 & n & n-1 & \cdots & 2 \\ 0 & 0 & n & \cdots & 3 \\ \cdots\cdots\cdots\cdots\cdots\cdots \\ 0 & 0 & 0 & \cdots & n \end{pmatrix}$;

k) $\begin{pmatrix} 1 & 0 & 0 & \cdots & 0 \\ 1 & 2 & 0 & \cdots & 0 \\ 1 & 2 & 3 & \cdots & 0 \\ \cdots\cdots\cdots\cdots\cdots \\ 1 & 2 & 3 & \cdots & n \end{pmatrix}$;

l) $\begin{pmatrix} 0 & 1 & 0 & 0 & \cdots & 0 \\ 0 & 0 & 1 & 0 & \cdots & 0 \\ 0 & 0 & 0 & 1 & \cdots & 0 \\ \cdots\cdots\cdots\cdots\cdots\cdots \\ 0 & 0 & 0 & 0 & \cdots & 1 \\ 1 & 0 & 0 & 0 & \cdots & 0 \end{pmatrix}$;

m) $\begin{pmatrix} \alpha & a_{12} & a_{13} & \cdots & a_{1n} \\ 0 & \alpha & a_{23} & \cdots & a_{2n} \\ 0 & 0 & \alpha & \cdots & a_{3n} \\ \cdots\cdots\cdots\cdots\cdots\cdots \\ 0 & 0 & 0 & \cdots & \alpha \end{pmatrix}$,

where $a_{12}, a_{23}, \ldots, a_{n-1,n} \neq 0$.

4102. Prove that the Jordan form of the matrix $A + \alpha E$ is equal to $A_j + \alpha E$ where A_j is the Jordan form of the matrix A.

4103. Let A be a Jordan box of size n with an element α at the principal diagonal.

a) Find the matrix $f(A)$ where $f(x)$ is a polynomial;

b) find the Jordan form of the matrix A^2.

4104. Find the Jordan form of the matrix

$$\begin{pmatrix} \alpha & 0 & 1 & 0 & \cdots & 0 & 0 \\ 0 & \alpha & 0 & 1 & \cdots & 0 & 0 \\ 0 & 0 & \alpha & 0 & \cdots & 0 & 0 \\ 0 & 0 & 0 & \alpha & \cdots & 0 & 0 \\ \cdots\cdots\cdots\cdots\cdots\cdots\cdots \\ 0 & 0 & 0 & 0 & \cdots & \alpha & 0 \\ 0 & 0 & 0 & 0 & \cdots & 0 & \alpha \end{pmatrix}.$$

4105. Find the Jordan form of the matrix:

a) A^2,

b) A^{-1} (A is a nonsingular matrix),

if A has the Jordan form A_j.

4106. Find the Jordan form of the matrix A, and find out the geometrical meaning of the corresponding linear operator $\mathcal{A}$, if

a) $A^2 = E$; b) $A^2 = A$.

4107. Prove that any periodic complex matrix is similar to a diagonal matrix. Find the form of this diagonal matrix.

4108. Prove that a matrix is nilpotent if and only if all its eigenvalues are equal to zero.

4109. Prove that for any linear operator $\mathcal{A}$ of rank 1 on a complex vector space there exists an integer k such that $\mathcal{A}^2 = k\mathcal{A}$.

4110. Find the Jordan form of the matrix of the linear operator $\mathcal{A}$ and a basis $(f_1, \ldots, f_n)$, with which $\mathcal{A}$ has this matrix if $\mathcal{A}$ is given with a basis $(e_1, \ldots, e_n)$ by the matrix:

a) $\begin{pmatrix} 3 & 2 & -3 \\ 4 & 10 & -12 \\ 3 & 6 & -7 \end{pmatrix}$; b) $\begin{pmatrix} 1 & 1 & -1 \\ -3 & -3 & 3 \\ -2 & -2 & 2 \end{pmatrix}$;

c) $\begin{pmatrix} 0 & 1 & -1 & 1 \\ -1 & 2 & -1 & 1 \\ -1 & 1 & 1 & 0 \\ -1 & 1 & 0 & 1 \end{pmatrix}$; d) $\begin{pmatrix} 6 & -9 & 5 & 4 \\ 7 & -13 & 8 & 7 \\ 8 & -17 & 11 & 8 \\ 1 & -2 & 1 & 3 \end{pmatrix}$.

4111. Find the Jordan form of a matrix of a linear operator on a complex vector space which has only one invariant line.

4112. Prove that the maximal number of linearly independent eigenvectors of a linear operator $\mathcal{A}$ with an eigenvalue λ is equal to the number of boxes with λ as a diagonal entry in the Jordan form of the matrix of $\mathcal{A}$.

4113. Prove that the set of linear operators on a n-dimensional complex vector space commuting with the given operator $\mathcal{A}$ is a vector space of dimension $\geq n$.

4114. Prove that if a linear operator $\mathcal{B}$ on a complex vector space commutes with any linear operator commuting with the operator $\mathcal{A}$, then $\mathcal{B}$ is a polynomial in $\mathcal{A}$.

4115. Prove that if matrices A and B satisfy the relation $AB - BA = B$ then the matrix B is nilpotent.

4116. The operator $\mathcal{A} : f(x, y) \rightarrow f(x+1, y+1)$ acts on the space of complex polynomials of degree at most two in x and y. Find the Jordan form of $\mathcal{A}$.

4117. The operator $\mathcal{A} = \dfrac{\partial}{\partial x} + \dfrac{\partial}{\partial y}$ acts on the space of complex polynomials of degree n at most. Find the Jordan form of $\mathcal{A}$.

4118. Prove that any matrix is similar to its transpose.

4119. Consider the linear operator $L_A(X) = AX$ on the space $M_2(\mathbb{C})$, where $X \in M_2(\mathbb{C})$ and A is a fixed matrix from $M_2(\mathbb{C})$. Find the Jordan form of L_A in terms of the Jordan form of A.

4120. Prove that for any nonsingular square complex matrix A and any natural number k the equation $X^k = A$ has a solution.

4121. Solve the equations:

a) $X^2 = \begin{pmatrix} 3 & 1 \\ -1 & 5 \end{pmatrix}$;

b) $X^2 = \begin{pmatrix} 6 & 2 \\ 3 & 7 \end{pmatrix}$.

4122. Using the Jordan form and Exercises 1707–1709 calculate

a) $\begin{pmatrix} 1 & 1 \\ -1 & 3 \end{pmatrix}^{50}$;

b) $\begin{pmatrix} 7 & -4 \\ 14 & -8 \end{pmatrix}^{64}$.

4123. Find the minimal polynomial of a diagonal matrix with distinct entries at the principal diagonal.

4124. Find the minimal polynomial of a Jordan box of size n with a number α at the principal diagonal.

4125. Prove that the minimal polynomial of a block-diagonal matrix is equal to the least common multiple of the minimal polynomials of its blocks.

4126. Find the minimal polynomial:

a) of the identity operator;

b) of the zero operator;

c) of the projection operator of a n-dimensional space V onto its k-dimensional subspace L $(0 < k < n)$;

d) of a reflection operator;

e) of a nilpotent operator of index k.

f) of the operator $\mathcal{A}$ from Exercise 4031a;

g) of the operator $\mathcal{A}$ from Exercise 4031b;

h) of the operator $\mathcal{A}$ from Exercise 4032a;

i) of the operator $\mathcal{A}$ from Exercise 4032b;

j) of the operator L_A from Exercise 4040;

k) of the operator $\mathcal{A}$ from Exercise 4116.

4127. Find the minimal polynomial of the matrix:

a) $\begin{pmatrix} 3 & -1 & -1 \\ 0 & 2 & 0 \\ 1 & 1 & 1 \end{pmatrix}$;

b) $\begin{pmatrix} 4 & -2 & 2 \\ -5 & 7 & -5 \\ -6 & 6 & -4 \end{pmatrix}$.

4128. Let a linear operator $\mathcal{A}$ with a basis (e_1, e_2, e_3) of a space V have the matrix

$$\begin{pmatrix} 1 & 0 & 0 \\ 1 & 2 & 1 \\ -1 & 0 & 1 \end{pmatrix}.$$

Find the minimal polynomial $g(t)$ of $\mathcal{A}$, and decompose the space V into a direct sum of invariant subspaces according to the factorization of the minimal polynomial into coprime factors.

4129. Prove that the minimal polynomial of a matrix of size ≥ 2 and of rank 1 has degree 2.

4130. What can be said about the Jordan form of a matrix of a linear operator $\mathcal{A}$ on a complex space if $\mathcal{A}^3 = \mathcal{A}^2$?

4131. Prove that some power of the minimal polynomial of a matrix is divisible by its characteristic polynomial.

4132. Prove that for similarity of two matrices it is necessary, but not sufficient that they have the same characteristic and minimal polynomials.

4133. Prove that if the degree of the minimal polynomial of a linear operator $\mathcal{A}$ is equal to the dimension of the space then any operator commuting with $\mathcal{A}$ is a polynomial in $\mathcal{A}$.

4134. A linear operator is termed *semisimple* if any invariant subspace has an invariant complementary subspace. Prove that

a)   the restriction of a semisimple operator to an invariant subspace is also a semisimple operator (see Exercise 4037);

b)   a linear operator is semisimple if and only if the space is the direct sum of the minimal invariant subspaces;

c)   a linear operator $\mathcal{A}$ is semisimple on the whole space if there exists a decomposition of the space into the direct sum of invariant subspaces, such that on each of them the restriction of $\mathcal{A}$ is semisimple.

4135. Prove that if the minimal polynomial of a linear operator $\mathcal{A}$ on a space V is the product of coprime polynomials $g_1(x)$ and $g_2(x)$, then V can be decomposed into the direct sum of two invariant subspaces such that the restrictions of $\mathcal{A}$ to these subspaces have minimal polynomials $g_1(x)$ and $g_2(x)$, respectively.

4136. Prove that for any linear operator there exists a decomposition of the space into a direct sum of invariant subspaces such that minimal polynomials of its restrictions to these subspaces are powers of distinct irreducible polynomials.

4137. Prove that if the minimal polynomial of a linear operator $\mathcal{A}$ is an irreducible polynomial of degree k, then for any $x \neq 0$ the vectors $x, \mathcal{A}x, \ldots, \mathcal{A}^{k-1}x$ form a basis of a minimal invariant subspace.

4138. Prove that a linear operator is semisimple if and only if its minimal polynomial has no multiple irreducible factors.

4139. Prove that a linear operator on a vector space over a field K of characteristic 0 is semisimple if and only if it has an eigen-basis over some extension of K.

4140. Prove that the sum of two commuting semisimple linear operators over a field of characteristic 0 is a semisimple operator.

4141. Let $\mathcal{A}$ be a linear operator on a vector space over a field K of characteristic 0 and $K[\mathcal{A}]$ be the ring of linear operators represented by polynomials in $\mathcal{A}$. Prove that if the minimal polynomial of $\mathcal{A}$ is a power of an irreducible polynomial $p(x)$ then

   a) elements of $K[\mathcal{A}]$ which are divisible in this ring by the element $p(\mathcal{A})$ form an ideal I different from $K[\mathcal{A}]$;

   b) for any operator $\mathcal{B} \in I$ the minimal polynomial of the operator $\mathcal{A} + \mathcal{B}$ is divisible by $p(x)$;

   c) there exists an operator $\mathcal{B} \in I$ such that the minimal polynomial of the operator $\mathcal{A} + \mathcal{B}$ is equal to $p(x)$.

4142. Prove that any linear operator $\mathcal{A}$ on a vector space over a field of characteristic 0 can be represented as the sum of a semisimple and of a nilpotent operator which are polynomials in $\mathcal{A}$.

4143. Prove that any linear operator $\mathcal{A}$ can be uniquely represented as a sum of commuting semisimple and nilpotent operators.

4144. Let $\mathcal{A}$ be a linear operator on a vector space V over a field K with the minimal polynomial $g(x)$. Assume that $g(x)$ is a power of a polynomial irreducible over K and the degree of $g(x)$ is equal to the dimension of V. Prove that

   a) V cannot be decomposed into the direct sum of two invariant subspaces;

   b) V is cyclic with respect to $\mathcal{A}$.

4145. Let $\lambda_1, \ldots, \lambda_n$ be eigenvalues of a matrix $A \in \mathbf{M}_n(\mathbb{C})$. Prove that

a) for any natural number k

$$\operatorname{tr} A^k = \lambda_1^k + \cdots + \lambda_n^k;$$

b) the coefficients of the characteristic polynomial of A are polynomials in $\operatorname{tr} A, \ldots, \operatorname{tr} A^n$;

c) if $\operatorname{tr} A = \operatorname{tr} A^2 = \cdots = \operatorname{tr} A^n = 0$ then the matrix A is nilpotent.

$$* \quad * \quad *$$

4146. Let

$$B = \begin{pmatrix} 0 & \cdots & 1 \\ \vdots & \ddots & \vdots \\ 1 & \cdots & 0 \end{pmatrix} \in \mathbf{M}_n(\mathbb{C})$$

and $S = \dfrac{1}{\sqrt{2}}(E + iB)$. For a Jordan box $J(n, \lambda) \in \mathbf{M}_n(\mathbb{C})$, prove the equality

$$SJ(n, \lambda)S^{-1} = \lambda E + \frac{1}{2}\sum_{k=1}^{n-1}[E_{k,k+1} + E_{k+1,k} + i(E_{n-k,k+1} - E_{n-k,k-1})].$$

4147. Prove that each complex matrix is similar to a symmetric one.

42 Normed vector spaces and algebras. Non-negative matrices

4201. Let K be a normed field (see Exercise 6535) with a norm $|x|$. Prove that the following functions on K^n are norms:

a) $\|(a_1, \ldots, a_n)\| = |a_1| + \cdots + |a_n|$;

b) $\|(a_1, \ldots, a_n)\| = \max(|a_1|, \ldots, |a_n|)$;

c) $\|(a_1, \ldots, a_n)\| = \sqrt{|a_1|^2 + \cdots + |a_n|^2}$.

4202. Let K be a locally compact normed field, and V be a finite-dimensional vector space over K. Prove that any two norms $\|x\|_1, \|x\|_2$ on V are equivalent, i.e. there exist positive real numbers C_1, C_2 such that

$$C_1\|x\|_1 \le \|x\|_2 \le C_2\|x\|_1$$

for all $x \in V$.

4203. Let K be a normed field, and V be a finite-dimensional vector space with a basis $(e_1, \ldots, e_m)$. Let $x_n = x_{n1}e_1 + \cdots + x_{nm}e_m \in V$, $x_{ij} \in K$, $n \geq 1$. Prove that a sequence of vectors x_n converges if and only if the sequences x_{ni}, $i = 1, \ldots, m$, converge.

4204. Let K be a complete normed field and V be a finite-dimensional normed vector space over K. Prove that V is a complete normed space.

4205. Let K be a normed field and V be a normed vector space over K. Denote by $L(V)$ the set of all linear operators $\mathcal{A}$ on V such that the number $\|\mathcal{A}(x)\|$ is bounded provided $\|x\| = 1$. Prove that

a) $L(V)$ is a subspace of the space of all linear operators on V;

b) $L(V)$ is a normed algebra with the norm

$$\|\mathcal{A}\| = \sup_{\|x\|=1} \|\mathcal{A}(x)\|;$$

c) if V is finite-dimensional then $L(V)$ is the space of all operators on V.

4206. Let K be a normed field and let norms a), b) from Exercise 4201 be given on K^n. Prove that the corresponding norms on $\mathbf{M}_n(K) = L(V)$, defined in Exercise 4205, are of the form

a) $\|A\| = \max\limits_{1 \leq j \leq n} \left(\sum\limits_{i=1}^{n} |a_{ij}| \right);$ b) $\|A\| = \max\limits_{1 \leq i \leq n} \left(\sum\limits_{j=1}^{n} |a_{ij}| \right).$

4207. Let K be a normed field. For $A = (a_{ij}) \in \mathbf{M}_n(K)$ put

a) $\|A\|_1 = \sum\limits_{i,j=1}^{n} |a_{ij}|;$ b) $\|A\|_2 = \sqrt{\sum\limits_{i,j=1}^{n} |a_{ij}|^2};$

c) $\|A\|_3 = n \cdot \max\limits_{1 \leq i,j \leq n} |a_{ij}|.$

Prove that each of these functions induces on $\mathbf{M}_n(K)$ the structure of a normed algebra.

4208. Prove that for any matrix $A \in \mathbf{M}_n(\mathbb{C})$ the matrices e^A, $\sin A$, $\cos A$ are well-defined.

4209. Let $A \in \mathbf{M}_n(\mathbb{C})$. Prove that

a) $\sin 2A = 2 \sin A \cos A;$ b) $e^{iA} = \cos A + i \sin A;$

c) $\sin A = \dfrac{1}{2}(e^{iA} - e^{-iA});$ d) $\cos A = \dfrac{1}{2}(e^{iA} + e^{-iA}).$

4210. If $A, B \in \mathbf{M}_n(\mathbb{C})$ and $AB = BA$ then $e^{A+B} = e^A e^B$.

4211. Let A be a normed algebra over a complete normed field K. Prove that if $x \in A$, then there exists a limit

$$\rho(x) = \lim_n \|x^n\|^{1/n}.$$

*   *   *

4212. Let x be an element of a Banach algebra A over a complete normed field K. Prove that the radius of convergence of the series $\sum_{n \geq 0} t^n x^n \in A[[t]]$ is equal to $\rho(x)^{-1}$ (see Exercise 4211).

4213. Let x be an element of a Banach algebra A over a complete normed field K. A *spectrum* $\mathrm{Sp}(x)$ is the set of all $\lambda \in K$ such that the element $x - \lambda$ is not invertible in A. Prove that

    a)   the radius of the least disc in K with the centre at the origin which contains $\mathrm{Sp}(x)$ is equal to $\rho(x)$;

    b)   the set $\mathrm{Sp}(x)$ is compact in K.

4214. Let $A \in \mathbf{M}_n(\mathbb{C})$ and $\lambda_1, \ldots, \lambda_n$ be all the eigenvalues of A. Prove that

$$\max_{1 \leq i \leq n} |\lambda_i| = \rho(A) \leq \|A\|,$$

where $\|A\|$ is an arbitrary norm on the algebra $\mathbf{M}_n(\mathbb{C})$ induced by virtue of Exercise 4205b by some norm on $\mathbb{C}^n$.

4215. Let x be an element of a Banach algebra A over a complete normed field K and let $f(t) \in K[[t]]$. Prove that if $\rho(x)$ is less than the radius of convergence of $f(t)$, then the series $f(x)$ converges.

4216. Let x be an element of a Banach algebra A over a complete normed field K and let g be an invertible element in A. Let $f(t) \in K[[t]]$. Prove that the series $f(x)$ converges if and only if $f(gxg^{-1})$ converges and $f(gxg^{-1}) = gf(x)g^{-1}$.

4217. Let a be an element of a Banach algebra over a complete normed field K. Assume that $f(t) \in K[[t]]$ and the series $f(a)$ converges. Let a have an annihilator polynomial of degree n. Prove that $f(a) \in \langle 1, a, a^2, \ldots, a^{n-1} \rangle$.

4218. Let $A \in \mathbf{M}_n(\mathbb{C})$ and $\lambda_0, \ldots, \lambda_n$ be all roots of the characteristic polynomial of A of multiplicities $k_0, \ldots, k_n$. Assume that $\lambda_0, \ldots, \lambda_n$ lie inside the disc of convergence of a series $u(t) \in \mathbb{C}[[t]]$. Prove that

$$u(A) = \sum_{i=0}^{n} G_i(A) \sum_{k=0}^{k_i} \sum_{l=0}^{k} \frac{1}{l!} \left[\frac{d^l u}{dx^l}(\lambda_i) \cdot \frac{d^l}{dx^l} G_i(x)^{-1} \Big|_{x=\lambda_i} (A - \lambda_i E)^k \right]$$

where $G_i(t) = \prod_{j \neq i} (t - \lambda_j)^{k_j}$.

4219. Calculate

a) $e^{\begin{pmatrix} 3 & -1 \\ 1 & 1 \end{pmatrix}}$;

b) $e^{\begin{pmatrix} 4 & -2 \\ 6 & -3 \end{pmatrix}}$;

c) $e^{\begin{pmatrix} 4 & 2 & -5 \\ 6 & 4 & -9 \\ 5 & 3 & 7 \end{pmatrix}}$;

d) $\ln \begin{pmatrix} 4 & -15 & 6 \\ 1 & -4 & 2 \\ 1 & -5 & 3 \end{pmatrix}$;

e) $\sin \begin{pmatrix} \pi - 1 & 1 \\ -1 & \pi + 1 \end{pmatrix}$.

4220. Find the determinant of the matrix e^A, where A is a square matrix of size n.

4221. Let $A \in \mathbf{M}_n(\mathbb{C})$ and

$$x(t) = \begin{pmatrix} x_1(t) \\ \vdots \\ x_m(t) \end{pmatrix}$$

be a continuously differentiable vector-function. Prove that the solution of the system of differential equations with constant coefficients

$$\frac{d}{dt} x(t) = Ax(t),$$

with initial condition $x(t_0)$, is the vector function $x(t) = e^{At} x(t_0)$.

4222. Let A be a non-negative matrix, and A^k for some natural number k be a positive matrix. Prove that $\rho(A) > 0$.

4223. Find an example of a non-negative 2×2 matrix such that A^2 is a positive matrix.

4224. Let a non-negative matrix A have a positive eigenvector. Then A is similar to a non-negative matrix in which the sums of entries of each row are equal.

4225. Let A, B be positive matrices and suppose that the matrix $A - B$ is also positive. Prove that $\rho(A) > \rho(B)$.

4226. Let A be a nonsingular non-negative matrix and suppose that the inverse matrix A^{-1} is also non-negative. Then $A = DP$ where D is a diagonal non-negative invertible matrix, P is a permutative matrix.

4227. Let A be a non-negative matrix, x be a nonzero complex vector such that $(Ax - \alpha x)$ is a non-negative vector for some real number α. Prove that $\rho(A) > \alpha$.

4228. Let A be a non-negative matrix and $^t A$ have a positive eigenvector. If $Ax - \rho(A)x$ is a non-negative vector for some nonzero vector x, then $Ax = \rho(A)x$.

4229. Let A be a non-negative matrix and suppose that the matrix A^k is positive for some natural number k. Prove that

a) A has a positive eigenvector;

b) $\rho(A)$ is an eigenvalue of A of multiplicity 1.

4230. Let a non-negative matrix A have an eigenvector $x = (x_1, \ldots, x_n)$ where $x_1, \ldots, x_r > 0$, $x_{r+1} = \cdots = x_n = 0$. Then there exists a permutative matrix P such that

$$P^{-1}AP = \begin{pmatrix} B & C \\ 0 & D \end{pmatrix},$$

where $B \in \mathbf{M}_r$, $D \in \mathbf{M}_{n-r}$, and B has a positive eigenvector.

4231. Let A be a non-negative matrix. Prove that there exists a positive matrix B commuting with A if and only if there exist positive eigenvectors of matrices A and $^t A$.

4232. Let A be a non-negative tridiagonal matrix. Prove that all eigenvalues of A are real.

4233. Let A be a non-negative matrix. Then $\rho(E + A) = 1 + \rho(A)$.

4234. Find $\rho(A)$ and non-negative eigenvectors x for the non-negative matrix A:

a) $\begin{pmatrix} 2 & 1 \\ 1 & 2 \end{pmatrix}$;

b) $\begin{pmatrix} 3 & 4 \\ 5 & 2 \end{pmatrix}$;

c) $\begin{pmatrix} 1 & 4 & 3 \\ 0 & 2 & 5 \\ 1 & 0 & 0 \end{pmatrix}$;

d) $\begin{pmatrix} 3 & 2 & 3 & 1 \\ 0 & 2 & 0 & 5 \\ 5 & 7 & 1 & 6 \\ 0 & 1 & 0 & 4 \end{pmatrix}$.

CHAPTER 10

Metric vector spaces

43 Geometry of metric spaces

4301. What vector spaces with bilinear forms in Exercise 3701 are metric?

4302. Prove that the real part $f(x, y)$ and the imaginary part $g(x, y)$ of an Hermitian function on a complex vector space V are invariant under multiplication by i; i.e. for any vectors $x, y \in V$

$$f(ix, iy) = f(x, y), \quad g(ix, iy) = g(x, y).$$

4303. Prove that a metric vector space is a direct sum of a subspace L and its orthocomplement $L^{\perp}$ if and only if the scalar product on L is nondegenerate and that in this case the scalar product on $L^{\perp}$ is also nondegenerate.

4304. Let $\mathbf{M}_n(\mathbb{C})$ be a space with an Hermitian scalar product

$$(X, Y) = \mathrm{tr}\, X^t \bar{Y}.$$

Find the orthocomplement of the subspace:

  a) of all matrices with zero trace;

  b) of all Hermitian matrices;

  c) of all skew-Hermitian matrices;

  d) of all upper-triangular matrices.

4305. Show that Hermitian and Euclidean spaces are normed.

4306. What norms on the spaces $\mathbb{R}^n$, $\mathbb{C}^n$ in Exercise 4201 are induced by Euclidean or Hermitian metric?

4307. Complete the system of vectors in Euclidean and Hermitian spaces to an orthogonal basis:

a) $((1, -2, 2, -3),\quad (2, -3, 2, 4))$;

b) $((1, 1, 1, 2),\quad (1, 2, 3, -3))$;

c) $\left(\left(\dfrac{2}{3}, \dfrac{1}{3}, \dfrac{2}{3}\right),\quad \left(\dfrac{1}{3}, \dfrac{2}{3}, -\dfrac{2}{3}\right)\right)$;

d) $\left(\left(\dfrac{1}{2}, \dfrac{1}{2}, \dfrac{1}{2}, \dfrac{1}{2}\right),\quad \left(\dfrac{1}{2}, \dfrac{1}{2}, -\dfrac{1}{2}, -\dfrac{1}{2}\right)\right)$;

e) $((1, 1 - i, 2),\quad (2, -1 + 3i, 3 - i))$;

f) $((-i, 2, -4 + i),\quad (4 - i, -1, i))$.

4308. Find the orthogonal projection of a vector x in Euclidean (Hermitian) space into the linear span of the orthonormal system of vectors $(e_1, \ldots, e_k)$.

4309. Prove that it is possible to choose orthonormal bases $(e_1, \ldots, e_k)$ and $(f_1, \ldots, f_l)$ of any two subspaces of Euclidean (Hermitian) space such that $(e_i, f_j) = 0$ if $i \neq j$ and $(e_i, f_i) \geq 0$.

4310. Let $(e_1, \ldots, e_k)$ and $(f_1, \ldots, f_l)$ be orthonormal bases of subspaces L and M of Euclidean (Hermitian) space, and let $A = \big((e_i, f_j)\big)$ be the matrix of size $k \times l$. Prove that all characteristic numbers of the matrix ${}^t A \cdot A$ belong to the segment $[0, 1]$ and do not depend on the choice of bases of the subspaces L and M.

4311. Prove that any real symmetric matrix of rank $\leq n$ with non-negative (positive) principal minors is the Gram matrix of some (linearly independent, respectively) system of vectors of n-dimensional Euclidean space.

Prove the similar statement for a Hermitian matrix and Hermitian space.

4312. Prove that the sum of squares of lengths of projections of vectors of any orthonormal basis of Euclidean (Hermitian) space into k-dimensional subspace is equal to k.

4313. Let G be the matrix of the scalar product in a basis $(e_1, \ldots, e_n)$ of Euclidean space V. Find the matrix of the change of the base to the dual one $(f_1, \ldots, f_n)$ and the matrix of the scalar product in the dual basis.

4314. Let S be the matrix of a change of the basis $\mathbf{e}$ to the basis $\mathbf{e}'$. Find the matrix of the change of the basis $\mathbf{e}'$ dual to $\mathbf{e}$ to the basis $\mathbf{f}'$ dual to $\mathbf{f}$:

a)   in Euclidean space;

b)   in Hermitian space.

4315. Construct, with the help of the orthonormalization process, an orthogonal basis of the linear span of the system of vectors in Euclidean (Hermitian) space:

a) $((1, 2, 2, -1), (1, 1, -5, 3), (3, 2, 8, -7))$;

b) $((1, 1, -1, -2), (5, 8, -2, -3), (3, 9, 3, 8))$;

c) $((2, 1, 3, -1), (7, 4, 3, -3), (1, 1, -6, 0), (5, 7, 7, 8))$;

d) $((2, 1, -i), (1 - i, 2, 0), (-i, 0, 1 - i))$;

e) $((0, 1 - i, 2), (-i, 2 + 3i, i), (0, 0, 2i))$.

4316. Find a basis of the orthocomplement of the linear span of the system of vectors in Euclidean (Hermitian) space:

a) $((1, 0, 2, 1), (2, 1, 2, 3), (0, 1, -2, 1))$;

b) $((1, 1, 1, 1), (-1, 1, -1, 1), (2, 0, 2, 0))$;

c) $((0, 1 + 2i, -i), (1, -1, 2 - i))$.

4317. Prove that the systems of linear equations determining a linear subspace of $\mathbb{R}^n$ and its orthocomplement are connected as follows: the coefficients of the linearly independent system determining one of these subspaces are coordinates of vectors of a basis of the other subspace.

4318. Find the equations determining the orthocomplement of the subspaces given by the system of equations:

a)
$$\begin{aligned} 2x_1 + x_2 + 3x_3 - x_4 &= 0, \\ 3x_1 + 2x_2 \qquad\quad -2x_4 &= 0, \\ 3x_1 + x_2 + 4x_3 - x_4 &= 0; \end{aligned}$$

b)
$$\begin{aligned} 2x_1 - 3x_2 + 4x_3 - 3x_4 &= 0, \\ 3x_1 - x_2 + 11x_3 - 13x_4 &= 0, \\ 4x_1 + x_2 + 18x_3 - 23x_4 &= 0; \end{aligned}$$

c)
$$\begin{aligned} x_1 + (1 - i)x_2 - ix_3 &= 0, \\ -ix_1 \qquad\quad +4x_2 \qquad &= 0. \end{aligned}$$

4319. Find the projection of the vector x on the subspace L and the orthogonal component of x where:

a) $L = \langle (1, 1, 1, 1), (1, 2, 2, -1), (1, 0, 0, 3) \rangle$,
$x = (4, -1, -3, 4)$;

b) $L = \langle (2, 1, 1, -1), (1, 1, 3, 0), (1, 2, 8, 1) \rangle$,
$x = (5, 2, -2, 2)$;

c) L is given by the system of equations

$$\begin{aligned} 2x_1 + x_2 + x_3 + 3x_4 &= 0, \\ 3x_1 + 2x_2 + 2x_3 + x_4 &= 0, \\ x_1 + 2x_2 + 2x_3 - 4x_4 &= 0, \end{aligned}$$

$x = (7, -4, -1, 2);$

d) $L = \langle(-i, 2+i, 0), (3, -i+1, i)\rangle,$
    $x = (0, 1+i, -i);$

e) L is given by the system of equations

$$(2+i)x_1 \qquad -ix_2+2x_3+ix_4 = 0,$$
$$(2+i)x_1 \qquad -ix_2+2x_3+ix_4 = 0,$$
$$5x_1+(-1+i)x_2 +x_3 \qquad = 0,$$

$x = (i, 2-i, 0).$

4320. Let the orthogonalization process transform the system of vectors $a_1, \ldots, a_n$ into the system $b_1, \ldots, b_n$. Prove that the vector b_k is the orthogonal component of the vector a_k with respect to the linear span of the system $a_1, \ldots, a_{k-1}$ $(k > 1)$.

4321. Find the distance between the vector x and the subspace given by the system of equations:

a) $x = (2, 4, 0, -1);$
    $2x_1 + 2x_2 + x_3 + x_4 = 0,$
    $2x_1 + 4x_2 + 2x_3 + 4x_4 = 0;$

b) $x = (3, 3, -4, 2);$
    $x_1 + 2x_2 + x_3 - x_4 = 0,$
    $x_1 + 3x_2 + x_3 - 3x_4 = 0;$

c) $x = (3, 3, -1, 1, -1);$
    $2x_1 - 2x_2 + 3x_3 - 2x_4 + 2x_5 = 0;$

d) $x = (3, 3, -1, 1, -1);$
    $x_1 - 3x_2 + 2x_4 - x_5 = 0;$

e) $x = (0, -i, 1+i);$
    $x_1 + ix_2 - (2-i)x_3 = 0;$

f) $x = (1, -1, i);$
    $x_1 + (5+4i)x_2 - ix_3 = 0.$

4322. *Bessel inequality. Parceval equality.* Let $(e_1, \ldots, e_k)$ be an orthonormal system of vectors in n-dimensional Euclidean (Hermitian) space V. Prove that for any vector x the inequality holds in

$$\sum_{i=1}^{k} |(x, e_i)|^2 \leq \|x\|^2.$$

Prove that the equality holds for any x if and only if $k = n$, i.e. the given system of vectors is an orthonormal basis of V (the *Parceval equality*).

4323. Applying the Cauchy inequality prove that

$$\left| \sum_{i=1}^{k} a_i b_i \right|^2 \le \sum_{i,j=1}^{k} |a_i|^2 |b_i|^2$$

for any complex numbers $a_1, \ldots, a_k, b_1, \ldots, b_k$.

4324. Prove that the square of the distance between a vector x in Euclidean (Hermitian) space and a subspace with a basis $(e_1, \ldots, e_k)$ is equal to the ratio of the Gram determinants of the systems of vectors $(e_1, \ldots, e_k, x)$ and $(e_1, \ldots, e_k)$.

4325. Prove that the Gram determinant of any system of vectors

a)   does not change under the orthogonalization process;

b)   is non-negative;

c)   is equal to zero if and only if the system is linearly dependent;

d)   does not surpass the product of squares of lengths of vectors of the system, and equality takes place if and only if either the vectors are pairwise orthogonal or one of them is equal to zero.

4326. Prove that the determinant of the matrix of a positively definite quadratic form does not surpass the product of entries of its principal diagonal.

4327. *Hadamard inequality.* Prove that for any real square matrix $A = (a_{ij})$ of size n the inequality

$$(\det A)^2 \le \prod_{i=1}^{n} \left(\sum_{j=1}^{n} a_{ij}^2 \right),$$

holds and the equality takes place if and only if either

$$\sum_{k=1}^{n} a_{ik} a_{jk} = 0 \quad (i, j = 1, \ldots, n; \quad i \ne j),$$

or the matrix A has a zero row.

Formulate and prove the similar statement for a complex matrix A.

4328. Find the lengths of the sides and the internal angles in the triangle abc in the space $\mathbb{R}^5$:

a) $a = (2, 4, 2, 4, 2)$,
   $b = (6, 4, 4, 4, 6)$,
   $c = (5, 7, 5, 7, 2)$;

b) $a = (1, 2, 3, 2, 1),$
    $b = (3, 4, 0, 4, 3),$
    $c = \left(1 + \dfrac{5}{26}\sqrt{78}, 2 + \dfrac{5}{13}\sqrt{78}, 3 + \dfrac{10}{13}\sqrt{78}, 2 + \dfrac{5}{13}\sqrt{78}, 1 + \dfrac{5}{26}\sqrt{78}\right).$

4329. Prove with the help of the scalar product of vectors that

a) the sum of squares of the diagonals in a parallelogram is equal to the sum of the squares of its sides;

b) the square of a side of a triangle is equal to the sum of the squares of two other sides minus the double product of these sides by the cosine of the angle between them.

4330. Solve the system of linear equations by the method of least squares:

a)
$$x_1 + x_2 - 3x_3 = -1,$$
$$2x_1 + x_2 - 2x_3 = 1,$$
$$x_1 + x_2 + x_3 = 3,$$
$$x_1 + 2x_2 - 3x_3 = 1;$$

b)
$$2x_1 - 5x_2 + 3x_3 + x_4 = 5,$$
$$3x_1 - 7x_2 + 3x_3 - x_4 = -1,$$
$$5x_1 - 9x_2 + 6x_3 + 2x_4 = 7,$$
$$4x_1 - 6x_2 + 3x_3 + x_4 = 8.$$

4331. *n-dimensional Pythagorean theorem.* Prove that the square of a diagonal of an n-dimensional rectangular parallelepiped is equal to the sum of squares of its edges outgoing from one vertex.

4332. Find the number of the diagonals in an n-dimensional cube which are orthogonal to a given diagonal.

4333. Find the length of a diagonal and the angles between the diagonals of the cube and its edges in an n-dimensional cube with an edge a.

4334. Find the radius R of a sphere circumscribed around an n-dimensional cube with an edge a, and solve the inequality $R < a$.

4335. Prove that the length of an orthogonal projection of an edge in an n-dimensional cube onto any of its diagonals is equal to $1/n$ of the length of the diagonal.

4336. Calculate the volume of an n-dimensional parallelepiped with sides:

a) $(1, -1, 1, -1),\quad (1, 1, 1, 1),\quad (1, 0, -1, 0),\quad (0, 1, 0, -1);$

b) $(1, 1, 1, 1),\quad (1, -1, -1, 1),\quad (2, 1, 1, 3),\quad (0, 1, -1, 0);$

c) $(1, 1, 1, 2, 1),\quad (1, 0, 0, 1, -2),\quad (2, 1, -1, 0, 2),\quad (0, 7, 3, -4, -2),$
    $(39, -37, 51, -29, 5);$

d) $(1, 0, 0, 2, 5),\quad (0, 1, 0, 3, 4),\quad (0, 0, 1, 4, 7),\quad (2, -3, 4, 11, 12),$
    $(0, 0, 0, 0, 1).$

4337. Prove, for the volume of a parallelepiped, the inequality

$$V(a_1, \ldots, a_k, b_1, \ldots, b_l) \leq V(a_1, \ldots, a_k) \cdot V(b_1, \ldots, b_l),$$

and show that the equality takes place if and only if $(a_i, b_j) = 0$ for all i and j.

4338. Find the angle between the vector x and the subspace L:

a) $L = \langle(3, 4, -4, -1), (0, 1, -1, 2)\rangle,$ $x = (2, 2, 1, 1);$

b) $L = \langle(5, 3, 4, -3), (1, 1, 4, 5), (2, -1, 1, 2)\rangle,$ $x = (1, 0, 3, 0);$

c) $L = \langle(1, 1, 1, 1), (1, 2, 0, 0), (1, 3, 1, 1)\rangle,$ $x = (1, 1, 0, 0);$

d) $L = \langle(0, 0, 0, 1), (1, -1, -1, 1), (-3, 3, 3, 0)\rangle,$ $x = (1, 2, 3, 0).$

4339. Prove that if the angle between any two of k distinct vectors in Euclidean space V is equal to $\pi/3$, then $k \leq \dim V$.

4340. Prove that if the angle between any two of k distinct vectors in Euclidean space V is obtuse, then $k \leq 1 + \dim V$.

4341. Find the angle between a diagonal of an n-dimensional cube and its k-dimensional face.

4342. Find the angle between two-dimensional sides $a_0a_1a_2$ and $a_0a_3a_4$ in the regular four-dimensional simplex $a_0a_1a_2a_3a_4$.

4343. Find the angle between the subspaces

$$\langle(1, 0, 0, 0), (0, 1, 0, 0)\rangle \quad \text{and} \quad \langle(1, 1, 1, 1), (1, -1, 1, -1)\rangle.$$

4344. Polynomials of the type

$$P_0(x) = 1, \quad P_k(x) = \frac{1}{2^k k!} \frac{d^k}{dx^k}\left[(x^2 - 1)^k\right] \quad (k = 1, 2, \ldots, n)$$

are called *Legendre polynomials*.

a) Prove that Legendre polynomials form an orthogonal basis of Euclidean space $\mathbb{R}[x]_n$ with the scalar product $\int_{-1}^1 f(x)g(x)\,dx$.

b) Find the explicit form of the polynomials $P_k(x)$ for $k \leq 4$.

c) Prove that $\deg P_k(x) = k$ and find the expansion of $P_k(x)$ for all k.

d) Calculate the length of the Legendre polynomial $P_k(x)$.

e) Calculate the value of $P_k(1)$.

f) Prove that the orthogonalization process applied to the basis $(1, x, x^2, \ldots, x^n)$ of the space $\mathbb{R}[x]_n$ gives us a basis which differs only by constant multipliers from appropriate Legendre polynomials. Find these multipliers.

g) Prove that the integral $\int_{-1}^{1} f(x)^2 \, dx$, where $f(x)$ is a polynomial of degree n with real coefficients and with the leading coefficient 1, achieves the minimum

$$\frac{2^{2n+1}}{(2n+1)\binom{2n}{n}^2} \quad \text{if} \quad f(x) = \frac{2^n}{\binom{2n}{n}} P_n(x).$$

4345. Find in the space $\mathbb{R}[x]_n$ with the scalar product $\int_{-1}^{1} f(x)g(x) \, dx$:

a) the volume of the parallelepiped $P(1, x, \ldots, x^n)$;

b) the distance between the vector x^n and the subspace $\mathbb{R}[x]_{n-1}$.

4346. Let L be the space of continuous functions in the segment $[-\pi, \pi]$ with the scalar product

$$(f, g) = \frac{1}{\pi} \int_{-\pi}^{\pi} f(t)g(t)dt.$$

Find the projection of the function t^m into the subspace

$$V = \langle 1, \cos t, \sin t, \ldots, \cos nt, \sin nt \rangle.$$

4347. Let V be a pseudoeuclidean space of signature (p, q) and let W be a subspace of V. Prove that

a) if the scalar product in W is definite positive, then $\dim W \leq p$;

b) if $(x, x) = 0$ for any $x \in W$, then $\dim W \leq \min(p, q)$.

4348. Let a nonsingular scalar product of signature (p, q) in a vector space be given, such that its restriction to a subspace W is a nonsingular scalar product of signature (p', q'). Prove that the restriction of the scalar product to $W^{\perp}$ is nondegenerate and has the signature $(p - p', q - q')$.

4349. Prove that a pseudoeuclidean space of signature (p, q), where p and q are distinct from zero, has a basis consisting of isotropic vectors.

44 Adjoint and normal operators

4401. Prove the following properties of the operation of passing to adjoint operators on metric space:

a) $A^{**} = A$;

b) $(A + B)^* = A^* + B^*$;

c) $(AB)^* = B^* A^*$;

d) $(\lambda A)^* = \bar{\lambda} A^*$;

e) $A^* A$ and $A A^*$ are selfadjoint operators;

f) if an operator A is nonsingular, then $(A^{-1})^* = (A^*)^{-1}$.

4402. Find the matrix of an operator A^*, with a basis e, of a metric vector space V if the operator A has the matrix A with this basis and the scalar product has the matrix G.

4403. Let (e_1, e_2) be an orthonormal basis of a metric vector space and an operator A have, with a basis $(e_1, e_1 + e_2)$, the matrix $\begin{pmatrix} 1 & 2 \\ 1 & -1 \end{pmatrix}$. Find the matrix of the operator A^* with this basis.

4404. Find the adjoint operator to the projection of the coordinate plane on the abscissa axis in parallel with the bisectrix of the first and third quadrants.

4405. Let A be a projection of a metric vector space V on a subspace V_1 in parallel with a subspace V_2. Prove that

a) $V = V_1^{\perp} \oplus V_2^{\perp}$;

b) A^* is a projection of the space V onto $V_2^{\perp}$ in parallel with $V_1^{\perp}$.

4406. Prove that if a subspace of a metric vector space is invariant under a linear operator A, then its orthocomplement is invariant under the operator A^*.

4407. Prove that the kernel and the image of the adjoint operator A^* are the orthocomplements to the image and the kernel of the operator A, respectively.

4408. Prove that if x is an eigenvector of operators A and A^* on a metric vector space with eigenvalues λ and μ, respectively, then $\mu = \bar{\lambda}$.

4409. Let V be the space of real infinitely differentiable periodic functions of a period $h > 0$ with the scalar product $\int_{-h}^{h} f(x) g(x)\, dx$.

a) Find the adjoint operator to the operator of differentiation $\mathcal{D}$.

b) Prove that mappings A and B given by the rule

$$A(f) = \sum_{i=0}^{n} u_i \mathcal{D}^i (f), \quad B(f) = \sum_{i=0}^{n} (-1)^i \mathcal{D}^i (u_i f),$$

where $u_0, u_1, \ldots, u_n \in V$ are fixed functions, are linear operators on V and $B = A^*$.

c) Prove that the operator given by the rule

$$A(f) = \sin^2 \frac{2\pi}{h} x \mathcal{D}^2(f) + \frac{2\pi}{h} \sin \frac{4\pi}{h} x \mathcal{D}(f),$$

is selfadjoint.

4410. Let V be the space of real infinitely differentiable functions on the segment $[a, b]$ with the scalar product $\int_a^b f(x)g(x)\,dx$. Prove that

a) if the functions $u_0, \ldots, u_n \in V$ satisfy conditions

$$\mathcal{D}^i(u_j)(a) = \mathcal{D}^i(u_j)(b) = 0 \quad (j = 1, \ldots, n; i = 0, 1, \ldots, j-1),$$

then the mappings A and B given by the rules

$$A(f) = \sum_{i=0}^{n} u_i \mathcal{D}^i(f), \quad B(f) = \sum_{i=0}^{n} (-1)^i \mathcal{D}^i(u_i f),$$

are linear operators on V and $B = A^*$;

b) the linear operator A given by the rule

$$A(f) = (x-a)^2(x-b)^2 \mathcal{D}^2(f) + 2(x-a)(x-b)\mathcal{D}(f),$$

is selfadjoint.

4411. Let the linear operators A and B on the space $\mathbb{R}[x]$ with the scalar product $\int_a^b f(x)g(x)\,dx$ are given by the rules

$$A(f) = \int_a^b P(x, y)f(y)\,dy, \quad B(f) = \int_a^b P(x, y)f(y)\,dy,$$

where $P(x, y) \in \mathbb{R}[x, y]$. Prove that $B = A^*$.

4412. Prove that if A is a selfadjoint operator then the function $f(x, y) = (Ax, y)$ is Hermitian.

4413. Prove that if A and B are selfadjoint operators on a metric vector space V and $(Ax, x) = (Bx, x)$ for all $x \in V$ then $A = B$.

4414. Prove that an operator A on Euclidean or Hermitian space V is normal if and only if $|Ax| = |A^*x|$ for all $x \in V$.

4415. Prove that if x is an eigenvector of a normal operator A in Euclidean or Hermitian space with an eigenvalue λ, then x is an eigenvector of the operator A^* with the eigenvalue $\bar{\lambda}$.

4416. Prove that eigenvectors of normal operators in metric vector spaces with distinct eigenvalues are orthogonal.

4417. Prove that

a) the orthocomplement of a linear span of an eigenvector of a normal operator $\mathcal{A}$ on a metric vector space is invariant under $\mathcal{A}$;

b) an operator on Hermitian space is normal if and only if it has an orthonormal eigenbasis;

c) an operator on Euclidean or metric space is normal if and only if each of its eigenvectors is an eigenvector of the adjoint operator.

4418. Prove that any set of commuting normal operators on Hermitian space has a common orthonormal eigenbasis.

4419. Prove that if a normal operator $\mathcal{A}$ on Hermitian space commutes with an operator $\mathcal{B}$ then $\mathcal{A}$ commutes with $\mathcal{B}^*$.

4420. Let $\mathcal{A}, \mathcal{B}$ be normal operators on Hermitian space and the characteristic polynomials of these operators be equal. Prove that the matrices of the operators $\mathcal{A}$ and $\mathcal{B}$ with any basis are similar.

4421. Let $\mathcal{A}$ be a normal nilpotent operator on Hermitian space. Prove that $\mathcal{A} = 0$.

4422. Prove that an operator $\mathcal{A}$ on Hermitian space is normal if and only if $\mathcal{A}^* = p(\mathcal{A})$ for some polynomial $p(t)$.

4423. Put $\overline{f(x)} = \sum_{i=0}^{n} \bar{a}_i x^i$ for any polynomial $f(x) = \sum_{i=0}^{n} a_i x^i \in K[x]$. Let $\mathcal{A}$ be an operator on a metric space. Prove that

a) $f(\mathcal{A})^* = \overline{f(\mathcal{A}^*)}$;

b) if $f(\mathcal{A}) = 0$, then $\overline{f(\mathcal{A}^*)} = 0$.

4424. Let $\mathcal{A}$ be a normal operator on a metric vector space V and $f(x) \in K[x]$. Prove that

a) the kernel $\operatorname{Ker} f(\mathcal{A})$ is invariant under $\mathcal{A}^*$;

b) $\operatorname{Ker} \overline{f(\mathcal{A}^*)} = \operatorname{Ker} f(\mathcal{A})$;

c) if $f(x) = f_1(x) f_2(x)$ where $f_1(x)$ and $f_2(x)$ are coprime then $\operatorname{Ker} f(\mathcal{A})$ is the orthogonal direct sum of subspaces $\operatorname{Ker} f_1(\mathcal{A})$ and $\operatorname{Ker} f_2(\mathcal{A})$;

d) if $(f(\mathcal{A}))^n = 0$ then $f(\mathcal{A}) = 0$.

4425. Let A be a normal operator on Euclidean space V and $A^2 = -E$. Prove that $A^* = -A$.

4426. Let $p(t) = t^2 + at + b$ be a real irreducible polynomial. Assume that A is a normal operator on Euclidean space and $p(A) = 0$. Prove that $A^* = -A - aE$.

4427. Let A be a normal linear operator on Euclidean space V and U be a two-dimensional subspace of V invariant under A. Assume that A has no eigenvectors in U. Prove that

a) U is invariant under A^*;

b) $U^\perp$ is invariant under A and A^*.

4428. Let A be a normal linear operator on two-dimensional Euclidean space U such that A has no eigenvectors. Let $\mathbf{e} = (e_1, e_2)$ be an orthonormal basis. Prove that the matrix of A in the basis $\mathbf{e}$ has the form

$$\begin{pmatrix} a & -b \\ b & a \end{pmatrix}.$$

4429. Let A be a normal operator on Euclidean space V. Prove that there exists an orthonormal basis V in which the matrix of A has a block-diagonal form

$$\begin{pmatrix} A_1 & & 0 \\ & \ddots & \\ 0 & & A_n \end{pmatrix}$$

in which the size of a block A_i is two at most, and blocks A_i of size two have the form

$$\begin{pmatrix} a_i & -b_i \\ b_i & a_i \end{pmatrix}.$$

4430. Prove that any operator on Euclidean (Hermitian) space is the sum of a symmetric and a skew-symmetric (of Hermitian and skew-Hermitian, respectively) operator.

4431. Prove that an operator A on Euclidean space V is skew-symmetric if and only if the vectors x and Ax are orthogonal for any $x \in V$.

4432. Prove that for any skew-symmetric operator on Euclidean space there exists an orthonormal basis in which its matrix has a block-diagonal form and the entries of the principal diagonal either equal zero or have the form

$$\begin{pmatrix} 0 & -\alpha \\ \alpha & 0 \end{pmatrix}, \quad (\alpha \in \mathbb{R}).$$

45 Selfadjoint operators. Reduction of quadratic functions to principal axes

4501. Prove that the product of two selfadjoint operators on a metric vector space is a selfadjoint operator if and only if these operators commute.

4502. Let A and B be selfadjoint operators on a metric vector space. Prove that

a) the operator $AB + BA$ is selfadjoint;

b) if $\bar{\lambda} = -\lambda$ then the operator $\lambda(AB - BA)$ is selfadjoint.

4503. Prove that the projection of a metric space $L_1 \oplus L_2$ onto the subspace L_1 in parallel with L_2 is a selfadjoint operator if and only if L_1 and L_2 are orthogonal.

4504. Find an orthonormal eigenbasis, and the matrix in this basis, of the operator given in some orthonormal basis by the matrix:

a) $\begin{pmatrix} 2 & 1 \\ 1 & 2 \end{pmatrix}$;

b) $\begin{pmatrix} 11 & 2 & -8 \\ 2 & 2 & 10 \\ -8 & 10 & 5 \end{pmatrix}$;

c) $\begin{pmatrix} 17 & -8 & 4 \\ -8 & 17 & -4 \\ 4 & -4 & 11 \end{pmatrix}$;

d) $\begin{pmatrix} 5 & -1 & -1 \\ -1 & 5 & -1 \\ -1 & -1 & 5 \end{pmatrix}$;

e) $\begin{pmatrix} 0 & 0 & 1 \\ 0 & 1 & 0 \\ 1 & 0 & 0 \end{pmatrix}$;

f) $\begin{pmatrix} 0 & 0 & 0 & 1 \\ 0 & 0 & 1 & 0 \\ 0 & 1 & 0 & 0 \\ 1 & 0 & 0 & 0 \end{pmatrix}$;

g) $\begin{pmatrix} 1 & 1 & 1 & 1 \\ 1 & 1 & -1 & -1 \\ 1 & -1 & 1 & -1 \\ 1 & -1 & -1 & 1 \end{pmatrix}$.

4505. Prove that the functions

$$\frac{1}{\sqrt{2}}, \cos x, \sin x, \ldots, \cos nx, \sin nx$$

constitute an orthonormal eigenbasis for the symmetric operator $\dfrac{d^2}{dx^2}$ on the space

$$V_n = \{a_0 + a_1 \cos x + b_1 \sin x + \cdots + a_n \cos nx + b_n \sin nx \mid a_i, b_i \in \mathbb{R}\}$$

with the scalar product $\dfrac{1}{\pi} \displaystyle\int_{-\pi}^{\pi} f(x)g(x)\, dx$.

4506. Prove that *Legendre polynomials* (see Exercise 4343) constitute an eigenbasis for the selfadjoint operator given by the rule

$$(A(f))(x) = (x^2 - 1) f''(x) + 2x f'(x),$$

on the space of polynomials of degree $\leq n$ with the scalar product $\int_{-1}^{1} f(x)g(x)\,dx$.

4507. Find an orthonormal eigenbasis and the matrix in this basis of the Hermitian operator given in some orthonormal basis by the matrix:

a) $\begin{pmatrix} 3 & 2+2i \\ 2-2i & 1 \end{pmatrix}$;   b) $\begin{pmatrix} 3 & -i \\ i & 3 \end{pmatrix}$;   c) $\begin{pmatrix} 3 & 2-i \\ 2+i & 7 \end{pmatrix}$.

4508. In the space of matrices $\mathbf{M}_n(\mathbb{C})$ let

$$(A, B) = \operatorname{tr}(A \cdot {}^t\bar{B}).$$

Prove that

a) $\mathbf{M}_n(\mathbb{C})$ is an Hermitian space;

b) any unitary matrix in this space has the length $\sqrt{n}$;

c) the operators $X \mapsto AX$ and $X \mapsto {}^t\bar{A}X$ on $\mathbf{M}_n(\mathbb{C})$ are adjoint;

d) the operator $X \mapsto AX$ is unitary if A is an unitary matrix.

4509. Prove that selfadjoint operators on Eucidean or Hermitian spaces are commuting if and only if they have a common orthonormal eigenbasis.

4510. Prove that a selfadjoint linear operator on Euclidean or Hermitian space

a) is non-negative if and only if all its eigenvalues are non-negative;

b) is positive if and only if all its eigenvalues are positive.

4511. Let $\mathcal{A}$ be an operator on Euclidean or Hermitian space. Prove that $\mathcal{A}^*\mathcal{A}$ is a non-negative selfadjoint operator. Prove that $\mathcal{A}^*\mathcal{A}$ is positive if and only if $\mathcal{A}$ is invertible.

4512. Prove that if two non-negative selfadjoint operators on Euclidean or Hermitian space commute, then their product is a non-negative selfadjoint operator.

4513. Prove that for any non-negative (positive) selfadjoint operator $\mathcal{A}$ on Euclidean or Hermitian space there exists a non-negative (positive, respectively) operator $\mathcal{B}$ such that $\mathcal{B}^2 = \mathcal{A}$.

4514. Let an operator $\mathcal{A}$ on three-dimensional Euclidean space in some orthonormal basis be given by the matrix

$$\begin{pmatrix} 13 & 14 & 4 \\ 14 & 24 & 18 \\ 4 & 18 & 29 \end{pmatrix}.$$

Find, in this basis, the matrix of a positive selfadjoint operator $\mathcal{B}$ such that $\mathcal{B}^2 = \mathcal{A}$.

4515. Prove that the eigenvalues of the product of two non-negative selfadjoint operators on Euclidean or Hermitian space, one of which is invertible, are real and non-negative.

4516. Prove that a non-negative selfadjoint operator of rank r on Euclidean or Hermitian space is a sum of r non-negative selfadjoint operators of rank 1.

4517. Prove that any linear operator A on Hermitian space has a unique decomposition $A = A_1 + iA_2$ where A_1 and A_2 are Hermitian operators.

4518. Let A be a real Jacobi matrix, i.e. a matrix of the form

$$\begin{pmatrix} \alpha_1 & \beta_1 & 0 & \cdots & 0 \\ \beta_1 & \alpha_2 & \beta_2 & \cdots & 0 \\ 0 & \beta_2 & \alpha_3 & \cdots & 0 \\ \vdots & \vdots & \vdots & \ddots & \vdots \\ 0 & 0 & 0 & \cdots & \beta_{n-1} \\ 0 & 0 & 0 & \cdots & \alpha_n \end{pmatrix},$$

where $\beta_1 \cdot \ldots \cdot \beta_{n-1} \neq 0$. Prove that A has no multiple eigenvalues.

4519. Find an orthogonal transformation reducing the quadratic function to the principal axes:

a) $6x_1^2 + 5x_2^2 + 7x_3^2 - 4x_1x_2 + 4x_1x_3$;

b) $11x_1^2 + 5x_2^2 + 2x_3^2 + 16x_1x_2 + 4x_1x_3 - 20x_2x_3$;

c) $x_1^2 + x_2^2 + 5x_3^2 - 6x_1x_2 - 2x_1x_3 + 2x_2x_3$;

d) $x_1^2 + x_2^2 + x_3^2 + 4x_1x_2 + 4x_1x_3 + 4x_2x_3$;

e) $x_1^2 - 5x_2^2 + x_3^2 + 4x_1x_2 + 2x_1x_3 + 4x_2x_3$;

f) $2x_1x_2 - 6x_1x_3 - 6x_2x_4 + 2x_3x_4$;

g) $3x_1^2 + 8x_1x_2 - 3x_2^2 + 4x_3^2 - 4x_3x_4 + x_4^2$;

h) $x_1^2 + 2x_1x_2 + x_2^2 - 2x_3^2 - 4x_3x_4 - 2x_4^2$;

i) $9x_1^2 + 5x_2^2 + 5x_3^2 + 8x_4^2 + 8x_2x_3 - 4x_2x_4 + 4x_3x_4$;

j) $4x_1^2 - 4x_2^2 - 8x_2x_3 + 2x_3^2 - 5x_4^2 + 6x_4x_5 + 3x_5^2$.

4520. Prove that if $f(x) = \sum_{i=1}^{r} \lambda_i x_i^2$ then

$$\max(|\lambda_1|, \ldots, |\lambda_r|) = \max_{|x|=1} |f(x)|.$$

4521. Reduce the Hermitian quadratic function to the principal axes:

a) $5|x_1|^2 + i\sqrt{3}x_1\bar{x}_2 - i\sqrt{3}\bar{x}_1x_2 + 6|x_2|^2.$

b) $2|x_1|^2 + |x_2|^2 + 2ix_1\bar{x}_2 - 2i\bar{x}_1x_2 + 2i\bar{x}_2x_3 - 2ix_2\bar{x}_3.$

c) $|x_1|^2 + 2|x_2|^2 + 3|x_3|^2 - 2\bar{x}_1x_2 + 2ix_1\bar{x}_2 + 2i\bar{x}_2x_3 - 2ix_2\bar{x}_3.$

46   Orthogonal and unitary operators. Polar factorization

4601. Prove that orthogonal (unitary) operators form a group with respect to multiplication.

4602. Prove that if an operator on Euclidean (Hermitian) space preserves the lengths of vectors then it is orthogonal (unitary, respectively).

4603. Prove that if vectors x and y on Euclidean (Hermitian) space have the same length, then there exists an orthogonal (unitary, respectively) operator which maps x to y.

4604. Let $x_1, \ldots, x_k$ and $y_1, \ldots, y_k$ be two systems of vectors in Euclidean (Hermitian) space. Prove that there exists an orthogonal (unitary, respectively) operator which maps x_i to y_i ($i = 1, \ldots, k$), if and only if $(x_i, x_j) = (y_i, y_j)$ for all i and j from 1 to k.

4605.

a) Let w be a nonzero vector in Euclidean (Hermitian) space. For any vector x we put $U_w(x) = x - 2\dfrac{(x, w)}{(w, w)}w$. Prove that $U_w(w) = -w$ and $U_w(y) = y$ if $x \in \langle w \rangle^{\perp}$.

b) Let x, y be nonzero vectors in Euclidean (Hermitian) space and $y \notin \langle x \rangle$. Prove that there exists a vector w such that $U_w(x) = \dfrac{\|x\|}{\|y\|}y$.

4606. Find a canonical basis, and the matrix in this basis, of the orthogonal operator given in some orthonormal basis by the matrix:

a) $\dfrac{1}{3}\begin{pmatrix} 2 & 2 & -1 \\ 2 & -1 & 2 \\ -1 & 2 & 2 \end{pmatrix};$

b) $\dfrac{1}{2}\begin{pmatrix} 1 & 1 & -\sqrt{2} \\ 1 & 1 & \sqrt{2} \\ \sqrt{2} & -\sqrt{2} & 0 \end{pmatrix};$

c) $\dfrac{1}{3}\begin{pmatrix} 2 & -1 & 2 \\ 2 & 2 & -1 \\ -1 & 2 & 2 \end{pmatrix};$

d) $\dfrac{1}{4}\begin{pmatrix} 3 & 1 & -\sqrt{6} \\ 1 & 3 & \sqrt{6} \\ \sqrt{6} & -\sqrt{6} & 2 \end{pmatrix};$

e) $\dfrac{1}{2}\begin{pmatrix} 1 & -\sqrt{2} & -1 \\ 1 & \sqrt{2} & -1 \\ \sqrt{2} & 0 & \sqrt{2} \end{pmatrix}$;

f) $\dfrac{1}{2}\begin{pmatrix} 1 & 1 & 1 & 1 \\ 1 & 1 & -1 & -1 \\ 1 & -1 & 1 & -1 \\ 1 & -1 & -1 & 1 \end{pmatrix}$;

g) $\dfrac{1}{2}\begin{pmatrix} 1 & 1 & 1 & 1 \\ 1 & 1 & -1 & -1 \\ -1 & 1 & -1 & 1 \\ -1 & 1 & 1 & -1 \end{pmatrix}$;

h) $\dfrac{1}{3}\begin{pmatrix} 2 & 2 & -1 \\ -1 & 2 & 2 \\ 2 & -1 & 2 \end{pmatrix}$;

i) $\dfrac{1}{9}\begin{pmatrix} 1 & -8 & 4 \\ 4 & 4 & 7 \\ -8 & 1 & 4 \end{pmatrix}$;

j) $\dfrac{1}{7}\begin{pmatrix} 3 & -2 & 6 \\ 6 & 3 & -2 \\ -2 & 6 & 3 \end{pmatrix}$;

k) $\begin{pmatrix} \dfrac{1}{\sqrt{2}} & 0 & -\dfrac{1}{\sqrt{2}} \\ \dfrac{1}{3\sqrt{2}} & \dfrac{4}{3\sqrt{2}} & \dfrac{1}{3\sqrt{2}} \\ \dfrac{2}{3} & -\dfrac{1}{3} & \dfrac{2}{3} \end{pmatrix}$;

l) $\begin{pmatrix} \dfrac{3}{4} & \dfrac{1}{4} & +\dfrac{\sqrt{6}}{4} \\ \dfrac{1}{4} & \dfrac{3}{4} & -\dfrac{\sqrt{6}}{4} \\ -\dfrac{\sqrt{6}}{4} & +\dfrac{\sqrt{6}}{4} & \dfrac{1}{2} \end{pmatrix}$.

4607. Find an orthonormal eigenbasis, and the matrix in this basis, of the unitary operator given in some orthonormal basis by the matrix:

a) $\begin{pmatrix} \cos\alpha & -\sin\alpha \\ \sin\alpha & \cos\alpha \end{pmatrix}$ $(\alpha \neq k\pi)$;

b) $\dfrac{1}{\sqrt{3}}\begin{pmatrix} 1+i & 1 \\ -1 & 1-i \end{pmatrix}$;

c) $\dfrac{1}{9}\begin{pmatrix} 4+3i & 4i & -6-2i \\ -4i & 4-3i & -2-6i \\ 6+2i & -2-6i & 1 \end{pmatrix}$;

d) $\dfrac{1}{4}\begin{pmatrix} 2+3i & -\sqrt{3} \\ \sqrt{3} & 2-3i \end{pmatrix}$;

e) $\dfrac{1}{\sqrt{2}}\begin{pmatrix} i & 1 \\ -1 & -i \end{pmatrix}$.

4608. Prove that a unitary matrix of size 2 with determinant 1 is similar to a real orthogonal matrix.

4609. Let A be a unitary operator on Hermitian space and the operator $A - \mathcal{E}$ be invertible. Prove that the operator $i(A - \mathcal{E})^{-1}(A + \mathcal{E})$ is Hermitian.

4610. Let A be an Hermitian operator. Prove that

a) the operator $A - i\mathcal{E}$ is invertible;

b) the operator $B = (A - i\mathcal{E})^{-1}(A + i\mathcal{E})$ is unitary;

c) the operator $B - \mathcal{E}$ is invertible;

d) $A = i(B - \mathcal{E})^{-1}(B + \mathcal{E})$.

4611. Prove that for any Hermitian operator $\mathcal{A}$ the operator $e^{i\mathcal{A}}$ is unitary, and conversely, any unitary operator is presentable in the form $e^{i\mathcal{A}}$ for some Hermitian operator $\mathcal{A}$.

4612. Let V be Euclidean space with a basis (e_1, e_2, e_3) and $\mathcal{A}$ be an orthogonal operator on V with determinant 1. Prove that $\mathcal{A} = \mathcal{A}_\varphi \mathcal{B}_\theta \mathcal{A}_\psi$ where $\mathcal{A}_\varphi$ and $\mathcal{A}_\psi$ are rotations in the plane $\langle e_1, e_2 \rangle$ on angles φ and ψ, respectively, and $\mathcal{B}_\theta$ is a rotation in the plane $\langle e_2, e_3 \rangle$ on the angle θ.

4613. Let V be the space of Hermitian matrices of size 2 over the field $\mathbb{R}$ with zero trace and $(A, B) = \operatorname{tr} AB$ $(A, B \in V)$. Prove that

a) V is Euclidean space with the orthonormal basis

$$e_1 = \frac{1}{\sqrt{2}} \begin{pmatrix} 1 & 0 \\ 0 & -1 \end{pmatrix}, \quad e_2 = \frac{1}{\sqrt{2}} \begin{pmatrix} 0 & 1 \\ 1 & 0 \end{pmatrix}, \quad e_3 = \frac{1}{\sqrt{2}} \begin{pmatrix} 0 & i \\ -i & 0 \end{pmatrix};$$

b)   the operator given by the rule $X \mapsto A X^t \bar{A}$ $(X \in V)$, where A is a unitary matrix, is orthogonal;

c)   for any orthogonal operator $\mathcal{A}$ on V, there exists a unitary matrix A of size 2 with determinant 1 such that $\mathcal{A}(X) = A X^t \bar{A}$ for all $X \in V$.

4614. Prove that any orthogonal operator $\mathcal{A}$ on Euclidean space is a product of reflections with respect to hyperplanes, and the minimal number of these factors is equal to the codimension of the subspace $\operatorname{Ker}(\mathcal{A} - \mathcal{E})$.

4615. Prove that if $\mathcal{A}$, $\mathcal{B}$ are positive selfadjoint operators, $\mathcal{A} = \mathcal{B}\mathcal{C}$ and the operator $\mathcal{C}$ is orthogonal (unitary), then $\mathcal{C} = \mathcal{E}$.

4616. Factorize an operator, given in some orthonormal basis by the matrix:

a) $\begin{pmatrix} 2 & -1 \\ 2 & 1 \end{pmatrix};$ b) $\begin{pmatrix} 1 & -4 \\ 1 & 4 \end{pmatrix};$ c) $\begin{pmatrix} 4 & -2 & 2 \\ 4 & 4 & -1 \\ -2 & 4 & 2, \end{pmatrix}.$

into the product of a positive selfadjoint and an orthogonal operator.

4617. Prove that the factorization $\mathcal{A} = \mathcal{B}\mathcal{C}$ of an operator on Euclidean (Hermitian) space where $\mathcal{B}$ is a non-negative selfadjoint symmetric (Hermitian, respectively) operator, and $\mathcal{C}$ is an orthogonal (unitary, respectively) operator, is unique.

4618. Prove that for any unitary operator $\mathcal{A}$ and for any natural number k, there exists a unitary operator $\mathcal{B}$ which is a polynomial in $\mathcal{A}$ such that $\mathcal{B}^k = \mathcal{A}$.

$$* \quad * \quad *$$

4619. Prove that a selfadjoint operator A is positive when the coefficients $c_1, \ldots, c_n$ of its characteristic polynomial $t^n + c_1 t^{n-1} + \cdots + c_n$ are non-zero and have alternating signs.

4620. Let A, B be selfadjoint operators and A be positive. Prove that the eigenvalues of the operator AB are real.

4621. Let A be a positive and B be a non-negative operators. Prove that the eigenvalues of AB real and non-negative.

4622. Let A be selfadjoint operator. Prove that the following conditions are equivalent:

a) all eigenvalues of A belong to a segment $[a, b]$;

b) the operator $A - \lambda \mathcal{E}$ is negative for $\lambda > b$ and is positive for $\lambda < a$.

4623. Let A, B be selfadjoint operators whose eigenvalues belong to segments $[a, b]$ and $[c, d]$, respectively. Prove that the eigenvalues of $A + B$ belong to the segment $[a + c, b + d]$.

4624. Let A be a selfadjoint operator. Prove that the operator e^A is positive and selfadjoint.

4625. Let $A = B\mathcal{U}$ be a polar factorization of an operator A where B is a non-negative selfadjoint operator, and $\mathcal{U}$ is a unitary operator. Prove that A is normal if and only if $B\mathcal{U} = \mathcal{U}B$.

4626. Let $A = B\mathcal{U}$ be a polar factorization of an operator A, where B is a non-negative selfadjoint operator, and $\mathcal{U}$ is a unitary operator. Assume that $\lambda_1 \geq \cdots \geq \lambda_n \geq 0$ are the eigenvalues of B. Consider the norm on the space of operators corresponding to the norm on the Hermitian space given in Exercise 4205b. Prove that

a) $\|A\| = \lambda_1$;

b) if the operator A is invertible then $\lambda_n > 0$ and $\|A^{-1}\| = \dfrac{1}{\lambda_n}$.

4627. Let A be a nonsingular square complex matrix of size n. Consider the system of linear equations $AX = b$. Let X_0 be its solution and X_1 be its approximation, and $r = b - AX_1$ be the vector of asidual error. Prove that

$$\frac{\|X_0 - X_1\|}{\|X_0\|} \leq \|A\| \cdot \|A^{-1}\| \cdot \frac{\|r\|}{\|b\|}.$$

4628. Let A be a square complex matrix. Prove that $A = U_1 D U_2$, where U_1, U_2 are unitary matrices, and D is a diagonal matrix. Prove that the entries of the principal diagonal of D are the square roots of the eigenvalues of the matrix $A \cdot {}^t A$.

4629. Let $A = (a_{ij})$ be a complex square matrix of size n. Prove that

a) $\det(A \cdot {}^t\bar{A}) \le \left(\sum\limits_{i=1}^{n} |a_{1i}|^2 \right) \cdots \left(\sum\limits_{i=1}^{n} |a_{ni}|^2 \right)$;

b) $|\det A| \le n^{\frac{n}{2}} \cdot (\max\limits_{ij} |a_{ij}|)^n$;

c)   the estimate mentioned in b) is sharp.

4630. Let $A \in \mathbf{M}_n(\mathbb{C})$. Prove that $A = UR$, where U is a unitary matrix, and R is an upper-triangular one. If $A \in \mathbf{M}_n(\mathbb{R})$ then $A = QR$, where Q is an orthogonal and R is a real upper-triangular matrix.

4631. Let $A \in \mathbf{M}_n(\mathbb{C})$. Prove that ${}^t\bar{A}A = {}^t\bar{R}R$ where R is an upper-triangular matrix. If $A \in \mathbf{M}_n(\mathbb{R})$ then R can be chosen in $\mathbf{M}_n(\mathbb{R})$.

4632. Prove that any unitary matrix is the product of a real orthogonal and a complex symmetric matrix.

4633. Let V be a complex vector space with a scalar product (we take the identity automorphism of the field $\mathbb{C}$). Prove that for any symmetric operator $\mathcal{A}$ on V, there exists a Jordan basis in which the matrix of the scalar product is block-diagonal with blocks of the same sizes as the Jordan boxes of the matrix of $\mathcal{A}$. These blocks are of the form

$$\begin{pmatrix} 0 & 0 & \ldots & 0 & 1 \\ 0 & 0 & \ldots & 1 & 0 \\ \multicolumn{5}{c}{\ldots\ldots\ldots\ldots\ldots} \\ 0 & 1 & \ldots & 0 & 0 \\ 1 & 0 & \ldots & 0 & 0 \end{pmatrix}.$$

CHAPTER 11

Tensors

47 Basic concepts

In this section V is n-dimensional vector space, $n \geq 2$, $(e_1, \ldots, e_n)$ is a basis of V, and $(e^1, \ldots, e^n)$ is the dual basis of the space V^*.

4701. Which tensors, given by their coordinates, are decomposable:

a) $t_{ij} = ij$;

b) $t^i_j = \delta_{1i} j$;

c) $t^{ij} = i + j$;

d) $t^k_{ij} = 2^{i+j+k^2}$;

e) $t^{ij}_k = \delta_{ij} \delta_{jk}$;

f) $t_{ijk} = \delta_{ij} \delta_{jk} \delta_{k1}$?

4702. Find the value $F(v, f)$ of the tensor

$$F = e^1 \otimes e_2 + e^2 \otimes (e_1 + 3e_3) \in \mathbf{T}^1_1(V),$$

where $v = e_1 + 5e_2 + 4e_3$, $f = e^1 + e^2 + e^3$.

4703. Find the value of the tensor $A \otimes B - B \otimes A \in \mathbf{T}^0_5(V)$ at the tuple $(v_1, \ldots, v_5)$:

a) $A = e^1 \otimes e^2 + e^2 \otimes e^3 + e^2 \otimes e^2 \in \mathbf{T}^0_2(V)$,
$B = e^1 \otimes e^1 \otimes (e^1 - e^3) \in \mathbf{T}^0_3(V)$,
$v_1 = e_1$, $v_2 = e_1 + e_2$, $v_3 = e_2 + e_3$, $v_4 = v_5 = e_2$;

b) $A = e^1 \otimes e^2 + e^2 \otimes e^3 + e^3 \otimes e^1 \in \mathbf{T}^0_2(V)$,
$B \in \mathbf{T}^0_3(V)$, all coordinates of the tensor B are equal to 1, and
$v_1 = e_1 + e_2$, $v_2 = e_2 + e_3$, $v_3 = e_3 + e_1$, $v_4 = v_5 = e_2$.

4704. Find the value $F(v, v, v, f, f)$ of the tensor $F =\in T_3^2(V)$ if all coordinates of F are equal to 3, and $v = e_1 + 2e_2 + 3e_3 + 4e_4$, $f = e^1 - e^4$.

4705. Find the coordinate $\tilde{t}_{123}^{12}$ of the tensor $T \in T_3^2(V)$ with the basis

$$(\tilde{e}_1, \tilde{e}_2, \tilde{e}_3) = (e_1, e_2, e_3) \begin{pmatrix} 1 & 2 & 3 \\ 0 & 1 & 2 \\ 0 & 0 & 1 \end{pmatrix},$$

if all its coordinates with the basis (e_1, e_2, e_3) are equal to 2.

4706. Find the coordinates with indices 1, 2, 3, 3, 3 of the products $A \otimes B$ and $B \otimes A$ of the tensors

$$A = e^1 \otimes e^2 + e^3 \otimes e^3 \in T_2^0(V),$$
$$B = B(v_1, v_2, v_3) \in T_3^0(V),$$

where $B(v_1, v_2, v_3)$ is the determinant composed of the coordinates v_1, v_2, v_3 with the basis (e_1, e_2, e_3).

4707. Find the coordinates:

a) $\tilde{t}_{21}^1$ of the tensor $e^1 \otimes e^2 \otimes (e_1 + e_2) \in T_2^1(V)$ with the basis

$$(\tilde{e}_1, \tilde{e}_2) = (e_1, e_2) \begin{pmatrix} 1 & 1 \\ 2 & 3 \end{pmatrix};$$

b) $\tilde{t}_1^{12}$ of the tensor $T \in T_1^2(V)$ with the basis

$$(\tilde{e}_1, \tilde{e}_2) = (e_1, e_2) \begin{pmatrix} 1 & 2 \\ 2 & 5 \end{pmatrix};$$

if all its coordinates are equal to 1 with the basis (e_1, e_2);

c) $\tilde{t}_{31}^{12}$ of the tensor $e^2 \otimes e^1 \otimes e_3 \otimes e_1 + e^3 \otimes e^3 \otimes e_1 \otimes e_2 \in T_2^2(V)$ with the basis

$$(\tilde{e}_1, \tilde{e}_2, \tilde{e}_3) = (e_1, e_2, e_3) \begin{pmatrix} 1 & 0 & 0 \\ 2 & 1 & 0 \\ 3 & 2 & 1 \end{pmatrix}.$$

4708. Find the coordinates of the tensors:

a) $(e_1 + e_2) \otimes (e_1 - e_2)$;

b) $(e_1 + e_2) \otimes (e_1 + e_2)$;

c) $(e_1 + 2e_2) \otimes (e_1 + e_2) - (e_1 + e_2) \otimes (e_1 + 2e_2)$;

d) $(e_1 + 2e_2) \otimes (e_3 + e_4) - (e_1 - 2e_2) \otimes (e_3 - e_4)$.

4709. Let $n = 4$, $T = e^1 \otimes e_2 + e^2 \otimes e_3 + e^3 \otimes e_4 \in T_1^1(V)$. Find all

a) $f \in V^*$ such that $T(v, f) = 0$ for any $v \in V$;

b) $v \in V$ such that $T(v, f) = 0$ for any $f \in V^*$.

4710. Let $n = 3$, the field $K = Z_p$ and $T = e^1 \otimes e_2 + e^2 \otimes e_3 \in T_1^1(V)$. Find the number of pairs $(v, f) \in V \times V^*$, for which $T(v, f) = 0$.

4711. Find the rank of the bilinear functions:

a) $(e^1 + e^2) \otimes (e^1 + e^3) - e^1 \otimes e^1 - e^2 \otimes e^2$;

b) $(e^1 - 2e^3) \otimes (e^1 + 3e^2 - e^4) + (e^1 - 2e^3) \otimes e^4$;

c) $(e^1 + e^3) \otimes (e^2 + e^4) - (e^2 - e^4) \otimes (e^1 - e^3)$.

4712. Prove that

a) the rank of the bilinear function $u \otimes v$ is equal to 1, if the elements $u, v \in V^*$ are different from 0;

b) the rank of the bilinear function $\sum_{i=1}^k u_i \otimes v_i$ does not exceed k, if $u_i, \dots, u_k, v_1, \dots, v_k \in V^*$.

4713. Find the total contraction of the tensors:

a) $(e_1 + 3e_2 - e_3) \otimes (e^1 - 2e^3 + 3e^4) - (e_1 + e_3) \otimes (e^1 - 3e^3 + e^4)$;

b) $(e_1 + 2e_2 + 3e_3) \otimes (e^1 + e^2 - 2e^3) - (e_1 - e_2 + e_4) \otimes (e^2 - 2e^3 - 3e^4)$;

c) $e_1 \otimes (e^1 + e^2 + e^3 + e^4) + e_2 \otimes (e^1 + 2e^2 + 2e^3 + 4e^4) + 2e_3 \otimes (e^1 - e^2 - e^4)$.

4714. Let $\alpha : V^* \otimes V \rightarrow L(V)$ be a canonical isomorphism. Calculate $\alpha(t)v$ for $n = 4$, where

a) $t = e^1 \otimes e_3$, $\quad v = e_1 + e_2 + e_3 + e_4$;

b) $t = (e^1 + e^2) \otimes (e_3 + e_4)$, $\quad v = 2e_1 + 3e_2 + 2e_3 + 3e_4$.

4715. Find $x \in V^* \otimes V$, such that $\alpha(x) = \alpha(t)^2$ for t equal to

a) $(2e^1 - e^3) \otimes (e_1 + e_2)$;

b) $e^1 \otimes e_2 + (e^1 + 2e^2) \otimes e_3$.

4716. Suppose that on a space V there is given a scalar product with the matrix

$$\begin{pmatrix} 2 & 1 & 0 & 0 \\ 1 & 1 & 0 & 0 \\ 0 & 0 & 1 & 1 \\ 0 & 0 & 1 & 2 \end{pmatrix}.$$

Make the lowering and the rising indices of tensors:

a) $e^1 \otimes e_3 + e^2 \otimes e_4$;

b) $(e^1 + e^2) \otimes (e_3 + e_4) - (e^1 + e^3) \otimes e_3$;

c) $t^i_j = \delta_{2i} + \delta_{4j}$;

d) $t^i_j = i\delta_{ij}$.

4717. Prove that if an operator $\mathcal{A}$ is diagonalizable then the operator $\mathcal{A}^{\otimes k}$ is also diagonalizable.

4718. Let a be the trace of an operator $\mathcal{A}$ and d be its determinant. Find:

a) $\mathrm{tr}(\mathcal{A} \otimes \mathcal{A})$;   b) $\mathrm{tr}(\mathcal{A}^{\otimes k})$;   c) $\det(\mathcal{A} \otimes \mathcal{A})$.

4719. Find the Jordan form of the matrix of the operator $\mathcal{A} \otimes \mathcal{B}$, if the matrices of $\mathcal{A}$ and $\mathcal{B}$ respectively, have the Jordan forms:

a) $\begin{pmatrix} 1 & 1 \\ 0 & 1 \end{pmatrix}$, $\begin{pmatrix} 1 & 0 & 0 \\ 0 & 2 & 0 \\ 0 & 0 & 3 \end{pmatrix}$;

b) $\begin{pmatrix} 1 & 1 \\ 0 & 1 \end{pmatrix}$, $\begin{pmatrix} 2 & 1 \\ 0 & 2 \end{pmatrix}$;

c) $\begin{pmatrix} 1 & 1 \\ 0 & 1 \end{pmatrix}$, $\begin{pmatrix} 0 & 1 & 0 \\ 0 & 0 & 1 \\ 0 & 0 & 0 \end{pmatrix}$.

48 Symmetric and skew-symmetric tensors

4801. Establish an isomorphism between the spaces $(T^p_q(V))^*$ and $T^q_p(V)$.

4802. Prove the following properties of the operators Sym and Alt on the space $T^q_0(V)$:

a) the intersection of the kernels KerSym and KerAlt is equal to zero if $q = 2$ and is different from zero if $q > 2$;

b) $\text{Sym} \cdot \text{Alt} = \text{Alt} \cdot \text{Sym} = 0$;

c)   the operator $\mathcal{P} = (\mathcal{E} - \text{Sym})(\mathcal{E} - \text{Alt})$ is a projection.

Find the rank of the operator $\mathcal{P}$ for $q = 3$.

4803. Prove that if the basic field has characteristic 0 then the span of tensors of the form v^k ($v \in V$) coincides with $S^k(V)$.

4804. Establish an isomorphism between:

a) $S^q(V_1 \oplus V_2)$ and $\oplus_{i=1}^{q} S^i(V_1) \otimes S^{q-i}(V_2)$;

b) $\Lambda^q(V_1 \oplus V_2)$ and $\oplus_{i=1}^{q} \Lambda^i(V_1) \otimes \Lambda^{q-i}(V_2)$.

4805. Prove that if $\dim V > 2$ then the spaces $\Lambda^2(\Lambda^2(V))$ and $\Lambda^4(V)$ do not coincide.

4806. Prove that for any nonsingular bilinear function f on a space V there exists a nonsingular bilinear tensor F in the space $\Lambda^2 V$ such that

$$F(v_1 \wedge v_2, v_3 \wedge v_4) = \det \begin{pmatrix} f(v_1, v_3) & f(v_1, v_4) \\ f(v_2, v_3) & f(v_2, v_4) \end{pmatrix}.$$

4807. Find the trace of an operator $\Lambda^q(\mathcal{A})$, where the operator $\mathcal{A}$ is given by the matrix:

a) $\begin{pmatrix} 1 & 1 & 0 \\ 0 & 2 & 2 \\ 0 & 0 & 3 \end{pmatrix}$ $(q = 2)$;   b) $\begin{pmatrix} 1 & -2 & 0 & 0 \\ 1 & 4 & 0 & 0 \\ 0 & 0 & -4 & 4 \\ 0 & 0 & -3 & 1 \end{pmatrix}$ $(q = 4)$;

c) $\begin{pmatrix} 1 & 0 & 1 & 2 \\ 0 & 2 & 1 & 0 \\ 1 & 0 & 1 & 0 \\ 0 & 0 & 1 & 3 \end{pmatrix}$ $(q = 2, 3)$.

4808. Find the Jordan form of the matrix of the operator $\Lambda^2(\mathcal{A})$ if the matrix of $\mathcal{A}$ has the Jordan form:

a) $\begin{pmatrix} 1 & 1 & 0 & 0 \\ 0 & 1 & 1 & 0 \\ 0 & 0 & 1 & 1 \\ 0 & 0 & 0 & 1 \end{pmatrix}$;   b) $\begin{pmatrix} 2 & 1 & 0 & 0 \\ 0 & 2 & 0 & 0 \\ 0 & 0 & 3 & 1 \\ 0 & 0 & 0 & 3 \end{pmatrix}$;   c) $\begin{pmatrix} 2 & 0 & 0 & 0 & 0 \\ 0 & -2 & 0 & 0 & 0 \\ 0 & 0 & -2 & 0 & 0 \\ 0 & 0 & 0 & 1 & 1 \\ 0 & 0 & 0 & 0 & 1 \end{pmatrix}$.

4809. Prove that if $\text{tr} \Lambda^q(\mathcal{A}) = 0$ for all $q > 0$ then the operator $\mathcal{A}$ is nilpotent.

4810. Let $\mathcal{A}$ be a nonzero operator on n-dimensional space V. Prove that a nonzero operator $\Lambda^{n-1}(\mathcal{A})$ on $\Lambda^{n-1}(V)$ either is nonsingular or has rank 1.

4811. Prove that k-dimensional subspace $W \subseteq V$ is invariant under a linear operator $\mathcal{A}$ if and only if $\Lambda^k W$ is invariant under $\Lambda^k(\mathcal{A})$.

4812. Prove that for any bivector $\xi \in \Lambda^2(V)$ there exists a basis $(e_1, \ldots, e_n)$ of V such that

$$\xi = e_1 \wedge e_2 + e_3 \wedge e_4 + \cdots + e_{k-1} \wedge e_k$$

for some even integer k.

4813. *Cartan's lemma.* Let a system $v_1, \ldots, v_k$ of vectors of a space V be linearly independent and $t_1, \ldots, t_k \in V$. Prove that $v_1 \wedge t_1 + \cdots + v_k \wedge t_k = 0$ if and only if $t_1, \ldots, t_k \in \langle v_1, \ldots, v_k \rangle$ and the matrix composed by elements α_{ij} such that $t_i = \sum_{j=1}^{k} \alpha_{ij} v_j$, is symmetric.

4814. Prove that a bivector ξ is decomposable if and only if $\xi \wedge \xi = 0$.

4815. Prove that for $\xi \in \Lambda^p(V)$, $x \in V$, $x \neq 0$, the equality $\xi \wedge x = 0$ holds if and only if $\xi = x \wedge \theta$ for some $\theta \in \Lambda^{p-1}(V)$.

4816. Let $\xi \in \Lambda^p(V)$ be a nonzero p-vector and $W = \{x \in V | \xi \wedge x = 0\}$. Prove that

a) $\dim W \leq p$;

b) $\dim W = p$ if and only if ξ is decomposable;

c) the least subspace whose pth power contains the p-vector ξ is equal to $U = \{\xi(v, \ldots, v_{p-1}) | v_i \in V^*\}$;

d) $\dim U \geq p$ and $\dim U = p$ if and only if ξ is decomposable.

4817. Prove that the operation of internal multiplication $i(v^*)$ where $v^* \in V^*$, is a *differentiation of the algebra* $S(V)$.

4818. Prove that the operators of internal multiplication $i(v_1^*)$ and $i(v_2^*)$ $(v_1^*, v_2^* \in V^*)$ commute in the algebra $S(V)$ and anticommute in the algebra $\Lambda(V)$.

CHAPTER 12

Affine, Euclidean,
and projective geometry

49 Affine spaces

4901. Prove that $\overline{ab} + \overline{bc} = \overline{ac}$ for any points a, b, c of an affine space.

4902. Prove that if $\sum_{i=1}^{k} \lambda_i = 0$, then the vector $\sum_{i=1}^{k} \lambda_i \overline{aa_i}$ does not depend on the point a for any points $a_1, \ldots, a_k$ of an affine space.

4903. Prove that if $\sum_{i=1}^{k} \lambda_i = 1$, then the point $a + \sum_{i=1}^{k} \lambda_i \overline{aa_i}$ (denoted by $\sum_{i=1}^{i} \lambda_i a_i$) does not depend on the point a for any points $a_1, \ldots, a_k$ of an affine space.

4904. Let (P, U) be an affine subspace (a plane) of an affine space. Prove that

a) $U = \{\overline{pq} \mid p, q \in P\}$;

b) $P = p + U$ for any point $p \in P$.

4905. Prove that the intersection of any family of planes in an affine space is either the empty set or a plane.

4906. Let S be a nonempty subset of an affine space A. Prove that

a) the subset $\langle S \rangle = a + \langle \overline{ax} \mid x \in S \rangle$, where $a \in S$, does not depend on a and it is the least plane, containing S;

b) $\langle S \rangle = \left\{ \sum_{i=1}^{k} \lambda_i a_i \;\middle|\; \sum_{i=1}^{k} \lambda_i = 1, \; a_i \in S, \; k \in \mathbb{N} \right\}$.

4907. Prove that a subset of an affinely independent set is affinely independent.

4908. Prove that any maximal affinely independent subset of a set S in an affine space contains $k + 1$ points, where $k = 1 + \dim \langle S \rangle$.

4909. Let, in an affine space (A, V), two systems of affine coordinates be given: $(a, e_i, \ldots, e_n)$, $(a', e'_1, \ldots, e'_n)$. Suppose that $(a_1, \ldots, a_n)$ are the coordinates of the point a' in the first system, $B = (b_{ij})$ is the matrix of the change of the basis $(e_1, \ldots, e_n)$ to the basis $(e'_1, \ldots, e'_n)$ in the vector space V. Express the coordinates $(x_1, \ldots, x_n)$ of a point $x \in A$ in the first system via its coordinates $(x'_1, \ldots, x'_n)$ in the second system and vice versa.

4910. Find a system of equations and parametric equations determining the affine hull of the set:

a) $(-1, 1, 0, 1)$, $(0, 0, 2, 0)$, $(-3, -1, 5, 4)$, $(2, 2, -3, -3)$;

b) $(1, 1, 1, -1)$, $(0, 0, 6, -7)$, $(2, 3, 6, -7)$, $(3, 4, 1, -1)$.

4911. Let $a_i = (a_{i1}, \ldots, a_{in})$ $(i = 1, \ldots, s)$ be points in n-dimensional affine space. Prove the inequalities

$$\mathrm{rk}(a_{ij}) - 1 \leq \dim\langle a_1, \ldots, a_s\rangle \leq \mathrm{rk}(a_{ij}).$$

Under what conditions does each inequality turn into an equality.

4912. Prove that any two lines in affine space are contained in a three-dimensional plane.

4913. Let $P_1 = a_1 + L_1$, $P_2 = a_2 + L_2$ be two planes in affine space. Prove that

a) $P_1 \cap P_2 = \emptyset$ if and only if $\overline{a_1 a_2} \notin L_1 + L_2$;

b) if $P_1 \cap P_2 \neq \emptyset$, then

$$\dim\langle P_1 \cup P_2\rangle = \dim P_1 + \dim P_2 - \dim(P_1 \cap P_2);$$

c) if $P_1 \cap P_2 = \emptyset$, then

$$\dim\langle P_1 \cup P_2\rangle = \dim P_1 + \dim P_2 - \dim(L_1 \cap L_2) + 1.$$

4914. Prove that for any planes $P_1, \ldots, P_s$ of affine space

$$\dim\langle P_1 \cup \cdots \cup P_s\rangle \leq \dim P_1 + \cdots + \dim P_s + s - 1.$$

4915. Prove that the *degree of parallelism* of two disjoint planes P_1, P_2 is equal to:

a) the greatest number k for which there exist parallel planes $Q_1 \subseteq P_1$ and $Q_2 \subseteq P_2$ of dimension k;

b) the greatest dimension of a plane contained in P_1 and parallel to P_2, if $\dim P_1 \leq \dim P_2$.

4916. Find the dimension of the affine hull of the union of planes P_1 and P_2 and the dimension of their intersections or the degree of parallelism, if

a) $\quad P_1 : 3x_1 + 2x_2 + 2x_3 + 2x_4 = 2, \quad P_2 : 2x_1 + 2x_2 + 3x_3 + 4x_4 = 5,$
$\qquad\quad 2x_1 + 3x_2 + 2x_3 + 5x_4 = 3, \qquad\quad 5x_1 - \ x_2 + 3x_3 - 5x_4 = 2;$

b) $\quad P_1 : 2x_1 + 3x_2 + 4x_3 + 5x_4 = 6, \quad P_2 : x_1 = 1 - t_1,$
$\qquad\quad 6x_1 + 5x_2 + 4x_3 + 3x_4 = 2, \qquad\quad x_2 = 1 + 2t_1 + t_2,$
$\qquad\qquad\qquad\qquad\qquad\qquad\qquad\qquad\quad x_3 = 1 - 2t_1 + 2t_2,$
$\qquad\qquad\qquad\qquad\qquad\qquad\qquad\qquad\quad x_4 = 1 + t_1 + t_2;$

c) $\quad P_1 : 3x_1 = 1 + 2t_1, \qquad\qquad\quad P_2 : x_1 = -6 + 4t,$
$\qquad\qquad x_2 = 3 + 2t_2, \qquad\qquad\qquad\quad x_2 = 2 + 3t,$
$\qquad\qquad x_3 = 5 + 4t_2, \qquad\qquad\qquad\quad x_3 = 2 + 7t,$
$\qquad\qquad x_4 = 4 + 3t_1 + 2t_2, \qquad\qquad x_4 = -2 + 5t,$
$\qquad\qquad x_5 = 2 + t_1 + 2t_2, \qquad\qquad\quad x_5 = -3 + 3t.$

4917. Let $P_1 = a_1 + L_1$ and $P_2 = a_2 + L_2$ be two disjoint planes. Prove that the minimal dimension of a plane, containing P_1 and parallel to P_2 is equal to

$$\dim P_1 + \dim P_2 - \dim(L_1 \cap L_2).$$

4918. Let P_1, P_2 be two planes in affine space A over a field K, $\langle P_1 \cup P_2 \rangle = A$, $P_1 \cap P_2 = \emptyset$ and let λ be the fixed element of K, $\lambda \neq 0, 1$. Find the locus of all points $\lambda a_1 + (1 - \lambda)a_2$, where a_1 and a_2 run over P_1 and P_2, respectively.

4919. Let $P_1 = a_1 + L_1$ and $P_2 = a_2 + L_2$ be skew planes in affine space. Prove that for any point $b \notin P_1 \cup P_2$ there exists at most one line passing through b, intersecting P_1 and P_2. This line exists if and only if $b \in \langle P_1 \cup P_2 \rangle$, but $\overline{a_1 b} \notin L_1 + L_2$ and $\overline{a_2 b} \notin L_1 + L_2$.

4920. Find a line passing through the point b and intersecting the planes P_1 and P_2:

a) $\qquad b = (6, 5, 1, -1),$
$\quad P_1 : -x_1 + 2x_2 + x_3 = 1, \qquad\qquad P_2 : x_1 = 4 + t,$
$\qquad\qquad x_1 + \qquad\ x_4 = 1, \qquad\qquad\qquad\quad x_2 = 4 + 2t,$
$\qquad\qquad\qquad\qquad\qquad\qquad\qquad\qquad\qquad x_3 = 5 + 3t,$
$\qquad\qquad\qquad\qquad\qquad\qquad\qquad\qquad\qquad x_4 = 4 + 4t;$

b)
$$b = (5, 9, 2, 10, 10),$$

$P_1 : x_1 - x_2 - x_4 + x_5 = 2,$ $P_2 : x_1 = 3,$

      $x_1 - x_3 - x_4 + x_5 = 1,$ $x_2 = 2 + 6t_1 + 5t_2,$

      $5x_1 + 3x_2 - 2x_3 - x_5 = 0,$ $x_3 = 0,$

                                    $x_4 = 5 + 4t_1 + 3t_2,$

                                    $x_5 = 6 + t_1 + 2t_2;$

c)
$$b = (6, -1, -5, 1),$$

$P_1 : x_1 = 3 + 2t,$ $P_2 : -6x_1 + 2x_2 - 5x_3 + 4x_4 = 1,$

      $x_2 = 5 - t,$ $9x_1 - x_2 + 6x_3 - 6x_4 = 5.$

      $x_3 = 3 - t,$

      $x_4 = 6 + t,$

4921. Let $a_0, a_1, \ldots, a_n$ be affinely independent points in n-dimensional affine space A. Prove that every point $a \in A$ has a unique decomposition $a = \sum_{i=0}^{n} \lambda_i a_i$, where $\sum_{i=0}^{n} \lambda_i = 1$.

4922. Let $(a_0, e_1, \ldots, e_n)$ be an affine system of coordinates in affine space (A, V), $a_i = a_0 + e_i$ $(i = 1, \ldots, n)$. Find the *barycentric coordinates* of a point $x = (x_1, \ldots, x_n)$ with respect to the system of points $a_0, a_1, \ldots, a_n$.

4923. Let (A, V) be an affine space over a field K, $|K| \geq 3$, and P be a nonempty subset of A. Prove that P is a plane if and only if P contains a line $\langle a, b \rangle$ for any two distinct points $a, b \in P$. Is this statement correct for a field K with two elements?

4924. Prove that an affine transformation has a fixed point, provided 1 is not an eigenvalue of its differential.

4925. Prove that for any two points a, b in an affine space (A, V) and any nonsingular linear operator $\mathcal{A}$ in V there exists a unique affine transformation f of (A, V) such that $f(a) = b$ and $Df = \mathcal{A}$.

4926. Prove that for any affine transformations f and g

$$D(fg) = Df \cdot Dg.$$

4927. Let f be an affine mapping of an affine space A into an affine space B over a field K, and $a_1, \ldots, a_s \in A$, $\alpha_1, \ldots, \alpha_s \in K$. Prove that

a)   if $\sum_{i=1}^{s} \alpha_i = 1,$ then $f\left(\sum_{i=1}^{s} \alpha_i a_i\right) = \sum_{i=1}^{s} \alpha_i f(a_i);$

b)   if $\sum_{i=1}^{s} \alpha_i = 0,$ then $Df\left(\sum_{i=1}^{s} \alpha_i a_i\right) = \sum_{i=1}^{s} \alpha_i f(a_i).$

4928. Let f be an affine transformation of finite order n in affine space (A, V) over a field K. Prove that if $\operatorname{char} K \nmid n$, then f has a fixed point. Is this statement correct when $\operatorname{char} K \mid n$?

4929. Prove that if G is a finite group of affine transformations over a field K and $\operatorname{char} K \nmid |G|$, then transformations from G have a common fixed point.

4930. Let $a_0, a_1, \ldots, a_n$ and $b_0, b_1, \ldots, b_n$ be two sets of affinely independent points in n-dimensional affine space A. Prove that there exists a unique affine transformation $f : A \to A$ for which $f(a_i) = b_i$ $(i = 0, 1, \ldots, n)$.

4931. Find all points, lines and planes in three-dimensional affine space that are invariant under the affine transformation which maps the points a_0, a_1, a_2, a_3 to the points b_0, b_1, b_2, b_3, respectively:

a) $a_0 = (1, 3, 4), \quad a_1 = (2, 3, 4), \quad a_2 = (1, 4, 4), \quad a_3 = (1, 3, 5),$
$b_0 = (3, 4, 3), \quad b_1 = (8, 9, 9),$
$b_2 = (-2, -2, -6), \quad b_3 = (5, 7, 8);$

b) $a_0 = (3, 2, 3), \quad a_1 = (4, 2, 3), \quad a_2 = (3, 3, 3), \quad a_3 = (3, 2, 4),$
$b_0 = (2, 4, 6), \quad b_1 = (1, 8, 12),$
$b_2 = (-1, -5, -1), \quad b_3 = (6, 12, 11);$

c) $a_0 = (2, 5, 1), \quad a_1 = (3, 5, 1), \quad a_2 = (2, 6, 1), \quad a_3 = (2, 5, 2),$
$b_0 = (3, 7, 3), \quad b_1 = (6, 11, 6),$
$b_2 = (5, 17, 9), \quad b_3 = (0, -5, -4);$

d) $a_0 = (2, 5, 4), \quad a_1 = (3, 5, 4), \quad a_2 = (3, 6, 4), \quad a_3 = (2, 5, 5),$
$b_0 = (1, 6, 6), \quad b_1 = (8, 16, 18),$
$b_2 = (-11, -13, -18), \quad b_3 = (7, 16, 19).$

4932. Prove that two *configurations* P_1, P_2 and Q_1, Q_2 in affine space are affine congruent if and only if

$$\dim P_1 = \dim Q_1, \qquad \dim P_2 = \dim Q_2,$$
$$\dim \langle P_1 \cup P_2 \rangle = \dim \langle Q_1 \cup Q_2 \rangle,$$

and both pairs have simultaneously an empty or a nonempty intersection.

4933. Does there exist an affine transformation mapping the points a, b, c to the points a_1, b_1, c_1, respectively, and the line l to the line l_1, if

a) $a = (1, 1, 1, 1), \quad b = (2, 3, 2, 3), \quad c = (3, 2, 3, 2),$
$l = (1, 2, 2, 2) + (0, 1, 0, 1)t,$
$a_1 = (-1, 1, -1, 1), \quad b_1 = (0, 4, 0, 4), \quad c_1 = (2, 2, 2, 2),$
$l_1 = (-1, 2, 0, 3) + (1, -5, 1, -5)t;$

b) $a = (2, -1, 3, -2),$ $b = (3, 1, 6, -1),$ $c = (5, 1, 4, 1),$
$l = (2, 0, 4, -1) + (0, 1, 2, 0)t,$
$a_1 = (1, -2, 3, 5),$ $b_1 = (2, 1, 8, 7),$ $c_1 = (3, 2, 10, -6),$
$l_1 = (1, -1, 5, -2) + (0, 2, 3, -3)t;$

c) $a = (2, -1, 2, 2,),$ $b = (5, -4, 0, 3),$ $c = (4, 4, 6, 8),$
$l = (7, 4, 10, 9) + (4, 4, 5, 6)t,$
$a_1 = (1, 3, 2, -2),$ $b_1 = (4, -2, 0, 0),$ $c_1 = (-3, 10, 6, 2),$
$l_1 = (5, -6, -1, 5) + (2, -6, -3, 2)t?$

4934. Does there exist an affine transformation mapping the points a, b, c, d to the points a_1, b_1, c_1, d_1, respectively, and the line l to the line l_1, if

a)
$a = (1, 2, 3, 4),$	$a_1 = (1, -1, 4, 2),$
$b = (1, 3, 3, 4),$	$b_1 = (2, -2, 5, 3),$
$c = (1, 2, 2, 4),$	$c_1 = (2, 0, 3, 3),$
$d = (1, 2, 3, 3),$	$d_1 = (2, 0, 5, 1)$
$l = (-3, 2, 4, 1) + (2, 1, -1, -2)t,$	$l_1 = (1, -5, 2, -12) + (1, 1, 1, 1)t;$

b)
$a = (-3, 0, 2, 4),$	$a_1 = (-1, 1, 2, 3),$
$b = (-3, 1, 3, 5),$	$b_1 = (1, -4, 3, 5),$
$c = (-2, 0, 3, 5),$	$c_1 = (-4, 8, 1, 7),$
$d = (-2, 1, 2, 5),$	$d_1 = (4, -8, 4, 10),$
$l = (-1, 5, 5, 6) + (1, 1, 1, 0)t,$	$l_1 = (4, 5, -1, 1) + (4, -6, 1, 2)t?$

4935. Let an affine space A be equal to $\langle P_1 \cup P_2 \rangle$, where P_1 and P_2 are skew planes, and let G be a subgroup of the affine group of the space A consisting of transformations under which P_1 and P_2 are invariant. Find the orbits of the action of G in A.

4936. Let (A, V) be an affine space over a field K. A bijective mapping $f : A \to A$ is a *collineation* if, for any three points $a, b, c \in A$ belonging to one line, the points $f(a), f(b), f(c)$ also belong to one line. Prove that if $|K| \geq 3$, then the image and the preimage of a plane $P \subseteq A$ under a collineation $f : A \to A$ are planes of the same dimension as P. Is the statement correct for $|K| = 2$?

4937. Let V be a vector space over a field K. The mapping $\varphi : V \to V$ is called *semilinear* with respect to some automorphism σ of K if

$$\varphi(x + y) = \varphi(x) + \varphi(y), \quad \varphi(\alpha x) = \sigma(\alpha)\varphi(x), \quad \text{where} \quad x, y \in V, \ \alpha \in K.$$

A *semiaffine transformation* of an affine space (A, V) is a pair (f, Df), where $f : A \to A$, $Df : V \to V$ satisfy the conditions:

(1) Df is a bijective semilinear mapping with respect to some automorphism of K;

(2) $f(a + v) = f(a) + Df(v)$ for every $a \in A$, $v \in V$.

Prove that

a) a semiaffine transformation is a collineation;

b) if (A, V) is an affine space over K and if $|K| \geq 3$, then any collineation $f : A \to A$ is a semiaffine transformation.

4938. Let (B, U) be a plane in an affine space (A, V) and W be a subspace of the space V complemented to U. Prove that any point $a \in A$ has a unique decomposition $a = b + w$, where $b \in B$, $w \in W$ and that the mapping $a \mapsto b$ of the *projection* onto B in parallel with W is an affine mapping of the space (A, V) into the space (B, U).

50 Convex sets

5001. Prove that any plane in an affine space is an intersection of finitely many half-spaces.

5002. Prove that a subset of a plane P in an affine space A is a *convex polyhedron* in P if and only if it is a convex polyhedron in A.

5003. Let a convex polyhedron M in an affine space be given by a system of linear inequalities

$$f_i(x) \geq 0 \quad (i = 1, \ldots, k; \quad f_i \neq \text{const}).$$

For any nonempty subset $J \subseteq \{1, \ldots, k\}$ let M^J denote the nonempty set given by conditions $f_i(x) = 0$ if $i \in J$, $f_i(x) \geq 0$ if $i \notin J$. Prove that nonempty M^J is a side of the polyhedron M and, conversely, any side of the polyhedron M coincides with M^J for some set $J \subseteq \{1, \ldots, k\}$.

5004. Let $a_0, a_1, \ldots, a_n$ be points in n-dimensional affine space in the *general position*, and let H_i $(i = 0, 1, \ldots, n)$ be a hyperplane passing through all these points, except for a_i, and H_i^+ be the half-space containing a_i and bounded by this hyperplane. Prove that

$$\text{conv}\{a_0, a_1, \ldots, a_n\} = \cap_{i=0}^n H_i^+.$$

5005. Prove that the sides of the n-dimensional simplex $\text{conv}\{a_0, a_1, \ldots, a_n\}$ are convex hulls of all possible proper subsets of the set $\{a_0, a_1, \ldots, a_n\}$.

5006. Find the sides of an n-dimensional parallelepiped given in some system of affine coordinates by inequalities $0 \leq x_i \leq 1$ ($i = 1, 2, \ldots, n$).

5007. Find the vertices and describe the form of the convex polyhedron in three-dimensional affine space, given by the inequalities

$$x_1 \le 1, \quad x_2 \le 1, \quad x_3 \le 1,$$
$$x_1 + x_2 \ge -1, \quad x_1 + x_3 \ge -1, \quad x_2 + x_3 \ge -1.$$

5008. Let a four-dimensional parallelepiped be given by the inequalities $0 \le x_i \le 1$ $(i = 1, 2, 3, 4)$. Find its vertices and describe the form of its sections by the planes:

a) $x_1 + x_2 + x_3 + x_4 = 1$;

b) $x_1 + x_2 + x_3 + x_4 = 2$;

c) $x_1 + x_2 + x_3 = 1$;

d) $x_1 + x_2 = x_3 + x_4 = 1$.

5009. Prove that the closure of a convex set is convex.

5010. Prove that the open kernel M^0 of *a solid* convex set M is convex and its closure contains M.

5011. Prove that the image and the preimage of a convex set under an affine map are convex sets.

5012. Prove that under a surjective affine map

a) the preimage of a hyperplane is a hyperplane;

b) the preimage of a half-space is a half-space.

5013. Prove that the convex hull of a set S consists of all possible combinations $\sum_{i=1}^{k} \lambda_i a_i$, where

$$a_i \in S, \quad \lambda_i \ge 0 \quad (i = 1, \ldots, k), \quad \sum_{i=1}^{k} \lambda_i = 1.$$

5014. Let M be a convex set and $a \in M$. Prove that

$$\mathrm{conv}(M \cup a) = \cup_{b \in M} \overline{ab}.$$

5015. Let S be a subset of n-dimensional affine space A. Prove that if $\langle S \rangle = A$, then $\mathrm{conv}\, S$ is a union of n-dimensional simplices whose vertices belong to S.

5016. Prove that the convex hull of a compact set is compact.

5017. Let M be a convex subset of two-dimensional affine space and $a \notin M^0$ (see Exercise 5010). Prove that it is possible to draw a line passing through a such that M belongs to one of the half-planes associated with this line.

5018. Let M be a convex subset of n-dimensional affine space A and a be a point not belonging to M^0 (see Exercise 5010). Prove that it is possible to draw a hyperplane through a such that M belongs to one of the half-planes associated with the hyperplane.

5019. Prove that it is possible to draw a *hyperplane of support* through any point of a closed convex set, not belonging to its open kernel.

5020. Prove that any closed convex set M is equal to an intersection (in general of infinitely many) half-spaces.

5021. Prove that any closed *convex* cone in a vector space is equal to an intersection (in general of infinitely many) half-spaces whose boundaries contain the origin.

5022. Let f_i $(i = 1, \ldots, k)$ be affine linear functions on an affine space A. Prove that the system of inequalities $f_i(x) \leq 0$ $(i = 1, \ldots, k)$ is incompatible if and only if there exist numbers $\lambda_i \geq 0$ such that $\sum_{i=1}^{k} \lambda_i f_i$ is a positive constant.

5023. Let M be a compact convex set containing a neighborhood of the origin in a vector space V, considered as an affine space, and let

$$M^* = \{f \in V^* \mid f(x) \leq 1 \text{ for any } x \in M\}.$$

Prove

  a) M^* is a compact convex set in the space V^* containing a neighborhood of origin;

  b) $M^{**} = M$ under the canonical identification of the space V^{**} with V.

5024. Prove that any compact convex set coincides with the convex hull of the set of its *extreme* points.

5025. Prove that the maximum of an affine linear function on a compact convex set is reached at some extreme point, but possibly it is also reached at some other points.

5026. Prove that the extreme points of a convex polyhedron are its vertices.

5027. Prove that any bounded convex polyhedron coincides with the convex hull of the set of its vertices.

5028. Prove that a convex hull of finitely many points is a convex polyhedron.

5029. Write down a system of linear inequalities defining the convex hull of the given points in four-dimensional affine space and find three-dimensional sides of this convex polyhedron:

  a) $O = (0,0,0,0),\ a = (1,0,0,0),\ b = (0,1,0,0),\ c = (1,1,0,0),$
    $d = (0,0,1,0),\ e = (0,0,0,1),\ f = (0,0,1,1);$

b) $O = (0, 0, 0, 0)$, $a = (1, 0, 0, 0)$, $b = (0, 1, 0, 0)$, $c = (0, 0, 1, 0)$,
       $d = (1, 1, 0, 0)$, $e = (1, 0, 1, 0)$, $f = (0, 1, 1, 0)$, $g = (1, 1, 1, 0)$,
       $h = (0, 0, 0, 1)$.

5030. Let M and N be convex sets of an affine space (A, V). Prove that

a)   the midpoints of intervals connecting points of M with points of N, form
     a convex set in A;

b)   the vectors connecting points of M with points of N form a convex set
     in V.

5031. Let M and N be disjoint closed convex sets in an affine space A, one of
which is bounded. Prove that there exists an affine linear function f on the space
A such that $f(x) < 0$ for all $x \in M$ and $f(y) > 0$ for all $y \in N$.

5032. Let M be a compact convex set of an affine space A and N be a compact
convex set of a vector space L of all affine linear functions on A. Let there exist
for each point $a \in M$ a function $f \in N$ such that $f(a) \geq 0$. Prove that there exists
a function $f_0 \in N$ such that $f_0(x) \geq 0$ for all $x \in M$.

5033. *Duality theorem in linear programming.* Let F be an affine bilinear
function on the direct product of affine spaces A and B. Let M and N be compact
convex subsets of A and B, respectively. Prove that

a) $$\max_{x \in M} \min_{y \in N} F(x, y) = \min_{y \in N} \max_{x \in M} F(x, y);$$

b)   there exist points $x_0 \in M$, $y_0 \in N$ such that for all $x \in M$, $y \in N$

$$F(x, y_0) \leq F(x_0, y_0) \leq F(x_0, y).$$

5034. Prove that

a)   the maximal number of parts (convex polyhedrons) on which n-
     dimensional real affine space can be decomposed by k hyperplanes, is
     equal to

$$\binom{k+1}{n} + \binom{k+1}{n-2} + \binom{k+1}{n-4} + \dots;$$

b)   the number of these parts is maximal if and only if the intersection of
     any m given hyperplanes is an $(n-m)$-dimensional plane (the empty set
     for $m > n$);

c)   if the number of these parts is maximal, then the number of bounded parts
     is equal to $\binom{k-1}{n}$.

5035. Find out if the polyhedrons given by the following inequalities are bounded:

a) $-3x_1 + 5x_2 \leq 10$, $\quad 5x_1 + 2x_2 \leq 35$, $\quad x_1 \geq 0$, $\quad x_2 \geq 0$;

b) $-x_1 + x_2 \leq 2$, $\quad 5x_1 - x_2 \leq 10$;

c) $3x_1 - x_2 \geq 4$, $\quad -x_1 + 3x_2 \geq 4$;

d) $-3x_1 + 4x_2 \leq 17$, $\quad 3x_1 + 4x_2 \leq 47$, $\quad x_1 - x_2 \leq 4$, $\quad x_1 + x_2 \geq 0$;

e) $-x_1 + 2x_2 \leq 6$, $\quad 5x_1 - 2x_2 \leq 26$, $\quad x_1 + 2x_2 \geq 10$;

f) $5x_1 - 2x_2 \geq 6$, $\quad 5x_1 - 2x_2 \leq 36$, $\quad 2 \leq x_1 \leq 7$.

5036. Find the vertices of the polyhedrons:

a) $x_1 + 2x_2 + x_3 + 3x_4 + x_5 = 5$,
$x_1 + x_3 - 2x_4 = 3$,
$x_1 \geq 0$, $\quad x_2 \geq 0$, $\quad x_3 \geq 0$, $\quad x_4 \geq 0$, $\quad x_5 \geq 0$;

b) $x_1 + x_2 - x_3 = 10$,
$x_1 - x_2 + 7x_3 = 7$,
$x_1 \geq 0$, $\quad x_2 \geq 0$, $\quad x_3 \geq 0$;

c) $4x_1 + 5x_2 + x_3 + x_4 = 29$,
$6x_1 - x_2 - x_3 + x_4 = 11$,
$x_1 \geq 0$, $\quad x_2 \geq 0$, $\quad x_3 \geq 0$, $\quad x_4 \geq 0$;

d) $x_1 + 2x_2 + x_3 = 4$,
$2x_1 + 2x_2 + 5x_3 = 5$,
$x_1 \geq 0$, $\quad x_2 \geq 0$, $\quad x_3 \geq 0$.

5037. Find the maximal and minimal values of the linear functions z on the bounded polyhedrons:

a) $x_1 + 2x_2 + x_3 + 3x_4 + x_5 = 5$,
$2x_1 + x_3 - 2x_4 = 3$,
$x_1 \geq 0$, $\quad x_2 \geq 0$, $\quad x_3 \geq 0$, $\quad x_4 \geq 0$, $\quad x_5 \geq 0$,
$z = x_1 - 2x_2 + x_3 + 3x_5$;

b) $3x_1 - x_2 + 2x_3 + x_4 + x_5 = 12$,
$x_1 - 5x_2 - x_4 + x_5 = -4$,
$x_1 \geq 0$, $\quad x_2 \geq 0$, $\quad x_3 \geq 0$, $\quad x_4 \geq 0$, $\quad x_5 \geq 0$,
$z = 4x_1 - x_2 + 2x_3 + x_5$;

c) $5x_1 + 2x_2 - x_3 + x_4 + x_5 = 42$,
$4x_1 - 4x_2 + x_3 + x_4 = 16$,
$x_1 \geq 0$, $\quad x_2 \geq 0$, $\quad x_3 \geq 0$, $\quad x_4 \geq 0$, $\quad x_5 \geq 0$,
$z = x_1 - 2x_2 + 4x_4 - x_5$;

d) $x_1 - 3x_2 + x_3 + 2x_5 = 8,$
     $4x_2 - 3x_4 - x_5 = 3,$
     $x_1 \geq 0, \quad x_2 \geq 0, \quad x_3 \geq 0, \quad x_4 \geq 0, \quad x_5 \geq 0,$
     $z = x_1 - 2x_2 + x_3 - x_5.$

51 Euclidean spaces

5101. Find the necessary and sufficient conditions under which a given set $\binom{n}{2}$ of non-negative real numbers is a set of distances between

a) *n affinely independent points* in Euclidean space;

b) *n arbitrary points* in Euclidean space.

5102. Does there exist a set of points a_1, a_2, a_3, a_4, a_5, in Euclidean space such that A is the matrix of distances $(\rho(a_i, a_j))$? What is the smallest dimension of a space in which this set can be placed:

a) $A = \begin{pmatrix} 0 & 1 & 2 & 2 & 2\sqrt{2} \\ 1 & 0 & \sqrt{5} & \sqrt{5} & 3 \\ 2 & \sqrt{5} & 0 & 2\sqrt{2} & 2 \\ 2 & \sqrt{5} & 2\sqrt{2} & 0 & 2\sqrt{3} \\ 2\sqrt{2} & 3 & 2 & 2\sqrt{3} & 0 \end{pmatrix};$

b) $A = \begin{pmatrix} 0 & 3 & \sqrt{5} & \sqrt{5} & 2\sqrt{2} \\ 3 & 0 & \sqrt{14} & \sqrt{14} & \sqrt{17} \\ \sqrt{5} & \sqrt{14} & 0 & \sqrt{2} & \sqrt{17} \\ \sqrt{5} & \sqrt{14} & \sqrt{2} & 0 & 3 \\ 2\sqrt{2} & \sqrt{17} & \sqrt{17} & 3 & 0 \end{pmatrix};$

c) $A = \begin{pmatrix} 0 & 1 & 2 & \sqrt{5} & 1 \\ 1 & 0 & \sqrt{5} & 2 & \sqrt{2} \\ 2 & \sqrt{5} & 0 & \sqrt{17} & 1 \\ \sqrt{5} & 2 & \sqrt{17} & 0 & \sqrt{10} \\ 1 & \sqrt{2} & 1 & \sqrt{10} & 0 \end{pmatrix};$

d)
$$A = \begin{pmatrix} 0 & \sqrt{5} & \sqrt{5} & \sqrt{5} & \sqrt{5} \\ \sqrt{5} & 0 & 2\sqrt{5} & 2\sqrt{2} & 2 \\ \sqrt{5} & 2\sqrt{5} & 0 & 2 & 2\sqrt{2} \\ \sqrt{5} & 2\sqrt{2} & 2 & 0 & 2\sqrt{5} \\ \sqrt{5} & 2 & 2\sqrt{2} & 2\sqrt{5} & 0 \end{pmatrix}.$$

5103. Prove the equivalence of the following two properties of a pair of planes $\{P, Q\}$ in Euclidean space:

a) any line belonging to one of these planes is perpendicular to any line belonging to the other plane;

b) the planes P, Q are perpendicular and either they are skew or they have only one common point.

5104. Let $Q \subset P$ be planes in n-dimensional Euclidean space E. Prove that any plane $P' \subset E$, which is perpendicular to P and $P \cap P' = Q$, has the dimension $\leq n - \dim P + \dim Q$. There exists a unique plane of dimension $n - \dim P + \dim Q$ with this property.

5105. Let P be a plane in Euclidean space and $a \notin P$. Prove that

a) there exists a unique line passing through a, intersecting P and perpendicular to P;

b) if c is a point in P and z is the orthogonal component of the vector $\overline{ac}$ with the respect to the directing subspace of the plane P, then $a + \langle z \rangle$ is the line mentioned in a) and $a + z$ is the point of the intersection of this line with P;

c) $\rho(a, P) = |z|$.

5106. Find the line in Euclidean space, passing through the point a, intersecting the plane P and perpendicular to P, if

a) $a = (5, -4, 4, 0)$, $\quad P = (2, -1, 2, 3) + \langle (1, 1, 1, 2), (2, 2, 1, 1) \rangle$;

b) $a = (5, 0, 2, 11)$, $\quad P : x_1 + 5x_2 + x_4 = 10,$
$$5x_1 + x_2 + 3x_3 + 8x_4 = -1.$$

5107. Find in Euclidean space the distance between the point a and the plane P, if

a) $a = (4, 1, -4, -5)$, $\quad P = (3, -2, 1, 5) + \langle (2, 3, -2, -2), (4, 1, 3, 2) \rangle$;

b) $a = (1, 1, -2, -3, -2)$, $P = (3, 7, -5, 4, 1) + \langle(1, 1, 2, 0, 1),$
$(2, 2, 1, 3, 1)\rangle$;

c) $a = (2, 1, -3, 4)$, $P : 2x_1 - 4x_2 - 8x_3 + 13x_4 = -19$,
$x_1 + x_2 - x_3 + 2x_4 = 1$;

d) $a = (1, -3, -2, 9, -4)$, $P : x_1 - 2x_2 - 3x_3 + 3x_4 + 2x_5 = -2$,
$x_1 - 2x_2 - 7x_3 + 5x_4 + 3x_5 = 1$.

5108. Find in n-dimensional Euclidean space the distance between the point $(b_1, \ldots, b_n)$ and the hyperplane $\sum_{i=1}^{n} a_i x_i = c$.

5109. In the space of polynomials with scalar product

$$(f, g) = \int_{-1}^{1} f(x)g(x)\, dx$$

find the distance between the polynomial x^n and the subspace of polynomials of degree less than n.

5110. In the space of trigonometrical polynomials with scalar product

$$(f, g) = \int_{-\pi}^{\pi} f(x)g(x)\, dx$$

find the distance between the function $\cos^{n+1} x$ and the subspace

$$\langle 1, \cos x, \sin x, \ldots, \cos nx, \sin nx \rangle.$$

5111. Let P be a plane in n-dimensional Euclidean space E. Prove that there exists a unique plane Q of dimension $n - \dim P$ passing through a point $a \in E$, perpendicular to P and intersecting it at only one point.

5112. Find in Euclidean space the plane of the maximal dimension, passing through a point a, perpendicular to the plane P and intersecting it at only one point, if

a) $a = (2, -1, 3, 5)$, $P = (7, 2, -3, 4) + \langle(-1, 3, 2, 1), (1, 2, 3, -1)\rangle$;

b) $a = (3, -2, 1, 4)$, $P : 2x_1 + 3x_2 - x_3 - 2x_4 = 4$,
$3x_1 + 2x_2 - 5x_3 + x_4 = 5$.

5113. Let $P_1 = c_1 + L_1$ and $P_2 = c_2 + L_2$ be two disjoint planes in Euclidean space, y and z be the orthogonal projection and the orthogonal component of the vector $\overline{c_1 c_2}$ with respect to the subspace $L_1 + L_2$, respectively, and let $y = y_1 + y_2$, where $y_1 \in L_1$, $y_2 \in L_2$.

a) Prove that the line $c_1 + y_1 + \langle z \rangle$ is perpendicular to the planes P_1, P_2 and intersects P_1 at the point $c_1 + y_1$, and P_2 at the point $c_2 - y_2$.

b) Find the distance $\rho(P_1, P_2)$.

c) Establish the bijective correspondence between $L_1 \cap L_2$ and the set of all lines which are perpendicular to P_1 and P_2 and intersect both planes.

d) Show that all lines described in c) are parallel and that their union is a plane of dimension $\dim(L_1 \cap L_2) + 1$.

5114. Find in Euclidean space the distance between the planes P_1 and P_2, if

a) $\qquad P_1 : x_1 + 3x_2 + x_3 + x_4 = 3,$
$\qquad\qquad x_1 + 3x_2 - x_3 + 2x_4 = 6,$
$\qquad P_2 = (0, 2, 6, -5) + \langle(-7, 1, 1, 1), (-10, 1, 2, 3)\rangle;$

b) $\qquad P_1 : -x_1 + x_2 + x_3 + x_4 = 3,$
$\qquad\qquad -3x_2 + 2x_3 - 4x_4 = 4,$
$\qquad P_2 = (1, 3, -3, -1) + \langle(1, 0, 1, 1)\rangle;$

c) $\qquad P_1 : x_1 + x_3 + x_4 - 2x_5 = 2,$
$\qquad\qquad x_2 + x_3 - x_4 - x_5 = 3,$
$\qquad\qquad x_1 - x_2 + 2x_3 - x_5 = 3,$
$\qquad P_2 = (1, -2, 5, 8, 2) + \langle(0, 1, 2, 1, 2), (2, 1, 2, -1, 1)\rangle;$

d) $\qquad P_1 : x_1 - 2x_2 + x_3 - x_4 + 3x_5 = 6,$
$\qquad\qquad x_1 - x_3 - x_4 + 3x_5 = 0,$
$\qquad P_2 = (-4, 3, -3, 2, 4) + \langle(2, 0, 1, 1, 1), (-5, 1, 0, 1, 1)\rangle.$

5115. The distance between any two of the points $a_0, a_1, \ldots, a_n$ in Euclidean space is equal to d. Find the distance between the planes $\langle a_0, a_1, \ldots, a_k \rangle$ and $\langle a_{k+1}, \ldots, a_n \rangle$.

5116. Prove that the configurations of two planes

$$\{a_1 + L_1, a_2 + L_2\} \text{ and } \{a_1' + L_1', a_2' + L_2'\}$$

in Euclidean space are metrically congruent if and only if $\rho(a_1 + L_1, a_2 + L_2) = \rho(a_1' + L_1', a_2' + L_2')$ and the configurations of the subspaces L_1, L_2, L_1', L_2' are orthogonally congruent in the corresponding Euclidean vector space.

5117. Find out if the given pairs of planes in Euclidean spaces are metrically congruent:

$$P_1 = (0, 9, 8, -12, 11) + \langle (0, 2, 2, 2, 1), (3, 1, 1, 1, -1) \rangle,$$
$$P_2 = (-3, -4, -5, 11, -12) + \langle (7, 5, -5, -1, -5), (3, 5, -1, 11, 13) \rangle;$$

$$Q_1 = (2, -5, -11, -8, -10) + \langle (2, -1, 1, -1, 1), (2, -2, 1, 0, 1) \rangle,$$
$$Q_2 = (8, 8, 10, 9, 11) + \langle (0, 3, 4, -4, -3), (14, -2, -5, 3, 4) \rangle;$$

$$R_1 = (7, -3, -9, -14, 5) + \langle (0, 0, 0, 1, 2), (2, -1, 2, 0, -6) \rangle,$$
$$R_2 = (0, 10, 9, 14, -5) + \langle (1, 7, 2, 0, 6), (4, -1, 0, 2, -2) \rangle.$$

5118. Find the locus of points in Euclidean space such that there exists a line passing through these points and intersecting planes P_1 and P_2, perpendicular to P_1 and P_2:

a) $P_1 = (1, 2, -1, -9, -13) + \langle (2, 3, 7, 10, 13), (3, 5, 11, 16, 21) \rangle,$
$$P_2 : 3x_1 - 5x_2 + 2x_3 - x_4 + x_5 = -22,$$
$$2x_1 + 4x_2 + 3x_3 - x_4 - 3x_5 = -4,$$
$$9x_1 + 3x_2 + x_3 - 2x_4 - 2x_5 = -138;$$

b) $P_1 = (3, 7, 2, 4, -3) + \langle (2, 5, 4, 5, 3), (4, 5, 6, 3, 3) \rangle,$
$$P_2 : -3x_1 + 2x_2 + x_3 - 2x_4 + x_5 = -14,$$
$$6x_1 - x_2 - 4x_3 + 2x_4 - x_5 = 16,$$
$$2x_1 - x_2 + 2x_4 - 3x_5 = 26.$$

5119. Prove that

a) if an isometry of Euclidean space has two skew invariant planes, then it has a fixed point;

b) an isometry f of n-dimensional Euclidean space with a fixed point has two skew invariant planes of positive dimensions if either f is proper and $n \geq 5$ is odd or f is improper and $n \geq 4$ is even.

5120. Let $a_0, a_1, \ldots, a_s$ and $b_0, b_1, \ldots, b_s$ be two sets of points in Euclidean space. Prove that there exists an isometry, mapping each one of the points a_i to the point b_i, respectively, if and only if

$$\rho(a_i, a_j) = \rho(b_i, b_j) \ (i, j = 1, \ldots, s).$$

5121. Prove that, for any isometry f of Euclidean space, the set of points a such that the distance $\rho(a, f(a))$ is minimal, form a plane invariant under f and the restriction of f to this plane is a parallel transfer.

5122. Prove that if two tetrahedrons in three-dimensional Euclidean space have equal corresponding dihedral angles, then these tetrahedrons are similar.

5123. Find the geometrical description of the proper isometry f of a Euclidean plane if

a) $$Df = \begin{pmatrix} 0 & 1 \\ -1 & 0 \end{pmatrix}, \qquad\qquad f(O) = (-2, 4);$$

b) $$Df = \frac{1}{\sqrt{2}} \begin{pmatrix} 1 & -1 \\ 1 & 1 \end{pmatrix}, \qquad\qquad f(O) = (1, 1).$$

5124. Find the geometrical description of the improper isometry f of a Euclidean plane if

a) $$Df = \begin{pmatrix} 0 & 1 \\ 1 & 0 \end{pmatrix}, \qquad\qquad f(O) = (1, 0);$$

b) $$Df = \frac{1}{2} \begin{pmatrix} 1 & \sqrt{3} \\ \sqrt{3} & -1 \end{pmatrix}, \qquad\qquad f(O) = (1, -\sqrt{3}).$$

5125. Find the geometrical description of the proper isometry f of three-dimensional Euclidean space if

a) $$Df = \frac{1}{3} \begin{pmatrix} 2 & -1 & 2 \\ 2 & 2 & -1 \\ -1 & 2 & 2 \end{pmatrix}, \qquad f(O) = (1, 0, -1);$$

b) $$Df = \frac{1}{9} \begin{pmatrix} 4 & 1 & -8 \\ 7 & 4 & 4 \\ 4 & -8 & 1 \end{pmatrix}, \qquad f(O) = (-1, -7, 2);$$

c) $$Df = \frac{1}{7} \begin{pmatrix} -2 & 3 & 6 \\ 6 & -2 & 3 \\ 3 & 6 & -2 \end{pmatrix}, \qquad f(O) = (-2, 4, 1).$$

5126. Find the geometrical description of the improper isometry f of three-dimensional Euclidean space if

a) $$Df = -\frac{1}{9} \begin{pmatrix} 4 & 1 & -8 \\ 7 & 4 & 4 \\ 4 & -8 & 1 \end{pmatrix}, \qquad f(O) = (1, 1, -2);$$

b) $$Df = \frac{1}{3} \begin{pmatrix} 2 & 2 & -1 \\ 2 & -1 & 2 \\ -1 & 2 & 2 \end{pmatrix}, \qquad f(O) = (4, 0, 2);$$

c) $Df = \dfrac{1}{3}\begin{pmatrix} -1 & 2 & 2 \\ -2 & 1 & -2 \\ 2 & 2 & -1 \end{pmatrix},$ $f(O) = (2, 0, 0);$

d) $Df = \dfrac{1}{7}\begin{pmatrix} -2 & 3 & 6 \\ 3 & 6 & -2 \\ 6 & -2 & 3 \end{pmatrix},$ $f(O) = (-3, 1, 2).$

52   Hypersurfaces of second degree

The notations and concepts used in the exercises in this section can be found in the section Theoretical material II.

5201. Prove that for any $x, y \in V$ the equality

$$Q(a_0 + x + y) = q(y) + 2f(x, y) + l(y) + Q(a_0 + x)$$

holds.

5202. Prove that if $b = a_0 + v$ ($v \in V$) is a central point of a quadratic function Q, then $Q(b + x) = Q(b - x)$ for all $x \in V$ and the linear function $y \mapsto 2f(v, y) + l(y)$ is zero.

5203. Prove that the set of central points (the *center*) of a quadratic function Q is given by the system of equations $\dfrac{\partial Q}{\partial x_i} = 0$ $(i = 1, \ldots, n)$.

5204. Let the passing from an affine system of coordinates $(a_0, e_1, \ldots, e_n)$ to a system of coordinates $(a_0, e'_i, \ldots, e'_n)$ be given by the formula

$$\begin{pmatrix} x_1 \\ \vdots \\ x_n \end{pmatrix} = T\begin{pmatrix} x'_1 \\ \vdots \\ x'_n \end{pmatrix} + \begin{pmatrix} t_1 \\ \vdots \\ t_n \end{pmatrix}.$$

Prove that the matrices of quadratic forms Q and q in the new system of coordinates are connected with their matrices in the old system by the formulae

$$\tilde{A}'_Q = {}^t\tilde{T}\tilde{A}_Q\tilde{T}, \quad A'_q = {}^tTA_qT,$$

where

$$\tilde{T} = \left(\begin{array}{c|c} T & \begin{matrix} t_1 \\ \vdots \\ t_n \end{matrix} \\ \hline 0\ldots0 & 1 \end{array}\right)$$

is the matrix of the affine change of coordinates.

5205. Prove that the points of intersection of an affine line $x_k = x_k^0 + r_k t$ $(k = 1, \ldots, n)$ with a quadric $Q(x_1, \ldots, x_n) = 0$ are determined by the values t which satisfy the equation

$$At^2 + 2Bt + C = 0,$$

where

$$A = q(r) = \sum_{i,j=1}^{n} a_{ij} r_i r_j, \quad C = Q(x_1^0, \ldots, x_n^0),$$

$$B = \sum_{i=1}^{n} \frac{\partial Q}{\partial x_i}(x_1^0, \ldots, x_n^0) r_i = \sum_{i,j=1}^{n} (a_{ij} x_j^0 + b_i) r_i.$$

5206. Find the center of the quadratic function over the field $\mathbb{R}$ given in some affine system of coordinates:

a) $$2 \sum_{1 \le i < j \le n} x_i x_j + 2 \sum_{i=1}^{n} x_i + 1;$$

b) $$x_1^2 + 2 \sum_{1 \le i < j \le n} x_i x_j + 2 \sum_{i=1}^{n} x_i + 1;$$

c) $$\sum_{i=1}^{n-1} x_i x_{i+1} + x_1 + x_n + 1;$$

d) $$\sum_{i=1}^{n} x_i^2 + 2 \sum_{1 \le i < j \le n} x_i x_j + x_1.$$

5207. Two quadratic functions $Q_i : A \to K$ $(i = 1, 2)$ are *equivalent* if there exists an affine transformation $f : A \to A$, such that $Q_2(x) = \lambda Q_1(f(x))$ for some $\lambda \in K^*$ and for all $x \in A$. Find the number of classes of equivalent quadratic functions over the fields $\mathbb{Z}_3$ if

a) the dimension of A is equal to two;

b) the dimension of A is equal to three.

5208. Find the number of classes of equivalent quadratic functions on n-dimensional affine space:

a) over the field $\mathbb{C}$;

b) over the field $\mathbb{R}$.

5209. Let a point a_0 in affine space (A, V) belong to a quadric X and a vector $u \in V$ determine an asymptotic direction. Prove that the line $x = a_0 + tu$ either lies entirely on the surface X, or intersects the quadric at only one point.

5210. Let $u \in V$ be a nonasymptotic vector for a quadric X_Q, that is $q(u) \neq 0$. Prove that the midpoints of chords of X_Q, which are parallel to u belong to the same hyperplane. Find its equation.

5211. Prove that the direction u is not asymptotic for a quadric X given by the equation in affine coordinates. Find the equation of the hyperplane conjugate to this direction:

a) $u = (1, 1, 1, 1),$ $X : x_1 x_2 + x_2 x_3 + x_3 x_4 - x_1 - x_4 = 0;$

b) $u = (1, 0, \ldots, 0, 1),$ $\displaystyle\sum_{1 \leq i < j \leq n} x_i x_j + x_1 + x_n = 1.$

5212. Prove that if a center of a quadric is nonempty then it is contained in a hyperplane conjugate to any nonasymptotic direction.

5213. Prove that the set of singular points of a quadric is equal to its intersection with its center.

5214. Prove that singular points of a quadric, if they exist, form a plane. Write down its equations.

5215. Find the points of the intersection of the quadric with the line:

a) $x_3^2 + x_1 x_2 - x_2 x_3 - x_1 = 0,$

$$x_1 = \frac{x_2 - 5}{3} = \frac{x_3 - 10}{7};$$

b) $5x_1^2 + 9x_2^2 + 9x_3^2 - 12x_1 x_2 - 6x_1 x_3 + 12x_1 - 36x_3 = 0,$

$$\frac{x_1}{3} = \frac{x_2}{2} = x_3 - 4;$$

c) $x_1^2 - 2x_2^2 + x_3^2 - 2x_1 x_2 - x_2 x_3 + 4x_1 x_3 + 3x_1 - 5x_3 = 0,$

$$\frac{x_1 + 3}{2} = x_2, \quad x_3 = 0.$$

5216. Find all lines which belong to the quadric

$$x_1^2 + x_2^2 + 5x_3^2 - 6x_1 x_2 + 2x_2 x_3 - 2x_1 x_3 - 12 = 0$$

and which are parallel to the line

$$\frac{x_1 - 1}{2} = \frac{x_2 + 3}{1} = -x_3.$$

5217. Find all lines passing through the origin and lying on the complex quadric

$$x_1^2 + 3x_1x_2 + 2x_2x_3 - x_1x_3 + 3x_1 + 2x_3 = 0.$$

5218. Find the equation of the quadric Q after the shift of the origin to the point O':

a) $Q : x_1^2 + 5x_2^2 + 4x_3^2 + 4x_1x_2 - 2x_2x_3 - 4x_1x_3 - 2x_1 - 10x_2 + 4x_3 = 0$,
$O' = (3, 0, 1)$;

b) $Q : x_1^2 + 2x_2^2 + x_3^2 - 4x_1x_2 + 6x_2x_3 - 2x_1x_3 + 10x_1 - 5 = 0$,
$O' = (-1, 1, 2)$.

5219. Find the affine type of the curve which is the intersection of the quadric and the plane:

a)
$$3x_2^2 + 4x_3^2 + 24x_1 + 12x_2 - 72x_3 + 360 = 0,$$
$$x_1 - x_2 + x_3 = 1;$$

b)
$$x_1^2 + 5x_2^2 + x_3^2 + 2x_1x_2 + 2x_2x_3 + 6x_1x_3 - 2x_1 + 6x_2 + 2x_3 = 0,$$
$$2x_1 - x_2 + x_3 = 0;$$

c)
$$x_1^2 - 3x_2^2 + x_3^2 - 6x_1x_2 + 2x_2x_3 - 3x_2 + x_3 - 1 = 0,$$
$$2x_1 - 3x_2 - x_3 + 2 = 0;$$

d)
$$x_1^2 + x_2^2 + x_3^2 - 6x_1 - 2x_2 + 9 = 0,$$
$$x_1 + x_2 - 2x_3 - 1 = 0.$$

5220. Find the affine and the metric type of the quadric, given in Euclidean space $\mathbb{R}^{n+1}$ by the equations:

a)
$$\sum_{i=1}^{n} x_i^2 + \sum_{1 \le i < j \le n} x_i x_j + x_1 + x_{n+1} = 0;$$

b)
$$\sum_{1 \le i < j \le n} x_i x_j + x_1 + x_2 + \cdots + x_n = 0.$$

5221. Determine the affine type of the quadric and its location with respect to the initial system of coordinates by transforming the left-hand side of its equations. Find the center.

a) $4x_1^2 + 2x_2^2 + 12x_3^2 - 4x_1x_2 + 8x_2x_3 + 12x_1x_3 + 14x_1 - 10x_2 + 7 = 0$;

b) $5x_1^2 + 9x_2^2 + 9x_3^2 - 12x_1x_2 - 6x_1x_3 + 12x_1 - 36x_3 = 0$;

c) $5x_1^2 + 2x_2^2 + 2x_3^2 - 2x_1x_2 - 4x_2x_3 + 2x_1x_3 - 4x_2 - 4x_3 + 4 = 0;$

d) $x_1^2 - 2x_2^2 + x_3^2 + 6x_2x_3 - 4x_1x_3 - 8x_1 + 10x_2 = 0;$

e) $x_1^2 + 2x_1x_2 + x_2^2 - x_3^2 + 2x_3 - 1 = 0;$

f) $3x_1^2 + 3x_2^2 + 3x_3^2 - 6x_1 + 4x_2 - 1 = 0;$

g) $3x_1^2 + 3x_2^2 - 6x_1 + 4x_2 - 1 = 0;$

h) $3x_1^2 + 3x_2^2 - 3x_3^2 - 6x_1 + 4x_2 + 4x_3 + 3 = 0;$

i) $4x_1^2 + x_2^2 - 4x_1x_2 - 36 = 0;$

j) $x_1^2 + 4x_2^2 + 9x_3^2 - 6x_1 + 8x_2 - 36x_3 = 0;$

k) $4x_1^2 - x_2^2 - x_3^2 + 32x_1 - 12x_3 + 44 = 0;$

l) $3x_1^2 - x_2^2 + 3x_3^2 - 18x_1 + 10x_2 + 12x_3 + 14 = 0;$

m) $6x_1^2 + 6x_3^2 + 5x_1 + 6x_2 + 30x_3 - 11 = 0.$

5222. Determine the metric type of the following surfaces in Euclidean space and the location with respect to the initial system of coordinates. Find out whether the surface is a surface of rotation.

a) $x_3^2 = 2x_1x_2;$

b) $x_3 = x_1x_2;$

c) $x_3^2 = 3x_1 + 4x_2;$

d) $x_3^2 = 3x_1^2 + 4x_1x_2;$

e) $x_3^2 = x_1^2 + 2x_1x_2 + x_2^2 + 1;$

f) $x_1^2 + 4x_2^2 + 5x_3^2 + 4x_1x_2 + 4x_3 = 0;$

g) $x_1^2 + 2x_1 + 3x_2 + 4x_3 + 5 = 0;$

h) $x_3 = x_1^2 + 2x_1x_2 + x_2^2 + 1;$

i) $x_1^2 - 2x_2^2 + x_3^2 + 4x_1x_2 - 8x_1x_3 - 4x_2x_3 - 14x_1 - 14x_2 + 14x_3 + 18 = 0;$

j) $5x_1^2 + 8x_2^2 + 5x_3^2 - 4x_1x_2 + 8x_1x_3 + 4x_2x_3 - 6x_1 + 6x_2 + 6x_3 + 10 = 0;$

k) $2x_1x_2 + 2x_1x_3 + 2x_2x_3 + 2x_1 + 2x_2 + 2x_3 + 1 = 0;$

l) $3x_1^2 + 3x_2^2 + 3x_3^2 - 2x_1x_2 - 2x_1x_3 - 2x_2x_3 - 2x_1 - 2x_2 - 2x_3 - 1 = 0;$

m) $2x_1^2 + 6x_2^2 + 2x_3^2 + 8x_1x_3 - 4x_1 - 8x_2 + 3 = 0;$

n) $4x_1^2 + x_2^2 + 4x_3^2 - 4x_1x_2 - 8x_1x_3 + 4x_2x_3 - 28x_1 + 2x_2 + 16x_3 + 45 = 0;$

o) $2x_1^2 + 5x_2^2 + 2x_3^2 - 2x_1x_2 - 4x_1x_3 + 2x_2x_3 + 2x_1 - 10x_2 - 2x_3 - 1 = 0$;

p) $7x_1^2 + 7x_2^2 + 16x_3^2 - 10x_1x_2 - 8x_1x_3 - 8x_2x_3 - 16x_1 - 16x_2 - 8x_3 + 72 = 0$;

q) $4x_1^2 + 4x_2^2 - 8x_3^2 - 10x_1x_2 + 4x_1x_3 + 4x_2x_3 - 16x_1 - 16x_2 + 10x_3 - 2 = 0$;

r) $2x_1^2 - 7x_2^2 - 4x_3^2 + 4x_1x_2 - 16x_1x_3 + 20x_2x_3 + 60x_1 - 12x_2 + 12x_3 - 90 = 0$;

s) $2x_1x_2 + 2x_1x_3 - 2x_1x_4 - 2x_2x_3 + 2x_2x_4 + 2x_3x_4 - 2x_2 - 4x_3 - 6x_4 + 5 = 0$;

t) $3x_1^2 + 3x_2^2 + 3x_3^2 + 3x_4^2 - 2x_1x_2 - 2x_1x_3 - 2x_1x_4 - 2x_2x_3 - 2x_2x_4 - 2x_3x_4$
$= 36$.

5223. For what values of the parameter a is the quadric

$$x_1^2 + x_2^2 + x_3^2 + 2ax_1x_2 + 2ax_1x_3 + 2ax_2x_3 = 4a$$

an ellipsoid?

5224. What are the necessary and sufficient conditions in order that two hyperboloids have a common asymptotic cone?

5225. Find the affine and the metric type of the quadric, given in Euclidean space $\mathbb{R}^{n+1}$ by the equation

$$a \sum_{i=1}^{n} x_i^2 + 2b \sum_{1 \le i < j \le n} x_i x_j + 2c \sum_{i=1}^{n+1} x_i = 0,$$

depending on the values of parameters a, b and c.

5226. A quadric is *k-planar* if for any of its points there exists a k-dimensional plane passing through this point and entirely belonging to the quadric but each $(k + 1)$-dimensional plane is not contained in the quadric. Prove that

a) a quadric of the type $I'_{n,s}$ over $\mathbb{R}$ is k-planar, where $k = \min(s, n - s)$;

b) a nonsingular quadric of the type $I_{n,s}$ over $\mathbb{R}$ is $(s - 1)$-planar, if $0 \le s \le n/2$, and is $(n - s)$-planar, if $s > n/2$;

c) a nonsingular quadric of the type $II_{n,s}$ over $\mathbb{R}$ is s-planar, if $0 \le s \le n/2$, and is $(n - 1 - s)$-planar, if $s > n/2$.

5227. For what values of parameters $a, b, c \neq 0$ does a plane of maximal dimension lie on the quadric

$$a \sum_{i=1}^{n} x_i^2 + 2b \sum_{1 \le i < j \le n} x_i x_j + 2c \sum_{i=1}^{n+1} x_i = 0$$

in the space $\mathbb{R}^{n+1}$? Find the dimension of this plane.

* * *

5228. Let $(e_1, \ldots, e_n)$ be a basis of a vector space V over a field K of characteristic different from 2.

a) Prove that if $n = 4$ then all decomposable elements $v_1 \wedge v_2$ in $\bigwedge^r V$ satisfy a nonsingular homogeneous quadratic equation $Q(x_0, \ldots, x_5) = 0$ (*Plücker quadric*).

b) Prove that all decomposable vectors in the spaces $\bigwedge^r V, 2 \le r \le n - 2$, satisfy a system of homogeneous quadratic equations $Q_i \left(x_0, \ldots, x_{\binom{n}{r}} \right) = 0$.

c) Let Q be a nondegenerate quadratic form in the space V. Then it is possible to introduce in the space $\wedge^p V$ a quadratic form $Q^{(p)}$ by the formulae

$$Q^{(0)} = 1,$$
$$Q^{(p)}(v_1 \wedge \cdots \wedge v_p) = \det \begin{pmatrix} Q(v_1, v_1) & \cdots & Q(v_1, v_p) \\ \cdots\cdots\cdots\cdots\cdots\cdots\cdots \\ Q(v_1, v_1) & \cdots & Q(v_p, v_p) \end{pmatrix}.$$

Prove that the obtained extension of Q to the algebra $\Lambda(V)$ is a nonsingular quadratic form in $\Lambda(V)$.

d) An *orientation* of n-dimensional vector space with a nonsingular quadratic form Q is the element $d \in \wedge^n V$, for which $Q^{(n)}(d) = 1$. Prove that if $\det Q$ is a square in the field K, then in V there are exactly two orientations and for any of them, say, d, it is possible to define an isomorphism of vector spaces $\lambda_d : V \to \wedge^{n-1} V$ satisfying the relation

$$v \wedge x = Q(v, \lambda_d^{-1} x)d = Q^{(n-1)}(\lambda_d v, v)d.$$

e) Applying the isomorphism λ_d from d), let there be defined in the case $\dim V = 3$ a bilinear mapping $V \times V \to V$ by the formula

$$[x, y] = \lambda_d^{-1}(x \wedge y) \quad (x, y \in V).$$

Prove that this multiplication induces in V a structure of a Lie algebra over K.

53   Projective spaces

5301. Find the projective transformation of the plane which maps the first pair of lines to the second pair:

a) $\qquad\qquad x = 0 \mapsto x = 0, \qquad\qquad y = 0 \mapsto x = 1;$

b) $\qquad\qquad x + y = 1 \mapsto x = 1, \qquad\qquad x + y = 0 \mapsto y = 0.$

5302. Find the projective transformation of the plane which maps the given curve to the other:

a) $\qquad\qquad\qquad x^2 + y^2 = 1 \mapsto y = x^2;$

b) $\qquad\qquad\qquad x^2 - y^2 = 1 \mapsto x^2 + y^2 = 1.$

5303. Find the projective transformation of the plane which maps the circle $x^2 + y^2 = 1$ to itself and

a)   the point $(0, 0)$ to the point $\left(\dfrac{1}{2}, 0\right)$;

b)   the line $x = 2$ to the point at infinity.

5304. Find the projective transformation of the space which maps the given quadric to the other:

a) $\qquad\qquad x^2 + y^2 + z^2 = 1 \mapsto xy - z^2 = 1;$

b) $\qquad\qquad\qquad xy = z \mapsto x^2 + y^2 - z^2 = 1;$

c) $\qquad\qquad\qquad xy = z^2 \mapsto y = x^2.$

5305. Find the maximal dimension of the planes which are contained in the quadric:

a) $\qquad\qquad x_1^2 + \cdots + x_k^2 - x_{k+1}^2 \cdots - x_n^2 = 1;$

b) $\qquad\qquad x_1^2 + \cdots + x_k^2 - x_{k+1}^2 \cdots - x_n^2 = x_n.$

5306. Prove that over the field of complex numbers any projective transformation has at least one fixed point.

5307. Prove that in real projective space of even dimension any projective transformation has a fixed point.

5308. Prove that if a projective transformation of n-dimensional projective space over an infinite field has finitely many fixed points, then the number of these points does not exceed $n + 1$.

5309. Prove that for any finite set of points A in the projective space over an infinite field there exists an affine chart containing A.

5310. Prove that any $(k-1)$-dimensional subspace of $\mathbf{P}^n$ can be covered by k affine charts and it cannot be covered by a smaller number of affine charts.

5311. Find the number of points in n-dimensional projective space over a field with q elements.

5312. Find the number of k-dimensional subspaces of n-dimensional projective space over a field with q elements.

5313. Find the number of projective transformations of n-dimensional projective space over a field with q elements.

5314. Let M_1 and M_2 be disjoint planes in $\mathbf{P}^n$, and L_1 and L_2 be disjoint planes of the same dimensions as M_1 and M_2, respectively. Prove that there exists a projective transformation which maps M_1 to L_1 and M_2 to L_2.

5315. Prove that if a projective transformation maps some affine chart to itself then it induces an affine transformation of this chart.

5316. Prove that any bijective transformation of a two-dimensional projective plane is projective if it maps lines to lines and preserves the cross-ratio of points on each line.

5317. Prove that any four lines in a projective plane, any three of which have no common point, can be mapped with the help of a suitable projective transformation to any four lines with the same property.

5318. Prove that there exists a projective transformation of a plane which preserves a given triangle and which maps a given internal point to any other internal given point.

5319. Prove that there exists a projective transformation of a plane which preserves a circle and which maps a given internal point to another given internal point.

5320. Prove that it is impossible to construct the center of a given circle with the help of a straight edge.

5321. Prove with the help of projective transformations that the segments connecting the vertices of a triangle with points A, B, C on opposite sides, have a common point if and only if A, B, C are points of tangency of some ellipse inscribed in the triangle.

5322. An avenue of trees is represented on a picture. Let l denote the distance from the first tree to the line of the horizon along the line of the avenue, and a_k the distance between the kth and $(k+1)$th trees. Express:

a) a_3 in terms of a_1 and a_2;

b) a_2 in terms of l and a_1.

5323. A projective transformation of a plane is a *homology*, if it preserves all points, lying on some line (the *axis of homology*) and all lines, passing through some point (the *center of homology*). Prove that

a) there exists a unique homology with the given axis l and the given center O which maps the given point $A \neq O$, $A \notin l$, to the given point $A' \neq O$, $A' \notin l$, lying on the line OA;

b) any projective transformation of a plane is the product of two homologies.

5324. Prove that there exists a unique projective transformation of a plane preserving the circle $x^2 + y^2 = 1$ and mapping three given points on this circle to three other given points also lying on this circle.

5325. *Desargues theorem.* Prove that if the lines AA', BB', CC' have a common point, then the points of intersections of the lines AB and $A'B'$, BC and $B'C'$, AC and $A'C'$ lie on one line.

5326. *Pascal theorem.* Prove that the points of intersection of opposite sides of a hexagon, inscribed in a circle, lie on one line.

5327. *Pappus theorem.* Prove that the points of intersection of opposite sides of a hexagon, whose vertices belong in turn to the two given lines, lie on one line.

5328. Let a_1, a_2, a_3, a_4 be lines on a plane, passing through the point O, and l be a line not passing through O. Prove that the cross-ratio of the points of intersection of the lines a_1, a_2, a_3, a_4 with the line l does not depend on l (the *cross-ratio of lines* a_1, a_2, a_3, a_4).

5329. Let f be a nonsingular bilinear function on an $(n+1)$-dimensional vector space V. Associate with each $(k + 1)$-dimensional subspace, $U \subset V$, a $(n - k)$-dimensional subspace

$$U^\perp = \{y \in V \mid f(x, y) = 0 \text{ for every } x \in U\}.$$

This correspondence in a projective space $\mathbf{P}(V)$ induces a mapping K_f, which associates an $(n-k-1)$-dimensional plane (*correlation* with respect to the function f) with each k-dimensional plane. Prove that

a) a correlation preserves incidence:

$$U_1 \subset U_2 \leftrightarrow K_f(U_1) \supset K_f(U_2);$$

b) if the function f is symmetric or skew-symmetric then the correlation K_f is involutive:

$$K_f(K_f(U)) = U;$$

    c)   a composition of a correlation with a projective transformation is a correlation;

    d)   any correlation is a composition of a fixed correlation with some projective transformation.

5330. Prove that any correlation of a projective line acts on its points as some projective transformation.

5331. Prove that a correlation of a projective plane preserves the cross-ratio.

5332. Prove that a correlation of a projective plane with respect to a symmetric bilinear function f maps each point of the curve $f(x, x) = 0$ onto the tangent to this curve at this point.

5333. Formulate the *Brianchon theorem*, obtained from the Pascal theorem (see Exercise 5326) by application of a correlation.

5334. Let a circle be given. For any chord passing through a fixed internal point, consider the point of intersection of the tangents to this circle, conducted through the ends of this chord. Prove with help of the notion of correlation that all these points of intersection lie on one line.

PART THREE
BASIC ALGEBRAIC STRUCTURES

CHAPTER 13

Groups

54 Algebraic operations. Semigroups

5401. Find out if the operation $*$ on the set M is associative:

a) $M = \mathbb{N}$, $x * y = x^y$;

b) $M = \mathbb{N}$, $x * y = \gcd(x, y)$;

c) $M = \mathbb{N}$, $x * y = 2xy$;

d) $M = \mathbb{Z}$, $x * y = x - y$;

e) $M = \mathbb{Z}$, $x * y = x^2 + y^2$;

f) $M = \mathbb{R}$, $x * y = \sin x \cdot \sin y$;

g) $M = \mathbb{R}^*$, $x * y = x \cdot y^{x/|x|}$?

5402. Let S be the semigroup of matrices $\begin{pmatrix} x & y \\ 0 & 0 \end{pmatrix}$, where $x, y \in \mathbb{R}$, with the operation of multiplication. Find in this semigroup the left and right neutral elements. Find elements which are left invertible and right invertible with respect to these neutral ones.

5403. The operation $\circ$ is defined on a set M by the rule $x \circ y = x$. Prove that $(M, \circ)$ is a semigroup. What can be said about neutral and invertible elements of this semigroup? In what case it is a group?

5404. Let M be a set. Define an operation $\circ$ on M^2 by the rule $(x, y) \circ (z, t) = (x, t)$. Is M^2 a semigroup with respect to this operation? Does there exist a neutral element in M^2?

5405. How many elements does the semigroup contain which consists of all powers of the matrix

$$\begin{pmatrix} -1 & 0 & 0 \\ 0 & 0 & 1 \\ 0 & 0 & 0 \end{pmatrix}?$$

Find out whether this semigroup is a group.

5406. Prove that the semigroups $(2^M, \cup)$ and $(2^M, \cap)$ are isomorphic.

5407. How many nonisomorphic semigroups of order 2 exist?

5408. Prove that in any finite semigroup there exists an *idempotent*.

5409. A semigroup is *monogenic* if it consists of positive powers of some of its elements (these elements are called *generators*). Prove that

   a)   a monogenic semigroup is finite if and only if it contains an idempotent;

   b)   a finite monogenic semigroup either is a group or has only one generator;

   c)   any two infinite monogenic semigroups are isomorphic;

   d)   any finite monogenic semigroup is isomorphic to a semigroup $S(n, k)$ which is defined on the set $\{a_1, \ldots, a_n\}$ as follows:

$$a_i + a_j = \begin{cases} a_{i+j}, & \text{if } i+j \le n, \\ a_{k+l+1}, & \text{if } i+j > n, \end{cases}$$

where l is the remainder of the number $i + j - n - 1$ divided by $n - k$.

55   Concept of a group. Isomorphisms of groups

5501. Find whether the following sets of numbers with the given operations are groups:

   a) $(A, +)$, where A is one of the sets $\mathbb{N}, \mathbb{Z}, \mathbb{Q}, \mathbb{R}, \mathbb{C}$;

   b) $(A, \cdot)$, where A is one of the sets $\mathbb{N}, \mathbb{Z}, \mathbb{Q}, \mathbb{R}, \mathbb{C}$;

   c) $(A_0, \cdot)$, where A is one of the sets $\mathbb{N}, \mathbb{Z}, \mathbb{Q}, \mathbb{R}, \mathbb{C}$, and $A_0 = A \setminus \{0\}$;

   d) $(n\mathbb{Z}, +)$, where n is a natural number;

   e) $(\{-1, 1\}, \cdot)$;

   f)   the set of powers of the given real number $a \neq 0$ with integer exponents, with respect to multiplication;

   g)   the set of all complex roots of 1 of the fixed order n, with respect to multiplication;

h) the set of complex roots of 1 of all orders, with respect to multiplication;

i) the set of complex numbers with the fixed absolute value r, with respect to multiplication;

j) the set of nonzero complex numbers with absolute values bounded by the fixed number r, with respect to multiplication;

k) the set of nonzero complex numbers located on rays outgoing from the origin and forming angles $\varphi_1 \varphi_2, \ldots, \varphi_n$ with the ray OX, with respect to multiplication?

5502. Prove that the semi-interval $[0, 1)$ with the operation $\oplus$, where $\alpha \oplus \beta$ is the fractional part of $\alpha + \beta$, is a group. To which group from Exercise 5501 is it isomorphic? Prove that any of its finite subgroups are cyclic.

5503. Prove that 2^M is a group with respect to the operation of symmetric difference $\triangle$ (see Exercise 104).

5504. Find out whether the following mappings from the set $M = \{1, 2, \ldots, n\}$ to itself form a group, with respect to multiplication:

a) the set of all mappings;

b) the set of all injective mappings;

c) the set of all surjective mappings;

d) the set of all bijective mappings;

e) the set of all even permutations;

f) the set of all odd permutations;

g) the set of all transpositions;

h) the set of all permutations, fixing elements of some subset $S \subseteq M$;

i) the set of all permutations such that the images of all elements of some subset $S \subseteq M$ belong to this subset;

j) the set $\{E, (12)(34), (13)(24), (14)(23)\}$;

k) the set $\{E, (13), (24), (12)(34), (13)(24), (14)(23), (1234), (1432)\}$?

5505. Find out whether the following sets of square real matrices of fixed size form a group:

a) the set of symmetric (skew-symmetric) matrices with respect to addition;

b) the set of symmetric (skew-symmetric) matrices with respect to multiplication;

c)   the set of nonsingular matrices with respect to addition;

d)   the set of nonsingular matrices with respect to multiplication;

e)   the set of matrices with a fixed determinant d with respect to multiplication;

f)   the set of diagonal matrices with respect to addition;

g)   the set of diagonal matrices with respect to multiplication;

h)   the set of diagonal matrices with nonzero diagonal entries with respect to multiplication;

i)   the set of *upper triangular matrices* with respect to addition;

j)   the set of *upper niltriangular* matrices with respect to multiplication;

k)   the set of upper niltriangular matrices with respect to addition;

l)   the set of *upper unitriangular* matrices with respect to multiplication;

m)   the set of all orthogonal matrices with respect to multiplication;

n)   the set of all matrices of the form $f(A)$, where A is a fixed nilpotent matrix, and $f(t)$ is an arbitrary polynomial with nonzero constant term, with respect to multiplication;

o)   the set of upper niltriangular matrices with respect to the operation $X \circ Y = X + Y - XY$;

p)   the set of nonzero matrices $\begin{pmatrix} x & y \\ -y & x \end{pmatrix}$ $(x, y \in \mathbb{R})$ with respect to multiplication;

q)   the set of nonzero matrices $\begin{pmatrix} x & y \\ \lambda y & x \end{pmatrix}$ $(x, y \in \mathbb{R})$, where λ is a fixed real number, with respect to multiplication;

r)   the set of matrices

$$\left\{ \pm \begin{pmatrix} 1 & 0 \\ 0 & 1 \end{pmatrix}, \ \pm \begin{pmatrix} i & 0 \\ 0 & -i \end{pmatrix}, \ \pm \begin{pmatrix} 0 & 1 \\ -1 & 0 \end{pmatrix}, \ \pm \begin{pmatrix} 0 & i \\ i & 0 \end{pmatrix} \right\}$$

with respect to multiplication.

5506. Prove that the set of upper niltriangular matrices of size 3 is a group with respect to the operation

$$X \circ Y = X + Y + \frac{1}{2}[X, Y].$$

5507. Let X be a set of points of the curve $y = x^3$, l be a line passing through points a, $b \in X$ (tangent to X if $a = b$) and c be the third point of its intersection with X. Let m be a line passing through the origin O and through the point c (tangent to X if $c = 0$).

Put $a \oplus b = d$ where d is the third point of the intersection of m and X or O if m is tangent to X at the point O. Prove that $(X, \oplus)$ is a commutative group.

5508. Prove that the set of functions $y(x) = (ax + b)/(cx + d)$, where a, b, c, $d \in \mathbb{R}$ and $ad - bc \neq 0$, is a group with respect to the operation of composition of functions.

5509. Prove that the commutator $[x, y] = xyx^{-1}y^{-1}$ of elements x, y of a group G has the following properties:

a) $[x, y]^{-1} = [y, x]$;

b) $[xy, z] = x[y, z]x^{-1}[x, z]$;

c) $[z, xy] = [z, x]x[z, y]x^{-1}$.

5510. Do the following equalities hold identically in the group S_3:

a) $x^6 = 1$;   b) $[[x, y], z] = 1$;   c) $[x^2, y^2] = 1$?

5511. Prove that the group of upper unitriangular matrices of size 3 satisfies the identity

$$(xy)^n = x^n y^n [x, y]^{-n(n-1)/2} \quad (n \in \mathbb{N}).$$

5512. Prove that if a group G satisfies the identity $[[x, y], z] = 1$, then G satisfies the identity

$$[x, yz] = [x, y][x, z], \quad [xy, z] = [x, z][y, z].$$

5513. Prove that if a group G satisfies the identity $x^2 = 1$, then G is commutative.

5514. Find out whether the following mappings of the groups $f : \mathbb{C}^* \mapsto \mathbb{R}^*$ are homomorphisms:

a) $f(z) = |z|$;   b) $f(z) = 2|z|$;   c) $f(z) = \dfrac{1}{|z|}$;

d) $f(z) = 1 + |z|$;   e) $f(z) = |z|^2$;   f) $f(z) = 1$;

g) $f(z) = 2$?

5515. Find out, for the groups G, whether the mapping $f : G \mapsto G$ defined by the rule:

a) $f(x) = x^2$,   b) $f(x) = x^{-1}$,

is a homomorphism? Under what condition is this mapping an isomorphism?

5516. Associate with each matrix $\begin{pmatrix} a & b \\ c & d \end{pmatrix} \in \mathbf{GL}(2, \mathbb{C})$ the function $y = \dfrac{ax + b}{cx + d}$ (see Exercise 5508). Is this mapping a homomorphism?

5517. Decompose the following groups into classes of isomorphic groups: $\mathbb{Z}$; $n\mathbb{Z}$; $\mathbb{Q}$; $\mathbb{R}$; $\mathbb{Q}^*$; $\mathbb{R}^*$; $\mathbb{C}^*$; $\mathbf{UT}_2(A)$, where A is one of the rings $\mathbb{Z}$, $\mathbb{Q}$, $\mathbb{R}$, $\mathbb{C}$.

5518. Find all isomorphisms between groups $(\mathbb{Z}_4, +)$ and $(\mathbb{Z}_5^*, \cdot)$.

5519. Prove that a group of order 6 is either commutative or isomorphic to the group S_3.

5520. Prove that if $a \neq 0$ is a rational number then the mapping $\varphi : x \mapsto ax$ is an automorphism of the group $\mathbb{Q}$. Find all automorphisms of the group $\mathbb{Q}$.

5521. Let G be a nonzero additive group consisting of real numbers such that each bounded interval contains only finitely many elements of G. Prove that $G \simeq \mathbb{Z}$.

5522. Find examples of planar geometrical figures whose groups of symmetries are isomorphic to:

a) $\mathbf{Z}_2$;   b) $\mathbf{Z}_3$;   c) $\mathbf{S}_3$;   d) $\mathbf{V}_4$.

5523. Find out whether some of the following groups are isomorphic:

– the group $\mathbf{D}_4$ of symmetries of a square;

– the group of quaternions $\mathbf{Q}_8$;

– the group from Exercise 5504k;

– the group from Exercise 5505r.

5524. Prove that the groups of proper symmetries of a tetrahedron, of a cube and of an octahedron are isomorphic to the groups $\mathbf{A}_4, \mathbf{S}_4, \mathbf{S}_4$, respectively.

5525. Prove that

a) the set of all automorphisms of an arbitrary group is a group with respect to composition;

b) the mapping $\sigma : x \mapsto axa^{-1}$, where a is a fixed element of a group G, is an (*inner*) automorphism of G;

c) the set of all inner automorphisms of an arbitrary group is a group with respect to composition.

5526. Find groups of automorphisms of the groups:

a) $\mathbb{Z}$;   b) $\mathbf{Z}_p$;   c) $\mathbf{S}_3$;   d) $\mathbf{V}_4$;   e) $\mathbf{D}_4$;   f) $\mathbf{Q}_8$.

5527. Prove that the mapping $a \mapsto \sigma$, which associates with each element a of a group G the permutation $\sigma : x \mapsto ax$ of the set G, is an injective homomorphism of the group G into the group S_G.

5528. Find subgroups of groups S_n which are isomorphic to the groups:

a) $\mathbf{Z}_3$;   b) $\mathbf{D}_4$;   c) $\mathbf{Q}_8$.

5529. For any permutation σ of degree n let $A_\sigma = (\delta_{i\sigma(j)})$ denote the square matrix of size n. Prove that if G is a group of permutations of degree n then the set of matrices A_σ, where $\sigma \in G$, forms a group isomorphic to G.

5530. Find subgroups of the groups of matrices $\mathbf{GL}_n(\mathbb{C})$ which are isomorphic to the groups:

a) $\mathbf{Z}_3$;   b) $\mathbf{D}_4$;   c) $\mathbf{Q}_8$.

5531. Find a subgroup of the group of real matrices of size 4 which is isomorphic to the group $\mathbf{Q}_8$.

5532. Prove that the group $\mathbf{U}_{p^\infty}$ cannot be mapped homomorphically onto a nontrivial finite group.

5533. Are the following groups isomorphic:

a) $\mathbf{SL}_2(3)$;   b) $\mathbf{S}_4$;   c) $\mathbf{A}_5$?

56 Subgroups. Orders of elements of groups. Cosets

5601. Prove that in any group

a) the intersection of any set of subgroups is a subgroup;

b) the union of two subgroups is a subgroup if and only if one of these subgroups is contained in the other;

c) if a subgroup C is contained in the union of subgroups A and B then either $C \subseteq A$ or $C \subseteq B$.

5602. Prove that a finite subsemigroup of any group is a subgroup. Is this statement valid for infinite subsemigroups?

5603. Find the order of the element of the group:

a) $\begin{pmatrix} 1 & 2 & 3 & 4 & 5 \\ 2 & 3 & 1 & 5 & 4 \end{pmatrix} \in S_5$;   b) $\begin{pmatrix} 1 & 2 & 3 & 4 & 5 & 6 \\ 2 & 3 & 4 & 5 & 1 & 6 \end{pmatrix} \in S_6$;

c) $\dfrac{-\sqrt{3}}{2} + \dfrac{1}{2}i \in \mathbb{C}^*$;

d) $\dfrac{1}{\sqrt{2}} - \dfrac{1}{\sqrt{2}}i \in \mathbb{C}$;

e) $\begin{pmatrix} 0 & 1 & 0 & 0 \\ 0 & 0 & 1 & 0 \\ 0 & 0 & 0 & 1 \\ 1 & 0 & 0 & 0 \end{pmatrix} \in \mathbf{GL}_4(\mathbb{R})$;

f) $\begin{pmatrix} 0 & i \\ 1 & 0 \end{pmatrix} \in \mathbf{GL}_2(\mathbb{C})$;

g) $\begin{pmatrix} -1 & a \\ 0 & 1 \end{pmatrix} \in \mathbf{GL}_2(\mathbb{C})$;

h) $\begin{pmatrix} 0 & -1 \\ 1 & -1 \end{pmatrix} \in \mathbf{GL}_2(\mathbb{C})$;

i) $\begin{pmatrix} \lambda_1 & * & \cdots & * \\ 0 & \lambda_2 & * & \cdots \\ & \cdots\cdots\cdots\cdots & & \\ 0 & \cdots & 0 & \lambda_n \end{pmatrix} \in \mathbf{GL}_n(\mathbb{C})$,

where $\lambda_1, \ldots, \lambda_n$ are distinct roots of 1 of order k.

5604. Let p be a prime odd number, X an integer square matrix of size n, and the matrix $E + pX$ lie in $\mathbf{SL}_n(\mathbb{Z})$ and have a finite order. Prove that $X = 0$.

5605. Prove that

a) the element $\dfrac{3}{5} + \dfrac{4}{5}i$ of the group $\mathbb{C}^*$ has infinite order;

b) the number $\dfrac{1}{\pi} \arctan \dfrac{4}{3}$ is irrational.

5606. How many elements of order 6 are contained in the group:

a) $\mathbb{C}^*$;   b) $\mathbf{D}_2(\mathbb{C})^*$;   c) S_5;   d) A_5?

5607. Prove that in any group

a) the elements x and yxy^{-1} have the same order;

b) the elements ab and ba have the same order;

c) the elements xyz and zyx can have different orders.

5608. Let elements x and y of a group G have finite orders and $xy = yx$. Prove that

a) if the orders of the elements x and y are coprime, then the order of the product xy is equal to the product of their orders;

b) there exist exponents k and l such that the order of the product $x^k y^l$ is equal to the least common multiple of orders of x and y.

Are these statements valid for noncommuting elements x and y?

5609. Prove that

a) if an element x of a group G has infinite order then $x^k = x^l$ if and only if $k = l$;

b) if an element x of a group G has order n then $x^k = x^l$ if and only if $n|(k - l)$;

c) if an element x of a group G has order n then $x^k = e$ if and only if $n|k$.

5610. Prove that in the group S_n

a) the order of an odd permutation is an even number;

b) the order of any permutation is equal to the least common multiple of the lengths of disjoint cycles occurring in its decomposition.

5611. Find the order of an element x^k, if the order of x is equal to n.

5612. Let G be a finite group, $a \in G$. Prove that $G = \langle a \rangle$ if and only if the order of a is equal to $|G|$.

5613. Find the number of elements of order p^m in the cyclic group $\mathbf{Z}_{p^n}$, where p is a prime, $0 < m \le n$.

5614. Let $G = \langle a \rangle$ be a cyclic group of order n. Prove that

a) elements a^k and a^l have the same order if and only if the greatest common divisors (k, n) and (l, n) are equal;

b) an element a^k is a generator of G if and only if k and n are coprime;

c) any subgroup $H \subseteq G$ is generated by an element a^d, where $d|n$;

d) for any divisor d of the number n there exists a unique subgroup $H \subseteq G$ of order d.

5615. In a cyclic group $\langle a \rangle$ n find all elements g satisfying the condition $g^k = e$ and all elements of order k if:

a) $n = 24,\quad k = 6$; b) $n = 24,\quad k = 4$; c) $n = 100,\quad k = 20$;

d) $n = 100,\quad k = 5$; e) $n = 360,\quad k = 30$; f) $n = 360,\quad k = 12$;

g) $n = 360,\quad k = 7$.

5616. Find all the subgroups of the cyclic group of order:

a) 24; b) 100; c) 360; d) 125; e) p^n (p is a prime).

5617. Let G be a finite group and $d(G)$ be the least natural number s such that $g^s = e$ for any element $g \in G$ (the *period* of the group G).

Prove that

a) the period $d(G)$ divides $|G|$ and it is equal to the least common multiple of orders of elements of G;

b) if G is commutative then there exists an element $g \in G$ of order $d(G)$;

c) a finite commutative group is cyclic if and only if $d(G) = |G|$.

Are the statements b) and c) valid for noncommutative groups?

5618. Does there exist an infinite group in which all elements have finite orders?

5619. The *periodic part* of a group G is the set of all its elements of finite orders.

a) Prove that the periodic part of a commutative group is a subgroup.

b) Is statement a) valid for a noncommutative group?

c) Find the periodic part of the groups $\mathbb{C}^*$ and $D_n(\mathbb{C})^*$.

d) Prove that if a commutative group G has elements of infinite order and all of them are contained in a subgroup H, then H coincides with G.

5620. Prove that in a commutative group the set of elements, whose orders divide a fixed number n, is a subgroup. Is this statement valid for a noncommutative group?

5621. Find all finite groups in which there exists the greatest proper subgroup.

5622. The set of all subgroups of a group G forms a chain, if for any two its subgroups one of them is contained in the other.

a) Prove that subgroups of the cyclic group of order p^n, where p is a prime, form a chain.

b) Find all finite groups in which subgroups form a chain.

c) Find all groups in which subgroups form a chain.

5623. Represent the group $\mathbb{Q}$ as a union of the ascending chain of cyclic subgroups.

5624. Establish an isomorphism between the group U_n of complex roots of order n of 1 and the group $\mathbb{Z}_n$ of residues modulo n.

5625. Find out if some of subgroups $\langle g \rangle$ generated by elements $g \in G$ are isomorphic:

a) $\qquad$ $G = \mathbb{C}^*,$ $\qquad$ $g = -\dfrac{1}{\sqrt{2}} + \dfrac{1}{\sqrt{2}} i;$

b) $\qquad$ $G = \mathbf{GL}_2(\mathbb{C}),$ $\qquad$ $g = \begin{pmatrix} 0 & 1 \\ i & 0 \end{pmatrix};$

c) $\qquad$ $G = \mathbf{S}_6,$ $\qquad$ $g = (32651);$

d) $\qquad$ $G = \mathbb{C}^*,$ $\qquad$ $g = 2 - i;$

e) $\qquad$ $G = \mathbb{R}^*,$ $\qquad$ $g = 10;$

f) $\qquad$ $G = \mathbb{C}^*,$ $\qquad$ $g = \cos\dfrac{6\pi}{5} + i \sin\dfrac{6\pi}{5};$

g) $\qquad$ $G = \mathbb{Z},$ $\qquad$ $g = 3?$

5626. Prove that

a)   any group of even order contains an element of order 2;

b)   a group is commutative if all its nontrivial elements have order 2.

5627. Prove that any finite subgroup of the group $\mathbb{C}^*$ is cyclic.

5628. Prove that any proper subgroup of the group $\mathbf{U}_{p^\infty}$ is a finite cyclic group.

5629. Prove that

a)   in the multiplicative group of a field for any natural number n there exists at most one subgroup of order n;

b)   any finite subgroup of the multiplicative group of a field is cyclic;

c)   the multiplicative group of a finite field is cyclic.

5630. Find all subgroups of the groups:

a) $\mathbf{S}_3;$ $\qquad$ b) $\mathbf{D}_4;$ $\qquad$ c) $\mathbf{Q}_8;$ $\qquad$ d) $\mathbf{A}_4.$

5631. Prove that if a subgroup H of the group $\mathbf{S}_n$ contains one of the sets

$$\{(12), (13), \ldots, (1n)\} \quad \{(12), (123\ldots n)\},$$

then $H = \mathbf{S}_n.$

5632. Find all elements of the group G commuting with the given element $g \in G$ (the *centralizer* of g), if:

a) $\qquad$ $G = \mathbf{S}_4,$ $\qquad$ $g = (12)(34);$

b)
$$G = SL_2(\mathbb{R}), \qquad g = \begin{pmatrix} a & 0 \\ 0 & b \end{pmatrix};$$

c)
$$G = S_n, \qquad g = (123\ldots n).$$

5633. Put

$$G_f = \{\sigma \in S_4 \mid f(x_{\sigma(1)}, x_{\sigma(2)}, x_{\sigma(3)}, x_{\sigma(4)}) = f(x_1, x_2, x_3, x_4)\}$$

for a polynomial f in variables x_1, x_2, x_3, x_4.

Prove that G_f is a subgroup of S_4 and find this subgroup for the polynomials:

a) $f = x_1 x_2 + x_3 x_4$;

b) $f = x_1 x_2 x_3$;

c) $f = x_1 + x_2$;

d) $f = x_1 x_2 x_3 x_4$;

e) $f = \prod_{1 \le j < i \le 4} (x_i - x_j)$.

5634. Find the cosets

a) of the additive group $\mathbb{Z}$ with respect to the subgroup $n\mathbb{Z}$, where n is a natural number;

b) of the additive group $\mathbb{C}$ with respect to the subgroup $\mathbb{Z}[i]$ of Gaussian integers, i.e. numbers $a + bi$ with integer a, b;

c) of the additive group $\mathbb{R}$ with respect to the subgroup $\mathbb{Z}$;

d) of the additive group $\mathbb{C}$ with respect to the subgroup $\mathbb{R}$;

e) of the multiplicative group $\mathbb{C}^*$ with respect to the subgroup U of numbers with absolute value 1;

f) of the multiplicative group $\mathbb{C}^*$ with respect to the subgroup $\mathbb{R}^*$;

g) of the multiplicative group $\mathbb{C}^*$ with respect to the subgroup of positive real numbers;

h) of the group of permutations S_n with respect to the stationary subgroup of the element n;

i) of the additive group of real (3×2) matrices with respect to the subgroup of all matrices (a_{ij}) such that $a_{31} = a_{32} = a_{22} = 0$;

j) of the additive group of all polynomials of degree at most 5 with complex coefficients with respect to the subgroup of polynomials of degree at most 3;

k) of the cyclic group $\langle a \rangle_6$ with respect to the subgroup $\langle a^4 \rangle$.

5635. Let g be a nonsingular matrix from $\mathbf{GL}_n(\mathbb{C})$ and $H = \mathbf{SL}_n(\mathbb{C})$. Prove that the coset gH consists of all matrices $a \in \mathbf{GL}(n, \mathbb{C})$ such that their determinant is equal to the determinant of g.

5636. Let H be a subgroup of a group G. Prove that the mapping $xH \mapsto Hx^{-1}$ induces a bijection between the set of left cosets and the set of right cosets of G with respect to H.

5637. Let g_1, g_2 be elements of a group G, and H_1, H_2 be subgroups of G. Prove that the following properties are equivalent:

a) $g_1 H_1 \subseteq g_2 H_2$;

b) $H_1 \subseteq H_2$ and $g_2^{-1} g_1 \in H_2$.

5638. Let g_1, g_2 be elements of a group G and H_1, H_2 be subgroups of G. Prove that a nonempty set $g_1 H_1 \cap g_2 H_2$ is a left coset of G with respect to the subgroup $H_1 \cap H_2$.

5639. Let H_1, H_2 be subgroups in a group G and $H_1 \subseteq H_2$. If the index H_1 in H_2 is equal to n, and the index of H_2 in G is equal to m, then the index of H_1 in G is equal to mn.

5640. Prove that all axial symmetries form a coset in the dihedral group with respect to the subgroup of rotations.

57 Action of a group on a set. Relation of conjugacy

5701. Find all orbits of the group G of nonsingular linear operators acting on n-dimensional space V, if:

a) G is the group of all nonsingular linear operators;

b) G is the group of all orthogonal operators;

c) G is the group of operators whose matrices in the basis $(e_1, \ldots, e_n)$ are diagonal;

d) G is the group of operators whose matrices in the basis $(e_1, \ldots, e_n)$ are upper triangular.

5702. Find the stabilizer G_a of the vector $a = e_1 + e_2 + \cdots + e_n$ if;

a) G is the group from Exercise 5701c;

b) G is the group from Exercise 5701d.

5703. Find the stabilizer G_x and the orbit of an arbitrary vector x if:

a) G is the group of all orthogonal operators on three-dimensional Euclidean space;

b) G is the group of all proper orthogonal operators on two-dimensional Euclidean space.

5704. Let G be the group of all nonsingular linear operators on n-dimensional vector space V and X be the set of all subspaces of dimension k of V.

a) Find the orbits of G in X.

b) Let $e_1, \ldots, e_n$ be a basis in V such that $e_1, \ldots, e_k$ is a basis of some subspace U. Find the matrices of the operators of the stabilizer G_U with the basis $e_1, \ldots, e_n$.

5705. Let G be a group of all nonsingular linear operators on n-dimensional vector space V and F be a set of *flags* in V, i.e. sets $f = (V_0, V_1, \ldots, V_n)$ of subspaces of V where $0 = V_0 < V_1 < \cdots < V_n = V$.

a) Find the orbits of G in F.

b) Let $e_i \in V_i \setminus V_{i-1}, i = 1, \ldots, n$. Prove that $e_1, \ldots, e_n$ is a basis in V.

c) Find the matrices of the operators of the stabilizer G_f with the basis $e_1, \ldots, e_n$.

5706. Let G be the group of all nonsingular linear operators on n-dimensional vector space V, and X (Y, respectively) be the set of all nonzero decomposable q-vectors in $\Lambda^q V$ (in $S^q(V)$, respectively).

a) Find the orbits of the action G in X and Y.

b) Find the stabilizer G_a of a decomposable q-vector a (a vector from $S^q(V)$).

5707. Let G be the group of all nonsingular linear operators on n-dimensional real (complex) space V, and B be the set of all symmetric (Hermitian) bilinear functions on V. Put $g(b)(x, y) = b(g^{-1}x, g^{-1}y)$ for $g \in G$ and $b \in B$.

a) Prove that G acts on B.

b) Determine the orbits of G in B. Find their number.

c) Determine the stabilizer G_b of a positively definite function b.

5708. Let G be the group of all nonsingular linear operators on n-dimensional complex space V and $L(V)$ be the set of all linear operators on V. Put $g(f) = gfg^{-1}$, if $g \in G$ and $f \in L(V)$.

a) Prove that G acts on $L(V)$.

b) Determine the orbits of G in $L(V)$.

5709. Find all the orbits in the set $\{1, 2, \ldots, 10\}$ and all stabilizers in the group G if G is generated by the permutation

a)
$$g = \begin{pmatrix} 1 & 2 & 3 & 4 & 5 & 6 & 7 & 8 & 9 & 10 \\ 5 & 8 & 3 & 9 & 4 & 10 & 6 & 2 & 1 & 7 \end{pmatrix} \in S_{10};$$

b)
$$g = \begin{pmatrix} 1 & 2 & 3 & 4 & 5 & 6 & 7 & 8 & 9 & 10 \\ 7 & 4 & 6 & 1 & 8 & 3 & 2 & 9 & 5 & 10 \end{pmatrix} \in S_{10};$$

c) $g = (169)(2, 10)(34578) \in S_{10}$.

5710. Consider the diamond with vertices, in a rectangular system of coordinates,
$$A = (0, 1), \quad B = (2, 0), \quad C = (0, -1), \quad D = (-2, 0).$$

a) Find the matrices of the orthogonal transformations of the plane mapping the diamond into itself.

b) Prove that these matrices, with respect to multiplication, form a group G, isomorphic to the group S_4.

c) Find the orbits of the action of G on the set of vertices of the diamond and find their stabilizers.

5711. Find the order of the dihedral group D_n.

5712. Find the order:

a) of the group of rotations of a cube;

b) of the group of rotations of a tetrahedron;

c) of the group of rotations of a dodecahedron.

5713. Prove that:

a) the group of rotations of an icosahedron is isomorphic to the group A_5;

b) the group of symmetries of a tetrahedron is isomorphic to S_4.

5714. Find the order of the stabilizer of a vertex in the group of rotations:

a) of an octahedron;

b) of an icosahedron;

c) of a tetrahedron;

d) of a cube;

e) of a dihedron.

5715. Let G be a group of affine transformations of n-dimensional affine space X. Suppose that Y is the family of all sets with $n + 1$ points $(A_0, \ldots, A_n)$ in the general position.

a) Find the orbits of G in Y;

b) Find the stabilizer G_a of a set $a \in Y$.

5716. Let G be a group of affine transformations of n-dimensional affine real (complex) space X. Denote by Q the set of all bilinear functions on X. Put $g(h) = h(g^{-1}x)$ for $g \in G, h \in Q$ and $x \in X$.

a) Prove that G acts on Q.

b) Determine the orbits of G in Q.

c) Determine the stabilizer G_h of a nonsingular function $h \in Q$.

5717. Let G be a group of linear-fractional transformations of the unit disc with the center O from Exercise 2422. Find:

a) the stabilizer of the point O;

b) the orbit of the point O;

c) the intersection of the stabilizers of two distinct points of the unit disc.

5718. Let a group G act on a set X and x, y be two elements of one orbit of G in X. Prove that all $g \in G$, such that $g(x) = y$, form a left coset in G with respect to the stabilizer G_x and form a right coset with respect to the stabilizer G_y.

5719. Let a commutative group G act on a set M. Prove that if $gm_0 = m_0$ for some $g \in G$ and $m_0 \in M$, then $gm = m$ for any point m, belonging to the orbit of the point m_0.

5720. Let H be a subgroup of a group $G, a \in G$. Prove that

a) the mapping $\sigma_a : gH \mapsto agH$ is a permutation on the set M of all left cosets of G with respect to H;

b) the mapping $f : a \mapsto \sigma_a$ defines an action of G on M;

c) σ_a is the identical permutation if and only if a belongs to the intersection of all subgroups conjugate to H in G.

5721. Enumerating left cosets of the group G, with respect to the subgroup H, find all permutations σ_a (see Exercise 5720) if:

a) $G = \mathbf{Z}_4$ and H is the identity subgroup;

b) $G = \mathbf{D}_4$, H is a subgroup consisting of the identity transformation and of some axial symmetry of a square.

5722. Prove that for any group G

a) a conjugation induces an action

$$m \mapsto g \cdot m = gmg^{-1} \quad (g, m \in G)$$

of G on the set G;

b) the stabilizer of a point m (the centralizer of the element m) coincides with the set of all elements of G commutating with m.

5723. Find the centralizer:

a) of the permutation $(12)(34)$ in the group $\mathbf{S}_4$;

b) of the permutation $(123\ldots n)$ in the group $\mathbf{S}_n$.

5724. In the group $\mathbf{GL}_2(\mathbb{R})$ find the centralizer of the matrix:

a) $\begin{pmatrix} 1 & 0 \\ 0 & -1 \end{pmatrix}$;   b) $\begin{pmatrix} 2 & 0 \\ 0 & 2 \end{pmatrix}$;   c) $\begin{pmatrix} 1 & 2 \\ 3 & 4 \end{pmatrix}$;   d) $\begin{pmatrix} 1 & 1 \\ 0 & 1 \end{pmatrix}$.

5725. In the group $\mathbf{GL}_n(\mathbb{R})$ find the centralizer of the matrix $\mathrm{diag}(\lambda_1, \ldots, \lambda_n)$ if

a) all elements of the diagonal are distinct;

b) $\lambda_1 = \cdots = \lambda_k = a$, $\lambda_{k+1} = \cdots = \lambda_n = b$ and $a \neq b$.

5726. Determine if some of the following three matrices are conjugate in the group $\mathbf{GL}_2(\mathbb{C})$:

$$A_1 = \begin{pmatrix} 1 & 1 \\ 0 & 1 \end{pmatrix}, \quad A_2 = \begin{pmatrix} \frac{1}{2} & 0 \\ 0 & 2 \end{pmatrix}, \quad A_3 = \begin{pmatrix} 1 & 0 \\ 2 & 1 \end{pmatrix}?$$

5727. Let F be a field. In the group $\mathbf{SL}_n(F)$ find

a) the centralizer C_{ij} of an elementary matrix $E + E_{ij}$ where $1 \leq i \neq j \leq n$;

b) the intersection of C_{ij} for all i, j, where $1 \leq i \neq j \leq n$;

c) the class of conjugate elements containing $E + E_{ij}$.

Prove that any two elementary matrices $E + \alpha E_{ij}$ and $E + \beta E_{pq}$, where $1 \leq i \neq j$, $p \neq q \leq n$, and $\alpha, \beta \in F^*$ are conjugate.

5728. In the group $\emptyset_2(\mathbb{R})$ of orthogonal operators find

a) the centralizer of the operator of rotation on an angle $q \neq k\pi$;

b) the centralizer of the symmetry relative to the axis OX.

5729. Prove that in the group $\emptyset_2(\mathbb{R})$ any two symmetries are conjugate.

5730. Find classes of conjugate elements in the groups:

a) S_3;   b) A_4;   c) D_4.

5731. Find all finite groups in which the number of classes of conjugate elements is equal to:

a) 1;   b) 2;   c) 3.

5732. In the group S_4 find the class of conjugate elements:

a) of the permutation $(12)(34)$;

b) of the permutation (124).

5733. Prove that two permutations are conjugate in the group S_n if and only if they have the same cyclic structure, i.e. their decompositions into the product of disjoint cycles for any integer k contain the same number of cycles of length k.

5734. Find the number of classes of conjugate elements in the groups:

a) S_4;   b) S_5;   c) S_6;   d) D_4.

5735. The *canonical form* of a matrix $A \in SO_3(\mathbb{R})$ is the matrix

$$\begin{pmatrix} 1 & 0 & 0 \\ 0 & \cos\varphi & -\sin\varphi \\ 0 & \sin\varphi & \cos\varphi \end{pmatrix}$$

conjugate to A. Prove that the matrices A_1 and A_2 are conjugate in $SO_2(\mathbb{R})$ if and only if their canonical forms are connected either by the relation $\varphi_1 + \varphi_2 = 2\pi k$ or by the relation $\varphi_1 - \varphi_2 = 2\pi k$ for some integer k.

5736. Prove that

a) if H and K are conjugate subgroups of a finite group and $K \subseteq H$ then $K = H$;

b) subgroups

$$H = \left\{ \begin{pmatrix} 1 & n \\ 0 & 1 \end{pmatrix} \ (n \in \mathbb{Z}) \right\}, \quad K = \left\{ \begin{pmatrix} 1 & 2n \\ 0 & 1 \end{pmatrix} \ (n \in \mathbb{Z}) \right\}$$

are conjugate in the group $GL_2(\mathbb{R})$ and $K \subset H$.

5737. Find the *normalizer* $N(H)$ of the subgroup H in the group G, if

a) $G = \mathbf{GL}_2(\mathbb{R})$ and H is the subgroup of diagonal matrices;

b) $G = \mathbf{GL}_2(\mathbb{R})$ and H is the subgroup of matrices

$$\begin{pmatrix} 1 & a \\ 0 & 1 \end{pmatrix} \quad (a \in \mathbb{R});$$

c) $G = \mathbf{S}_4, \quad H = \langle (1234) \rangle$.

5738. Find the automorphism group of:

a) the group $\mathbf{Z}_5$;

b) the group $\mathbf{Z}_6$.

5739. Prove that

a) $\mathrm{Aut}\mathbf{S}_3 \simeq \mathbf{S}_3$ and all automorphisms of the group $\mathbf{S}_3$ are inner;

b) $\mathrm{Aut}\mathbf{V}_4 \simeq \mathbf{S}_3$ and only the identical automorphism is inner in $\mathbf{V}_4$.

5740. Find out whether the automorphism groups of the following groups are cyclic:

a) $\mathbf{Z}_9$;

b) the group $\mathbf{Z}_8$.

5741. Find the order of the group $\mathrm{Aut}\mathrm{Aut}\mathrm{Aut}\mathbf{Z}_9$.

5742. Construct an external automorphism in the group $\mathbf{S}_6$.

5743. Prove that in the group $\mathbf{S}_n$, $n \neq 6$ all automorphisms are inner.

5744. Prove that the group of automorphisms of $\mathbf{D}_4$ is isomorphic to $\mathbf{D}_4$. Find the subgroup of inner automorphisms of $\mathbf{D}_4$.

5745. Find the group of automorphisms of $\mathbf{D}_n$ and the subgroup of its inner automorphisms.

58 Homomorphisms and normal subgroups. Factorgroups, centers

5801. Prove that the subgroup H of the group G is normal if

a) G is a commutative group and H is an arbitrary subgroup;

b) $G = \mathbf{GL}_n(\mathbb{R})$ and H is the subgroup of matrices with determinants equal to 1;

c) $G = S_n$, $H = A_n$;

d) $G = S_4$, $H = V_4$;

e) G is the group of nonsingular complex upper triangular matrices and H is the group of matrices $E + \sum\limits_{\substack{1 \le i < j \le n \\ j - i \ge k}} \alpha_{ij} E_{ij}$, $\alpha_{ij} \in \mathbb{C}$.

5802. Prove that any subgroup of index 2 is normal.

5803. Find all nontrivial normal subgroups in the groups:

a) S_3;   b) A_4;   c) S_4.

5804. Using the example of the group A_4, show that a normal subgroup K of a normal subgroup H in a group G is not necessarily normal in G.

5805. Let A and B be normal subgroups of a group G and $A \cap B$ be the identity subgroup. Prove that $xy = yx$ for any $x \in A$, $y \in B$.

5806. Let H be a subgroup of G of index 2, C be a class of conjugate elements in G and $C \subset H$. Prove that C is either a class of conjugate elements in H or a union of two classes of conjugate elements in H, consisting of the same number of elements.

5807. Find the number of classes of conjugate elements in the group A_5 and find the number of elements in each class.

5808. Prove that the group A_5 is simple.

5809.

a) Prove that every subgroup of the group of quaternions Q_8 is normal.

b) Find the center and all classes of conjugate elements in the group Q_8.

c) Prove that the complex matrices

$$\pm E = \pm \begin{pmatrix} 1 & 0 \\ 0 & 1 \end{pmatrix}, \qquad \pm I = \pm \begin{pmatrix} i & 0 \\ 0 & -i \end{pmatrix},$$

$$\pm J = \pm \begin{pmatrix} 0 & 1 \\ -1 & 0 \end{pmatrix}, \qquad \pm K = \pm \begin{pmatrix} 0 & i \\ i & 0 \end{pmatrix},$$

form a group with respect to multiplication of matrices. This group is isomorphic to Q_8.

5810. Find all normal subgroups of the dihedral group D_n.

5811. Let F be a field and G be a subgroup of $\mathbf{GL}_n(F)$, containing $\mathbf{SL}_n(F)$. Prove that G is normal in $\mathbf{GL}_n(F)$.

5812. Associate with each matrix $\begin{pmatrix} a & c \\ b & d \end{pmatrix} \in \mathbf{GL}_2(\mathbb{C})$ a linear-fractional transformation

$$f(z) = \frac{az + b}{cz + d}.$$

Find the kernel of this homomorphism.

5813. Let n, $m \geq 2$ be natural numbers and $\mathbf{SL}_n(\mathbb{Z}; m\mathbb{Z})$ be a subset of $\mathbf{SL}_n(\mathbb{Z})$, consisting of matrices $E + Xm$, where X is an integer square matrix of size n. Prove that

a) $\mathbf{SL}_n(\mathbb{Z}; m\mathbb{Z})$ is a normal subgroup in $\mathbf{SL}_n(\mathbb{Z})$;

b) if $m = p$ is a prime then $\mathbf{SL}_n(\mathbb{Z})/\mathbf{SL}_n(\mathbb{Z}; p\mathbb{Z}) \simeq \mathbf{SL}_n(\mathbb{Z}_p)$;

c) the group $\mathbf{SL}_n(\mathbb{Z}; m\mathbb{Z})$ does not contain elements of finite order if $m \geq 3$;

d) if G is a finite subgroup of $\mathbf{SL}_n(\mathbb{Z})$ then the order of G divides $\frac{1}{2}(3^n - 1)(3^n - 3) \ldots (3^n - 3^{n-1})$.

5814. Prove that for any group G the set of all its inner automorphisms is a normal subgroup in the group $\mathrm{Aut}\,G$ of all automorphisms of G.

5815. Prove that any subgroup, containing a commutant of a group, is normal.

5816. Find the center of the group:

a) $\mathbf{S}_n$;

b) $\mathbf{A}_n$;

c) $\mathbf{D}_n$.

5817. Prove that the center of a group of order p^n, where p is a prime ($n \in \mathbb{N}$), contains more than one element.

5818. Let G be the set of all upper unitriangular matrices of size 3 with elements from the field $\mathbf{Z}_p$.

a) Prove that G is a noncommutative group of order p^3 with respect to multiplication.

b) Find the center of G.

c) Find all classes of conjugate elements in G.

5819. Find the center of the group:

a) $\mathbf{GL}_n(\mathbb{R})$;   b) $\mathbf{O}_2(\mathbb{R})$;   c) $\mathbf{SO}_2(\mathbb{R})$;   d) $\mathbf{SO}_3(\mathbb{R})$;

e) $\mathbf{SU}_2(\mathbb{C})$;   f) $\mathbf{SU}_n(\mathbb{C})$;   g) of upper-triangular matrices.

5820. Find the center:

a)   of the group of all linear-fractional transformations of the complex plane;

b)   of the group of all transformations of the unit disc from Exercise 2422.

5821. Prove that the group H is a homomorphic image of a finite cyclic group G if and only if H is cyclic and its order divides the order of G.

5822. Prove that if a group G is homomorphically mapped onto a group H and $a \mapsto a'$ then

a)   the order of a is divisible by the order of a';

b)   the order of G is divisible by the order of H.

5823. Find all the homomorphisms:

a) $Z_6 \to Z_6$;   b) $Z_6 \to Z_{18}$;   c) $Z_{18} \to Z_6$;

d) $Z_{12} \to Z_{15}$;   e) $Z_6 \to Z_{25}$.

5824. Prove that the additive group of rational numbers cannot be mapped homomorphically onto the additive group of integers.

5825. Find factorgroups:

a) Z/nZ;   b) U_{12}/U_3;   c) $4Z/12Z$;   d) R^*/R_+.

5826. Let F^n be the additive group of n-dimensional linear space over a field F and H be the subgroup of vectors of a k-dimensional subspace. Prove that the factorgroup F^n/H is isomorphic to F^{n-k}.

5827. Let H_n be the set of complex numbers with arguments $2\pi k/n$ ($k \in Z$). Prove that

a) $R/Z \simeq U$;   b) $C^*/R^* \simeq U$;   c) $C^*/U \simeq R_+$;

d) $U/U_n \simeq U$;   e) $C^*/U_n \simeq C^*$;   f) $C^*/H_n \simeq U$;

g) $H_n/R_+ \simeq U_n$;   h) $H_n/U_n \simeq R_+$.

5828. Let $G = GL_n(R)$, $H = GL_n(C)$, $P = SL_n(R)$, $Q = SL_n(C)$,

$$A = \{X \in G \mid |\det X| = 1\}, \qquad B = \{X \in H \mid |\det X| = 1\},$$
$$B = \{X \in G \mid \det X > 1\}, \qquad N = \{X \in H \mid \det X > 0\}.$$

Prove that

a) $G/P \simeq \mathbb{R}^*$;   b) $H/Q \simeq \mathbb{C}^*$;   c) $G/M \simeq \mathbf{Z}_2$;

d) $H/N \simeq \mathbf{U}$;   e) $G/A \simeq \mathbb{R}_+$;   f) $H/B \simeq \mathbb{R}_+$.

5829. Let G be the group of affine transformations of n-dimensional space, H be the subgroup of parallel transfers, and K be the subgroup of transformations fixing the given point O. Prove that

a) H is a normal subgroup in G;

b) $G/H \simeq K$.

5830. Prove that the factorgroup of the group S_4 with respect to the normal subgroup $\{e, (12)(34), (13)(24), (14)(23)\}$ is isomorphic to the group S_3.

5831. Prove that if H is a subgroup of index k in a group G then H contains some subgroup which is normal in G and whose index in G divides $k!$.

5832. Prove that a subgroup is normal if its index is the least prime divisor of the order of the group.

5833. Prove that the factorgroup of the group $\mathbf{GL}_2(\mathbf{Z}_3)$ by its center is isomorphic to the group S_4.

5834. Prove that in the group $\mathbb{Q}/\mathbb{Z}$

a)   each element has finite order;

b)   for any natural number n there exists exactly one subgroup of order n.

5835. Prove that the group of inner automorphisms of a group G is isomorphic to the factorgroup of G by its center.

5836. Prove that the factorgroup of a noncommutative group by its center cannot be cyclic.

5837. Prove that a group of order p^2, where p is a prime, is commutative.

5838. Prove that the group of all automorphisms of a noncommutative group cannot be cyclic.

*   *   *

5839. Find the number of classes of conjugate elements and the number of elements in each class for a noncommutative group of order p^3, where p is a prime.

5840. A subgroup H is *maximal* in a group G if $H \neq G$ and any subgroup containing H coincides with H or with G. Prove that

a) the intersection of any two different maximal commutative subgroups is contained in the center of the group;

b) any finite simple noncommutative group has two different maximal subgroups whose intersection contains more than one element;

c) any finite simple noncommutative group has a proper noncommutative subgroup.

5841. Prove that the factorgroup of $SL_2(5)$ by its center is isomorphic to A_5.

5842. Prove that every finite subgroup in $SL_2(\mathbb{Q})$ is a subgroup in one of the following groups: D_3, D_4, D_6.

5843. Let F be a field, $n \geq 3$ and G be a normal subgroup in $GL_n(F)$. Prove that either $G \supseteq SL_n(F)$ or G consists of scalar matrices.

5844. Let F be a field containing at least four elements and G be a normal subgroup in $GL_2(F)$. Prove that either $G \supseteq SL_2(F)$ or G consists of scalar matrices.

5845. Prove that $SL_2(2) \simeq S_3$.

5846. Find all normal subgroups in $SL_2(3)$.

5847. Let G be a normal subgroup of finite index in $SL_n(\mathbb{Z})$, $n \geq 3$. Show that there exists a natural number m such that $G \subseteq SL_n(\mathbb{Z}, m\mathbb{Z})$.

5848. Let F be a field, $n \geq 3$ and φ be an automorphism of the group $GL_n(F)$. Prove that there exist a homomorphism of groups $\tau : GL_n(F) \rightarrow F^*$ and an automorphism τ of F, such that either $\varphi(x) = \tau(x)g\tau(x)g^{-1}$ or $\varphi(x) = \tau(x)g'\tau(x)^{1-}g^{-1}$, where $g \in GL_n(F)$.

59 Sylow subgroups. Groups of small orders

5901. Find the order of the groups:

a) $GL_n(q)$;

b) $SL_n(q)$;

c) of all nonsingular upper triangular matrices of size n over the finite field with q elements.

5902. Are the following groups isomorphic:

a) Q_8 and D_4;

b) S_4 and $SL_2(3)$?

5903. Find all Sylow 2-subgroups and 3-subgroups in the groups:

a) S_3;　　b) A_4.

5904. Indicate the conjugating elements for Sylow 2-subgroups and for Sylow 3-subgroups in the groups:

a) S_3;　　b) A_4.

5905. Prove that any Sylow 2-subgroup in the group S_4 is isomorphic to the dihedral group D_4.

5906. Which Sylow 2-subgroups of the group S_4 contain the permutations:

a) (1324);　　　　　　b) (13);　　　　　　c) (12)(34)?

5907. Prove that there exist exactly two noncommutative nonisomorphic groups of order 8, namely, the group of quaternions Q_8 and the dihedral group D_4.

5908. Prove that the Sylow 2-subgroup in the group $SL_2(3)$

a) is isomorphic to the group of quaternions;

b) is normal in $SL_2(3)$.

5909. How many different Sylow p-subgroups has the group A_5, if

a) $p = 2$;　　　　　　b) $p = 3$;　　　　　　c) $p = 5$?

5910. Find the order of Sylow p-subgroups in the group S_n.

5911. How many different Sylow p-subgroups has the group S_p, where p is a prime?

5912. Prove that Sylow p-subgroup in a group G is unique if and only if it is normal in G.

5913. Let

$$P = \left\{ \begin{pmatrix} 1 & a \\ 0 & 1 \end{pmatrix} \mid a \in Z_p, \quad p \text{ is a prime} \right\}.$$

a) Prove that P is a Sylow p-subgroup in the group $SL_2(p)$.

b) Find the normalizer of P in $SL_2(p)$.

c) Find the number of different Sylow p-subgroups in $SL_2(p)$.

d) Prove that P is a Sylow p-subgroup in $GL_2(p)$.

e) Find the normalizer of P in $GL_2(p)$.

f) Find the number of different Sylow p-subgroups in $GL_2(p)$.

5914. Prove that the subgroup of upper unitriangular matrices is a Sylow p-subgroup in $\mathbf{GL}_n(p)$.

5915. In the dihedral group $\mathbf{D}_n$, for each prime divisor p of the number $2n$, find:

a)   all Sylow p-subgroups;

b)   conjugating elements for Sylow p-subgroups.

5916. Prove that the image of a Sylow p-subgroup of a finite group G, under a homomorphism of G onto a group H, is a Sylow subgroup in H.

5917. Prove that any Sylow p-subgroup of the direct product of finite groups A and B is the product of Sylow p-subgroups of A and B.

5918. Let P be a Sylow p-subgroup of a finite group G and H be a normal subgroup in G.

a)   Prove that the intersection $P \cap H$ is a Sylow p-subgroup in H.

b)   Find an example showing that for non-normal subgroups H the statement a) is not valid.

5919. Prove that all Sylow subgroups of a group of order 100 are commutative.

5920. Prove that any group of order:

a)   15,   b)   35,   c)   185,   d)   255

is commutative.

5921. How many different Sylow 2-subgroups and Sylow 5-subgroups are there in a noncommutative group of order 20?

5922. Prove that there exist no simple groups of order:

a)   36,   b)   80,   c)   56,   d)   196,   e)   200.

5923. Let p and q be primes, $p < q$. Prove that

a)   if $q - 1$ is not divisible by p then any group of order pq is commutative;

b)   if $q-1$ is divisible by p then in the group of nonsingular matrices $\begin{pmatrix} a & b \\ 0 & 1 \end{pmatrix}$ $(a, b \in \mathbf{Z}_q)$ there exists a noncommutative subgroup of order pq.

5924. How many elements of order 7 has a simple group of order 168?

* * *

5925. Let K be a normal subgroup in p-group G. Prove that $K \cap Z(G) \neq 1$.

5926. Let V be a finite-dimensional vector space over a field of characteristic p and G be p-group of linear nonsingular operators on V. Prove that there exists a nonzero vector $x \in V$ such that $gx = x$ for all $g \in G$.

5927. Let P be a Sylow p-subgroup of a finite group G and H be a subgroup of G, containing normalizer $N_G(P)$. Prove that $N_G(H) = H$.

60 Direct products and direct sums. Abelian groups

6001. Prove that the groups $\mathbb{Z}$ and $\mathbb{Q}$ are not decomposable into a direct sum of nonzero subgroups.

6002. Do the following groups

a) S_3;   b) A_4;   c) S_4;   d) Q_8

have a nontrivial direct decomposition?

6003. Prove that a finite cyclic group is a direct sum of prime-power cyclic subgroups.

6004. Prove that the direct sum of cyclic groups $\mathbb{Z}_m \oplus \mathbb{Z}_n$ is a cyclic group if and only if m and n are coprime.

6005. Decompose into a direct sum the groups:

a) $\mathbb{Z}_6$;   b) $\mathbb{Z}_{12}$;   c) $\mathbb{Z}_{60}$.

6006. Prove that the multiplicative group of complex numbers is the direct product of the group of positive real numbers and of the group of all complex numbers with absolute value 1.

6007. Prove that if $n \geq 3$ then the multiplicative group of the residue ring $\mathbb{Z}_{2^n}$ is the direct product of the subgroup $\{\pm 1\}$ and the cyclic group 2^{n-2}.

6008. What is the order:

a)   of a direct product of finite groups;

b)   of an element of a direct product of finite groups?

6009. Prove that if subgroups $A_1, A_2, \ldots, A_k$ of an abelian group have coprime orders then their sum is direct.

6010. Let D be a subgroup of the direct product $A \times B$ of groups A and B of coprime orders. Prove that $D \simeq (D \cap A) \times (D \cap B)$.

6011. Let k be the greatest order of elements of a finite abelian group G. Prove that the order of any element of G divides k. Is this statement valid without the assumption about the commutativity of G?

6012. Find all direct decompositions of the group, consisting of the numbers of the form $\pm 2^n$.

6013. Let A be a finite abelian group. Find all direct decompositions of the group $\mathbb{Z} \oplus A$ such that one of its summands is an infinite cyclic group.

6014. Find classes of conjugate elements in the group $A \times B$ in terms of classes of conjugate elements in the groups A and B.

6015.

a) Prove that the center of the direct product $A \times B$ is equal to the direct product of the centers of A and B.

b) Let N be a normal subgroup in $A \times B$ and $N \cap A = N \cap B = 1$. Prove that N belongs to the center of $A \times B$.

6016. Prove that if a factorgroup A/B of an abelian group A by a subgroup B is a free abelian group, then $A = B \oplus C$ where C is a free abelian group.

6017. Prove that a subgroup A of an abelian group G is a direct summand of G if and only if there exists a surjective homomorphism $\pi G \to A$ such that $\pi^2 = \pi$.

6018. Let φ_1, φ_2 be homomorphisms from groups A_1, A_2 to an abelian group B. Prove that there exists a unique homomorphism $\varphi : A_1 \times A_2 \to B$ whose restrictions to A_1 and A_2 coincide with φ_1 and φ_2, respectively. Is the abelience of B essential in this statement?

6019. On a set of homomorphisms from an abelian group A to an abelian group B define the operation of addition by the rule

$$(\alpha + \beta)(x) = \alpha(x) + \beta(x).$$

Prove that the homomorphisms $A \to B$ form an abelian group $\mathrm{Hom}(A, B)$ with respect to this addition.

6020. Find the groups of homomorphisms:

a) $\mathrm{Hom}(\mathbb{Z}_{12}, \mathbb{Z}_6)$;

b) $\mathrm{Hom}(\mathbb{Z}_{12}, \mathbb{Z}_{18})$;

c) $\mathrm{Hom}(\mathbb{Z}_6, \mathbb{Z}_{12})$;

d) $\mathrm{Hom}(A_1 \oplus A_2, B)$;

e) $\mathrm{Hom}(A, B_1 \oplus B_2)$;

f) $\mathrm{Hom}(\mathbb{Z}_n, \mathbb{Z}_k)$;

g) $\mathrm{Hom}(\mathbb{Z}, \mathbb{Z}_n)$;

h) $\mathrm{Hom}(\mathbb{Z}_n, \mathbb{Z})$;

i) $\mathrm{Hom}(\mathbb{Z}, \mathbb{Z})$;

j) $\mathrm{Hom}(\mathbb{Z}_2 \oplus \mathbb{Z}_2, \mathbb{Z}_8)$;

k) $\mathrm{Hom}(\mathbb{Z}_2 \oplus \mathbb{Z}_3, \mathbb{Z}_{30})$.

6021. Prove that $\mathrm{Hom}(\mathbb{Z}, A) \simeq A$.

6022. Let A be an abelian group. Prove that all its endomorphisms, with respect to addition and usual multiplication of mappings, form the ring $\text{End}\,A$ with unit.

6023. Prove that the group of automorphisms of an abelian group coincides with the group of invertible elements of the endomorphism ring.

6024. Find the endomorphism rings of the groups:

a) $\mathbb{Z}$; b) $\mathbb{Z}_n$; c) $\mathbb{Q}$.

6025. Prove that the mapping $x \to nx$ $(n \in \mathbb{Z})$ in an abelian group is an endomorphism. For what groups it will be: a) injective; b) surjective?

6026. Prove that the endomorphism ring of a free abelian group of rank n is isomorphic to the ring $\mathbf{M}_n(\mathbb{Z})$.

6027. Find the groups of automorphisms of the groups:

a) $\mathbb{Z}$; b) $\mathbb{Q}$; c) $\mathbb{Z}_{2^n}$; d) the free abelian group of rank n.

6028. Prove that

a) $\text{Aut}\mathbb{Z}_{30} \simeq \text{Aut}\mathbb{Z}_{15}$; b) $\text{Aut}(\mathbb{Z} \oplus \mathbb{Z}_2) = \mathbb{Z}_2 \oplus \mathbb{Z}_2$.

6029. Prove that the ring $\text{End}(\mathbb{Z} \oplus \mathbb{Z}_2)$ is infinite and is noncommutative.

6030. Prove that the endomorphism ring of a finite abelian group is the direct sum of the endomorphism rings of its prime components.

6031. Prove that a subgroup of a finitely generated abelian group is also finitely generated.

6032. Prove that any homomorphism from a finitely generated abelian group onto itself is an automorphism.

Is a similar statement valid for the additive group of a ring of polynomials?

6033. Prove that the free abelian groups of ranks m and n are isomorphic if and only if $m = n$.

6034. Let A, B, C be finitely generated abelian groups and $A \oplus C \simeq B \oplus C$. Prove that $A \simeq B$.

6035. The order of a finite abelian group G is divisible by a number m. Prove there exists in G a subgroup of order m.

6036. Let A and B be finite abelian groups. Suppose that for any natural number m the numbers of elements of order m in A and in B are the same. Prove that $A \simeq B$.

6037. Let A and B be finitely generated abelian groups such that each of them is isomorphic to a subgroup of the other one. Prove that $A \simeq B$.

6038. Prove that a subgroup B of a free abelian group A is free and the rank of B does not exceed the rank of A.

6039. Applying the main theorem on finitely generated abelian groups, find up to isomorphism all abelian groups of the orders:

a) 2;   b) 6;   c) 8;   d) 12;   e) 16;   f) 24;   g) 36;   h) 48.

6040. We say that an abelian group is of the type $(n_1, n_2, \ldots, n_k)$ if it is a direct sum of cyclic groups of orders $n_1, n_2, \ldots, n_k$. Does an abelian group of the type $(2,16)$ contain a subgroup of the type:

a) (2,8);   b) (4,4);   c) (2,2,2)?

6041. Find the type of the group $(\langle a \rangle_9 \oplus \langle b \rangle_{27})/(3a + 9b)$.

6042. Are the following pairs of groups isomorphic:

a) $(\langle a \rangle_2 \oplus \langle b \rangle_4)/(3a + 9b)/2(\langle b \rangle$ and $(\langle a \rangle_2 \oplus \langle b \rangle_4)/\langle a + 2b \rangle$;

b) $\mathbf{Z}_6 \oplus \mathbf{Z}_{36}$ and $\mathbf{Z}_{12} \oplus \mathbf{Z}_{18}$;

c) $\mathbf{Z}_6 \oplus \mathbf{Z}_{36}$ and $\mathbf{Z}_9 \oplus \mathbf{Z}_{24}$;

d) $\mathbf{Z}_6 \oplus \mathbf{Z}_{10} \oplus \mathbf{Z}_{10}$ and $\mathbf{Z}_{60} \oplus \mathbf{Z}_{10}$?

6043. How many subgroups

a)   of orders 2 and 6 are there in a noncyclic abelian group of order 12;

b)   of orders 3 and 6 are there in a noncyclic abelian group of order 18;

c)   of orders 5 and 15 are there in a noncyclic abelian group of order 75?

6044. Find all direct decompositions of the groups:

a) $\langle a \rangle_2 \oplus \langle b \rangle_2$;   b) $\langle a \rangle_p \oplus \langle b \rangle_p$;   c) $\langle a \rangle_2 \oplus \langle b \rangle_4$.

6045.

a)   How many elements of orders 2, 4 and 6 are there in the group $\mathbf{Z}_2 \oplus \mathbf{Z}_4 \oplus \mathbf{Z}_3$?

b)   How many elements of orders 2, 4 and 5 are there in the group $\mathbf{Z}_2 \oplus \mathbf{Z}_4 \oplus \mathbf{Z}_4 \oplus \mathbf{Z}_5$?

6046. Using primary decomposition prove that a finite subgroup of the multiplicative group of a field is cyclic.

6047. Let F be a field whose multiplicative group F^* is finitely generated. Prove that F is finite.

6048. Prove that a finitely generated subgroup of the multiplicative group of complex numbers is the direct product of a free abelian group and a finite cyclic group.

6049. Let A be a free abelian group with a basis $e_1, \ldots, e_n$ and $x = m_1 e_1 + \cdots + m_n e_n \in A \setminus 0$, where $m_i \in \mathbb{Z}$. Prove that the cyclic group $\langle x \rangle$ is a direct summand in A if and only if the numbers $m_1, \ldots, m_n$ are coprime.

6050. Let A be a free abelian group with a basis $x_1, \ldots, x_n$. Prove that the elements

$$y_j = \sum_{i=1}^{n} a_{ij} x_i \quad (j = 1, \ldots, n)$$

form a basis of the group A if and only if $\det(a_{ij}) = \pm 1$.

6051. Let A be a free abelian group with a basis $x_1, \ldots, x_n$ and B be its subgroup generated by the elements

$$y_j = \sum_{i=1}^{n} a_{ij} x_i \quad (j = 1, \ldots, n).$$

Prove that the factorgroup A/B is finite if and only if $\det(a_{ij}) \neq 0$, and in this case $|A/B| = |\det(a_{ij})|$.

6052. Decompose, into a direct sum of cyclic groups, the factorgroup A/B where A is the free abelian group, with the basis x_1, x_2, x_3 and B is its subgroup generated by y_1, y_2, y_3:

a) $\quad y_1 = 7x_1 + 2x_2 + 3x_3,$
$\quad\quad y_2 = 21x_1 + 8x_2 + 9x_3,$
$\quad\quad y_3 = 5x_1 - 4x_2 + 3x_3;$

b) $\quad y_1 = 4x_1 + 5x_2 + 3x_3,$
$\quad\quad y_2 = 5x_1 + 6x_2 + 5x_3,$
$\quad\quad y_3 = 8x_1 + 7x_2 + 9x_3;$

c) $\quad y_1 = 5x_1 + 5x_2 + 2x_3,$
$\quad\quad y_2 = 11x_1 + 8x_2 + 5x_3,$
$\quad\quad y_3 = 17x_1 + 5x_2 + 8x_3;$

d) $\quad y_1 = 6x_1 + 5x_2 + 7x_3,$
$\quad\quad y_2 = 8x_1 + 7x_2 + 11x_3,$
$\quad\quad y_3 = 6x_1 + 5x_2 + 11x_3;$

e) $\quad y_1 = 4x_1 + 5x_2 + x_3,$
$\quad\quad y_2 = 8x_1 + 9x_2 + x_3,$
$\quad\quad y_3 = 4x_1 + 6x_2 + 2x_3;$

f) $\quad y_1 = 2x_1 + 6x_2 + 2x_3,$
$\quad\quad y_2 = 2x_1 + 8x_2 - 4x_3,$
$\quad\quad y_3 = 4x_1 + 12x_2 - 2x_3;$

g) $\quad y_1 = 6x_1 + 5x_2 + 4x_3,$
$\quad\quad y_2 = 7x_1 + 6x_2 + 9x_3,$
$\quad\quad y_3 = 5x_1 + 4x_2 + 4x_3;$

h) $\quad y_1 = x_1 + 2x_2 + 3x_3,$
$\quad\quad y_2 = 2y_1,$
$\quad\quad y_3 = 3y_1;$

i) $\quad y_1 = 4x_1 + 7x_2 + 3x_3,$ j) $\quad y_1 = 2x_1 + 3x_2 + 4x_3,$

$\quad\quad y_2 = 2x_1 + 3x_2 + 2x_3,$ $y_2 = 5x_1 + 5x_2 + 6x_3,$

$\quad\quad y_3 = 6x_1 + 10x_2 + 5x_3;$ $y_3 = 2x_1 + 6x_2 + 9x_3.$

6053. Find the order of the coset $(x_1 + 2x_3) + B$ in the factorgroup of the free abelian group A with the basis x_1, x_2, x_3 with respect to the subgroup B generated by $x_1 + x_2 + 4x_3$ and by $2x_1 - x_2 + 2x_3$.

6054. Find the order of the coset $32x_1 + 31x_2 + B$ in the factorgroup of the free abelian group A with the basis x_1, x_2, x_3, with respect to the subgroup B generated by $2x_1 + x_2 - 50x_3$ and by $4x_1 + 5x_2 + 60x_3$.

6055. Prove that the endomorphism ring of a finite abelian group is commutative if and only if each of its prime components is cyclic.

*   *   *

6056. An additive subgroup H in n-dimensional real space $\mathbb{R}^n$ is discrete if there exists a neighborhood U of the origin such that $U \cap H = 0$. Prove that a discrete subgroup in $\mathbb{R}^n$ is a free abelian group and its rank does not exceed n.

6057. Find all elements of finite order in $\mathbb{R}^n / \mathbb{Z}^n$.

6058. Let $H = \mathbb{Z}[i]$ be the subgroup of Gaussian integers in the additive group of complex numbers $\mathbb{C}$. Assume that $z = x + iy \in \mathbb{C} \setminus H$, where $x, y \in \mathbb{R}^*$, and let xy^{-1} be irrational. Prove that $\langle z \rangle + H$ is dense in $\mathbb{C}$.

6059. Let H be an additive closed subgroup in $\mathbb{R}^n$. Prove that $H = L \oplus H_1$, where L is a subspace of $\mathbb{R}^n$ and H_1 is a discrete subgroup in $\mathbb{R}^n$.

6060. Prove that if the order of an element a in the abelian group A is coprime with n then the equation $nx = a$ has a solution in A.

6061. An abelian group A is *divisible* if the equation $nx = a$ has a solution for any $a \in A$ and for any integer $n \neq 0$. Prove that a group is divisible if and only if the equation $px = a$ has a solution for any a and any prime p.

6062. Prove that the direct sum is divisible if and only if all its direct summands are divisible.

6063. Prove that the groups $\mathbb{Q}$ and $\mathbf{U}_{p^\infty}$ (p is a prime) are divisible.

6064. Prove that in a torsion-free group it is possible to define a structure of a linear space over the field $\mathbb{Q}$ if and only if the group is divisible.

6065. Let A be a divisible subgroup of a group G and let B be a maximal subgroup of G such that $A \cap B = \{0\}$ (this subgroup always exists). Prove that $G = A \oplus B$.

6066. Prove that any abelian group has a divisible subgroup such that the factorgroup by this subgroup has no divisible subgroups.

6067. Let A be a finitely generated abelian group and let B be a subgroup of A. Suppose that A/B is torsion-free. Then $A = B \oplus C$ where C is a free abelian group.

6068. Let A, B be free abelian groups and let $\varphi : A \to B$ be a homomorphism of groups. Prove that $\text{Ker}\varphi$ is a direct summand in A.

6069. Let A be a free abelian group with a base $e_1, \dots, e_n$ and let C be an integer square matrix of size n. Denote by B the set of vectors $x_1 e_1 + \cdots + x_n e_n \in A$ such that

$$C \begin{pmatrix} x_1 \\ \vdots \\ x_n \end{pmatrix} = 0.$$

Prove that B is a subgroup of A and a direct summand in A. Conversely, any direct summand in A is a set of solutions of a system of linear homogeneous integer equations.

61 Generators and defining relations

6101. Prove that

a)   the group S_n is generated by the transposition (12) and the cycle $(12 \dots n)$;

b)   the group A_n is generated by threefold cycles.

6102. Prove that

a)   the group $\text{GL}_n(K)$ over a field K is generated by matrices $E + a E_{ij}$, where $a \in K$, $1 \le i \ne j \le n$, and by matrices $E + b E_{11}$, where $b \in K$, $b \ne -1$;

b)   the group $\text{UT}_n(k)$ is generated by matrices $E + a E_{ij}$, where $a \in K$, $1 \le i < j \le n$.

6103. Prove that the special linear group $\text{SL}_n(K)$ over a field K is generated by *transvections*, i.e. by elementary matrices $E + \alpha E_{ij}$ $(i \ne j)$.

6104. Prove that

a)   any integer matrix with determinant 1 can be reduced to the identity matrix by elementary row transformations of additions to some row of another one multiplied by ± 1;

b)   the group $\text{SL}_n(\mathbb{Z})$ is finitely generated.

$$* \quad * \quad *$$

6105. Let $\mathbf{F}_q$ be the field with $q \neq 9$ elements and a a generator of the cyclic group $\mathbf{F}_q^*(q)$. Prove that $\mathbf{SL}_2(q)$ is generated by two matrices

$$\begin{pmatrix} 1 & 1 \\ 0 & 1 \end{pmatrix}, \quad \begin{pmatrix} 1 & 0 \\ a & 1 \end{pmatrix}.$$

6106.

a) Prove that $\mathbf{A}_5$ is generated by two permutations (254) and (12345).

b) Prove that $\mathbf{A}_n$ for even degree $n \geq 4$ is generated by two elements $a = (12)(n-1, n)$, $b = (1, 2, \ldots, n-1)$.

c) Prove that $\mathbf{A}_n$ for odd degree $n \geq 5$ is generated by two elements $a = (1, n)(2, n-1)$, $b = (1, 2, \ldots, n-2)$.

6107. Find all two-element sets generating the groups:

a) $\mathbf{Z}_6$;   b) $\mathbf{S}_3$;   c) $\mathbf{Q}_8$;   d) $\mathbf{D}_4$;   e) $\langle a \rangle_2 \oplus \langle b \rangle_2$.

6108. Prove that if d is the minimal number of generators in a finite abelian group A then the analogous number for the group $A \oplus A$ is equal to $2d$.

6109. Prove that the group $\mathbf{S}_2 \times \mathbf{S}_3$ is generated by two elements.

6110. Prove that if a group has a finite system of generators then any system of generators contains a finite subsystem generating the whole group.

6111. Determine whether a *normal closure* of the matrix $A = \begin{pmatrix} 1 & 1 \\ 0 & 1 \end{pmatrix}$ in the group G generated by matrices A and $B = \begin{pmatrix} 2 & 0 \\ 0 & 1 \end{pmatrix}$ is finitely generated.

6112. Prove that

a) each word in a free group is equivalent to a unique uncancelled word;

b) 'the free group' is, in fact, a group.

6113. Let F be a free group with free generators $x_1, \ldots, x_n$ and let G be an arbitrary group. Prove that for any elements $g_1, \ldots, g_n \in G$ there exists a unique homomorphism $\varphi : F \to G$ such that $\varphi(x_1) = g_1, \ldots, \varphi(x_n) = g_n$.

Deduce from this statement that any finitely generated group is isomorphic to a factorgroup of a suitable free group of a finite rank.

6114. Prove that a free group is torsion-free.

6115. Prove that any two commuting elements of a free group belong to one cyclic subgroup.

6116. Prove that a word w lies in the *commutant* of a free group with a system of free generators $x_1, \ldots, x_n$ if and only if for every $i = 1, \ldots, n$ the sum of exponents for all occurrences of x_i in w is equal to 0.

6117. Determine all words in a free group conjugate to the given word w.

6118. Prove that the factorgroup of a free group by its commutant is a free abelian group.

6119. Prove that free groups of ranks m and n are isomorphic if and only if $m = n$.

6120. How many subgroups of index 2 are there in the free group of rank 2?

6121. Let F be a free group of rank k.

a) Prove that all words in which the sum of all exponents in each variable is divisible by n form the normal subgroup N.

b) Prove that $F/N = \overbrace{\mathbf{Z}_n \oplus \cdots \oplus \mathbf{Z}_n}^{k \text{ times}}$.

6122. Prove that all surjective homomorphisms from the free groups of rank 2 onto the group $\mathbf{Z}_n \oplus \mathbf{Z}_n$ have the same kernel.

6123. How many homomorphisms are there from a free group of rank 2 into the group:

a) $\mathbf{Z}_2 \oplus \mathbf{Z}_2$;   b) $\mathbf{S}_3$?

6124. Prove that in $\mathbf{SL}_2(\mathbb{Z})$ the set of matrices

$$\begin{pmatrix} a & b \\ c & d \end{pmatrix}, \quad \text{where} \quad a \equiv d \equiv 1 \pmod 4, \quad b \equiv c \equiv 0 \pmod 2$$

form a group with two generators

$$\begin{pmatrix} 1 & 2 \\ 0 & 1 \end{pmatrix}, \quad \begin{pmatrix} 1 & 0 \\ 2 & 1 \end{pmatrix}.$$

6125. Prove that if a group G with generators $x_1, \ldots, x_n$ is given by defining relations $R_i(x_1, \ldots, x_n) = 1$ ($i \in I$), and in any group H for elements $h_1, \ldots, h_n \in H$ we have $R_i(h_1, \ldots, h_n) = 1$, then there exists a unique homomorphism $\varphi : G \to H$ such that $\varphi(x_1) = h_1, \ldots, \varphi(x_n) = h_n$.

6126. Prove that if elements a and b of some group satisfy the relations $a^5 = b^3 = 1, b^{-1}ab = a^2$ then $a = 1$.

6127. Show that a group generated by elements a, b with relations $a^2 = b^7 = 1$, $a^{-1}ba = b^{-1}$ is finite.

6128. Prove that the group given by generators x_1, x_2, with defining relations

a) $x_1^2 = x_2^3 = (x_1 x_2)^2 = 1$,

b) $x_1^2 = x_2^3 = 1$, $x_1^{-1} x_2 x_1 = x_2^2$,

is isomorphic to $\mathbf{S}_3$.

6129. Prove that the group given by generators x_1, x_2, and by defining relations

$$x_1^2 = x_2^n = 1, \quad x_1^{-1} x_2 x_1 = x_2^{-1},$$

is isomorphic to the dihedral group $\mathbf{D}_n$.

6130. Prove that the group given by generators x_1, x_2, and by defining relations

$$x_1^4 = 1, \quad x_1^2 = x_2^2, \quad x_2^{-1} x_1 x_2 = x_1^3,$$

is isomorphic to the group of quaternions $\mathbf{Q}_8$.

6131. Prove that the group given by generators x_1, x_2 and by defining relations, $x_1^2 = x_2^2 = 1$ is isomorphic to the group of matrices

$$\left\{ \begin{pmatrix} \pm 1 & n \\ 0 & 1 \end{pmatrix} \quad (n \in \mathbb{Z}) \right\}.$$

6132. Prove that the group given by generators x_1, x_2, and by defining relations $x_1^2 = x_2^2 = (x_1 x_2)^n = 1$, is isomorphic to the group of matrices

$$\left\{ \begin{pmatrix} \pm 1 & k \\ 0 & 1 \end{pmatrix} \quad (k \in \mathbb{Z}_n) \right\}.$$

6133. Let G be the group generated by elements x_{ij}, $1 \le i < j \le n$, with defining relations

$$x_{ij} x_{kl} = x_{kl} x_{ij}, \qquad\qquad 1 \le i < j \ne k < l \ne i \le n;$$
$$x_{ij} x_{jl} x_{ij}^{-1} x_{jl}^{-1} = x_{il}, \qquad 1 \le i < j < l \le n.$$

Prove that

a) each element of G has the form

$$x_{12}^{m_{12}} x_{13}^{m_{12}} \ldots x_{1n}^{m_{1n}} x_{23}^{m_{23}} \ldots x_{2n}^{m_{2n}} \ldots x_{n-1,n}^{m_{n-1,n}}$$

where $m_{ij} \in \mathbb{Z}$;

b) $G \simeq \mathbf{UT}_n(\mathbb{Z})$.

6134. Prove that if $G/H = \langle gH \rangle$ is an infinite cyclic group then $G = \langle g \rangle H$, $\langle g \rangle \cap H = \{e\}$.

6135. Determine, in terms of generators and defining relations, all groups which have an infinite cyclic normal subgroup with an infinite cyclic factorgroup.

6136. Let the group G be given by generators x_1, x_2 and by the defining relation $x_1 x_2 x_1^{-1} = x_2^2$. Find the smallest subgroup generated in G by x_2. Is this subgroup normal?

62 Solvable groups

6201. Find the commutator:

a) of nonsingular matrices $\begin{pmatrix} 0 & 1 \\ 1 & 0 \end{pmatrix}$ and $\begin{pmatrix} a & 0 \\ 0 & 1 \end{pmatrix}$;

b) $\begin{pmatrix} a & b \\ 0 & c \end{pmatrix}$ and $\begin{pmatrix} x & y \\ 0 & z \end{pmatrix}$;

c) of two transpositions in the symmetric groups S_n.

6202. Prove the following properties of the *commutant* G' of groups:

a) G' is a normal subgroup in G;

b) the factorgroup G/G' is commutative;

c) if N is normal in G and G/H is commutative then $G' \subseteq N$.

6203. Prove that $\varphi(G)' = H'$ under a surjective homomorphism $\varphi : G' \to H$.

6204. Find a bijective correspondence between the homomorphisms from a group G to commutative group A and homomorphisms from the factorgroup G/G' to A.

6205. Prove that the commutant of the group $GL_n(K)$ is contained in $SL_n(K)$.

6206. Prove that the commutant of a direct product is the direct product of the commutants of the factors.

6207. Find the commutants and the orders of factorgroups by commutants for the groups:

a) S_3;   b) A_4;   c) S_4;   d) Q_8.

6208. Find the commutants of the groups:

a) S_n;   b) D_n.

6209. Prove that the commutant of a normal subgroup is normal in the whole group.

6210. The *series of commutants* (or a *derived series*) in a group G is the series of subgroups

$$G = G^0 \supseteq G' \supseteq G'' \supseteq \ldots,$$

where $G^{i+1} = (G^i)'$. Prove that

    a)   all members of the series of commutants are normal in G;

    b) $\varphi(G^i) = H^i$ for any homomorphism φ from G onto a group H.

6211. Prove that

    a)   any subgroup of a solvable group is solvable;

    b)   any factorgroup of a solvable group is solvable;

    c)   if A and B are solvable groups, then the group $A \times B$ is solvable;

    d)   if $G/A \simeq B$ and A, B are solvable groups, then G is solvable.

6212. Prove that the groups:

    a) S_3;   b) A_4;   c) S_4;   d) Q_8;   e) D_n

are solvable.

6213. Let $UT_n(K)$ be the group of upper unitriangular matrices. Prove that

    a)   the set $UT_n^m(K)$ of matrices from $UT_n(K)$ with $m - 1$ zero diagonals above the principal one is a subgroup of $UT_n(K)$;

    b)   if $A \in UT_n^i(K)$, and $B \in UT_n^j(K)$, then $[A, B] \in UT_n^{(i+j)}(K)$;

    c)   the group $UT_n(K)$ is solvable.

6214. Prove that the group of nonsingular upper triangular matrices is solvable.

6215. Prove that a finite group G is solvable if and only if it has a series of subgroups

$$G = H_0 \supseteq H_1 \supseteq \cdots \supseteq H_k = \{e\}$$

such that H_{i+1} is normal in H_i and H_i/H_{i+1} are cyclic groups of prime orders.

6216. Prove that a finite p-group is solvable.

6217. Prove that a group of order pq where p, q are distinct primes is solvable.

6218. Prove that groups of orders:

a) 20;

b) 12;

c) p^2q, where p, q are distinct primes;

d) 42;

e) 100;

f) $n < 60$

are solvable.

6219. For transvections $t_{ij}(\alpha) = E + \alpha E_{ij}$ prove the formula $[t_{ik}(\alpha), t_{kj}(\beta)] = t_{ij}(\alpha\beta)$, where i, j, k are distinct.

6220. Let F be a field and $n \geq 3$. Prove that

a) $\mathbf{SL}_n'(F) = \mathbf{GL}_n'(F) = \mathbf{SL}_n(F)$;

b) the groups $\mathbf{SL}_n(F)$ and $\mathbf{GL}_n(F)$ are not solvable.

6221. Let F be a field containing at least four elements. Prove that

a) $\mathbf{SL}_2'(F) = \mathbf{GL}_2'(F) = \mathbf{SL}_2(F)$;

b) the groups $\mathbf{SL}_2(F)$ and $\mathbf{GL}_2(F)$ are not solvable.

* * *

6222. Let p, q, r be distinct primes. Prove that any group of order pqr is solvable.

6223. Let p, q, r be distinct primes. Prove that any nonsolvable group of order p^2qr is isomorphic to $\mathbf{A}_5$.

6224. If the order of a finite group G is square-free then G is a solvable group with a cyclic normal subgroup N such that G/N is a cyclic group.

6225. Let G be a finite group such that $G = G'$ and the center of G is of order 2. Suppose that the factorgroup by the center is isomorphic to $\mathbf{A}_5$. Prove that $G \simeq \mathbf{SL}_2(5)$.

6226. Let F be a field, V be an n-dimensional vector space over F and let G be the group of nonsingular linear operators on V such that if $g \in G$ then $g = 1 + h$, where $h^n = 0$. Prove that

a) there exists a vector $x \neq 0$ in V such that $gx = x$ for all $g \in G$;

b) there exists a basis $e_1, \ldots, e_n$ in V such that matrices of all operators g, $g \in G$ in this basis are upper-triangular;

c) the group G is solvable.

6227. Let p, q be primes and p divides $q - 1$. Prove that

a)   there exists an integer $r \not\equiv 1 \pmod{q}$ such that $r^p \equiv 1 \pmod{q}$;

b)   there exists a unique (up to isomorphism) noncommutative group of order pq.

6228. Prove that

a)   if elements a, b of a commutative group satisfy the relations $a^3 = b^5 = (ab)^7 = e$ then $a = b = e$;

b)   the subgroup S_7 generated by permutations (123) and (14567) is not solvable;

c)   the group generated by x_1, x_2 with defining relations $x_1^3 = x_2^5 = (x_1 x_2)^7 = e$ is not solvable.

6229. In what cases is a free group solvable?

CHAPTER 14

Rings

63 Rings and algebras

6301. Which of the following sets of numbers form a ring with respect to the usual operations of addition and multiplication:

a) the set $\mathbb{Z}$;

b) the set $n\mathbb{Z}$ $(n > 1)$;

c) the set of all non-negative integers;

d) the set $\mathbb{Q}$;

e) the set of rational numbers where the denominators divide a fixed integer $n \in \mathbb{N}$;

f) the set of rational numbers where the denominators are not divisible by a fixed prime p;

g) the set of rational numbers where the denominators are powers of a fixed prime p;

h) the set of real numbers of the form $x + y\sqrt{2}$, where $x, y \in \mathbb{Q}$;

i) the set of real numbers of the form $x + y\sqrt[3]{2}$, where $x, y \in \mathbb{Q}$;

j) the set of real numbers of the form $x + y\sqrt[3]{2} + z\sqrt[3]{4}$, where $x, y, z \in \mathbb{Q}$;

k) the set of complex numbers of the form $x + yi$, where $x, y \in \mathbb{Z}$;

l) the set of complex numbers of the form $x + yi$, where $x, y \in \mathbb{Q}$;

m) the set of all possible sums of the form $a_1 z_1 + a_2 z_2 + \cdots + a_n z_n$, where $a_1, a_2, \ldots, a_n$ are real numbers, $z_1, z_2, \ldots, z_n$ are complex roots of 1;

n) the set of complex numbers of the form $\dfrac{x + y\sqrt{D}}{2}$, where D is a fixed square-free integer, (i.e. not divisible by the square of a prime), and x and y are integers of the same parity?

6302. Which of the sets of matrices listed below form a ring with respect to matrix addition and multiplication:

a) the set of real symmetric matrices of size n;

b) the set of real orthogonal matrices of size n;

c) the set of upper triangular matrices of size $n \geq 2$;

d) the set of matrices of size $n \geq 2$, in which the two last rows consist of zeros;

e) the set of matrices of the form $\begin{pmatrix} x & y \\ Dy & x \end{pmatrix}$, where D is a fixed integer, and $x, y \in \mathbb{Z}$;

f) the set of matrices of the form $\begin{pmatrix} x & y \\ Dy & x \end{pmatrix}$, where D is a fixed element of a ring K, $x, y \in K$;

g) the set of matrices of the form $\dfrac{1}{2} \begin{pmatrix} x & y \\ Dy & x \end{pmatrix}$, where D is a fixed square-free integer, and x and y are integers of the same parity;

h) the set of complex matrices of the form $\begin{pmatrix} z & w \\ -\bar{w} & \bar{z} \end{pmatrix}$;

i) the set of real matrices of the form

$$\begin{pmatrix} x & -y & -z & -t \\ y & x & -t & z \\ z & t & x & -y \\ t & -z & y & x \end{pmatrix}?$$

6303. Which of the following sets of functions form a ring with respect to the usual operations of addition and multiplication of functions:

a) the set of functions in a real variable continuous on a segment $[a, b]$;

b) the set of functions having the second derivative on the interval (a, b);

c) the set of *entire* rational functions in a real variable;

d) the set of rational functions in a real variable;

e) the set of functions in a real variable, vanishing on a subset $D \subseteq \mathbb{R}$;

f) the set of trigonometrical polynomials

$$a_0 + \sum_{k=1}^{n}(a_k \cos kx + b_k \sin kx)$$

with real coefficients, where n is an arbitrary natural number;

g) the set of trigonometrical polynomials of the form $a_0 + \sum_{k=1}^{n} \cos kx$ with real coefficients, where n is an arbitrary natural number;

h) the set of trigonometrical polynomials of the form $a_0 + \sum_{k=1}^{n} a_k \sin kx$ with real coefficients, where n is an arbitrary natural number;

i) the set of functions, defined on some set D with values in a ring R;

j) all power series in one or several variables;

k) all univariate Laurent series?

6304. In the set of polynomials in one variable t consider the operation of multiplication given by the rule

$$(f \circ g)(t) = f(g(t)).$$

Does this set form a ring with respect to this multiplication and ordinary addition?

6305. Does the set of all subsets of some set form a ring with respect to symmetric difference and intersection, considered as addition and multiplication, respectively?

6306. Prove that the rings in Exercises:

a) 6301m and 6302g;

b) 6302h and 6302i.

are isomorphic.

6307. Which of the rings, mentioned in Exercises 6301–6305, contain zero divisors?

6308. Find the invertible elements in rings with unit elements from Exercises 6301–6305.

6309. Prove that one of the rings from Exercises 6303e and 6303f is isomorphic to the ring of polynomials $\mathbb{R}[x]$, while the other one is not isomorphic to $\mathbb{R}[x]$.

6310. Prove that all invertible elements of a ring with unity form a group with respect to multiplication.

6311. Find all invertible elements, all zero divisors and all nilpotent elements in the rings:

a) $\mathbf{Z}_n$;

b) $\mathbf{Z}_{p^n}$, where p is a prime;

c) $K[x]/f K[x]$, where K is a field;

d) of the upper triangular matrices over a field;

e) $\mathbf{M}_2(\mathbb{R})$;

f) of all functions, defined on some set S with the values in a field K;

g) of all univariate power series.

6312. Let R be a finite ring. Prove that

a) if R does not contain zero divisors then it has a unit element and all its nonzero elements are invertible;

b) if R has a unit element then an element is invertible if it has a one-sided inverse;

c) if R has a unit element then any left zero divisor is a right zero divisor.

Are the statements b) and c) valid for rings without unit?

6313. Prove that in a ring with unity and without zero divisors each one-sided invertible element is invertible.

6314. Let R be a ring with unity and $x, y \in R$. Prove that

a) if the products xy and yx are invertible then the elements x and y are also invertible;

b) if R has no zero divisors and the product xy is invertible then x and y are invertible;

c) without additional assumptions, on a ring R, invertibility of the product xy does not imply the invertibility of x and y.

6315. Let R be a *direct sum* of rings $R_1, \ldots, R_k$.

a) Under what conditions is the ring R commutative, does it have a unit element, and have no zero divisors?

b) Find all invertible elements, all zero divisors, and all nilpotent elements in R.

6316. Prove that

a) if numbers k and l are coprime then

$$\mathbf{Z}_{kl} = \mathbf{Z}_k \oplus \mathbf{Z}_l;$$

b)   if $n = p_1^{k_1} \ldots p_s^{k_s}$, where $p_1, \ldots, p_s$ are distinct primes then

$$\mathbf{Z}_n = \mathbf{Z}_{p_1^{k_1}} \oplus \cdots \oplus \mathbf{Z}_{p_s^{k_s}};$$

c)   if numbers k and l are coprime then

$$\phi(kl) = \phi(k)\phi(l),$$

where ϕ is the Euler function.

6317. Find all zero divisors in $\mathbb{C} \oplus \mathbb{C}$.

6318. Prove that

a)   a zero divisor in any (associative) algebra is not invertible;

b)   in a finite-dimensional algebra with unity any nonzero divisor is invertible;

c)   a finite-dimensional algebra without zero divisors is a skew-field (a *division algebra*).

6319. Prove that

a)   a finite-dimensional algebra with unity and without zero divisors over the field $\mathbb{C}$ is isomorphic to $\mathbb{C}$;

b)   over the field $\mathbb{C}$ any finite-dimensional division algebra is isomorphic to $\mathbb{C}$.

6320. Determine up to isomorphism all commutative two-dimensional algebras over $\mathbb{C}$:

a)   with unity;

b)   not necessarily with unity.

6321. Determine up to isomorphism all commutative two-dimensional algebras over $\mathbb{R}$:

a)   with unity;

b)   not necessarily with unity.

6322. Let $\mathbb{H}$ be the skew-field of quaternions.

a)   Is $\mathbb{H}$ an algebra over $\mathbb{C}$ in the sense that multiplication by a scalar $\alpha \in \mathbb{C}$ is a left-side multiplication by $\alpha \in \mathbb{H}$?

b)   Prove that the mapping

$$1 \mapsto \begin{pmatrix} 1 & 0 \\ 0 & 1 \end{pmatrix}, \quad i \mapsto \begin{pmatrix} i & 0 \\ 0 & -i \end{pmatrix}, \quad j \mapsto \begin{pmatrix} 0 & 1 \\ -1 & 0 \end{pmatrix}, \quad k \mapsto \begin{pmatrix} 0 & i \\ i & 0 \end{pmatrix}$$

his an isomorphism of $\mathbb{H}$ as an algebra over $\mathbb{R}$ with a subalgebra of the algebra of matrices $\mathbf{M}_2(\mathbb{C})$ over $\mathbb{R}$.

c)   Prove that the mapping $z \mapsto \begin{pmatrix} z & 0 \\ 0 & \bar{z} \end{pmatrix}$ is an isomorphic embedding of the field $\mathbb{C}$ into the algebra $\mathbb{H}$ which is considered as a subalgebra of the algebra $\mathbf{M}_2(\mathbb{C})$ over $\mathbb{R}$ (see b)).

d)   Solve in $\mathbb{H}$ the equation $x^2 = -1$.

6323. The *tensor algebra* $T(V)$ of a vector space V over a field K is an (infinite-dimensional) vector space

$$T(V) = \oplus_{k=0}^{\infty} T_k(V),$$

where $T_0(V) = K$, $T_k(V) = \underbrace{V \otimes \cdots \otimes V}_{k \text{ times}}$ for $k > 0$, with multiplication
$f \cdot g = f \otimes g$, where $f \in T_k(V)$, $g \in T_m(V)$. Prove that

a) $T(V)$ is an associative algebra with unity over K;

b) $T(V)$ has no zero divisors.

6324. The *Grassman algebra* $\Lambda(V)$ of a vector space V over a field K is the vector space
$$\Lambda(V) = \oplus_{k=0}^{\infty} \Lambda^k(V),$$

where $\Lambda^0(V) = K$, with multiplication

$$f \cdot g = f \wedge g, \quad \text{where} \quad f \in \Lambda^k(V), \quad g \in \Lambda^m(V)$$

for all $k, m > 0$. Prove that

a) $\Lambda(V)$ is an associative algebra with unity over K;

b)   each element of $I = \oplus_{k \geq 1} \Lambda^k(V)$ is nilpotent;

c)   each element of $\Lambda(V) \setminus I$ is invertible.

6325. The *symmetric algebra* $S(V)$ of a vector space V over a field K is the vector space
$$S(V) = \oplus_{k=0}^{\infty} S^k(V),$$

where $S^0(V) = k$, with multiplication $f \cdot g = \text{Sym}(f \otimes g)$, where $f \in S^k(V)$, $g \in S^m(V)$ for all $k, m > 0$. Prove that

a) $S(V)$ is an associative, commutative algebra over K;

b) if $x_1, \ldots, x_n$ is a basis of the space V then $S(V)$ is isomorphic to the algebra of polynomials in $x_1, \ldots, x_n$.

6326. Let A and B be algebras over a field K. The *tensor product* of algebras $C = A \otimes_K B$ is the tensor product of the vector spaces A and B over K with multiplication

$$(a' \otimes b') \cdot (a'' \otimes b'') = a'a'' \otimes b'b''.$$

Prove that there exists an algebra isomorphism over K:

a) $\mathbb{C} \otimes_R \mathbb{C} \simeq \mathbb{C} \oplus \mathbb{C}$ $(K = \mathbb{R})$;

b) $\mathbf{M}_n(K) \otimes_K \mathbf{M}_m(K) \simeq \mathbf{M}_{mn}(K)$;

c) $\mathbf{M}_n(K) \otimes_K A \simeq \mathbf{M}_n(A)$, where A is an arbitrary associative algebra over K;

d) $K[X_1, \ldots, X_n] \otimes_K K[Y_1, \ldots, Y_m] \simeq K[X_1, \ldots, X_n, Y_1, \ldots, Y_m]$;

e) $\mathbb{H} \otimes_\mathbb{R} \mathbb{C} \simeq \mathbf{M}_2(\mathbb{C})$ $(K = \mathbb{C}$ or $\mathbb{R})$;

f) $S(V) \otimes \Lambda(V) = T(V)$ if $\dim V = 2$;

g) $\mathbb{Q}(\sqrt{p}) \otimes \mathbb{Q}(\sqrt{q}) \simeq \mathbb{Q}(\sqrt{p} + \sqrt{q})$, where p and q are distinct primes.

6327. Let K be a field of characteristic zero and let $R = K[X_1, \ldots, X_n]$ be a ring of polynomials. Let p_i, q_i be linear operators on R as a vector space over K such that for $f \in R$

$$p_i(f) = x_i f, \quad q_i(f) = \frac{\partial}{\partial x_i} f.$$

Denote by $A_n(K)$ the subalgebra of the algebra of linear operators on R generated by $p_1, \ldots, p_n, q_1, \ldots, q_n$. This subalgebra is called the *Weyl algebra* or the *algebra of differential operators*. Prove that

a) $q_j p_i - p_i q_j = \delta_{ij}, \quad p_i p_j = p_j p_i, \quad q_i q_j = q_j q_i$;

b) the monomials

$$p_1^{l_1} \ldots p_n^{l_n} q_1^{t_1} \ldots q_n^{t_n}, \quad l_i, t_j \geq 0$$

form a basis of $A_n(K)$ as a vector space over K.

6328. Let $f = f(p_1, \ldots, p_n, q_1, \ldots, q_n)$ be an element of the Weyl algebra $A_n(K)$ (see Exercise 6327.) Prove that

$$p_i f = f p_i + \frac{\partial f}{\partial q_i};$$
$$q_i f = f q_i - \frac{\partial f}{\partial p_i}.$$

6329. Prove that the algebra of upper niltriangular matrices of size n is a nilpotent algebra of index n.

6330. Prove that

a) zero divisors in the ring of all functions on the segment $[0, 1]$ are exactly the functions vanishing at some point;

b) zero divisors in the ring of continuous functions on the segment $[0, 1]$ are exactly the functions vanishing on some segment $[a, b]$, where $a < b$.

64   Ideals, homomorphisms, factor-rings

6401. Find all ideals of the rings:

a) $\mathbb{Z}$;

b) $K[x]$, where K is a field.

6402. Prove that

a) $\mathbb{Z}[x]$;

b) $K[x, y]$, where K is a field,

are not principal ideal rings.

6403. Prove that in the ring of matrices over a field, any two-sided ideal is either zero or coincides with the whole ring.

6404. Prove that ideals in the ring of matrices $\mathbf{M}_n(R)$, with entries in an arbitrary ring R, are exactly the sets of matrices whose entries belong to a fixed ideal of R.

6405. Find all ideals in the ring of integer upper triangular matrices of size 2.

6406. Let I and J be sets of matrices of the form

$$\begin{pmatrix} 0 & g & h \\ 0 & 0 & 2k \\ 0 & 0 & 0 \end{pmatrix} \quad \text{and} \quad \begin{pmatrix} 0 & l & 2m \\ 0 & 0 & 2n \\ 0 & 0 & 0 \end{pmatrix}$$

with integers g, h, k, l, m, n. Prove that I is an ideal in the ring R of upper triangular matrices over $\mathbb{Z}$, and J is an ideal of I but J is not an ideal in R.

6407. Find all left-sided ideals in the algebra $M_2(\mathbb{Z}_2)$.

6408. Find all ideals in a two-dimensional algebra L over $\mathbb{R}$ with a basis $(1, e)$, where 1 is the unit in L, and

a) $e^2 = 0$;

b) $e^2 = 1$.

6409. Prove that an ideal of a ring containing an invertible element coincides with the whole ring.

6410. Determine whether non-invertible elements of the rings:

a) $\mathbb{Z}$;   b) $\mathbb{C}[x]$;   c) $\mathbb{R}[x]$;   d) $\mathbb{Z}_n$

form an ideal.

6411. Prove that the ring of integers has no minimal ideals.

6412. Find maximal ideals of the rings:

a) $\mathbb{Z}$;   b) $\mathbb{C}[x]$;   c) $\mathbb{R}[x]$;   d) $\mathbb{Z}_n$.

6413. Prove that the set I_S of continuous functions vanishing on a fixed subset $S \subseteq [a, b]$ is an ideal in the ring of continuous functions on $[a, b]$.

Does any ideal of this ring have the form I_S for some $S \subseteq [a, b]$?

6414. Let R be the ring of continuous functions on the segment $[0,1], 0 \le c \le 1$, and $I_c = \{f(x) \in R \| \ f(c) = 0\}$. Prove that

a) I_c is a maximal ideal in R;

b) any maximal ideal of R coincides with I_c for some c.

6415. Prove that a commutative ring with unity (different from zero) is a field if it has only two ideals: the zero ideal and the whole ring. Is this statement valid for rings without unity?

6416. Prove that a ring with nonzero multiplication and without proper one-sided ideals is a skew-field.

6417. Prove that a ring with a unit element and without zero divisors is a skew-field if any decreasing chain of its left-sided ideals is finite.

6418. Let K be a commutative ring without zero divisors. Suppose that $\delta : K \setminus \{0\} \to \mathbb{N}$ is a mapping such that for any elements $a, b \in K$, where $b \neq 0$, there exist elements $q, r \in K$ for which $a = bq + r$ and either $\delta(r) < \delta(b)$ or $r = 0$.

Prove that there exists a mapping $\delta_1 : K \setminus \{0\} \to \mathbb{N}$ satisfying the previous condition and the following property: $\delta_1(ab) \geq \delta(b)$ for any $a, b \in K$, where $ab \neq 0$.

6419. Prove that

a)   the ring of Gaussian integers $x + iy$ $(x, y \in \mathbb{Z})$ is Euclidean;

b)   the ring of complex numbers $x + iy\sqrt{3}$ $(x, y \in \mathbb{Z})$ is not Euclidean;

c)   the ring of complex numbers of the form $\dfrac{x + iy\sqrt{3}}{2}$, where x and y are integers of the same parity, is Euclidean.

6420. Prove that any rectangular matrix with entries in a Euclidean ring, with the help of elementary row and column transformations, can be reduced to the form

$$\begin{pmatrix} e_1 & 0 & \cdots & 0 & 0 & \cdots & 0 \\ 0 & e_2 & \cdots & 0 & 0 & \cdots & 0 \\ \multicolumn{7}{c}{\cdots\cdots\cdots\cdots\cdots\cdots\cdots\cdots} \\ 0 & 0 & \cdots & e_r & 0 & \cdots & 0 \\ 0 & 0 & \cdots & 0 & 0 & \cdots & 0 \\ \multicolumn{7}{c}{\cdots\cdots\cdots\cdots\cdots\cdots\cdots\cdots} \\ 0 & 0 & \cdots & 0 & 0 & \cdots & 0 \end{pmatrix}$$

where $e_1|e_2|\ldots|e_r$, $e_i \neq 0$ $(i = 1, 2\ldots, r)$.

6421. Prove that in Exercise 6420 the product $e_1 \ldots e_i$ for $i = 1, \ldots, r$ coincides with the greatest common divisor of all minors of size i of the original matrix.

6422. Prove that a ring which contains a principal ideal ring R, and which is contained in its field of fractions Q, is a principal ideal ring.

6423. Prove that a ring of polynomials $R[x]$ over a commutative ring R with a unit element and without zero divisors is a principal ideal ring if and only if R is a field.

6424. Find all ideals in the algebra $\mathbb{C}[[x]]$.

6425. Prove that the Weil algebra $A_n(K)$ (see Exercise 6327) is simple if K is a field of zero characteristic.

6426. *Chinese remainder theorem.* Let A be a commutative ring with unity. Prove that

a)   if I_1 and I_2 are ideals of A and $I_1 + I_2 = A$ then for any elements $x_1, x_2 \in A$ there exists $x \in A$ such that $x - x_1 \in I_1$, $x - x_2 \in I_2$;

b)   if $I_1, \ldots, I_n$ are ideals of A and $I_i + I_j = A$ for all $i \neq j$ then for any elements $x_1, \ldots, x_n \in A$ there exists $x \in A$ such that $x - x_k \in I_k$ $(k = 1, \ldots, n)$.

6427. Let R and S be rings with unit elements and $\varphi : R \to S$ be a homomorphism.

a) Find out if the image of the unit of R is the unit of S?

b) Is the statement a) valid for a surjective homomorphism φ?

6428. Let K be a field and $K[x_1, \ldots, x_n]$ be an algebra of polynomials. Assume that $f_1, \ldots, f_n \in K[x_1, \ldots, x_n]$. Prove that

a) a mapping φ for which

$$\varphi(g(x_1, \ldots, x_n)) = g(f_1, \ldots, f_n)$$

is an endomorphism of K-algebra $K[x_1, \ldots, x_n]$;

b) if φ is an automorphism of $K[x_1, \ldots, x_n]$ then the Jacobian

$$J = \det \left(\frac{\partial f_i}{\partial x_j} \right)$$

is not equal to zero;

c) if $h = h(x_2 \ldots x_n)$ then the mapping Ψ such that

$$\Psi(g(x_1, \ldots, x_n)) = g(x_1 + h, x_2, \ldots, x_n)$$

is an automorphism of $K[x_1, \ldots, x_n]$.

6429. Let K be a field and $f_1, \ldots, f_n \in K[[x_1 \ldots x_n]]$ have zero constant terms. Prove that

a) a mapping φ such that

$$\varphi(g(x_1 \ldots x_n)) = g(f_1, \ldots, f_n)$$

is an endomorphism of $K[[x_1, \ldots, x_n]]$;

b) the mapping φ is an automorphism if and only if the Jacobian

$$J = \det \left(\frac{\partial f_i}{\partial x_j} \right)$$

has a nonzero constant term.

6430. Let K be a field of zero characteristic and $h = h(q_1) \in A_n(K)$. Prove that the mapping φ such that

$$\varphi(f(p_1, \ldots, p_n, q_1, \ldots, q_n)) = f(p_1 + h, p_2, \ldots, p_n, q_1, \ldots, q_n)$$

is an automorphism of the K-algebra $A_n(K)$.

6431. Let φ be an automorphism of the $\mathbb{C}$-algebra $\mathbf{M}_n(\mathbb{C})$. Prove that

a)   the left annihilator of the matrix $\varphi(E_{nn})$ has a dimension $n(n-1)$;

b)   the Jordan canonical form of the matrix $\varphi(E_{nn})$ is equal to E_{11};

c)   there exists an invertible matrix Y such that $Y^{-1}\varphi(E_{nn})Y = E_{nn}$;

d)   the mapping $A \rightarrow Y^{-1}\varphi(A)Y$ is an automorphism of $\mathbf{M}_n(\mathbb{C})$, which maps $\mathbf{M}_{n-1}(\mathbb{C})$ into itself;

e)   there exists an invertible matrix X such that $\varphi(A) = XAX^{-1}$ for all matrices $A \in \mathbf{M}_n(\mathbb{C})$.

6432. Let K be a field.

a)   Prove that the linear mapping

$$\varphi : \mathbf{M}_n(K) \otimes \mathbf{M}_m(K) \rightarrow \mathbf{M}_{nm}(K),$$

where $1 \le i, j \le n$, $1 \le r, s \le m$ and $\varphi(E_{ij} \otimes E_{rs}) = E_{i+nr,j+ns}$ is an isomorphism of K-algebras.

b)   Prove that the linear mapping

$$\Psi : \mathbf{M}_n(K) \rightarrow \mathbf{M}_n(K) \otimes \mathbf{M}_n(K),$$

where

$$\Psi(E_{ij}) = E_{ij} \otimes E_{ij},$$

is a homomorphism of K-algebras. Find $\mathrm{Ker}\Psi$.

6433. Prove that the image of a commutative ring under a homomorphism is a commutative ring.

6434. Prove that the mapping $\varphi : f(x) \rightarrow f(c)$ ($c \in \mathbb{R}$) is a homomorphism from the ring of real functions defined on $\mathbb{R}$ to $\mathbb{R}$.

6435. Find all homomorphisms of rings:

a) $\mathbb{Z} \rightarrow 2\mathbb{Z}$;   b) $2\mathbb{Z} \rightarrow 2\mathbb{Z}$;   c) $2\mathbb{Z} \rightarrow 3\mathbb{Z}$;   d) $\mathbb{Z} \rightarrow \mathbf{M}_2(\mathbb{Z}_2)$.

6436. Find all homomorphisms

a)   from the group $\mathbb{Z}$ to the group $\mathbb{Q}$;

b)   from the ring $\mathbb{Z}$ to the field $\mathbb{Q}$.

6437. Prove that any homomorphism from a field to a ring is either a zero one or is an isomorphic mapping onto some subfield of the ring.

6438. Let K be a field and $R = K[x_1, \ldots, x_n]$ be an algebra of polynomials in $x_1, \ldots, x_n$ over K. Construct a bijection between the space of rows K^n and the set of all K-algebra homomorphisms $R \to K$.

6439. Prove that

a) $F[x]/\langle x - \alpha \rangle \simeq F$ (F is a field);

b) $\mathbb{R}[x]/\langle x^2 + 1 \rangle \simeq \mathbb{C}$;

c) $\mathbb{R}[x]/\langle x^2 + x + 1 \rangle \simeq \mathbb{C}$.

6440. Find all elements a and b such that the factor-rings

$$\mathbb{Z}_2[x]/\langle x^2 + ax + b \rangle$$

a) are isomorphic;

b) are fields.

6441. Find whether these factor-rings are isomorphic:

$$\mathbb{Z}[x]/\langle x^3 + 1 \rangle \quad \text{and} \quad \mathbb{Z}[x]/\langle x^3 + 2x^2 + x + 1 \rangle.$$

6442. Find whether these factor-rings are isomorphic:

$$\mathbb{Z}[x]/\langle x^2 - 2 \rangle \quad \text{and} \quad \mathbb{Z}[x]/\langle x^2 - 3 \rangle.$$

6443. Let a and b be different elements of a field F. Prove that $F[x]$-modules $F[x]/\langle x - a \rangle$ and $F[x]/\langle x - b \rangle \simeq F$ are not isomorphic but the factor-rings $F[x]/\langle x - a \rangle$ and $F[x]/\langle x - b \rangle$ are isomorphic.

6444. Let $a \neq b$ and $c \neq d$ be elements of a field F. Prove that the factor-rings $F[x]/\langle (x - a)(x - b) \rangle F$ and $F[x]/\langle (x - c)(x - d) \rangle$ are isomorphic.

6445. Which of the following algebras are isomorphic over $\mathbb{C}$:

$$A_1 = \mathbb{C}[x, y]/\langle x - y, xy - 1 \rangle, \quad A_2 = \mathbb{C}[x]/\langle (x - 1)^2 \rangle,$$
$$A_3 = \mathbb{C} \oplus \mathbb{C}, \quad A_4 = \mathbb{C}[x, y], \quad A_5 = \mathbb{C}[x]/\langle x^2 \rangle?$$

6446. Find whether the algebras A and B over $\mathbb{C}$ are isomorphic:

a) $\quad A = \mathbb{C}[x, y]/\langle x^n - y \rangle, \qquad B = \mathbb{C}[x, y]/\langle x - y^m \rangle$;

b) $\quad A = \mathbb{C}[x, y]/\langle x^2 - y^2 \rangle, \qquad B = \mathbb{C}[x, y]/\langle (x - y)^2 \rangle$?

6447. Find if the following algebras over $\mathbb{R}$ are isomorphic:

a) $\quad A = \mathbb{R}[x]/\langle x^2 + x + 1 \rangle, \qquad B = \mathbb{R}[x]/\langle 2x^2 - 3x + 3 \rangle$;

b) $\qquad A = \mathbb{R}[x]/\langle x^2 + 2x + 1\rangle, \qquad B = \mathbb{R}[x]/\langle (x^2 - 3x + 2)\rangle$?

6448. Prove that the element f of the algebra $K[x]/\langle x^{n+1}\rangle$ (K is a field) is invertible if and only if $f(0) \neq 0$.

6449. Let K be a field and $f \in K[x]$ has degree n. Prove that the dimension of K-algebra $K[x]/fK[x]$ is equal to n.

6450. Let K be a field. Prove that

a) if the polynomials $f, g \in K[x]$ are coprime then

$$K[x]/fgK[x] \simeq K[x]/fK[x] \oplus K[x]/gK[x];$$

b) if $f = p_1^{k_1} \dots p_s^{k_s}$, where $p_1, \dots, p_s$ are coprime irreducible polynomials then

$$K[x]/fK[x] \simeq K[x]/p_1^{k_1}K[x] \oplus \cdots \oplus K[x]/p_s^{k_s}K[x].$$

6451. Prove that a factor-ring R/I of a commutative ring with unity is a field if and only if I is a maximal ideal of R.

6452. Prove that the ideal I of a commutative ring R is prime if and only if I is the kernel of a homomorphism from R to a field.

6453. Prove that

a) the factor-ring $\mathbb{Z}[i]/\langle 2\rangle$ is not a field;

b) the factor-ring $\mathbb{Z}[i]/\langle 3\rangle$ is a field with 9 elements;

c) $\mathbb{Z}[i]/\langle n\rangle$ is a field if and only if n is a prime which is not a sum of two squares of integers.

6454. Prove that the factor-ring $\mathbb{Z}[x]/\langle n\rangle$ for any integer $n > 1$ is isomorphic to $\mathbf{Z}_n[x]$.

6455. Let $f(x)$ be an irreducible polynomial of degree n in the ring $\mathbf{Z}_p[x]$ Prove that the factor-ring $\mathbf{Z}_p[x]/\langle f(x)\rangle$ is a finite field. Find the number of its elements.

6456. Prove that

a) any ring is isomorphic to a subring of a ring with unity;

b) an n-dimensional algebra with unity over a field F is isomorphic to a subalgebra of an algebra with unity of dimension $n + 1$;

c) an n-dimensional algebra with unity over a field K is isomorphic to a subalgebra of the algebra $\mathbf{M}_n(K)$;

d) an n-dimensional algebra over K is isomorphic to a subalgebra of the algebra $\mathbf{M}_{n+1}(K)$.

6457. Let $I_1, \ldots, I_s$ be ideals of an algebra A with unity, $I_i + I_j = A$ if $i \neq j$. Prove that the mapping $f : A/\bigcap_{k=1}^{s} I_k \mapsto A/I_1 \oplus \cdots \oplus A/I_s$, given by the formula $f(a + \bigcap_{k=1}^{s} I_k) = (a + I_1, \ldots, a + I_s)$, is an isomorphism of algebras.

6458. Establish the isomorphism

$$\mathbb{Q}[x]/\langle x^2 - 1 \rangle \simeq \mathbb{Q} \oplus \mathbb{Q}.$$

6459. Prove that $\mathbb{Q}[x]/\langle x^2 - 2 \rangle \simeq \mathbb{Q}[\sqrt{2}]$.

6460. Let I be a maximal ideal of $\mathbb{Z}[x]$. Prove that $\mathbb{Z}[x]/I$ is a finite field.

6461. Let V be a vector space over a field K of zero characteristic. Prove that $S(V) \simeq T(V)/I$, where I is the ideal of $T(V)$, generated by all elements $x \otimes y - y \otimes x$, where $x, y \in V$.

6462. Let V be a vector space over a field K of zero characteristic. Prove that $\Lambda(V) \simeq T(V)/I$, where I is the ideal of $T(V)$, generated by all elements $x \otimes y + y \otimes x$, where $x, y \in V$.

6463. Let V be a vector space of dimension $2n$ with a basis $p_1 \ldots p_n, q_1 \ldots q_n$ over a field K of zero characteristic. Prove that $A_n(K) \simeq T(V)/I$, where I is the ideal of $T(V)$, generated by all elements $p_i \otimes q_j - q_j \otimes p_i - \delta_{ij}, \ p_i \otimes p_j - p_j \otimes p_i,$ $q_i \otimes q_j - q_j \otimes q_i$.

6464. Let $(e_1, \ldots, e_n)$ be a basis of a vector space V over a field K of characteristic different from 2, and let $\Lambda(V)$ be the external (or Grassman algebra) over V. Prove that

a) $\dim \Lambda(V) = 2^n$;

b)   if $x_1, \ldots, x_{n+1} \in \Lambda^1(V) \oplus \cdots \oplus \Lambda^n(V)$, then $x_1 \ldots x_{n+1} = 0$;

c)   the formula

$$\varphi(e_i) = \sum_{j=1}^{n} a_{ij} e_j + \omega_i, \quad i = 1, \ldots, n$$

where $\omega_i \in \Lambda^1(V) \oplus \cdots \oplus \Lambda^n(V)$, defines an automorphism of $\Lambda(V)$ if and only if $\det(a_{ij}) \neq 0$.

6465. Let R be a ring with unity. The *left annihilator* of a subset $M \subseteq R$ is the set

$$\{x \in R \mid xm = 0 \quad \text{for} \quad \text{any} \quad m \in M\}.$$

Prove that

a)   the left annihilator of any subset is a left ideal of R;

b)   the left annihilator of a right ideal of R, generated by an idempotent, is also generated (as a left ideal) by some idempotent.

6466. Prove that a sum of finitely many left ideals generated by orthogonal idempotents is generated by an idempotent.

6467. Let I_k $(k = 1, \ldots, n)$ be a set of matrices of size n over a field K, consisting of all matrices in which all columns except the kth are equal to 0. Prove that

a) I_k is a left ideal of $\mathbf{M}_n(K)$;

b) I_k is a minimal submodule in $\mathbf{M}_n(K)$ considered as the left module over itself;

c) $\mathbf{M}_n(K) = I_1 \oplus \cdots \oplus I_n$;

d)   the module $\mathbf{M}_2(K)$ has a decomposition into the direct sum of minimal submodules which is different from the decomposition c);

e)   there exists a module isomorphism between these two decompositions of the module $\mathbf{M}_2(K)$.

6468. Prove that the sets of matrices:

a) $I = \left\{ \begin{pmatrix} x & 2x \\ y & 2y \end{pmatrix} \ (x, y \in K) \right\}, \quad J = \left\{ \begin{pmatrix} x & 0 \\ y & 0 \end{pmatrix} \ (x, y \in K) \right\},$

b) $I = \left\{ \begin{pmatrix} -x & 3x \\ -y & 3y \end{pmatrix} \ (x, y \in K) \right\}, \quad J = \left\{ \begin{pmatrix} 0 & x \\ 0 & y \end{pmatrix} \ (x, y \in K) \right\}$

are submodules of the ring $\mathbf{M}_2(K)$ considered as a left module over itself and $\mathbf{M}_2(K)/I \simeq J$.

6469. Let $R = I_1 \oplus I_2$ be a decomposition of a ring with unity e into the direct sum of two-sided ideals I_1, I_2 and $e = e_1 + e_2$, where $e_1 \in I_1, e_2 \in I_2$. Prove that e_1 and e_2 are units of the rings I_1 and I_2, respectively.

6470. Prove that the rings $\mathbf{Z}_{mn}$ and $\mathbf{Z}_m \oplus \mathbf{Z}_n$ are isomorphic if and only if m and n are coprime.

6471. A ring is *completely right (left) reducible* if it is a direct sum of right (left, respectively) ideals which are simple modules over the ring. For what integers n is the residue ring $\mathbf{Z}_n$ completely right (left) reducible?

6472. Prove that the algebra of all upper triangular matrices $n \geq 2$ over a field is not completely reducible.

6473. Prove that in a commutative completely reducible ring with unity the number of idempotents and the number of ideals are finite.

6474. Prove that in any completely right reducible algebra the intersection of all maximal ideals is equal to zero.

6475. Prove that any commutative completely reducible ring with unity is isomorphic to a direct sum of fields.

6476. A module is *completely reducible* if it is a direct sum of *minimal* submodules. What cyclic groups are completely reducible as $\mathbb{Z}$-modules?

6477. Prove that if a ring R is completely left reducible (see 6471) and I is a left ideal of R then $R = I \oplus J$ for some left ideal J of R.

6478. Prove that any left ideal of a completely left reducible ring R

    a) is completely reducible as a left R-module;

    b) is generated by an idempotent.

6479. Let R be a completely left reducible ring with unity. Prove that

    a) if R has no idempotents different from 0 and 1 then R is a skew-field;

    b) if R has no zero divisors then R is a skew-field.

Are these statements valid for rings in which the existence of a unit element is not assumed?

6480. Prove that if $xy = 0$ for any two elements x, y of a left ideal I of a completely left reducible ring R with unity then $I = \{0\}$.

6481. Prove that if I is an ideal of a ring R with unity then the factor-ring R/I has a unit element.

6482. Prove that a factor-ring of a commutative *Noetherian* ring is also Noetherian.

6483. Prove that the residue ring $\mathbb{Z}_{p_1 \ldots p_m}$, where $p_1, \ldots, p_m$ are distinct primes, is a direct sum of fields.

6484. Find all submodules of a vector space with a basis $(e_1, \ldots, e_n)$ as a module over the ring of all diagonal matrices, if

$$\mathrm{diag}(\lambda_1, \ldots, \lambda_n) \circ (\alpha_1 e_1 + \cdots + \alpha_n e_n) = \lambda_1 \alpha_1 e_1 + \cdots + \lambda_n \alpha_n e_n.$$

6485. Let R be a commutative ring with unity and without zero divisors. Consider R as a module over itself. Prove that R is isomorphic to any of its nonzero submodules if and only if R is a principal ideal ring.

6486. Prove that the rule $h(x) \circ f = h(x^r) f$, where $h(x)$ is a fixed polynomial, transforms the ring of polynomials $F[x]$ over a field F into a free module of rank r over $F[x]$.

* * *

6487. Let K be a field of zero characteristic. Prove that the polynomial algebra $K[x_1, \ldots, x_n]$ is a simple module over the Weyl algebra $A_n(K)$ (see Exercise 6327).

6488. Let K be a field of zero characteristic. Prove that each nonzero module over the Weil algebra $A_n(K)$ has infinite dimension over K.

65 Special classes of algebras

6501. Prove that a ring of univariate polynomials over a commutative Noetherian ring with unity is Noetherian.

6502. Prove that an algebra of polynomials in a finite number of variables over a field is Noetherian.

6503. The *algebra* $A(\alpha, \beta)$ *of generalized quaternions* over a field F of characteristic different from 2, where $\alpha, \beta \in F^*$, is defined as the vector space over F with the basis $(1, i, j, k)$ and with the multiplication table

$$1 \cdot 1 = 1, \quad 1 \cdot i = i \cdot 1 = i,$$
$$1 \cdot j = j \cdot 1 = j, \quad 1 \cdot k = k \cdot 1 = k,$$
$$i^2 = -\alpha, \quad j^2 = -\beta, \quad ij = -ji = k.$$

Prove the following.

a) $A(\alpha, \beta)$ is an (associative) *central simple* algebra over F.

b) The mapping

$$x = x_0 + x_1 i + x_2 j + x_3 k \mapsto x_0 - x_1 i - x_2 j - x_3 k = \bar{x}$$

is an *involution* (i.e. $\overline{x+y} = \bar{x} + \bar{y}$, $\overline{xy} = \bar{y}\bar{x}$, $\bar{\bar{x}} = x$) for all $x, y \in A(\alpha, \beta)$.

c) For any $x \in A(\alpha, \beta)$,

$$x^2 - (\mathrm{tr}x)x + N(x) = 0,$$

where $\mathrm{tr}x = x + \bar{x}$ and $N(x) = x\bar{x}$ are elements of F.

d) The algebra $A(\alpha, \beta)$ is a skew-field if and only if the norm equation $N(x) = 0$ has only a zero solution.

e) The algebra $A(\alpha, \beta)$ is either a skew-field (if it has no zero divisors) or isomorphic to the algebra of matrices $M_2(F)$ (if it has zero divisors).

f) If the norm equation has a nonzero solution in $A(\alpha, \beta)$, then it has also a solution in the set of nonzero *pure* quaternions.

g) A subalgebra $F(a)$ of $A(\alpha, \beta)$ generated by an element a is a commutative algebra of dimension ≤ 2 over F, and if a is not a zero divisor then $F(a)$ is a field isomorphic to the splitting field of the polynomial $x^2 - (\text{tr}a)x + N(a)$.

i) *Witt's theorem.* The norm $N(x)$ is a quadratic form of rank 3 on the space of pure quaternions and, conversely, each quadratic form of rank 3 on a three-dimensional vector space W over F corresponds to an algebra of generalized quaternions, defined on a vector space $F \oplus W$ by multiplication

$$1 \cdot w = w \cdot 1,$$
$$w_1 \cdot w_2 = -Q(w_1, w_2) \cdot 1 + [w_1, w_2],$$

where Q is a bilinear form on W associated with the given quadratic form, and $S[w_1, w_2]$ is the vector product of elements of W.

j) The construction mentioned above establishes a bijective correspondence between quaternion algebras over F (up to isomorphism) and equivalence classes of quadratic forms of rank 3 in a three-dimensional vector space over F. (Note: the forms $Q : W \times W \to F$ and $Q' : W' \times W' \to F$ are equivalent if there exist an isomorphism $\alpha : W \to F$ and an element $\lambda \in F^*$ such that $Q'(\alpha(x), \alpha(y)) = \lambda Q(x, y)$ for all $x, y \in W$).

6504. A finite-dimensional algebra is *semisimple* if it does not contain nonzero nilpotent ideals. Prove that

a) a factor-algebra $\mathbb{C}[x]/\langle f(x) \rangle$ is semisimple if and only if the polynomial $f(x)$ has no multiple roots;

b) an algebra generated by the field $\mathbb{C}$ and a matrix A in the algebra $\mathbf{M}_n(\mathbb{C})$ is semisimple if and only if the minimal polynomial of A has no multiple roots;

c) a finite-dimensional algebra over a field is semisimple if and only if it is completely left reducible;

d) a commutative semisimple algebra with unit is isomorphic to a direct sum of fields;

e) if all idempotents of a semisimple algebra are central then the algebra is a direct sum of skew-fields.

6505. Let $H = (h_{ij})$ be a symmetric $(n \times n)$-matrix over a field F. A *Clifford algebra* is a $2n$-dimensional vector space $\mathbb{C}(F, H)$ over F with the basis consisting of symbols

$$e_{i_1\ldots i_k} \quad (1 \le i_1 < i_2 < \cdots < i_k \le n) \quad \text{and} \quad e_0 = 1,$$

and with multiplication determined by the rules

$$e_i e_i = h_{ii}, \quad e_0 e_i = e_i e_0 = e_i, \quad e_i e_j + e_j e_i = h_{ij},$$
$$e_{i_1\ldots i_k} = e_{i_1} \ldots e_{i_k} \quad (1 \le i_1 < \cdots < i_k \le n).$$

Let V be an n-dimensional vector space with a basis $(e_1, \ldots, e_n)$ and Q be a quadratic form on V. The Clifford algebra $\mathbb{C}_Q(F)$ of the quadratic form Q is defined as the algebra $\mathbb{C}(F, H)$, where $h_{ij} = Q(e_i, e_j)$.

a)   Prove that if $H = 0$ then $\mathbb{C}(F, H) \simeq \Lambda(V)$.

b)   The *even Clifford algebra* $\mathbb{C}^+(F, H)$ (or $\mathbb{C}_Q^+(F)$) is the subalge-
     bra of the Clifford algebra generated by the elements $e_{i_1} \cdot \ldots \cdot$
     $e_{i_{2m}}$ $\left(m = 0, 1, \ldots, \left[\frac{n}{2}\right]\right)$. Prove that the even Clifford algebra of the
     quadratic form

$$Q(x_1, x_2, x_3) = h_{11}x_1^2 + h_{12}x_1 x_2 + h_{22}x_2^2,$$

which is not a product of linear factors in F, is a quadratic extension of F.
This extension is isomorphic to the splitting field $F\left(\sqrt{h_{12}^2 - 4h_{11}h_{22}}\right)$
of the form Q.

c)   Prove that if $\operatorname{char} F \ne 2$ then the even Clifford algebra of a quadratic form
     Q on a three-dimensional vector space V is isomorphic to an algebra of
     generalized quaternions of the form $Q^{(2)}$ on the three-dimensional vector
     space $W = \Lambda^2 V$ (see Exercise 6503).

d)   Prove that under the assumptions of c) the quadratic form $N(x) = x\bar{x}$ on
     the space of pure quaternions is equivalent to the form λQ ($\lambda \in F^*$).

6506. Let $A = A_0 \oplus A_1$ be a 2-graded associative algebra over a field K, i.e.
$A_i A_j \subset A_{i+j}$ (addition of indices modulo 2). Define on A a new operation setting
$[x, y] = xy - (-1)^{ij} yx$, where $x \in A_i$, $y \in A_j$.

a)   Prove that

$$[x, y] = (-1)^{ij}[y, x],$$
$$[x, [y, z]] + [y, [z, x]] + (-1)^{ij+1}[z, [x, y]] = 0,$$

for all homogeneous elements $x \in A_i$, $y \in A_j$, $z \in A$. The algebra with
2-grading for which the homogeneous elements satisfy the given relations
is called the *Lie superalgebra*.

b) Let V be an n-dimensional vector space with a basis $(e_1, \ldots, e_n)$ over K of characteristic not equal to 2 and $\Lambda(V)$ be an external algebra on V. Let I be the identity operator on V, $L_0 = K \cdot I$ and let L_1 be the linear span of operators φ_i and ψ_i, where

$$\varphi_i(w) = w \wedge e_i,$$

$$\psi_i(e_{i_1} \wedge \cdots \wedge e_{i_p}) = \begin{cases} (-1)^{p-k} e_{i_1} \wedge \cdots \wedge \hat{e}_{i_k} \wedge \cdots \wedge e_{i_p}, & \text{if } i_k = i, \\ 0, & \text{if } i_k \neq i \text{ for all } k = 1, \ldots, p. \end{cases}$$

Prove that $L = L_0 \oplus L_1$ is a Lie superalgebra with respect to the operation introduced in a).

6507. Let K be an extension of the field $\mathbb{Q}$ of degree n. Prove that

a) for any polynomial $f(x) \in \mathbb{Q}[x]$ of degree n, there exists a matrix A of size n such that $f(A) = 0$;

b) the algebra $\mathbf{M}_n(\mathbb{Q})$ contains a subalgebra which is isomorphic to K;

c) if L is a subalgebra of $\mathbf{M}_n(\mathbb{Q})$ which is a field, then $[L : \mathbb{Q}] \leq n$.

6508. Does the $\mathbb{C}$-algebra of analytical functions defined on a domain $U \subseteq \mathbb{C}$ have zero divisors?

6509. A function of a complex variable is *entire* if it is analytical in the complex plane. Prove that any finitely generated ideal of the algebra of entire functions is a principal one.

6510. A *differentiation* of a ring R is a mapping $D : R \to R$ satisfying the conditions:

(1) $$D(x + y) = D(x) + D(y);$$
(2) $$D(xy) = D(x)y + xD(y) \quad (x, y \in R).$$

Find all differentiations of the rings:

a) $\mathbb{Z}$;

b) $\mathbb{Z}[x]$;

c) $\mathbb{Z}[x_1, x_2, \ldots, x_n]$.

6511. Let L be a set with operations of addition and multiplication. Suppose that L is a commutative group with respect to addition, and multiplication $\circ$ is connected with addition by distributivity. L is a *Lie ring*, if for any $x, y, z \in L$ we have:

(1) $$x \circ x = 0;$$
(2) $$(x \circ y) \circ z + (y \circ z) \circ x + (z \circ x) \circ y = 0 \quad (\textit{Jacobi identity}).$$

Prove that

a)   the Lie ring satisfies the identity $x \circ y = -y \circ x$;

b)   vectors in three-dimensional space form a Lie ring with respect to addition and vector multiplication;

c)   any ring R is a Lie ring with respect to addition and multiplication $x \circ y = xy - yx$;

d)   a set of all differentiations of a ring R is a Lie ring with respect to addition and multiplication $D_1 \circ D_2 = D_1 D_2 - D_2 D_1$.

6512. Let K be a field and D be a differentiation of the K-algebra of matrices $\mathbf{M}_n(K)$.
Prove that there exists a matrix $A \in \mathbf{M}_n(K)$ such that $D(X) = AX - XA$ for all X.

6513. Let K be a field of zero characteristic and let D be a differentiation of the Weyl algebra $A_n(K)$. Prove that there exists an element $f \in A_n(K)$ such that $D(g) = fg - gf$ for all $g \in A_n(K)$.

6514. Prove that the semigroup ring $R[S]$ of an ordered semigroup S has no zero divisors if and only if R has no zero divisors.

6515. Let p be a prime and $\mathbb{Z}_p$ be the ring of p-adic integers, i.e. the set of all formal series $\sum_{i \geq 0} a_i p^i$, where $a_i \in \mathbb{Z}$ and $0 \leq a_i < p$. Put

$$\sum_{i \geq 0} a_i p^i + \sum_{i \geq 0} b_i p^i = \sum_{i \geq 0} c_i p^i$$

$$\left(\sum_{i \geq 0} a_i p^i \right) \left(\sum_{i \geq 0} b_i p^i \right) = \sum_{i \geq 0} d_i p^i,$$

if for any $n \geq 0$ in $\mathbf{Z}_{p^n}$

$$\sum_{i=0}^{n-1} a_i p^i + \sum_{i=0}^{n-1} b_i p^i = \sum_{i=0}^{n-1} c_i p^i$$

$$\left(\sum_{i=0}^{n-1} a_i p^i \right) \left(\sum_{i=0}^{n-1} b_i p^i \right) = \sum_{i=0}^{n-1} d_i p^i.$$

Prove that

a) $\mathbb{Z}_p$ is a domain containing $\mathbb{Z}$;

b)   an element $\sum_{i \geq 0} a_i p^i$ is invertible in $\mathbb{Z}_p$ if and only if $a_0 = 1, 2, \dots, p - 1$;

c) the natural homomorphism of the groups of invertible elements $Z_p^* \to Z_{p^n}^*$ is surjective for all n;

d) every ideal of $\mathbb{Z}_p$ is principal and it is of the form (p^n), $n \geq 0$.

Find all prime elements in $\mathbb{Z}_p$.

6516.

a) Prove that the field of p-adic numbers $\mathbb{Q}_p$, i.e. the field of fractions of $\mathbb{Z}_p$, consists of elements of the form $p^m h$, where $m \in \mathbb{Z}$, $h \in \mathbb{Z}_p$.

b) Show that $\mathbb{Q}$ is a subfield of $\mathbb{Q}_p$.

c) Prove that the element $p^m (\sum_{i \geq 0} a_i p^i)$ from $\mathbb{Q}_p$, where $0 \leq a_i \leq p - 1$, belongs to $\mathbb{Q}$ if and only if elements a_i, $i \geq N$, starting from some N, form a periodic sequence.

d) Find in $\mathbb{Q}_5$ the images of $\dfrac{2}{7}$ and $\dfrac{1}{3}$.

6517. Let K be a field and p be an irreducible polynomial in one variable X with coefficients in K. Construct, by analogy with Exercise 6515, the ring $K[X]_p$ and its field of fractions $K(X)_p$. Show that if p has degree 1 then $K[X]_p \simeq K[[X]]$.

6518. Find all subrings of the field of rational numbers $\mathbb{Q}$ which contain the unit element.

66   Fields

6601. Which of the rings in Exercises 6301–6303 are fields?

6602. Which of the following sets of matrices form a field with respect to the usual matrix operations:

a) $\left\{ \begin{pmatrix} x & y \\ ny & x \end{pmatrix}; \quad x, y \in \mathbb{Q} \right\}$, where n is a fixed integer;

b) $\left\{ \begin{pmatrix} x & y \\ ny & x \end{pmatrix}; \quad x, y \in \mathbb{R} \right\}$, where n is a fixed number;

c) $\left\{ \begin{pmatrix} x & y \\ ny & x \end{pmatrix}; \quad x, y \in \mathbb{Z}_p \right\}$, where $p = 2, 3, 5, 7$?

6603. Let K be a field and F be a field of fractions of $K[[x]]$. Prove that every element in F is a product $x^{-s} h$ where $s \geq 0$ and $h \in K[[x]]$.

6604. Prove that an order of the unit of a field in its additive group is either infinite or a prime.

6605. For which of the numbers $n = 2, 3, 4, 5, 6, 7$ does there exist a field with n elements?

6606. Prove that the field with p^2 elements, where p is a prime, has a unique proper subfield.

6607. Prove that the fields $\mathbb{Q}$ and $\mathbb{R}$ have only identical automorphism.

6608. Find all automorphisms of the field $\mathbb{C}$ which fix each real number.

6609. Does the field $\mathbb{Q}(\sqrt{2})$ have a nonidentical automorphism?

6610. Prove that in a field F of characteristic p

a) $(x + y)^{p^m} = x^{p^m} + y^{p^m}$ (m is a natural number);

b) if F is finite, then the mapping $x \mapsto x^p$ is an automorphism.

6611. Prove that if a complex number z is not real then the ring $\mathbb{R}[z]$ coincides with $\mathbb{C}$.

6612. For which $m, n \in \mathbb{Z} \setminus \{0\}$ are the fields $\mathbb{Q}(\sqrt{m})$ and $\mathbb{Q}(\sqrt{n})$ isomorphic?

6613. Prove that the set of elements fixed by an automorphism φ of a field K is a subfield.

6614. Prove that any two fields with four elements are isomorphic.

6615. Does there exist a field properly containing the field of complex numbers?

6616. Prove that any finite field has a positive characteristic.

6617. Does there exist an infinite field with a positive characteristic?

6618. Solve, in the field $\mathbb{Q}(\sqrt{2})$, the equations:

a) $x^2 + (4 - 2\sqrt{2})x + 3 - 2\sqrt{2} = 0$;

b) $x^2 - x - 3 = 0$;

c) $x^2 + x - 7 + 6\sqrt{2} = 0$;

d) $x^2 - 2x + 1 - \sqrt{2} = 0$.

6619. Solve the system of equations

$$x + 2z = 1, \quad y + 2z = 2, \quad 2x + z = 1$$

a) in the field $\mathbb{Z}_3$;

b) in the field $\mathbb{Z}_5$.

6620. Solve the system of equations

$$3x + y + 2z = 1, \quad x + 2y + 3z = 1, \quad 4x + 3y + 2z = 1$$

in the residue fields modulo 5 and modulo 7.

6621. Find a polynomial $f(x)$ of degree at most 3 with coefficients in $\mathbf{Z}_5$ such that $f(0) = 3$, $f(1) = 3$, $f(2) = 5$, $f(4) = 4$.

6622. Find all polynomials $f(x)$ with coefficients in $\mathbf{Z}_5$ such that $f(0) = f(1) = f(4) = 1$, $f(2) = f(3) = 3$.

6623. Which of the equations

a) $x^2 = 5$,   b) $x^7 = 7$,   c) $x^3 = a$

have solutions in the field $\mathbf{Z}_{11}$?

6624. In the residue field modulo 11 solve the equations

a) $x^2 + 3x + 7 = 0$;

b) $x^2 + 5x + 1 = 0$;

c) $x^2 + 2x + 3 = 0$;

d) $x^2 + 3x + 5 = 0$.

6625. Prove that a field with n elements satisfies the identity $x^n = x$.

6626. Solve in a field $\mathbf{Z}_p$ the equation $x^p = a$.

6627. Prove that if $x^n = x$ for all elements x of a field K then K is finite and its characteristic divides n.

6628. Find all generators of the multiplicative group of the fields:

a) $\mathbf{Z}_7$;   b) $\mathbf{Z}_{11}$;   c) $\mathbf{Z}_{17}$.

6629. Let a, b be elements of a field of order 2^n where n is odd. Prove that if $a^2 + ab + b^2 = 0$ then $a = b = 0$.

6630. Let F be a field such that the group F^* is cyclic. Prove that F is finite.

6631. Solve the equations:

a) $f^4 = 1$;   b) $f^2 - f - x = 0$,

in the field of rational functions with real coefficients.

6632. Prove that in a field $\mathbf{Z}_p$

a) $\displaystyle\sum_{k=1}^{p-1} k^{-1} = 0 \quad (p > 2)$;

b) $\displaystyle\sum_{k=1}^{(p-1)/2} k^{-2} = 0 \quad (p > 3)$.

6633. Let $n \geq 2$ and $\zeta_1, \ldots, \zeta_m$ be all roots of 1 of order n in a field K. Prove that

a) $\{\zeta_1, \ldots, \zeta_m\}$ is a multiplicative group;

b) $\{\zeta_1, \ldots, \zeta_m\}$ are roots of 1 of order m;

c) m divides n;

d) if $k \in \mathbb{Z}$ then

$$\zeta_1^k + \cdots + \zeta_m^k = \begin{cases} 0, & \text{if } m \text{ does not divide } k \\ m, & \text{if } m \text{ divides } k. \end{cases}$$

6634. Let $m_k m_{k-1} \ldots m_0$ and $n_k n_{k-1} \ldots n_0$ be records of the natural numbers m and n in the number systems with the radix s, where s is a prime. Prove that

a) the numbers $\binom{m}{n}$ and $\binom{m_0}{n_0}\binom{m_1}{n_1} \ldots \binom{m_n}{n_n}$ have the same remainder after division by s;

b) $\binom{m}{n}$ is divisible by s if and only if $m_i < n_i$ for some i.

6635. An *absolute value* on a field K is a function $\|x\|$, $x \in K$, taking real non-negative values such that

(1) $\|x\| = 0$ if and only if $x = 0$;

(2) $\|xy\| = \|x\|\|y\|$;

(3) $\|x + y\| \leq \|x\| + \|y\|$.

Prove that the following functions on $\mathbb{Q}$ are absolute values:

a) $\|x\| = \begin{cases} 1, & x \neq 0, \\ 0, & x = 0; \end{cases}$

b) $\|x\| = |x|^s$ where s is a fixed number, $0 < s \leq 1$;

c) $\|x\| = |x|_p^s$ where p is a prime and $s < 1$ is a fixed positive number; here $|x|_p = p^{-r}$ if $x = p^r mn^{-1}$, where m, n are integers which are not divisible by p.

6636. Let $\|x\|$ be a valuation on $\mathbb{Q}$ and let y be an element such that $\|y\| \neq 0, 1$. Then $\|x\|$ is either of the form b) or of the form c) from Exercise 6635.

6637. Let K be a field and $K(x)$ be a field of rational functions in one variable x. Prove that the following functions on $K(x)$ are valuations:

a) $\|f\| = \begin{cases} 1, & \text{if } f \neq 0, \\ 0, & \text{if } f = 0; \end{cases}$

b) $\|hg^{-1}\| = c^{\deg h - \deg g}$, where $h, g \in K[x]$ and $0 < c < 1$;

c) if $p(x)$ is an irreducible polynomial, $h = p(x)^r u(x) v(x)$, where $u(x)$, $v(x)$ are polynomials coprime with $p(x)$, then $|h| = c^r$ where $0 < c < 1$.

6638. Prove that

a) the completion of $\mathbb{Q}$ with respect to the valuation from Exercise 6637b is equal to $\mathbb{R}$;

b) the completion of $\mathbb{Q}$ with respect to the valuation from Exercise 6637c is equal to $\mathbb{Q}_p$;

c) the completion of $\mathbb{Z}$ with respect to the valuation from Exercise 6637b is equal to $\mathbb{Z}_p$;

d) the completion of $\mathbb{C}[x]$ with respect to the valuation from Exercise 6639c with $p = x$ is equal to $\mathbb{C}[[x]]$.

6639. The sequence x_n, $n \geq 1$ of elements of $\mathbb{Q}_p$ converges in a metric $\|f\|$ from 6637c if and only if $\lim_{n \to \infty} \|x_n - x_{n+1}\|_p = 0$.

6640. For which $t \in \mathbb{Q}_p$ do the series:

a) $e^t = \sum \frac{t^n}{n!}$;

b) $\ln(1 + x) = \sum_{n \geq 1} \frac{1}{n} (-1)^n t^n$;

c) $\sum_{n \geq 0} x^n$

converge?

6641. Let $a \in \mathbb{Q}_p$ and $x_n = a^{p^n}$. Does there exist $\lim_{n \to \infty} x_n$?

6642. Let $f(x) \in \mathbb{Z}_p[x]$, $a_0 \in \mathbb{Z}_p$, and $\|f(a_0)/f'(a_0)^2\|_p < 1$. Put

$$a_{n+1} = a_n - \frac{f(a_n)}{f'(a_n)}.$$

Prove that $a = \lim a_n$ exists, $\|a - a_0\|_p < 1$ and $f(a) = 0$.

6643. Prove that any automorphism of $\mathbb{Q}_p$ is identical.

6644. Let $f(x) \in \mathbb{Z}_p[x]$ have degree n and the leading coefficient of $f(x)$ be equal to 1. Let the image $\overline{f(x)}$ of the polynomial $f(x)$ in $\mathbb{Z}_p[x]$ be factorized,

$\overline{f(x)} = g(x)h(x)$, where $g(x)$, $h(x)$ are coprime and have the leading coefficients 1. Suppose that $\deg g(x) = r$, $\deg h(x) = n - r$. Then $f(x) = u(x)v(x)$, where $\deg u(x) = r$, $\deg v(x) = n - r$, the leading coefficients of $u(x)$, $v(x)$ are equal to 1 and the images of $u(x)$, $v(x)$ in $\mathbb{Z}_p[x]$ are equal to $g(x)$ and $h(x)$, respectively.

6645. Let $f(x) \in \mathbb{Z}_p[x]$, $a \in \mathbb{Z}_p$, and

$$f(a) = 0, \quad f'(a) \neq 0$$

in $\mathbb{Z}_p$. Then there exists an element $b \in \mathbb{Z}_p$ such that $f(b) = 0$ and its image in $\mathbb{Z}_p$ is equal to a.

6646. Let m be a natural number which is not divisible by p, $a \in 1 + p\mathbb{Z}_p$. Then there exists $b \in \mathbb{Z}_p$ such that $b^m = a$.

6647. Let the fields $\mathbb{Q}_p$ and $\mathbb{Q}_{p'}$ be isomorphic. Prove that $p = p'$.

6648. Prove that the ring $\mathbb{Z}_p$ is compact in $\mathbb{Q}_p$ with respect to p-adic topology.

67   Extensions of fields. Galois theory

In this section all rings and algebra are assumed to be commutative and with unit elements.

6701. Let A be an algebra over a field K and let

$$K = K_0 \subset K_1 \subset K_1 \subset \cdots \subset K_s$$

be a tower of subfields in A. Prove that

$$(A : K) = (A : K_s)(K_s : K_{s-1}) \ldots (K_1 : K_0).$$

6702. Let A be an algebra over a field K and $a \in A$. Prove that

a)   if a is not algebraic over K then the subalgebra $K[a]$ is isomorphic to the ring of polynomials $K[x]$;

b)   if a is algebraic over K then $K[a] \simeq K[x]/\langle \mu_a(x) \rangle$, where $\mu_a(x)$ is a uniquely defined unitary polynomial (the *minimal polynomial* of a) over K;

c)   if A is a field and a is algebraic over K then $\mu_a(x)$ is irreducible in $K[x]$;

d)   if all elements in A are algebraic over K, and for any $a \in A$ the polynomial $\mu_a(x)$ is irreducible, then A is a field.

6703. Find minimal polynomials for:

a) $\sqrt{2}$ over $\mathbb{Q}$;

b) $\sqrt[7]{5}$ over $\mathbb{Q}$;

c) $\sqrt[105]{9}$ over $\mathbb{Q}$;

d) $2 - 3i$ over $\mathbb{R}$;

e) $2 - 3i$ over $\mathbb{C}$;

f) $\sqrt{2} + \sqrt{3}$ over $\mathbb{Q}$;

g) $1 + \sqrt{2}$ over $\mathbb{Q}(\sqrt{2} + \sqrt{3})$.

6704. Prove that

a) if A is a finite-dimensional algebra over K then any element of A is algebraic over K;

b) if $a_1, \ldots, a_s \in A$ are algebraic elements over K then the subalgebra $K[a_1, \ldots, a_s]$ is finite-dimensional over K.

6705. Prove that if A is a field and $a_1, \ldots, a_s \in A$ are algebraic elements over K then the extension $K(a_1, \ldots, a_s)$ coincides with the algebra $K[a_1, \ldots, a_s]$.

6706. Prove that the set of all elements of K-algebra A, algebraic over K, is a subalgebra of A. If A is a field then this subalgebra is a subfield.

6707. Prove that if in the tower of fields $K = K_0 \subset K_1 \subset K_1 \subset \cdots \subset K_s = L$ each stage $K_{i-1} \subset K_i$ $(i = 1, \ldots, s)$ is an algebraic extension, then L/K is an algebraic extension.

6708. Prove that any polynomial with coefficients in a field K has a root in some extension L/K.

* * *

6709. Let K be a field. Prove that

a) for an arbitrary polynomial from $K[x]$ there exists a *splitting field* of this polynomial over K;

b) for any finite set of polynomials from $K[x]$ there exists a splitting field over K.

6710. Let K be a field, $g(x) \in K[x]$, $h(x) \in K[x]$, $f(x) = g(h(x))$, and α be a root of $g(x)$ in some extension L/K. Prove that f is irreducible over K if and only if $g(x)$ is irreducible over K and $h(x) - \alpha$ is irreducible over $K[\alpha]$.

6711. Let K be a field, $a \in K$. Prove that

a)   if p is a prime then the polynomial $x^p - a$ either is irreducible or has a root in K;

b)   if the polynomial $x^n - 1$ can be factorized into linear factors in $K[x]$ then either the polynomial $x^n - a \in K[x]$ is irreducible or the polynomial $x^d - a$ has a root in K for some divisor $d \neq 1$ of n;

c)   the assumption about the factorization of $x^n - 1$ into linear factors is essential for the validity of statement b).

6712. Prove that the polynomial $f(x) = x^p - x - a$ either is irreducible over a field K of characteristic $p \neq 0$ or it can be factorized over K into a product of linear factors. Determine this factorization if $f(x)$ has a root x_0.

6713. Find the degree of the splitting field over $\mathbb{Q}$ for the polynomials:

a) $ax + b \quad (a, b \in \mathbb{Q}, \quad a \neq 0)$;

b) $x^2 - 2$;

c) $x^3 - 1$;

d) $x^3 - 2$;

e) $x^4 - 2$;

f) $x^p - 1 \quad (p \text{ is a prime})$;

g) $x^n - 1 \quad (n \in \mathbb{N})$;

h) $x^p - a \ (a \in \mathbb{Q} \text{ is not a } p\text{th power in } \mathbb{Q}, p \text{ is a prime})$;

i) $(x^2 - a_1) \ldots (x^2 - a_n) \ (a_1, \ldots, a_n \in \mathbb{Q}^* \text{ are distinct})$.

6714. Prove that a finite field extension L/K is simple if and only if the set of intermediate fields between K and L is finite. Find an example of a finite extension which is not simple.

6715. Let L/K be an algebraic field extension. Prove that the extension $L(x)/K(x)$ is also algebraic and $(L(x) : K(x)) = (L : K)$.

6716. Let L/K be a field extension. The elements $a_1, \ldots, a_s \in L$ are said to be *algebraically independent over* K, if $f(a_1, \ldots, a_s) \neq 0$ for any nonzero polynomial $f(x_1, \ldots, x_s) \in K[x_1, \ldots, x_s]$. Prove, that $a_1, \ldots, a_s \in L$ are algebraically independent over K if and only if the extension $K(a_1, \ldots, a_s)$ is K-isomorphic to the field of rational functions $K(x_1, \ldots, x_s)$.

6717. Let L/K be a field extension and $a_1, \ldots, a_m; b_1, \ldots, b_n$ two maximal algebraically independent over K systems of elements of L. Prove, that $m = n$ (the *transcendence degree* of L over K).

6718. Prove that

a) there exist finitely many maximal ideals in finite-dimensional commutative K-algebra A and their intersection coincides with the set $N(A)$ of all nilpotent elements of A, (the *nilradical* of A);

b) zero is the only nilpotent element in A^{red} (the algebra $A^{\text{red}} = A/N(A)$ is called *reduced*);

c) the algebra $A/N(A)$ is isomorphic to a direct product of fields $K_1, \ldots, K_s$ which are extensions of K;

d) $s \le (A : K)$;

e) the set of extensions K_i is defined for the algebra A up to isomorphism[1];

f) if B is a subalgebra of A then any component of B is an extension of one or several components of A;

g) if I is an ideal of A then the components of the algebra A/I are contained among the components of A.

6719. Let K be a field, $f(x) \in K[x]$, $p_1(x)^{k_1} \ldots p_s(x)^{k_s}$ be a factorization of $f(x)$ into a product of powers of distinct irreducible polynomials over K, $A = K[x]/\langle f(x)\rangle$. Prove that $A^{\text{red}} = A/N(A) \simeq \prod_{i=1}^{s} k[x]/\langle p_l(x)\rangle$.

6720. Let A be a K-algebra and L be an extension of the field K. Prove that

a) if $f_1, \ldots, f_n$ are distinct K-homomorphisms $A \to L$ then $f_1, \ldots, f_n$ are linearly independent as elements of the L-vector spaces of all K-linear mappings $A \to L$;

b) the number of distinct K-homomorphisms $A \to L$ does not exceed $(A : L)$.

Find all automorphisms of the fields $\mathbb{Q}(\sqrt{2})$, $\mathbb{Q}(\sqrt{2} + \sqrt{3})$, $\mathbb{Q}(\sqrt[3]{2})$.

6721. Let A be a finite-dimensional K-algebra and L be an extension of the field K. Let $A_L = L \otimes_K A$. Let $(e_1, \ldots, e_n)$ be a basis of A over K. Prove that

a) $(1 \otimes e_1, \ldots, 1 \otimes e_n)$ is a basis of A_L over L;

b) the image of A under the natural embedding of A into A_L is a K-subalgebra of A_L.

6722. Let A be a finite-dimensional K-algebra, L/K be a field extension. Prove that

---

[1]Extensions $K_1,\ldots,K_s$ together with canonical homomorphisms $A \to K_s$ are called the *components* of A.

a)   if B is a subalgebra of A then B_L is a subalgebra of A_L;

b)   if I is an ideal of A and I_L is the corresponding ideal of A_L then
     $(A/I)_L \simeq A_L/I_L$;

c)   if $A = \prod_{i=1}^{s} A_i$, $A_L \simeq \prod_{i=1}^{s}(A_i)_L$;

d)   if $K_1, \ldots, K_s$ is the set of components of A, then the set of the components
     of A_L coincides with the union of sets of the components of the algebras
     $(K_1)_L, \ldots, (K_s)_L$;

e)   if F/L is a field extension then $(A_L)_F \cong A_F$.

6723. Let A be a finite-dimensional K-algebra and let L/K be a field extension.
Suppose that B is L-algebra. Prove that

a)   each K-homomorphism $A \to B$ has a unique extension to L-
     homomorphism $A_L \to B_L$;

b)   the set of K-homomorphisms $A \to L$ is in bijective correspondence with
     the set of components of A_L which are isomorphic to L;

c)   the number of distinct K-homomorphisms $A \to L$ does not exceed
     $(A : K)$ (see Exercise 6720b).

6724. Let F/K and L/K be field extensions and suppose that F/K is finite.
Prove that there exists an extension E/K with embeddings of F and L into E such
that all elements of K are fixed.

6725. Let A be a finite-dimensional K-algebra and $A = K[a_1, \ldots, a_s]$. Prove
that the following properties of a field extension L/K are equivalent:

a)   all components of A_L are isomorphic to L;

b) L is the *splitting field* for the minimal polynomial of any $a \in A$ (the
     splitting field for K-algebra A).

6726. Prove that if L is a splitting field for K-algebra A and B is a subalgebra
of A then any K-homomorphism $B \to L$ can be extended to K-homomorphism
$A \to L$.

6727. A splitting field L for the finite-dimensional K-algebra A is the *field of
decomposition* for A if any of its proper subfields containing K is not a splitting
field A. Prove that

a)   if $A = K[a_1, \ldots, a_s]$, then L is the field of decomposition for A if and
     only if L is the field of decomposition for the minimal polynomials of
     the elements $a_1, \ldots, a_s$;

b)   any two fields of decomposition for A are K-isomorphic;

c) there exists K-embedding of the field of decomposition for A into any splitting field for A.

6728. Let A be a finite-dimensional K-algebra and let L be the field of decomposition for A. Prove that the number of components of the L-algebra A_L is the same for all splitting fields of A (the *separable degree* $(A : K)_s$ of A over K).

6729. Let A be a K-algebra and L/K be a field extension. Prove that

a) the number of the components of A_L does not exceed $(A : K)_s$;

b) the number of distinct K-homomorphisms $A \to L$ does not exceed $(A : K)_s$ and the equality takes place if and only if L is a splitting field for A.

6730. Prove that the following properties of finite field extension L/K are equivalent:

a) all components of the algebra L_L are isomorphic to L;

b) L has $(L : K)$ K-automorphisms;

c) for any K-embeddings $\varphi_i : L \to L'$ $(i = 1, 2)$ of L into any extension L'/K, we have $\varphi_1(L) = \varphi_2(L)$;

d) any irreducible polynomial in $K[x]$ with a root in L can be factorized into a product of linear factors over L;

e) L is a field of decomposition of a polynomial from $K[x]$. (An extension L/K satisfying these conditions is called *normal*).

6731. Let $K \subset L \subset F$ be a tower of finite extensions of a field K. Prove that

a) if the extension F/K is normal then the extension F/L is also normal;

b) if extensions L/K and F/L are normal then F/K is not necessarily normal;

c) any extension of degree 2 is normal.

6732. Let A be a finite-dimensional K-algebra and $a \in A$. The characteristic polynomial, the determinant and the trace of the linear operator $t \to at$ on A are denoted by $\chi_{A/K}(a, x)$, $N_{A/K}(a)$, $\mathrm{tr}_{A/K}(a)$, respectively, and are called the *characteristic polynomial*, the *norm* and the *trace*, respectively, of $a \in A$ over K.

Prove that if $K \subset L \subset F$ is a tower of finite field extensions and $a \in F$ then:

a) $\chi_{F/K}(a, x) = N_{L(x)/K(x)}(\chi_{F/L}(a, x))$, where $\chi_{F/L}(a, x)$ is considered as an element of the field of rational functions $K(x)$;

b) $N_{F/K}(a) = N_{L/K}(N_{F/L}(a))$;

c) $\mathrm{tr}_{F/K}(a) = \mathrm{tr}_{L/K}(\mathrm{tr}_{F/L}(a))$.

6733. Let L/K be a finite extension and $a \in L$. Prove that

a)   the minimal polynomial of a is equal to $\pm \chi_{K(a)/K}(a, x)$;

b) $\chi_{L/K}(a, x)$ is (up to a sign) a power of the minimal polynomial of a.

6734. Let L/K be a finite field extension. Prove that the K-bilinear form on L

$$(x, y) \mapsto \operatorname{tr}_{L/K}(xy)$$

either is nonsingular or $\operatorname{tr}_{L/K}(x) = 0$ for all $x \in L$.

6735. Prove that the following properties of a finite-dimensional K-algebra A are equivalent:

a)   for any field extension L/K the algebra A_L is reduced (see Exercise 6718);

b) $(A : K)_s = (A : K)$ (Exercise 6718);

c)   for some field extension L/K there exists $(A : K)$ K-homomorphisms $A \to L$;

d)   the bilinear form $(x, y) \mapsto \operatorname{tr}_{A/K}(xy)$ on A is nonsingular. (An algebra A satisfying these conditions is called *separable*.)

6736. Let L/K be a field extension. Prove that a finite-dimensional K-algebra A is separable if and only if A_L is a separable L-algebra.

6737. Prove that any subalgebra and any factor-algebra of a separable K-algebra is a separable K-algebra.

6738. Let A be a separable K-algebra, $(A : K) = n$, $\varphi_1, \ldots, \varphi_n$ be distinct K-homomorphisms of A into some its splitting field L. Prove that for any $a \in A$

$$\operatorname{tr}_{A/K}(a) = \sum_{i=1}^{n} \varphi_i(a), \quad N_{A/K}(a) = \prod_{i=1}^{n} \varphi_i(a),$$

$$\chi_{A/K}(a, x) = \prod_{i=1}^{n}(\varphi_i(a) - x).$$

6739. A finite field extension L/K is called *separable* if L is a separable K-algebra.

a)   Prove that a separable extension of a field is simple.

b)   Find out if $a = -\dfrac{1}{2} + i\dfrac{\sqrt{2}}{2}$ and $b = \sqrt{2} + i$ are primitive elements of the extension $\mathbb{Q}(\sqrt{2}, i)/\mathbb{Q}$?

6740. Prove that a finite-dimensional K-algebra is separable if and only if it is a direct product of separable extensions of the field K.

6741. Let $K = K_0 \subset K_1 \subset \cdots \subset K_s = L$ be a tower of finite field extensions. Prove that L/K is separable if and only if each extension K_i/K_{i-1} $(i = 1, \ldots, s)$ is separable.

6742. Let K be a field. A polynomial $f(x) \in K[x]$ is *separable* if it has no multiple roots in any extension of K. Prove that

    a)   if K has characteristics 0 then any irreducible polynomial from $K[x]$ is separable;

    b)   if K has characteristics $p \neq 0$ then an irreducible polynomial $f(x) \in K[x]$ is separable if and only if it cannot be represented in the form $g(x^p)$ where $g(x) \in K[x]$.

Find an example of a nonseparable irreducible polynomial.

6743. Let A be a finite-dimensional K-algebra. An element $a \in A$ is *separable* over K if $K[a]$ is a separable K-algebra. Prove that an element is separable if and only if its minimal polynomial is separable.

6744. Let $K \subset L \subset F$ be a tower of finite field extensions. Prove that

    a)   if an element $a \in F$ is separable over K then a is separable over L;

    b)   the converse statement is valid if the extension L/K is separable.

6745. Let A be a separable K-algebra and $f(x) \in K[x]$ be a separable polynomial. Prove that the algebra $B = A[x]/\langle f(x) \rangle$ is separable.

6746. Let $A = K[a_1, \ldots, a_s]$ be a finite-dimensional K-algebra. Prove that the following conditions are equivalent:

    a) A is a separable K-algebra;

    b)   any element $a \in A$ is separable;

    c)   elements $a_1, \ldots, a_s$ are separable.

6747. Prove that

    a)   a finite field extension K/F is separable if and only if either K has the characteristic 0 or the characteristic of K is equal to $p > 0$ and $K^p = K$;

    b)   any finite extension of a finite field is separable.

6748. A finite field extension L/K of characteristic $p > 0$ is *purely nonseparable* if L/K has no separable elements over K. Prove, that L/K is purely nonseparable if and only if $L^{p^k} \subseteq K$ for some $k \geq 1$.

6749. Let $K \subset K_0 \subset K_1 \cdots \subset K_s = L$ be a tower of finite extensions of fields. Prove that an extension L/K is purely nonseparable if and only if each extension K_i/K_{i-1} $(i = 1, \ldots, s)$ is purely nonseparable.

6750. Prove that the degree of a purely nonseparable extension of a field of characteristic $p > 0$ is a power of p and its separable degree is equal to 1.

6751. Let L/K be a finite field extension. Prove that

a) the set K_s of all elements of L which are separable over K is a subfield, separable over K;

b) L/K_s is a purely nonseparable extension;

c) $(K_s : K) = (L : K)_s$;

d) $(L : K) = (L : K)_s \cdot (L : K)_i$, where $(L : K)_i = (L : K_s)$, the *nonseparable* degree of L/K.

6752. Let $K \subset L \subset F$ be a tower of finite field extensions. Prove that

a) $(F : K)_s = (F : L)_s \cdot (L : K)_s$;

b) $(F : K)_i = (F : L)_i \cdot (L : K)_i$.

6753. Let L/K be a finite field extension, $n = (L : K)_s$ and let $\varphi_1, \ldots, \varphi_n$ be the set of all K-embeddings of L into some splitting field of L/K. Prove that for any $a \in L$

a) $\mathrm{tr}_{L/K}(a) = (L : K)_i \sum_{j=1}^{n} \varphi_j(a)$;

b) $N_{L/K}(a) = \left(\prod_{j=1}^{n} \varphi_j(a) \right)^{(L:K)_i}$;

c) $\chi_{L/K}(a, x) = \left(\prod_{j=1}^{n} (\varphi_j(a) - x) \right)^{(L:K)_i}$.

6754. A normal separable field extension L/K is called a *Galois extension*, a group of K-automorphisms of this extension is called a *Galois group* and it is denoted by $G(L/K)$. Prove that

a) $G(L/K)$ acts transitively on the set of roots in L for the minimal polynomial of any element of L;

b) the order of the group $G(L/K)$ is equal to the degree of L/K.

6755. Find the Galois group of the extensions:

a) $\mathbb{C}/\mathbb{R}$;

b) $\mathbb{Q}(\sqrt{2})/\mathbb{Q}$;

c) L/K, where $(L : K) = 2$;

d) $\mathbb{Q}(\sqrt{2} + \sqrt{3})/\mathbb{Q}$.

6756. The *Galois group* of a separable polynomial $f(x) \in K[x]$ over a field K is the Galois group of the field of decomposition for $f(x)$ over K. It is a group of permutations on the sets of roots of $f(x)$. Find Galois groups over $\mathbb{Q}$ of the polynomials from Exercise 6713.

6757. Let G be a finite group of automorphisms of a field L and $K = L^G$ be its subfield of fixed elements. Prove that L/K is a Galois extension and $G(L/K) = G$.

6758. Prove that if elements $a_1, \ldots, a_n$ are algebraically independent over a field K then the Galois group of the polynomial $x^n + a_1 x^{n-1} + \cdots + a_n$ over the field of rational functions $K(a_1, \ldots, a_n)$ is equal to $\mathbf{S}_n$.

6759. Prove that any finite group is a Galois group of some field extension.

6760. *Fundamental theorem of Galois theory.* Let L/K be a field extension and G be its Galois group. Prove that the juxtaposition of a subfield L^H of fixed elements to a subgroup $H \subset G$ determines a bijective correspondence between all subgroups of G and all intermediate subfields of L/K. Under this bijection an intermediate subfield F corresponds to the subgroup $H = G(L/F)$. An extension F/K is normal if and only if the subgroup H is normal in G and in this case the canonical map $G \to G(F/K)$ induces an isomorphism $G(F/K) \simeq G/H$.

6761. Using the fundamental theorem of Galois theory and the existence of a real root for any real polynomial of odd degree, prove that the field of complex numbers is algebraically closed.

6762. Prove that the Galois group of any finite field extension $L/\mathbf{F}_p$ is cyclic and is generated by the automorphism $x \mapsto x^p$ ($x \in L$).

6763. Prove that the Galois group of a separable polynomial $f(x) \in K[x]$ over a field K, considered as a subgroup of $\mathbf{S}_n$, is contained in the alternating group if and only if the discriminant $D = \prod_{i>j}(x_i - x_j)^2$ of $f(x)$ is a square in K, where $x_1, \ldots, x_n$ are the roots of $f(x)$ in its field of decomposition.

6764. Let L/K be a Galois extension with the cyclic Galois group $\langle \varphi \rangle_n$. Prove that there exists $a \in L$ such that elements $a, \varphi(a), \ldots, \varphi^{n-1}(a)$ form a basis of L over K.

6765. Let L/K be a separable extension of degree n and $\varphi_1, \ldots, \varphi_n$ be distinct K-embeddings of L into some splitting field for L. Prove that $a \in L$ is primitive in L/K if and only if the images $\varphi_1(a), \ldots, \varphi_n(a)$ are distinct.

6766. Find the automorphism group of K-algebra A which is the direct product of n copies of fields isomorphic to K.

6767. Let L/K be a Galois extension with Galois group G, $L = \prod L_\sigma$, where L_σ is a component of the algebra L_L such that the projection from L onto L_σ induces an automorphism σ of L_σ. Let e_σ be the unit of L_σ. Prove that extensions of automorphisms from G to L-automorphisms of L_L satisfy the property $\tau(e_\sigma) = e_{\sigma\tau^{-1}}$ $(\sigma, \tau \in G)$.

6768. Let L be a splitting field for a separable K-algebra A and $\varphi_1, \ldots, \varphi_n$ be the set of all K-homomorphisms $A \to L$. Prove that $y_1, \ldots, y_n \in A$ form a basis of A over K if and only if $\det(\varphi_i(y_j)) \neq 0$.

6769. *Normal basis theorem.* Prove that a Galois extension L/K with the Galois group G has an element $a \in L$ such that the set $\{\sigma(a) \mid \sigma \in G\}$ is a basis of L over K.

6770. Find the field of invariants $K(x_1, \ldots, x_n)^{A_n}$ for the group $\mathbf{A}_n$ acting in the field of rational functions by permutations of the variables.

6771. Let ε be a primitive complex root of 1 of order n and a group $G = \langle\sigma\rangle_n$ act on the field $\mathbb{C}(x_1, \ldots, x_n)$ by the rule $\sigma(x_1) = \varepsilon^i x_i$ $(i = 1, \ldots, n)$. Find the field of invariants $\mathbb{C}(x_1, \ldots, x_n)^G$.

6772. Find the field of invariants for the group G from Exercise 6771 acting on the field $\mathbb{C}(x_1, \ldots, x_n)$ by cyclic permutation of the variables.

6773. Let a field K contain all roots of 1 of order n. Suppose that an element $a \in K$ is not a dth power for any divisor $d > 1$ of n. Find the Galois group of the polynomial $x^n - a$ over K.

6774. Let a field K contain all roots of 1 of order n and L/K be a Galois extension with cyclic Galois group of order n. Prove that $L = K(\sqrt[n]{a})$ for some element $a \in K$.

6775. Let a field K contain all roots of 1 of order n. Prove that a finite field extension L/K is a Galois extension with an abelian Galois group of *period n* if and only if $L = K(\theta_1, \ldots, \theta_s)$ where $\theta_i^n = a_i \in K$ $(i = 1, \ldots, s)$ (i.e. L is the field of decomposition over K for the polynomial $\prod_{i=1}^s (x_i^n - a_i)^s$).

6776. Let a field K contain all roots of 1 of order n and $L = K(\theta_1, \ldots, \theta_s)$, where $\theta_i^n = a_i \in K^*$ $(i = 1, \ldots, s)$. Prove that

$$G(L/K) \simeq \langle (K^*)^n, a_1, \ldots, a_s\rangle/(K^*)^n.$$

6777. Let a field K contain all roots of 1 of order n. Establish a bijective correspondence between the set of all (up to K-isomorphism) Galois extensions with abelian Galois groups of period n and the set of all finite subgroups of the group $K^*/(K^*)^n$.

6778. Prove that any Galois extension L/K of degree p of a field K of characteristic $p > 0$ has the form $L = K(\theta)$, where θ is a root of the polynomial $x^p - x - a$ $(a \in K)$, and, conversely, any extension of this form is a Galois extension of degree 1 or p.

6779. Let K be a field of characteristic $p > 0$. Prove that a finite field extension L/K is a Galois extension of period p if and only if $L = K(\theta_1, \ldots, \theta_s)$, where θ_i is a root of the polynomial $x^p - x - a_i$ $(a_i \in K; i = 1, \ldots, s)$.

6780. Let K be a field of characteristic $p > 0$ and $L = K(\theta_1, \ldots, \theta_s)$, where θ_i is a root of the polynomial $x^p - x - a_i$ ($a_i \in K, i = 1, \ldots, s$). Prove that

$$G(L/K) \simeq \langle \rho(K), a_1, \ldots, a_s \rangle / K,$$

where $\rho : K \to K$ is the additive homomorphism $x \mapsto x^p - x$.

6781. Let K be a field of characteristic $p > 0$. Establish a bijective correspondence between the set of all (up to K-isomorphism) Galois extensions L/K with the abelian Galois group of period p and the set of all finite subgroups of the group $K/\rho(K)$.

68 Finite fields

6801. Prove that any finite extension of a finite field is simple.

6802. Prove that

a) a finite extension of a finite field is normal;

b) any two finite extensions of a finite field F of the same degree are F-isomorphic.

6803. Prove that

a) for any prime power q there exists a unique field $\mathbf{F}_q$ with q elements (up to isomorphism);

b) there exists an embedding of the field $\mathbf{F}_q$ into the field $\mathbf{F}_{q'}$ if and only if q' is a power of q;

c) if K and L are finite extensions of a finite field F then there exists F-embedding of K into L if and only if $(K : L)|(L : F)$;

d) if a polynomial $f(x)$ over a finite field F can be factorized into a product of irreducible factors of degrees $n_1, \ldots, n_s$ then the degree of the field of decomposition for $f(x)$ over F is equal to the least common multiple of $n_1, \ldots, n_s$.

6804. Let F be a finite field of odd order q. An element $a \in F^*$ is a *quadratic residue* in F if the binomial $x^2 - a$ has a root in F. Prove that

a) the number of quadratic residues is equal to $(q - 1)/2$;

b) a is a quadratic residue if $a^{(q-1)/2} = 1$, and it is not a quadratic residue if $a^{(q-1)/2} = -1$.

$$* \quad * \quad *$$

6805. Let F be a finite field, an let $\left(\dfrac{a}{F}\right)$ for an element $a \in F^*$ be equal to 1 if a is a quadratic residue in F and equal to -1 otherwise. Prove that

a)   the map $F^* \to \{-1, 1\}$ such that $a \mapsto \left(\dfrac{a}{F}\right)$ is a group homomorphism;

b) $\left(\dfrac{a}{F}\right) = \mathrm{sgn}\,\sigma_a$, where $\sigma_a : x \mapsto ax$ is a permutation on the set F.

6806. Let a and b be coprime integers and $\sigma : x \to ax$ be a permutation on the set of residue classes modulo b. Prove that

a)   if b is even then $\mathrm{sgn}\,\sigma = \begin{cases} 1 & \text{for } \quad b \equiv 2 \pmod 4, \\ (-1)^{(a-1)/2} & \text{for } \quad b \equiv 0 \pmod 4; \end{cases}$

b)   if b is odd, and $b = \displaystyle\prod_{i=1}^{s} p_i$ ($p_1, \ldots, p_s$ are primes), then $\mathrm{sgn}\,\sigma =$

$\displaystyle\prod_{i=1}^{s} \left(\dfrac{n}{p_i}\right)$, where $\left(\dfrac{a}{p_i}\right) = \left(\dfrac{a}{\mathbf{Z}_{p_i}}\right)$ is the *Legendre symbol* (in this case

$\mathrm{sgn}\,\sigma$ is denoted by $\left(\dfrac{a}{b}\right)$ and is called the *Jacobi symbol*;

c) $\left(\dfrac{a}{b_1 b_2}\right) = \left(\dfrac{a}{b_1}\right)\left(\dfrac{a}{b_2}\right),\quad \left(\dfrac{a_1 a_2}{b}\right) = \left(\dfrac{a_1}{b}\right)\left(\dfrac{a_2}{b}\right);$

d) $\left(\dfrac{-1}{b}\right) = (-1)^{(b-1)/2}.$

6807. Let G be an additive finite abelian group of odd order and σ be an automorphism of G, $\left(\dfrac{\sigma}{G}\right) = \mathrm{sgn}\,\sigma$, where σ is considered as a permutation on the set G. Prove that if G is a disjoint union $\{0\} \cup S \cup \{-S\}$ then $\left(\dfrac{\sigma}{G}\right) = (-1)^{|\sigma(S) \cap \sigma(-S)|}.$

6808. Let σ be an automorphism of the group G from Exercise 6807, G_1 be a subgroup of G invariant under σ, $G_2 = G/G_1$ and σ_1, σ_2 be automorphisms of G_1 and G_2 induced by σ. Prove that

$$\left(\dfrac{\sigma}{G}\right) = \left(\dfrac{\sigma_1}{G_1}\right)\left(\dfrac{\sigma_2}{G_2}\right),$$

and deduce from here Statement 6806b.

6809. The *Gauss lemma*. Prove that if N is the number of integers x in the segment $1 \le x \le \left(\dfrac{b-1}{2}\right)$ for which $ax \equiv r \pmod b$, $-\left(\dfrac{b-1}{2}\right) \le r \le 1$, then $\left(\dfrac{a}{b}\right) = (-1)^N.$

6810. Prove that $\left(\dfrac{2}{b}\right) = (-1)^{(b^2-1)/8}$.

6811. The *quadratic reciprocity law*. Prove that

$$\left(\frac{a}{b}\right)\left(\frac{b}{a}\right) = (-1)^{\frac{a-1}{2}\cdot\frac{b-1}{2}}.$$

for any coprime odd integers a and b.

6812. Let V be a finite-dimensional space over a finite field F of odd order, and A be a nonsingular linear operator on V. Prove that

$$\left(\frac{A}{V}\right) = \left(\frac{\det A}{F}\right).$$

6813. Let $F/\mathbf{F}_q$ be a finite field extension of degree n. Prove that there exists a basis of $\mathbf{F}_q$-vector space F of the form $x, x^q, \ldots, x^{q^{m-1}}$ for some $x \in F$.

6814. Prove that elements $x_1, \ldots, x_n \in \mathbf{F}_{q^n}$ form a basis of $\mathbf{F}_{q^n}$ over $\mathbf{F}_q$ if and only if

$$\det\begin{pmatrix} x_1 & x_2 & \cdots & x_n \\ x_1^q & x_2^q & \cdots & x_n^q \\ \vdots & \vdots & \ddots & \vdots \\ x_1^{q^{n-1}} & x_2^{q^{n-1}} & \cdots & x_n^{q^{n-1}} \end{pmatrix} \neq 0.$$

6815. Let $a \in \mathbf{F}_{q^n}$. Elements $a, a^q, \ldots, a^{q^{n-1}}$ form a basis of $\mathbf{F}_{q^n}$ over $\mathbf{F}_q$ if and only if the polynomials $x^n - 1$ and

$$ax^{n-1} + a^q x^{n-2} + \cdots + a^{q^{n-2}} x + a^{q^{n-1}}$$

in $\mathbf{F}_{q^n}[x]$ are coprime.

CHAPTER 15

Elements of representation theory

69 Representations of groups. Basic concepts

6901. Prove that the mapping $\rho : \mathbb{Z} \to \mathbf{GL}_2(\mathbb{C})$ for which

$$\rho(n) = \begin{pmatrix} 1 & n \\ 0 & 1 \end{pmatrix} \quad (n \in \mathbb{Z}),$$

is a reducible two-dimensional complex representation of the group $\mathbb{Z}$. It is not equivalent to a direct sum of two one-dimensional representations.

6902. Prove that the mapping $\rho : \langle a \rangle_p \to \mathbf{GL}_2(\mathbf{F}_p)$ (p is a prime), for which

$$\rho(a^k) = \begin{pmatrix} 1 & k \cdot 1 \\ 0 & 1 \end{pmatrix},$$

is a reducible two-dimensional representation of the cyclic group $\langle a \rangle_p$. It is not equivalent to a direct sum of two one-dimensional representations.

6903. Let $A \in \mathbf{GL}_n(\mathbb{C})$. Prove that the mapping $\rho_A : \mathbb{Z} \to \mathbf{GL}_n(\mathbb{C})$ for which $\rho_A(n) = A^n$ is a representation of the group $\mathbb{Z}$. Representations ρ_A and ρ_B are equivalent if and only if the Jordan normal forms of matrices A and B coincide (up to a permutation of Jordan boxes).

6904. Find out whether the mappings L given by the formulae

a) $(L(t)f)(x) = f(x - t)$;

b) $(L(t)f)(x) = f(tx)$;

c) $(L(t)f)(x) = f(e^t x)$;

d) $(L(t)f)(x) = e^t f(x)$;

289

e) $(L(t)f)(x) = f(x) + t$;

f) $(L(t)f)(x) = e^t f(x + t)$,

define linear representations of the additive group $\mathbb{R}$ in the space $C(\mathbb{R})$ of continuous functions on a real line.

6905. Which of these subspaces of $C(\mathbb{R})$ are invariant under the linear representation L from Exercise 6904a:

a)   the subspace of infinitely differentiable functions;

b)   the subspace of polynomials;

c)   the subspace of polynomials of degree $\leq n$;

d)   the subspace of even functions;

e)   the subspace of odd functions;

f)   the linear span of the functions $\sin x$ and $\cos x$;

g)   the subspace of the polynomials in $\cos x$ and $\sin x$;

h)   the linear span of the functions $\cos x, \cos 2x, \ldots, \cos nx$;

i)   the linear span of the functions $e^{c_1 t}, e^{c_2 t}, \ldots, e^{c_n t}$, where $c_1, \ldots, c_n$ are distinct fixed real numbers?

6906. Find subspaces of the space of polynomials which are invariant under the representation L from Exercise 6904a.

6907. Find matrices (in some basis) of the linear representation L from Exercise 6905 restricted to the subspace of polynomials of degree ≤ 2.

6908. Find matrices (in some basis) of the linear representation L from Exercise 6905 restricted to the linear span of the functions $\sin x$ and $\cos x$.

6909. Prove that each of the following formulae defines a linear representation of the group $\mathbf{GL}_n(F)$ in the space $\mathbf{M}_n(F)$:

a) $\Lambda(A) \cdot X = AX$;

b) $\mathrm{Ad}(A) \cdot X = AXA^{-1}$;

c) $\Phi(A) \cdot X = AX^t A$.

6910. Prove that the linear representation Λ (see Exercise 6909a) is completely reducible and its invariant subspaces coincide with left ideals of the algebra $\mathbf{M}_n(K)$.

6911. Prove that if $\mathrm{char}\, F$ does not divide n then the linear representation Ad (see Exercise 6909b) is completely reducible and its only nontrivial invariant subspaces are the space of matrices with zero trace and the spaces of scalar matrices.

6912. Prove that if $\operatorname{char} F \neq 2$ then the linear representation Φ (see Exercise 6909c) is completely reducible and its only nontrivial invariant subspaces are spaces of symmetric and skew-symmetric matrices.

6913. Let V be a two-dimensional space over a field F. Show that there exist representations ρ_1 and ρ_2 of the group S_3 in V for which, in some basis of the space V,

$$\rho_1((1\,2)) = \begin{pmatrix} 0 & 1 \\ 1 & 0 \end{pmatrix}, \quad \rho_1((1\,2\,3)) = \begin{pmatrix} 0 & -1 \\ 1 & -1 \end{pmatrix},$$

$$\rho_2((1\,2)) = \begin{pmatrix} 0 & 1 \\ 1 & 0 \end{pmatrix}, \quad \rho_2((1\,2\,3)) = \begin{pmatrix} 0 & 1 \\ -1 & -1 \end{pmatrix}.$$

Prove that these representations are isomorphic if and only if $\operatorname{char} F \neq 3$.

6914. Let V be a two-dimensional vector space over a field F. Show that there exist two representations ρ_1, ρ_2 of the group $\mathbf{D}_4 = \langle a, b \mid a^4 = b^2 = (ab)^2 = 1 \rangle$ in V for which, in some basis of the space V,

$$\rho_1(a) = \begin{pmatrix} 0 & 1 \\ -1 & 0 \end{pmatrix}, \quad \rho_1(b) = \begin{pmatrix} 0 & 1 \\ 1 & 0 \end{pmatrix},$$

$$\rho_2(a) = \begin{pmatrix} 0 & 1 \\ -1 & 0 \end{pmatrix}, \quad \rho_2(b) = \begin{pmatrix} 1 & \\ 0 & \\ 0 & -1 \end{pmatrix}.$$

Are these representations equivalent?

6915. Let ρ_1 and ρ_2 be representations of the groups S_3 and $\mathbf{D}_4$ from Exercises 6913 and 6914. Are these representations irreducible?

6916. Let V be the vector space over a field F with a basis $(e_1, \ldots, e_n)$. Define a mapping $\psi : S_n \to \mathbf{GL}(V)$

$$\psi_\sigma(e_i) = e_{\sigma(i)},$$

where $\sigma \in S_n$, $i = 1, \ldots, n$. Prove that

a) ψ is a representation of S_n;

b) the subspace W of vectors such that the sum of their coordinates in the basis $(e_1, \ldots, e_n)$ is equal to zero and the subspace U of vectors with equal coordinates are invariant under ψ;

c) if $\operatorname{char} F$ does not divide n then the restriction of ψ to W is an irreducible $(n - 1)$-dimensional representation of S_n.

6917. Let $P_{n,m}$ be the subspace of homogeneous polynomials of degree m in the algebra $F[x_1, \dots, x_n]$ and let $\operatorname{char} F = 0$. Define the mapping Θ: $\mathbf{GL}_n(F) \to \mathbf{GL}(P_{n,m})$ by setting

$$(\Theta_A f)(x_1, \dots, x_n) = f\left(\sum_{i=1}^{n} x_i a_{i1}, \dots, \sum_{i=1}^{n} x_i a_{in}\right)$$

for $f \in P_{n,m}$ and $A = (a_{ij}) \in \mathbf{GL}_n(F)$. Prove that Θ is an irreducible representation of $\mathbf{GL}_n(F)$ in $P_{n,m}$.

6918. Let V be n-dimensional space over a field F of zero characteristic. Define a mapping $\theta : \mathbf{GL}(V) \to \mathbf{GL}(\wedge^m V)$ by setting

$$\theta(f)(x_1 \wedge \dots \wedge x_m) = (f x_1) \wedge \dots \wedge (f x_m),$$

where $x_1, \dots, x_m \in V$ and $f \in \mathbf{GL}(V)$. Prove that θ is an irreducible representation of $\mathbf{GL}(V)$.

6919. Prove that

a) for any representation ρ of a group G there exists a representation $\rho^{\otimes m}$ of G in a space

$$V^{\otimes m} = \underbrace{V \otimes \dots \otimes V}_{m}$$

of m times contravariant tensors on the spaces V, such that

$$\rho^{\otimes m}(g)(v_1 \otimes \dots \otimes v_m) = (\rho(g)v_1) \otimes \dots \otimes (\rho(g)v_m)$$

for all $v_1, \dots, v_m \in V$, $g \in G$;

b) the subspaces of symmetric and of skew-symmetric tensors are invariant under $\rho^{\otimes m}$. Find the dimensions of these subspaces if $\dim V = n$.

6920. Let $\Phi : G \to \mathbf{GL}(V)$ be a representation over a field F and let $\xi : G \to F^*$ be a group homomorphism. Consider the mapping $\Phi_\xi : G \to \mathbf{GL}(V)$ given by the rule $\Phi_\xi(g) = \xi(g)\Phi(g)$, $g \in G$. Prove that Φ_ξ is a representation of G. Show that it is irreducible if and only if Φ is irreducible.

6921. Let Φ be a complex representation of a finite group G. Prove that each operator Φ_g ($g \in G$) is diagonalizable.

6922. Let $\rho : G \to \mathbf{GL}(V)$ be a finite-dimensional representation of a group G over a field F. Prove that for any $g \in G$ there exists a basis of V in which the matrix $\rho(g)$ is block-upper-triangular

$$\rho(g) = \begin{pmatrix} \rho_1(g) & & * \\ & \ddots & \\ 0 & & \rho_m(g) \end{pmatrix},$$

where ρ_i are irreducible representations of G.

6923. Let $\rho : G \to \mathbf{GL}(V)$ be a finite-dimensional representation of a group G and let $(e_1, \ldots, e_n)$ be a basis of V such that for any $g \in G$ the matrix $\rho(g)$ is block-upper-triangular as in Exercise 6922, where the size d_i of the square matrix $\rho_i(g)$ does not depend on g. Prove that

a) the linear span V_i of the vectors $e_{d_1 + \cdots + d_{i-1} + 1}, \ldots, e_{d_1 + \cdots + d_i}$ is a G-invariant subspace $(1 \leq i \leq m)$;

b) a mapping $g \mapsto \rho_i(g)$ is a matrix representation of G;

c) the linear representation of G corresponding to this matrix representation is isomorphic to the representation appearing in the factorspace V_i / V_{i-1} (by definition $V_0 = 0$).

6924. Let $\rho : G \to \mathbf{GL}(V)$ be a representation of a group G. Prove that

a) for any $v \in V$ the linear span $\langle \rho(g)v \mid g \in G \rangle$ is an invariant subspace for ρ;

b) any vector of V belongs to some invariant subspace of dimension $\leq |G|$;

c) the minimal invariant subspace containing a vector $v \in V$ coincides with $\langle \rho(g)v \mid g \in G \rangle$.

6925. Let $\rho : G \to \mathbf{GL}(V)$ be a representation of a group G and let H be a subgroup of G, $[G : H] = k < \infty$. Prove that if a subspace U is invariant under the restriction of ρ to H then the dimension of the minimal subspace containing U invariant under the representation ρ, does not exceed $k \cdot \dim U$.

6926. Let V be a vector space over the field $\mathbb{C}$ with a basis $(e_1, \ldots, e_n)$. Define in V the representation Φ of a cyclic group $\langle a \rangle_n$ by setting $\Phi(a)(e_i) = e_{i+1}$ if $i < n$ and $\Phi(a)(e_n) = e_1$. Let $n = 2m$. Find the dimension of the minimal invariant subspace which contains the vectors:

a) $e_1 + e_{m+1}$;

b) $e_1 + e_3 + \cdots + e_{2m-1}$;

c) $e_1 - e_2 + e_3 - \cdots - e_{2m}$;

d) $e_1 + e_2 + \cdots + e_m$.

6927. Prove that any set of commuting operators on a finite-dimensional complex vector space has a common eigenvector.

6928. Prove that any irreducible representation of an abelian group in a finite-dimensional vector space over the field $\mathbb{C}$ is one-dimensional.

6929. Let $G = \langle a \rangle_p \times \langle b \rangle_p$, where p is a prime and K is a field of characteristic p. Assume that V is a vector space over K with a basis $x_0, x_1, \ldots, x_n, y_1, \ldots, y_n$.

Define a mapping $\rho : G \to \mathbf{GL}(V)$ by setting

$$\rho(a)x_i = \rho(b)x_i = x_i, \quad 0 \le i \le n, \qquad \rho(a)y_i = x_i + y_i, \quad 1 \le i \le n;$$
$$\rho(b)y_i = y_i + x_{i-1}, \quad 1 \le i \le n.$$

Prove that ρ can be extended to a representation of G. Check whether this representation is indecomposable.

6930. Prove that irreducible complex representations of the group $\mathbf{U}_{p^\infty}$ are in one-to-one correspondence with sequences (a_n) of natural numbers such that $0 \le a_n \le p^n - 1$ and $a_n \equiv a_{n+1} \pmod{p^n}$ for all n.

6931. Prove that irreducible complex representations of the group $\mathbb{Q}/\mathbb{Z}$ are in one-to-one correspondence with sequences of natural numbers (a_n) such that $0 \le a_n \le n - 1$ and $a_n \equiv a_m \pmod{n}$, if n divides m.

70 Representations of finite groups

7001. Let $\mathcal{A}$ and $\mathcal{B}$ be two commuting operators on a finite-dimensional vector space V over $\mathbb{C}$ and let $\mathcal{A}^m = \mathcal{B}^n = \mathcal{E}$ for some natural numbers m and n. Prove that V is a direct sum of one-dimensional subspaces which are invariant under $\mathcal{A}$ and $\mathcal{B}$.

7002. List all irreducible complex representations of the groups:

a) $\langle a \rangle_2$;   b) $\langle a \rangle_4$;   c) $\langle a \rangle_2 \times \langle b \rangle_2$;   d) $\langle a \rangle_6$;

e) $\langle a \rangle_8$;   f) $\langle a \rangle_4 \times \langle b \rangle_2$;   g) $\langle a \rangle_2 \times \langle b \rangle_2 \times \langle c \rangle_2$;   h) $\langle a \rangle_6 \times \langle b \rangle_3$;

i) $\langle a \rangle_9 \times \langle b \rangle_{27}$.

7003. Let V be a vector space over a field F, $\mathcal{A} \in \mathbf{GL}(V)$ and $\mathcal{A}^n = \mathcal{E}$.

   a) Prove that the correspondence $a^k \mapsto \mathcal{A}^k$ induces a representation of the cyclic group $\langle a \rangle_n$ in V.

   b) Find all invariant subspaces of this representation in the cases:

$$n = 4, \quad A = \begin{pmatrix} 0 & 1 \\ -1 & 0 \end{pmatrix} \quad \text{and} \quad n = 6, \quad A = \begin{pmatrix} 0 & -1 \\ 1 & -1 \end{pmatrix}.$$

   c) Let $F = \mathbb{C}$ and $e_0, e_1, \ldots, e_{n-1}$ be a basis of V such that

$$A(e_i) = \begin{cases} e_{i+1}, & \text{if } i < n - 1 \\ e_0, & \text{if } i = n - 1. \end{cases}$$

Decompose this representation into the direct sum of irreducible representations.

d) Prove that the representation from c) is isomorphic to the regular representation of the group $\langle a \rangle_n$.

7004. Decompose the regular representation of the groups:

a) $\langle a \rangle_2 \times \langle b \rangle_2$;   b) $\langle a \rangle_2 \times \langle b \rangle_3$;   c) $\langle a \rangle_2 \times \langle b \rangle_4$

into the direct sum of one-dimensional representations.

7005. Let $H = \langle a \rangle_3$ be a cyclic subgroup of a group G. Let Φ be the regular representation of G and let Ψ be its restriction to H. Find the multiplicities of each irreducible representation of H in the decomposition of Ψ into the direct sum of irreducible representations:

a) $G = \langle b \rangle_6$, $a = b^2$;   b) $G = S_3$, $a = (1, 2, 3)$.

7006. Find all nonisomorphic one-dimensional real representations of the group $\langle a \rangle_n$.

7007. Prove that the dimension of an irreducible real representation of a cyclic group is at most two.

7008. Let $\rho_k : \langle a \rangle_n \to \mathbf{GL}_2(\mathbb{R})$ be a representation for which

$$\rho_k(a) = \begin{pmatrix} \cos \dfrac{2\pi k}{n} & -\sin \dfrac{2\pi k}{n} \\ \sin \dfrac{2\pi k}{n} & \cos \dfrac{2\pi k}{n} \end{pmatrix} \quad (0 < k < n).$$

Prove that

a) the representation ρ_k is irreducible if $k \neq n/2$;

b) the representations ρ_k and $\rho_{k'}$ are equivalent if and only if either $k = k'$ or $k + k' = n$;

c) any two-dimensional real irreducible representation of $\langle a \rangle_n$ is equivalent to ρ_k for some k.

7009. Find the number of nonequivalent irreducible real representations:

a) of the group $\mathbf{Z}_n$;

b) of all abelian groups of order 8.

7010. Find the number of nonequivalent two-dimensional complex representations of the groups:

a) $\mathbf{Z}_2$,   b) $\mathbf{Z}_4$,   c) $\mathbf{Z}_2 \oplus \mathbf{Z}_2$.

7011. Let G be an abelian group of order n. Prove that the number of non-equivalent k-dimensional complex representations of G is equal to the coefficient at t^k of the series $(1 - t)^{-n}$. Find this coefficient.

7012. Prove that the kernel of any one-dimensional representation of a group G contains the commutant of G.

7013. Let ρ be a representation of a group G in a space V. Suppose that there exists a basis in which all operators $\rho(g)$ $(g \in G)$ are diagonalizable. Prove that $\operatorname{Ker}\rho \supseteq G'$.

7014. Prove that all irreducible complex representations of a finite group are one-dimensional if and only if the group is commutative.

7015. Find all nonisomorphic one-dimensional complex representations of the groups S_3 and A_4.

7016. Find all one-dimensional complex representations of the groups S_n and D_n.

7017. Construct an irreducible two-dimensional complex representation of the group S_3.

7018. Using the homomorphism of the group S_4 onto the group S_3 construct an irreducible two-dimensional complex representation of S_4.

7019. Using the isomorphism of groups of permutations and groups of symmetries of a cube and of a tetrahedron (see Exercise 5713), construct:

a)   two irreducible three-dimensional matrix complex representations of S_4;

b)   an irreducible three-dimensional representation of A_4.

7020. Prove that if ε is a root of 1 of order n then the mapping

$$a \mapsto \begin{pmatrix} \varepsilon & 0 \\ 0 & \varepsilon^{-1} \end{pmatrix}, \quad b \mapsto \begin{pmatrix} 0 & 1 \\ 1 & 0 \end{pmatrix}$$

can be extended to a representation ρ_ε of the group D_n. Is it irreducible if $\varepsilon \neq \pm 1$?

7021. Let ρ_ε and $\rho_{\varepsilon'}$ be irreducible two-dimensional complex representations of the groups D_n from Exercise 7020. Prove that ρ_ε and $\rho_{\varepsilon'}$ are isomorphic if and only if $\varepsilon' = \varepsilon^{\pm 1}$.

7022. Let ρ be an irreducible complex representation of the group D_n. Prove that ρ is isomorphic to ρ_ε for some ε.

7023. Let ρ be the natural two-dimensional real representation of D_n by symmetries of a regular n-gon. Find ε such that ρ is isomorphic to ρ_ε.

7024. Using the realization of quaternions by complex matrices of size 2 (see Exercise 5809c) construct a two-dimensional complex representation of the group Q_8.

7025. Let a group G have an exact reducible two-dimensional representation. Prove that

a) the commutant of the group G' is abelian;

b) if G is finite and the ground field has a characteristic zero then G is commutative.

7026. Prove that an exact two-dimensional complex representation of a finite noncommutative group is irreducible.

7027. Let G be a finite group and let ρ be a finite-dimensional complex representation of G such that matrices of all operators $\rho(g)$ ($g \in G$) in some basis are upper-triangular. Prove that $\mathrm{Ker}(\rho) \supseteq G'$.

7028. Prove that if the ground field in Exercises 6922 and 6923 is the field of complex numbers and the group G is finite, then the representation ρ is equivalent to the direct sum of representations $\rho_1, \ldots, \rho_m$.

7029. Prove that if the ground field in Exercise 6922 is the field of complex numbers and the group G is finite, then there exists a nonsingular matrix C such that for all $g \in G$

$$C^{-1}\rho(g)C = \begin{pmatrix} \rho_1(g) & & 0 \\ & \ddots & \\ 0 & & \rho_m(g) \end{pmatrix}.$$

7030. Let G be a finite group of order n and ρ be its regular representation. Prove that

$$\mathrm{tr}\rho(g) = \begin{cases} 0, & g \neq 1, \\ n, & g = 1. \end{cases}$$

7031. Prove that for any nonidentical element of a finite group there exists an irreducible complex representation which maps this element to a nonidentical operator.

7032. Let $\mathcal{A}$, $\mathcal{B}$ be linear operators on a finite-dimensional vector space V over a field F of characteristic zero and let $\mathcal{A}^3 = \mathcal{B}^2 = \mathcal{E}$, $\mathcal{A}\mathcal{B} = \mathcal{B}\mathcal{A}^2$.

Prove that for any subspace U invariant under $\mathcal{A}$ and $\mathcal{B}$ there exists a subspace W invariant under $\mathcal{A}$, $\mathcal{B}$ such that $V = U \oplus W$.

7033. Find all nonequivalent two-dimensional complex representations of the groups a) A_4, b) S_3.

7034. Find the number and dimensions of irreducible complex representations of the groups:

a) S_3; b) A_4; c) S_4; d) Q_8; e) D_n; f) A_5.

7035. How many irreducible components are contained by the regular representation of the following groups:

a) $\mathbf{Z}_3$; b) $\mathbf{S}_3$; c) $\mathbf{Q}_8$; d) $\mathbf{A}_4$?

7036. Prove with the help of representation theory that a group of order 24 can not coincide with its commutant.

7037. Is it possible that a finite group has irreducible complex representations only of the following dimensions:

  a) three one-dimensional and four two-dimensional;

  b) two one-dimensional and two five-dimensional;

  c) five one-dimensional and one five-dimensional?

7038. Prove that the group $\mathbf{GL}_2(\mathbb{C})$ has no subgroups isomorphic to $\mathbf{S}_4$.

7039. Prove that any 8-dimensional complex representation of the group $\mathbf{S}_4$ contains a two-dimensional invariant subspace.

7040. Prove that any 5-dimensional representation of the group $\mathbf{A}_4$ contains an one-dimensional invariant subspace.

7041. Prove that the number of irreducible representations of a group G is greater than the number of irreducible representations of any of its factorgroups by a nontrivial normal subgroup.

7042. For what finite groups does the regular representation over the field $\mathbb{C}$ contain finitely many subrepresentations?

7043. Prove that any irreducible representation of a finite p-group over a field of characteristic p is identical.

7044. Let G be a finite p-group and ρ be its representation in a finite-dimensional space V over a field of characteristic p. Prove that there exists a basis of V such that for any $g \in G$ the matrix of the operator $\rho(g)$ is upper-unitriangular.

7045. Let H be a normal subgroup of a finite group G. Prove that the dimension of any irreducible representation of G over a field F does not exceed $[G : H]m$, where m is the greatest dimension of irreducible representations of H over F.

7046. Prove that in $\mathbf{GL}_n(\mathbb{C})$ there exist only finitely many nonconjugate subgroups of a fixed finite order.

7047. Let $\rho : G \rightarrow \mathbf{GL}_3(\mathbb{R})$ be an irreducible three-dimensional real representation of a finite group G and let a representation $\tilde{\rho} : G \rightarrow \mathbf{GL}_3(\mathbb{C})$ be the composition of the mapping ρ with the standard embedding $\mathbf{GL}_3(\mathbb{R}) \rightarrow \mathbf{GL}_3(\mathbb{C})$. Prove that the representation $\tilde{\rho}$ is irreducible.

7048. Prove that any irreducible complex representation of a group of order p^3 of dimension at least two is exact.

7049. Find the number and dimensions of irreducible complex representations of a noncommutative group of order p^3.

7050. Decompose the real representation Φ of the cyclic group $\langle a \rangle$ of order 4 for which

$$\Phi(a) = \begin{pmatrix} 0 & 0 & -1 \\ 0 & 1 & 0 \\ 1 & 0 & 0 \end{pmatrix}$$

into a direct sum of irreducible representations.

7051. Consider the real three-dimensional representation of the group

$$G = \langle a \rangle_2 \times \langle b \rangle_2,$$

where

$$\Phi(a) = \begin{pmatrix} 5 & -4 & 0 \\ 6 & -5 & 0 \\ 0 & 0 & 1 \end{pmatrix}, \qquad \Phi(b) = -E.$$

Decompose Φ into a direct sum of irreducible representations.

7052. Consider the two-dimensional complex representation Φ of the group $G = \langle a \rangle_2 \times \langle b \rangle_2$, where

$$\Phi(a) = \begin{pmatrix} 0 & 1 \\ 1 & 0 \end{pmatrix}, \qquad \Phi(b) = \begin{pmatrix} 0 & -1 \\ -1 & 0 \end{pmatrix}.$$

Decompose Φ into a direct sum of irreducible representations.

71 Group algebras and their modules

7101. Find out if the algebra of quaternions is a real group algebra:

  a) of the group of quaternions;

  b) of some group?

7102. Let V be a vector space over a field F with a basis (e_1, e_2, e_3), $\varphi : F[S_3] \to \operatorname{End} V$ be a homomorphism where $\varphi(\sigma)(e_i) = e_{\sigma(i)}$ for all $\sigma \in S_3$ $(i = 1, 2, 3)$. Find the dimensions of the kernel and of the image of φ.

7103. Find a basis of the kernel of the homomorphism $\varphi \colon \mathbb{C}(\langle a \rangle_n) \to \mathbb{C}$ such that $\varphi(a) = \varepsilon$, where ε is a root of 1 of order n.

7104. Let a group H be isomorphic to a factorgroup of a group G. Prove that $F[H]$ is isomorphic to a factoralgebra of the algebra $F[G]$.

7105. Let $G = G_1 \times G_2$. Prove that $F[G] \simeq F[G_1] \otimes F[G_2]$.

7106. Let G be a finite group and R be a set of mappings from G to a field F. Put in R

$$(\alpha f_1 + \beta f_2)(g) = \alpha f_1(g) + \beta f_2(g),$$
$$(f_1 f_2)(g) = \sum_{h \in G} f_1(h) f_2(h^{-1}g),$$

where $f_1, f_2 \in R$. Prove that R is an algebra over F and the mapping $f \mapsto \sum_{g \in G} f(g)g$ from R to $F[G]$ is an algebra isomorphism.

7107. Prove that if a group G contains elements of finite order then the group algebra $F[G]$ has zero divisors.

7108. Prove that any irreducible $F[G]$-module is isomorphic to a factormodule of the regular $F[G]$-module.

7109. Find all commutative two-sided ideals of the group algebra $\mathbb{C}[G]$ where

a) $G = S_3$;   b) $G = Q_8$;   c) $G = D_5$.

7110. Find all elements x of the group algebra $F[G]$ such that $xg = x$ for all $g \in G$.

7111. Find a basis of the center of the group algebra for the groups:

a) S_3;   b) Q_8;   c) A_4.

7112. Prove that the group algebra A of a free abelian group of rank r has no zero divisors. Show that the field of fractions for A is isomorphic to the field of rational fractions in r variables.

7113. Let A be a ring and A-module V have a decomposition $V = U \oplus W$ where U is an irreducible module and W has no submodules isomorphic to U. Prove that if α is an automorphism of V then $\alpha(U) = U$.

7114. Let A be a ring and let A-module V have a direct decomposition $V = U \oplus W$. Let $\varphi : U \to W$ be a homomorphism of A-modules. Prove that $U_1 = \{x + \varphi(x) \mid x \in U\}$ is A-submodule of V isomorphic to U and $V = U_1 \oplus W$.

7115. Let A be a semisimple finite-dimensional algebra over $\mathbb{C}$. Suppose that A-module V is a direct sum of nonisomorphic irreducible A-modules $V = V_1 \oplus \cdots \oplus V_k$. Find the group of automorphisms of the module V.

7116. Let A be a semisimple finite-dimensional algebra over $\mathbb{C}$. Suppose that A-module V is a direct sum of two isomorphic irreducible A-modules. Prove that the group of automorphisms of A-module V is isomorphic to $GL_2(\mathbb{C})$.

7117. Let A be a semisimple finite-dimensional algebra over $\mathbb{C}$ and let V be A-module finite dimensional over $\mathbb{C}$. Prove that V has finitely many A-submodules if and only if V is a direct sum of nonisomorphic irreducible A-modules.

7118. Let G be a finite group and F be a field of characteristic zero. Consider the group algebra $A = F[G]$ as a left module over A. Let U be a submodule of A. Prove that for any A-module homomorphism $\varphi : U \to A$ there exists an element $a \in A$ such that $\varphi(u) = ua$ for all $u \in U$.

7119. For what finite groups is the complex group algebra simple?

7120. Let $A = F[G]$ (F is a field) where G is a finite group of order $n > 1$. Put

$$e_1 = (n \cdot 1)^{-1} \sum_{g \in G} g, \quad e_2 = 1 - e_1$$

for $n \cdot 1 \neq 0$. Prove that Ae_1 and Ae_2 are proper two-sided ideals and $A = Ae_1 \oplus Ae_2$.

7121. Prove that the equality

$$xy = f(x, y) \cdot 1 + \sum_{g \in G\backslash 1} \alpha_g \cdot g \quad (\alpha_g \in F)$$

defines a symmetric bilinear function f on the group algebra $F[G]$ and the kernel of f is a two-sided ideal.

7122. Let G be a finite group and f be the bilinear function on $\mathbb{R}[G]$ from Exercise 7121. Prove that f is nondegenerate. Find the signature of f for the groups:

a) $\mathbf{Z}_2$;   b) $\mathbf{Z}_3$;   c) $\mathbf{Z}_4$;   d) $\mathbf{Z}_2 \oplus \mathbf{Z}_2$.

7123. Let H be a subgroup of a group G and $\omega(H)$ be the minimal left ideal of $F[G]$ containing $\{h - 1 \mid h \in H\}$. Prove that if H is a normal subgroup then the ideal $\omega(H)$ is two-sided.

7124. Decompose the group algebras of the group $\langle a \rangle_3$ over the fields of real and complex numbers into direct sums of fields.

7125. Prove that $\mathbb{Q}[\langle a \rangle_p]$ (p is a prime) is a direct sum of two two-sided ideals, one of which is isomorphic to $\mathbb{Q}$ and the other isomorphic to $\mathbb{Q}(\varepsilon)$ where ε is a primitive root of 1 of order p.

7126. Let G be a finite group and $\operatorname{char} F$ not divide $|G|$. Let I be an ideal in $F[G]$. Prove that $I^2 = I$.

7127. Find the idempotents and minimal ideals of the rings:

a) $\mathbf{F}_3[\langle a \rangle_2]$;   b) $\mathbf{F}_2[\langle a \rangle_2]$;   c) $\mathbb{C}[\langle a \rangle_2]$;   d) $\mathbb{R}[\langle a \rangle_3]$.

7128. Let G be a finite group. Prove that for any $a \in \mathbb{C}[G]$ the equation $a = axa$ is solvable in $\mathbb{C}[G]$.

7129. How many two-sided ideals are there in the algebras:

a) $\mathbb{C}[S_3]$;   b) $\mathbb{C}[Q_8]$?

7130. For what finite groups G is the group algebra $\mathbb{C}[G]$ a direct sum of $n = 1, 2, 3$ matrix algebras?

7131. Let G be a group and A be an algebra with unity over a field F. Suppose that $\varphi : G \to A^*$ is a group homomorphism. Prove that there exists a unique algebra homomorphism $F[G] \to A$ whose restriction to G coincides with φ.

7132. Prove that if char F does not divide the order of a finite group G then any two-sided ideal of the group algebra $F[G]$ is a ring with unity. Is this statement valid for an arbitrary algebra with unity?

7133. Let F be a field of characteristic $p > 0$ where p divides the order of a finite group G and $u = \sum_{g \in G} g \in F[G]$. Prove that $F[G]u$ is a submodule of the left regular module and it is not a direct summand of $F[G]$.

7134. Let $G = \langle a \rangle_p$ and F be a field of characteristic p. Let $\Phi : G \to \mathbf{GL}_2(F)$, where

$$\Phi(a^s) = \begin{pmatrix} 1 & s \cdot 1 \\ 0 & 1 \end{pmatrix}$$

is a representation of G. Find $F[G]$-submodule U of the regular representation $V = F[G]$ such that the representation of G in V/U is isomorphic to Φ. For which p is the representation Φ isomorphic to the regular one?

7135. Prove that the algebra $\mathbf{F}_2[\langle a \rangle_2]$ is not a direct sum of minimal left ideals.

7136. Let H be a p-group which is a normal subgroup of a finite group G and let F be a field of characteristic p.

a) Prove that the ideal $\omega(H)$ (see Exercise 7123) is nilpotent.

b) Find the index of nilpotency of $\omega(H)$ if $G = \langle a \rangle_2, H = \langle a \rangle_2$ and $F = \mathbf{F}_2$.

7137. Prove that all ideals of the group algebra of the infinite cyclic group are principal.

7138. Prove that a cyclic module over the algebra $F[\langle a \rangle_\infty]$ is either finite-dimensional over F or isomorphic to the left regular $F[\langle a \rangle_\infty]$-module.

7139. Let $A = \mathbb{C}[\langle g \rangle_\infty]$ and $P = Ax_1 \oplus Ax_2$ be a free A-module with a basis (x_1, x_2). Suppose that H is a submodule of P generated by elements h_1, h_2. Decompose P/H into a direct sum of cyclic A-modules and find their dimensions, if

a) $h_1 = gx_1 + x_2,$ $h_2 = x_1 - (g + 1)x_2;$

b) $h_1 = g^2x_1 + g^{-2}x_2,$ $h_2 = g^4x_1 + (1 - g)x_2;$

c) $h_1 = gx_1 + 2g^{-1}x_2,$ $h_2 = (1 + g)x_1 + 2(g^{-2} + g^{-1})x_2.$

7140. Let $\mathcal{A}, \mathcal{B}$ be the linear operators on $V = F[x]$, $\mathcal{A}(f(x)) = f'(x)$, $\mathcal{B}(f(x)) = xf(x)$. Prove that the mapping $\varphi : g \mapsto \mathcal{AB}$ can be extended to a homomorphism $F[\langle g \rangle_\infty] \to \text{End}\,V$. Find $\text{Ker}\varphi$.

7141. Let M be a maximal ideal of the algebra $A = F[\langle a \rangle_\infty]$ and $r = \dim_F (A/M)$. Prove that

a) if $F = \mathbb{C}$ then $r = 1$;

b) if $F = \mathbb{R}$ then either $r = 1$ or $r = 2$;

c) if $F = \mathbb{F}_2$ then r cannot be bounded.

7142. Prove that the group algebra of a free abelian group of finite rank is Noetherian.

7143. Prove that the group algebra of a free abelian group of finite rank is a unique factorization domain.

7144. Find, in the group algebra $A = \mathbb{C}[G]$ of the free abelian group G with a basis (g_1, g_2), the prime factorization of the elements:

a) $g_1g_2 + g_1^{-1}g_2^{-1};$

b) $1 + g_1^{-1}g_2 - g_1g_2^{-1} - g_1^{-2}g_2^2.$

7145. Let G be the free abelian group with a basis (g_1, g_2). Find the factor-algebra of the group algebra $A = F[G]$ by the ideal I generated by the elements:

a) $g_1g_2^{-1};$

b) $g_1 - g_2;$

c) $g_1 - 1$ and $g_2 - 2$.

7146. Prove that if G is a finite group and the group algebra $\mathbb{C}[G]$ has no nilpotent elements then G is commutative.

7147. Let H be a normal subgroup of a group G and V be some $F[G]$-module. Denote by $(H - 1)V$ the linear span of the elements $(h - 1)v$ where $h \in H, v \in V$. Prove that

a) $(H - 1)V$ is an $F[G]$-submodule of V;

b) if H is a Sylow (normal) p-subgroup of G and $\text{char}\,F = p$ and $(H - 1)V = V$, then $V = 0$.

7148. Prove that the complex group algebras of the groups D_4 and Q_8 are isomorphic.

7149. Find the number of nonisomorphic complex group algebras of dimension 12.

7150. Prove that the number of summands in the decomposition of the group algebra of the symmetric group S_n over the field $\mathbb{C}$ in a direct sum of matrix algebras is equal to the number of representations $n = n_1 + n_2 + \cdots + n_k$, where $n_1 \geq n_2 \geq \cdots \geq n_k > 0$.

72 Characters of representations

7201. Let an element g of a group G have order k and χ be an n-dimensional character of G. Prove that $\chi(g)$ is a sum of n (not necessarily distinct) roots of 1 of order k.

7202. Let Φ be a three-dimensional complex representation of the group $\langle a \rangle_3$ and $\chi_\Phi(g) = 0$ for some $g \in \mathbb{Z}_3$. Prove that Φ is equivalent to the regular representation.

7203. Let χ be a two-dimensional complex character of the group $G = \langle a \rangle_3 \times \langle b \rangle_3$. Prove that $\chi(g) \neq 0$ for all $g \in G$.

7204. Let χ be a two-dimensional complex character of a group of odd order. Prove that $\chi(g) \neq 0$ for all $g \in G$.

7205. Let Φ be an n-dimensional complex representation of a finite group G. Prove that $\chi_\Phi(g) = n$ if and only if g belongs to the kernel of Φ.

7206. Let A be the additive group of the n-dimensional vector space V over the field $\mathbf{F}_p$ and χ be an irreducible nontrivial complex character of A. Prove that $\{a \in A \mid \chi(\alpha) = 1\}$ is an $(n-1)$-dimensional subspace of V.

7207. Let χ be a complex character of a finite group G and let $m = \max\{|\chi(g)| \mid g \in G\}$. Prove that

$$H = \{g \in G \mid \chi(g) = m\} \quad \text{and} \quad K = \{g \in G \mid |\chi(g)| = m\}$$

are normal subgroups of G.

7208. Prove that a two-dimensional complex character χ of the group S_3 is irreducible if and only if $\chi((123)) = -1$.

7209. Let χ be a two-dimensional complex character of a finite group G and $g \in G'$. Prove that if $\chi(g) \neq 2$ then χ is irreducible.

7210. Find the 'mean value' $\dfrac{1}{|G|} \sum_{g \in G} \chi(g)$ of an irreducible character of a non-unit finite group G.

7211. Prove that for any element g of a non-unit finite group G there exists a nontrivial irreducible complex character χ of G such that $\chi(g) \neq 0$.

7212. Prove that a mapping of a group G into $\mathbb{C}$ is a one-dimensional character of G if and only if this mapping is a homomorphism of G into the group $\mathbb{C}^*$.

7213. Prove that a central function which is equal to a product of two one-dimensional characters of a group G is a one-dimensional character of G.

7214. Prove that the operation of multiplication of functions induces on a set of one-dimensional characters of a group G a structure of an abelian group $\hat{G}$, the *dual group of* G.

7215. Prove that for a finite cyclic group A, the group $\hat{A}$ is a finite cyclic group of the same order.

7216. Let a finite abelian group A have a direct decomposition $A = A_1 \times A_2$ and let $\alpha_1 \in \hat{A}_1$, $\alpha_2 \in \hat{A}_2$. Prove that the mapping $A \to \mathbb{C}^*$ sending (a_1, a_2) to $\alpha_1(a_1) \cdot \alpha_2(a_2)$ is a one-dimensional character of A and that $\hat{A} \simeq \hat{A}_1 \times \hat{A}_2$.

7217. Let B be a subgroup of a finite abelian group A and let $B^0 = \{\alpha \in \hat{A} \mid \alpha(b) = 1 \text{ for all } b \in B\}$. Prove that

a) B^0 is a subgroup of $\hat{A}$ and every subgroup of $\hat{A}$ coincides with B^0 for some B;

b) $\hat{B} \simeq \hat{A}/B^0$;

c) $B_1 \subset B_2$ if and only if $B_1^0 \supset B_2^0$;

d) $(B_1 \cap B_2)^0 = B_1^0 \cdot B_2^0$;

e) $(B_1 B_2)^0 = B_1^0 \cap B_2^0$.

7218. Let Φ be a homomorphism of a group G into $\mathbf{GL}_n(\mathbb{C})$. Prove that

a) the mapping $\Phi^* : g \mapsto (\Phi(g^{-1}))^t$ is a representation of G;

b) $\chi_\Phi(g) = \bar{\chi}_{\Phi^*}(g)$ for all $g \in G$;

c) the representations Φ and Φ^* are equivalent if and only if the character χ is real-valued.

7219. Let Φ be an irreducible complex representation of a group $\mathbf{S}_n$ and let $\Phi'(\sigma) = \Phi(\sigma)\text{sgn}\sigma$ ($\sigma \in \mathbf{S}_n$). Prove that Φ' is a representation of $\mathbf{S}_n$ and the following statements are equivalent:

a) $\Phi \sim \Phi'$;

b) the restriction of Φ to $\mathbf{A}_n$ is reducible;

c) $\chi_\Phi(\sigma) = 0$ for any odd permutation $\sigma \in \mathbf{S}_n$.

7220. In Exercise 5809 the group of matrices from $M_2(\mathbb{C})$ isomorphic to the group of quaternions Q_8 was given. Prove that the induced two-dimensional representation of Q_8 is irreducible. Find its character.

7221. Find the character of the representation of the group S_n in the space with basis $(e_1, \ldots, e_n)$ where

$$\Phi(\sigma)e_i = e_{\sigma(i)} \quad \text{for} \quad \sigma \in S_n.$$

7222. Find the character of the two-dimensional representation of the group D_n induced by the isomorphism of D_n with the group of symmetries of a fixed regular n-gon.

7223. Find the character of the three-dimensional representation of the group S_4 induced by the isomorphism of S_4 with the group of symmetries of a fixed regular tetrahedron.

7224. Find the character of the representation of the group S_4 induced by the isomorphism of S_4 with the group of rotations of the cube.

7225. Compose the table of irreducible characters of the groups:

a) $\langle a \rangle_2$; b) $\langle a \rangle_3$; c) $\langle a \rangle_4$; d) $\langle a \rangle_2 \times \langle b \rangle_2$; e) $\langle a \rangle_2 \times \langle b \rangle_2 \times \langle c \rangle_2$.

7226. Compose the table of characters of one-dimensional representations and calculate the groups of one-dimensional characters (Exercise 7214) for the groups:

a) S_3;   b) A_4;   c) Q_8;   d) S_n;   e) D_n.

7227. Find the absolute value of the determinant of the matrix whose rows coincide with rows of the table of irreducible characters of an abelian group of order n.

7228. Compose the table of irreducible characters of the groups:

a) S_3;   b) S_4;   c) Q_8;   d) D_4;   e) D_5;   f) A_4.

7229. Find out whether the character of a representation of some group of order 8 can take the values $(1, -1, 2, 0, 0, -2, 0, 0)$.

7230. Decompose the central function

$$(1, -1, i, -i, j, -j, k, -k) \mapsto (5, -3, 0, 0, -1, -1, 0, 0)$$

on Q_8 with respect to the basis of irreducible characters. Is this function a character of some representation?

7231. Determine which of the central functions on S_3,

$$f_1 : (e, (12), (13), (23), (123), (132)) \mapsto (6, -4, -4, -4, 0, 0),$$
$$f_2 : (e, (12), (13), (23), (123), (132)) \mapsto (6, -4, -4, -4, 3, 3),$$

are characters. Find the corresponding representation.

7232. Let A be the additive group of a finite-dimensional vector space V over the field $\mathbf{F}_p$ and let Ψ be a nontrivial irreducible (complex) character of the additive group of $\mathbf{F}_p$.

a) Prove that any irreducible character χ of A is of the form $\chi(a) = \Psi(l(a))$ for some linear function $l \in V^*$.

b) Establish an isomorphism between the dual group $\hat{A}$ (see Exercise 7214) and the additive group of the space V^*.

c) Construct an isomorphism between A and $\hat{\hat{A}}$.

7233. Assume that f in the hypothesis of the previous exercise is a complex-valued function on A. Define the function $\hat{f}$ on $\hat{A}$ by setting

$$\hat{f}(\chi) = \frac{1}{|A|} \sum_{a \in A} f(a) \chi(a) = (f, \chi)_A$$

for $\chi \in \hat{A}$.

a) Prove that

$$f = \sum_{\chi \in \hat{A}} \hat{f}(\chi) \cdot \chi.$$

b) Prove that

$$\widehat{fg}(\chi) = \sum_{\varphi \in \hat{A}} \hat{f}(\varphi) \hat{g}(\varphi^{-1} \cdot \chi).$$

c) Compare the function f on A and $\hat{\hat{f}}$ on $\hat{\hat{A}}$ using the isomorphism from Exercise 7232c.

7234. Let A be the additive group of the field $\mathbf{F}_p$. Consider the function f on A where

$$f(a) = \begin{cases} 0, & \text{if } a = 0, \\ 1, & \text{if } a = x^2 \text{ for some } x \in \mathbf{F}_p^*, \\ -1, & \text{in the other cases.} \end{cases}$$

Prove that if χ is an irreducible complex character of A then $|(f, \chi)_A| = p^{-1/2}$.

7235. Let G be a finite group and H be a subgroup. Prove that a central function on H which is the restriction to H of a character of G is a character of H.

7236. Let Φ be a matrix of an n-dimensional representation of a group G. Construct a representation Ψ of G in the space of square matrices of size n by setting

$$\Psi_g(A) = \Phi_g A^t \Phi_g$$

for $A \in \mathbf{M}_n(K)$. Express χ_Ψ in terms of χ_Φ.

7237. Find the irreducible components of the representations Ψ from Exercise 7236 and find their multiplicities if

a) Φ is a two-dimensional irreducible representation of $\mathbf{S}_3$;

b) Φ is the representation from Exercise 7223;

c) Φ is the two-dimensional representation of $\mathbf{Q}_8$ from Exercise 7220.

7238. Let Φ be a matrix n-dimensional representation of a group G. Construct the representation Ψ of G in the space of square matrices $\mathbf{M}_n(K)$ by setting

$$\Psi_g(A) = \Phi_g \cdot A.$$

Express χ_Ψ in terms of χ_Φ.

7239. Let $\rho : G \to \mathbf{GL}(V)$ be the regular complex representation of a group $G = \langle a \rangle_n$. Find the multiplicity of the identity representation of G as an irreducible constituent of $\rho^{\otimes m}$ (see Exercise 6919).

7240. Let ρ be a two-dimensional irreducible complex representation of the group $\mathbf{S}_3$. Decompose $\rho^{\otimes 2}$ and $\rho^{\otimes 3}$ into a direct sum of irreducible representations.

7241. Let $\rho : G \mapsto \mathbf{GL}(V)$ be the complex regular representation of the group $G = \langle a \rangle_n$. Find the multiplicity of the identity representation of G as an irreducible constituent of the representation induced on the space of skew-symmetric m-contravariant tensors on V (see Exercise 6919).

$$* \quad * \quad *$$

7242. Let χ be a character of a group G and f be the central function on G,

$$f(g) = \frac{1}{2}(\chi(g)^2 - \chi(g^2)).$$

Prove that f is a character of G.

7243. Let Φ be a representation of the group $G = \mathbf{S}_3$ in a space $\mathbb{C}(G)$ of all complex-valued functions on G:

$$(\Phi_\sigma f)(x) = f(\sigma^{-1}x) \ (f \in \mathbb{C}(G), x \in G, \sigma \in G)$$

and let $f_0 \in \mathbb{C}(G)$. Denote by V_0 the linear span of all elements $\Phi_\sigma f_0$ where $\sigma \in G$. Find the character of the restriction Φ to V_0 if

a) $f_0(\sigma) = \text{sgn}\sigma$;

b) $f_0(\sigma) = \begin{cases} 1, & \text{if } \sigma \in \{e, (12)\}, \\ 0, & \text{otherwise}; \end{cases}$

c) $f_0(\sigma) = \begin{cases} 1, & \text{if } \sigma \in \{e, (123), (132)\}, \\ 0, & \text{otherwise}; \end{cases}$

d) $f_0(\sigma) = \begin{cases} 1, & \text{if } \sigma \in \{e, (13), (23)\}, \\ -1, & \text{if } \sigma \in \{(12), (123), (132)\}. \end{cases}$

7244. Let Φ be a complex representation of a finite group G in a space V and Ψ be a representation of G in a space W. Denote by $T(\Phi, \Psi)$ the space of all linear mappings S from V to W such that $S \circ \Phi_g = \Psi_g \circ S$ for all $g \in G$. Prove that

$$\dim T(\Phi, \Psi) = (\chi_\Phi, \chi_\Psi)_G.$$

73 Initial information on representations of continuous groups

All representations considered in this section are assumed finite-dimensional if the opposite is not explicitly stated.

$$* \quad * \quad *$$

7301. Let F be either the field $\mathbb{R}$ or the field $\mathbb{C}$. Prove that

a) for any matrix $A \in \mathbf{M}_n(F)$ the mapping $P_A : t \mapsto e^{tA}$ $(t \in F)$ is a differentiable matrix representation of the additive group of F;

b) any differentiable matrix representation P of the additive group of F has the form P_A where $A = P'(0)$;

c) representations P_A and P_B are equivalent if and only if the matrices A and B are similar.

7302. Prove that P is a matrix representation of the additive group of the field $\mathbb{R}$ and find a matrix A such that $P = P_A$ (see Exercise 7301) if

a) $P(t) = \begin{pmatrix} \cos t & -\sin t \\ \sin t & \cos t \end{pmatrix}$;

b) $P(t) = \begin{pmatrix} \cosh t & \sinh t \\ \sinh t & \cosh t \end{pmatrix}$;

c) $P(t) = \begin{pmatrix} e^t & 0 \\ 0 & 1 \end{pmatrix}$;

d) $P(t) = \begin{pmatrix} e^t & 0 \\ 0 & e^{-t} \end{pmatrix}$;

e) $P(t) = \begin{pmatrix} 1 & t \\ 0 & 1 \end{pmatrix}$;

f) $P(t) = \begin{pmatrix} 1 & e^t - 1 \\ 0 & e^t \end{pmatrix}$.

7303. Which of the matrix representations of the group $\mathbb{R}$ from Exercise 7302 are equivalent?

7304. In what case are the representations P_A and P_{-A} equivalent if $F = \mathbb{C}$?

7305. Find all differentiable complex matrix representations of the groups:

a) $\mathbb{R}_+^*$;

b) $\mathbb{R}^*$;

c) $\mathbb{C}^*$;

d) $\mathbf{U}$ (in this case we suppose that the representation is differentiable with respect to the argument of a complex number z).

7306. Can any complex linear representation of the group $\mathbb{Z}$ be obtained by restriction to $\mathbb{Z}$ of some representation of the group $\mathbb{C}$?

7307. Find all subspaces of the space $\mathbb{C}^n$ invariant under the matrix representation P_A (see Exercise 7301) in the case when the characteristic polynomial of A has no multiple roots.

7308. Prove that the matrix representation P_A (see Exercise 7301) is completely reducible if and only if the matrix A is diagonalizable.

7309. Let R_n be the space of homogeneous polynomials of degree n in x and y with complex coefficients. Put

$$(\Phi_n(A)f)(x, y) = f(ax + cy, bx + dy).$$

for $A = \begin{pmatrix} a & b \\ c & d \end{pmatrix} \in \mathbf{SL}_2(\mathbb{C})$ and $f \in R_n$. Prove that the restriction of Φ_n to the subgroup $\mathbf{SU}_2(\mathbb{C})$ is irreducible.

7310. Let $G = \mathbf{GL}_2(\mathbb{C})$. A complex function on G is called *polynomial* if it is a polynomial in matrix entries.

a)   Let $t(A) = \mathrm{tr} A$ and $d(A) = \det A$. Prove that t and d are central polynomial functions on G.

b)   Prove that any central polynomial function on G is a polynomial in t and d.

c) Let $A = (a_{ij}) \in G$ and $R = \mathbb{C}[x, y]$. Denote by $\Psi(A)$ the homomorphism $R \to R$ for which

$$\Psi(A) : x \mapsto a_{11}x + a_{12}y,$$
$$\Psi(A) : y \mapsto a_{21}x + a_{22}y.$$

Prove that Ψ is a representation of G in the space R and subspaces of homogeneous polynomials of degree n are invariant under Ψ.

d) Prove that the restriction of $\Psi(A)$ to the subspace R_n for $A \in SL_2(\mathbb{C})$ coincides with the operator $\Phi_n(A)$ from Exercise 7309.

e) Let χ_n be the character of the restriction $\Psi|_{R_n}$. Prove that $\chi_n = t\chi_{n-1} - d\chi_{n-2}$.

7311. Let $\mathbb{H}$ be the space of all complex matrices $X = \begin{pmatrix} z & w \\ -\bar{w} & \bar{z} \end{pmatrix}$ with the structure of 4-dimensional Euclidean space $(X, X) = \det X$, and let $\mathbb{H}_0 = \{X \in \mathbb{H} | \operatorname{tr} X = 0\}$. Prove that

a) the mapping $P : SU_2 \to GL(\mathbb{H}_0)$, defined by the formula $P(A) : X \mapsto AXA^{-1}$, is a (real) linear representation of SU_2 such that $\operatorname{Ker} P = \pm E$ and $\operatorname{Im} P$ consists of all proper orthogonal transformations of $\mathbb{H}_0$;

b) the mapping $R : SU_2 \times SU_2 \to GL(\mathbb{H})$, defined by the formula $R(A, B) : X \mapsto AXB^{-1}$, is a (real) linear representation of $SU_2 \times SU_2$ such that $\operatorname{Ker} R = \{(E, E), (-E, -E)\}$ and $\operatorname{Im} R$ consists of all proper orthogonal transformations of $\mathbb{H}_0$;

c) the complexification of the linear representation P is isomorphic to the restriction of the representation Φ_2 of SL_2 from Exercise 7309 to SU_2.

7312. Let G be a topological connected solvable group and ρ be a continuous homomorphism from G to the group of nonsingular linear operators on a finite-dimensional complex space V. Prove that

a) there exists in V a nonzero common eigenvector for all operators $\rho(g)$, $g \in G$;

b) there exists a basis $e_1, \ldots, e_n$ of V such that all matrices $\rho(g)$, $g \in G$ in this basis are upper-triangular.

7313. Let F be an algebraically closed field and G be a solvable group of nonsingular linear operators on a finite-dimensional vector space V over F. Prove that there exists a basis $e_1, \ldots, e_n$ of V and a normal subgroup N of G of a finite index (depending only on n) such that N consists of upper-triangular matrices.

Answers and hints

103. *Hint.* Make use of induction on $|\bigcup_{i=1}^{r} A_i|$.

104. *Hint.* Induction on n.

105. 2^{2^n}. *Hint.* Let $X_1, \ldots, X_n$ be the given subsets and let X_i^ε denote either X_i or $\bar{X}_i$. Then any subset formed of $\{X_i\}$ can be written in the form

$$\bigcup_{(\varepsilon_1,\ldots,\varepsilon_n) \in \varepsilon} (X_1^{\varepsilon_1} \cap X_2^{\varepsilon_2} \cap \cdots \cap X_n^{\varepsilon_n}),$$

where ε is a subset of the set of all sequences $(\varepsilon_1, \ldots, \varepsilon_n)$. Let X be the set of all subsets consisting of n elements. Construct n subsets X_i of X such that any element of X can be presented in the form $X_1^{\varepsilon_1} \cap \cdots \cap X_n^{\varepsilon_n}$.

202. *Hint.* Represent X in the form $(Y \cup A) \cup (X \setminus (Y \cup A))$ where A is countable and $A \cap Y = \emptyset$, $X \setminus Y$ in the form $A \cup (X \setminus (Y \cup A))$ and use a bijection $Y \cup A \to A$.

204. 2^n.

205. a) $|Y^X| = n^m$.

b) $n(n-1) \ldots (n-m+1)$.

c) $n!$ if $m = n$ and 0 if $m \neq n$.

d) $n^m - n(n-1)^m + \cdots + (-1)^i \binom{n}{i}(n-i)^m + \cdots + (-1)^{n-1}\binom{n}{n-1}$. *Hint.* Apply Exercise 103.

206. $n(n-1) \ldots (n-m+1)/m!$.

207. 2^{n-1}.

209. $n!/(m_1! \ldots m_k!)$.

211. h), i), j). *Hint.* Induction on n.

212. *Hint.* Induction on m.

213. $\binom{n+k-1}{k-1}$.

Hint. Establish a bijection between the set of the given partitions and the set of combinations of $n + k - 1$ things $k - 1$ at a time. Apply Exercise 103.

301. a) $\begin{pmatrix} 1 & 2 & 3 & 4 & 5 \\ 2 & 1 & 3 & 4 & 5 \end{pmatrix}$ and $\begin{pmatrix} 1 & 2 & 3 & 4 & 5 \\ 1 & 2 & 5 & 4 & 3 \end{pmatrix}$.

b) $\begin{pmatrix} 1 & 2 & 3 & 4 & 5 & 6 \\ 6 & 5 & 3 & 2 & 1 & 4 \end{pmatrix}$ and $\begin{pmatrix} 1 & 2 & 3 & 4 & 5 & 6 \\ 1 & 3 & 5 & 6 & 4 & 2 \end{pmatrix}$.

c) $\begin{pmatrix} 1 & 2 & 3 & 4 & 5 \\ 5 & 4 & 3 & 1 & 2 \end{pmatrix}$ and $\begin{pmatrix} 1 & 2 & 3 & 4 & 5 & 6 \\ 4 & 1 & 6 & 5 & 3 & 2 \end{pmatrix}$.

302. a) (153)(247).

b) (1362)(47).

c) (1362745).

d) (1472365).

e) (12)(34)...(2n − 1, 2n).

f) (1 n + 1)(2 n + 2)...(n 2n).

303. a) $\begin{pmatrix} 1 & 2 & 3 & 4 & 5 & 6 & 7 \\ 3 & 4 & 6 & 7 & 5 & 1 & 2 \end{pmatrix}$.

b) $\begin{pmatrix} 1 & 2 & 3 & 4 & 5 & 6 & 7 \\ 6 & 3 & 7 & 2 & 4 & 5 & 2 \end{pmatrix}$.

c) $\begin{pmatrix} 1 & 2 & 3 & 4 & \dots & 2n-1 & 2n \\ 3 & 4 & 5 & 6 & \dots & 1 & 2 \end{pmatrix}$.

304. a) (1642573).

b) (26537).

305. a) 5.

b) 8.

c) 13.

d) 14.

e) $\dfrac{n(n-1)}{2}$.

f) $\dfrac{n(n+1)}{2}$.

g) $(n-k)(k-1)$.

h) $(k-1)(n-k) + \dfrac{(k-1)(k-2)}{2}$.

306. a) −1.

b) 1.

c) 1.

d) −1.

e) $(-1)^{\left[\frac{n+2}{2}\right]}$.

f) $(-1)^{\left[\frac{n+1}{4}\right]}$.

g) $(-1)^{\left[\frac{n}{2}\right]}$.

h) $(-1)^{\left[\frac{n}{2}\right]\left[\frac{n+1}{2}\right]}$.

307. a), b) -1 if k is even.

c) 1.

d) 1 if k is even.

e) 1 if $p + q + r + s$ is even.

308. $\binom{n}{2} - k$.

309. a) $\begin{pmatrix} 1 & 2 & \cdots & n \\ n & n-1 & \cdots & 1 \end{pmatrix}$.

b) $k - 1$.

c) $n - k$.

310. *Hint.* If the pair of numbers is distinct from the pairs $(q, q+1)$ and $(q+1, q)$ then it forms an inversion in both sequences simultaneously.

311. *Hint.* Apply Exercise 310.

312. *Hint.* Show that if $\sigma = (i_1, \ldots, i_k)$ then $\pi G \pi^{-1} = (\pi(i_1), \ldots, \pi(i_k))$.

313. If i, j occur in disjoint cycles then these cycles are merged into one; if i, j occur in one cycle then it is factorized into two cycles and other cycles do not change; the decrement increases or decreases by 1.

317. *Hint.* If g is another polynomial of the same type (for another choice of binomials) then $\sigma_f / \sigma_g = f/g$; then make use of $\prod_{i>j}(x_j - x_i)$.

318. *Hint.* a) If the graph is connected then the transpositions $(12), \ldots, (1n)$ have the presentation mentioned; if the graph is disconnected then only the cycles which are contained in one of the connected components have that presentation.

b) Make use of statement a).

319. *Hint.* Consider a series of sequences starting with $1, 2, \ldots, n$ which can be obtained as follows: at first 1 is located at the positions $2, 3, \ldots, n$, respectively; then 2 is located at all positions to the $(n-1)$th and so on. On each step the number of inversions is increased by 1 and it reaches the number $\binom{n}{2}$ at the last sequence $n, n-1, \ldots, 1$.

320. $\dfrac{n!}{2}\binom{n}{2}$.

Hint. Make use of Exercise 308.

321. a) sgnξ = (sgnσ)^n(sgnτ)^m. *Hint.* Put $X \times Y$ in lexicographic order and calculate the number of inversions.

b)   Lengths of cycles are equal to the least common multiple of (k_i, l_j), each of them occurs $GCD(k_i, l_j)$ times $(i = 1, \ldots, s; \ j = 1, \ldots, t)$;

Hint. First consider the case when σ, τ are cycles. Notice that the parity of ξ coincides with the parity of $|X| + |Y|$.

322. c) *Hint.* Make use of Exercise 312.

323. *Hint.* Factorize σ into a product of disjoint cycles. Each cycle of length at least three can be interpreted as a rotation of a regular n-gon. Represent the rotation as a product of two symmetries.

402. a) $u(n) = 3 \cdot 2^n - 5.$ b) $u(n) = (-1)^n (2n - 1).$

403. *Hint.* Induction on n.

404. a) $\dfrac{4^{n+1} - 1}{3} - n - 1.$

b) $-n^2 + 1.$

c) $2^{n+2}(n - 2) + 6.$

405. $u(n) = \dfrac{1}{\sqrt{5}} \left(\dfrac{1 + \sqrt{5}}{2} \right)^n - \dfrac{1}{\sqrt{5}} \left(\dfrac{1 - \sqrt{5}}{2} \right)^n.$

406. – 410. *Hint.* Induction on n.

411. *Hint.* a) – e) Induction on n.

f)   Follows from e).

g)   Follows from e), f) and the Euclidean algorithm.

412. *Hint.* Induction on n.

413. *Hint.* If m is an integer such that $rm \in \mathbb{Z}$ then $U_m(2 \cos r\pi) = 0$ by 412b. By 412b and 2801 the number $2 \cos r\pi$ is integer. Since $|\cos r\pi| \le 1$ then $2 \cos r\pi = 0, \pm 1, \pm 2.$

414. $\dfrac{n(n + 1)}{2} + 1.$

Hint. The addition of the nth line increases the number of areas by n.

501. a) $n(n + 1)(2n + 1)/6$; *Hint.* Consider the sum $(0 + 1)^3 + (1 + 1)^3 + \cdots + (n - 1) + 1)^3.$

b) $n^2(n + 1)^2/4.$

502. *Hint.* See the hint for Exercise 501.

503. *Hint.* Let T be a set of pairs (σ, i), where $\sigma \in S_n$, $\sigma(i) = i$; then

$$\sum_{\sigma \in S_n} N(\sigma)^{s+1} = \sum_{(\sigma, i) \in T} N(\sigma)^s = \sum_{i=1}^{n} \sum_{\sigma' \in S_{n-1}} (N(\sigma' + 1)^s.$$

504. *Hint.* Make use of Exercises 207 and 212a.

505. *Hint.* Represent one of the functions as a sum of the values of the other one and make use of Exercise 504.

506. *Hint.* Make use of Exercise 505.

601. $(1, 4, -7, 7)$.

602. a) $(0, 1, 2, -2)$.

b) $(1, 2, 3, 4)$.

603. a) Independent. b) Dependent. c) Independent. d) Independent. e) Dependent. f) Independent.

607. a) Independent. b) Dependent. c) Independent. d) Independent. e) Dependent if k is even. f) Dependent.

608. No.

609. a) $\lambda = 15$.

b) λ is any number.

c) λ is any number.

d) $\lambda \neq 12$.

e) There is no such λ.

610. a) (a_1, a_2), (a_2, a_3).

b) (a_1, a_3), (a_2, a_4), (a_1, a_4), (a_2, a_3).

c) Any two vectors form a basis.

d) (a_1, a_2, a_4), (a_2, a_3, a_4).

e) (a_1, a_2, a_4), (a_1, a_2, a_5), (a_2, a_3, a_4), (a_2, a_3, a_5).

611. When the system is either linearly independent or when it is obtained from a linearly independent system by addition of zero vectors.

612. a) (a_1, a_2, a_4), $a_3 = a_1 - a_2$.

b) (a_1, a_2, a_3), $a_4 = 2a_1 - 3a_2 + 4a_3$, $a_5 = a_1 + 5a_2 - 5a_3$.

c) (a_1, a_2, a_5), $a_3 = a_1 - a_2 + a_5$, $a_4 = 3a_1 + 4a_2 - 2a_5$.

d) (a_1, a_2), $a_3 = a_1 + 3a_2$, $a_4 = 2a_1 - a_2$.

e) (a_1, a_2, a_3).

f) (a_1, a_2), $a_3 = 2a_1 - a_2$;

g) (a_1, a_2), $a_3 = -a_1 + a_2$, $a_4 = -5a_1 + 4a_2$.

h) (a_1, a_2, a_3), $a_4 = a_1 + a_2 - a_3$.

i) (a_1, a_2, a_4), $a_3 = 2a_1 - a_2$.

j) (a_1, a_2), $a_3 = 3a_1 - a_2$, $a_4 = a_1 - a_2$.

k) (a_1, a_2, a_3), $a_4 = a_1 - a_2 - a_3$.

613. Any $k - 1$ different vectors form a basis.

618. a), b) $p = 3$.

701. a) 2. b) 3. d) 3. e) 3. f), g), h) 4. i), j) 5. k) n if n is odd and $n - 1$ if n is even.

702. a) 1 if $\lambda = 1$, 2 if $\lambda = -1$ and 3 if $\lambda \pm 1$.

b) 2 if $\lambda = 1$, 3 if either $\lambda = 2$ or $\lambda = 3$ and 4 in the other cases.

c) 2 if $\lambda = 0$ and 3 if $\lambda \neq 0$.

d) 2 if $\lambda = 3$ and 3 if $\lambda \neq 3$.

e) 3 if $\lambda = \pm 1, \pm 2$; 4 if $\lambda \neq \pm 1, \pm 2$.

f) 3 if $\lambda = 0, -2, -4$ and 4 in the other cases.

g) n if $\lambda = 1, 2, \ldots, n$, and $n + 1$ for the other values of λ.

h) n if $\lambda = \frac{1}{2}$ and $n + 1$ if $\lambda \neq \frac{1}{2}$.

704. *Hint.* The system of rows of a product of matrices is a linear combination of the system of rows of the second factor.

706. *Hint.* The system of rows of a sum of matrices is a linear combination of the union of systems of rows of these matrices.

707. *Hint.* If, for example, the system of rows of a matrix A of rank 2 is equal to $(a, b, \alpha a + \beta b, \gamma a + \delta b)$ then A is a sum of matrices with rows $(a, 0, \alpha a, \gamma a)$ and $(0, b, 0, \beta b, \delta b)$; then apply Exercise 706.

709. 0 if $r \leq n - 2$, 1 if $r = n - 1$ and n if $r = n$.

710. *Hint.* Apply elementary row and column transformations.

715. *Hint.* Make use of the reduction to the row-echelon form by elementary row transformations of the type II.

716. *Hint.* Induction on the number of columns.

719. *Hint.* Induction on the number of rows. For the proof of uniqueness consider the basis of the system of columns with the least possible indices.

801. a) $x_3 = (x_1 - 9x_2 - 2)/11$, $x_4 = (-5x_1 + x_2 + 10)/11$; $(0, 1, -1, 0)$.

b) $x_3 = -\frac{11}{8}x_4$, $x_1 = \frac{2}{3}x_2 - \frac{1}{24}x_4 + \frac{1}{3}$; $(\frac{1}{3}, 0, 0, 0)$.

c) The system has no solutions.

d) $x_3 = 1 - 4x_1 - 3x_2$, $x_4 = 1$, $(1, -1, 0, 1)$.

e) $x_3 = 6 + 10x_1 - 15x_2$; $x_4 = -7 - 12x_1 + 18x_2$, $(1, 1, 1, -1)$.

f) $x_1 = 3$, $x_2 = 0$, $x_3 = -5$, $x_4 = 11$.

g) $x_1 = 3$, $x_2 = 2$, $x_3 = 1$.

h) $x_3 = 13$, $x_5 = -34$, $x_4 = 19 - 3x_1 - 2x_2$.

802. a) If $\lambda = 0$ then the system is incompatible; if $\lambda \neq 0$ then

$$x_1 = \frac{1}{\lambda}, \quad x_3 = \frac{9\lambda - 16}{5\lambda} - \frac{8}{5}x_2, \quad x_4 = \frac{4 - \lambda}{5\lambda} - \frac{3}{5}x_2.$$

b) If $\lambda \neq 0$ then the system is incompatible; if $\lambda = 0$ then $x_1 = -\frac{1}{2}(7 + 19x_3 + 7x_4)$, $x_2 = -\frac{1}{2}(3 + 13x_1 + 5x_4)$.

c) If $\lambda = 1$ then the system is incompatible; if $\lambda \neq 1$ then $x_1 = \dfrac{43 - 8\lambda}{8 - 8\lambda} - \dfrac{9}{8}x_3$,
$x_2 = \dfrac{5}{4 - 4\lambda} + \dfrac{x_3}{4}$, $x_4 = \dfrac{5}{\lambda - 1}$.

d) If $\lambda = 8$ then $x_2 = 4 + 2x_1 - 2x_4$, $x_3 = 3 - 2x_4$; if $\lambda \neq 8$ then $x_2 = 4 - 2x_4$, $x_3 = 3 - 2x_4$.

e) If $\lambda = 8$ then $x_3 = -1$, $x_4 = 2 - x_1 - \frac{3}{2}x_2$; if $\lambda \neq 8$ then $x_2 = 4 - 2/3x_1$, $x_3 = -1$, $x_4 = 0$.

f) If $\lambda \neq 1, -2$ then $x_1 = x_2 = x_3 = 1/(\lambda + 2)$; if $\lambda = 1$ then $x_1 = 1 - x_2 - x_3$; if $\lambda = -2$ then the system is incompatible.

g) If $\lambda \neq 1, -3$ then $x_1 = x_2 = x_3 = x_4 = 1/(\lambda + 3)$; if $\lambda = 1$ then $x_1 = 1 - x_2 - x_3 - x_4$; if $\lambda = -3$ then the system is incompatible.

h) If $\lambda \neq 0, -3$, then

$$x_1 = \frac{2 - \lambda^2}{\lambda(\lambda + 3)}, \quad x_2 = \frac{2\lambda - 1}{\lambda(\lambda + 3)}, \quad x_3 = \frac{\lambda^3 + 3\lambda^2 - \lambda - 1}{\lambda(\lambda + 3)};$$

if either $\lambda = 0$ or $\lambda = -3$ then the system is incompatible.

i) If $\lambda \neq 0, -3$ then $x_1 = 2 - \lambda^2$, $x_2 = 2\lambda - 1$, $x_3 = \lambda^3 + 2\lambda^2 - \lambda - 1$; if $\lambda = 0$ then $x_1 = -x_2 - x_3$; if $\lambda = -3$ then $x_1 = x_2 = x_3$.

803. a) $^t(2, 3, 1)^{1)}$.

b) The set of vectors of the form

$$^t(0, 0, 2, -1) + \alpha^t(13, 0, 9, -1) + \beta^t(0, 13, -27, 3).$$

c) The set of vectors of the form

$$^t(2, 1, -1, 0, 1) + \alpha^t(1, 0, 4, 0, -1) + \beta^t(0, 1, -8, 0, 2).$$

---

1) In the answers the symbol $^t u$ denotes the vector-column obtained by transposition of a row u.

d)   The set of vectors of the form $'(2, -2, 3, -1) + \alpha'(-13, 8, -6, 7)$.

e)   Ø.

f)   The set of vectors of the form

$$'(1, 2, 22/5, 8/5) + \alpha'(5, 0, -17, -8) + \beta'(0, 5, 34, 16).$$

g)   The set of vectors of the form

$$'(-3, 1, 3/2, -1/2, -5/2) + \alpha'(1, 0, -2, -4, -4) + \beta'(0, 1, -1, -2, -2).$$

h) $'(3, 0, -5, 11)$.

   804. a) $x_1 = 8x_3 - 7x_4, x_2 = -6x_3 + 5x_4$; $('(8, -6, 1, 0), '(-7, 5, 0, 1))$.

b)   The system has only the zero solution.

c) $x_1 = x_4 = x_5, x_2 = x_4 - x_6, x_3 = x_4$; $('(1, 1, 1, 1, 0, 0), '(-1, 0, 0, 0, 1, 0), '(0, -1, 0, 0, 0, 1))$.

d)   If either $n = 3k$ or $n = 3k + 1$ then the system has only the zero solution; if $n = 3k + 2$ then the general solution is equal to $x_{3i} = 0, x_{3i+1} = -x_n, x_{3i+2} = x_n$ $(i = 1, \ldots, k)$; $('(-1, 1, 0, -1, 1, 0, \ldots, 0, -1, 1))$.

   805. a) $('(7, -5, 0, 2), '(-7, 5, 1, 0))$.

b) $('(-9, 3, 4, 0, 0)), '(-3, 1, 0, 2, 0))$.

c)   The kernel consists only of the zero vector.

d) $('(-9, -3, 11, 0, 0), '(3, 1, 0, 11, 0), '(-10, 4, 0, 0, 11))$.

e) $('(0, 1, 3, 0, 0), '(0, 0, 2, 0, 1))$.

f) $('(-3, 2, 1, 0, 0), '(-5, 3, 0, 0, 1))$.

   806. a) $x_1 = x_2 = 1$.

b) $x_3 = 3, x_2 = -1$.

c) $x_1 = \cos(\alpha - \beta), x_2 = \sin(\alpha - \beta)$.

d) $'(0, -3, 6)$.

e) $x_1 = 3, x_2 = 2, x_3 = 1$.

f) $x_1 = 3, x_2 = -2, x_3 = 2$.

   807. $x^2 + 3x + 4$.

   808. $x^3 + 3x^2 + 4x + 5$.

   809. $-x^4 - x + 1$.

   810. a) $'(2, 4, 2)$.

b) $'(15, 2, 4)$.

811. *Hint.* Deduce Cramer's formula $\Delta x_i = \Delta_i$ and multiply both sides of this formula by an integer u such that $\Delta u + mv = 1$.

812. *Hint.* If $d = a_{ij}$ and $a_{ik} = dq + r$ $(0 < r < |d|)$ then the given matrix can be reduced by elementary transformation to a matrix with an entry $r < |d|$; therefore all entries of the ith row and of the jth column are divisible by d and the matrix can be reduced to a matrix B where $b_{11} = d, b_{1i} = b_{k1} = 0$; if $b_2 = dq + s$ $(0 \le s < |d|)$, then after subtructing the second row from the first one and after adding the first column multiplied by q to the second one we obtain a matrix with an entry $-s$, i.e. $s = 0$.

813. *Hint.* Make use of Exercise 812 and its proof.

814. *Hint.* Apply Cramer's theorem. The converse statement is not correct: the system consisting of one equation $2x = 2$ is definite over the ring of integers and is indefinite modulo 2.

815. *Hint.* The system consisting of one equation $4x = 2$ has no integer solutions but is compatible modulo any prime.

816. a) There is a unique solution modulo $p \ne 3$; $x_1 = -1 + x_2 + x_3$ if $p = 3$.

b) There is a unique solution modulo $p \ne 3$; modulo 3 the system is incompatible.

c) There is a unique solution modulo $p \ne 2$; modulo $p = 2$ the system is incompatible.

819. *Hint.* Make use of the result of the previous exercise.

823. Apply Exercises 820 – 822.

824. a) $\{{}^t(1 - 3k - 2l, 2k, l) \mid k, l \in \mathbb{Z}\}$.

b) $\{{}^t(k, 0, 11(2k - 1), -8(2k - 1)) \mid k \in \mathbb{Z}\}$.

825. *Hint.* Make use of Exercise 819.

826. *Hint.* For a column X denote by $\|X\|$ the maximum of absolute values of its entries. Prove that $\|X_n - X_m\| < q\|X_{n-1} - X_{m-1}\|$ for any natural numbers n, m. Deduce that the sequence X_n converges to a solution of $AX = b$.

901. a) -1.

b) 0.

c) 1.

d) $\sin(\alpha - \beta)$.

e) 0.

f) 0.

g) $a^2 + b^2 + c^2 + d^2$.

902. a) -8.

b) -50.

c)   16.

d)   0.

e) $3abc - a^3 - b^3 - c^3$.

f)   0.

g) $\sin(\beta - \gamma) + \sin(\gamma - \alpha) + \sin(\alpha - \beta)$.

h) -2.

i)   0.

j) $3i\sqrt{3}$.

1001. a) The product occurs with the sign plus.

b)   The product occurs with the sign minus.

c)   The product does not occur.

1002. $i = 2, j = 3, k = 2$.

1003. $2x^4 - 5x^3 + \ldots$.

1004. a) $a_{11}a_{22}\ldots a_{nn}$.

b) $(-1)^{n(n-1)/2}a_{1n}a_{2,n-1}\ldots a_{n1}$.

c) $abcd$.

d) $abcd$.

e)   0.

1006. 1.

1101. a) The determinant is multiplied by $(-1)^n$.

b)   It is not changed.

c)   It is not changed; *Hint.* the transformation can be replaced by two symmetries with respect to horizontal and vertical medians and a symmetry with respect to the principal diagonal.

d)   It is not changed.

e)   It is multiplied by $(-1)^{n(n-1)/2}$.

1102. a) The determinant is multiplied by $(-1)^{n-1}$.

b)   It is multiplied by $(-1)^{n(n-1)/2}$.

1103. a), b) The determinant is not changed.

c)   It vanishes.

d)   The determinant of even size vanishes and of odd size doubles.

1104. *Hint.* Transpose the determinant and factor out -1 from each row.

1105. *Hint.* Make use, for example, of the equality $20604 = 2\cdot 10^4 + 6 \cdot 10^2 + 4$.

1106. 0. *Hint.* One row is equal to the half-sum of two others.

1107. 0.

1110. a) $a_1a_2 \ldots a_n + (a_1a_2 \ldots a_{n-1} + a_1 \ldots a_{n-2}a_n + \cdots + a_2a_3 \ldots a_n)x$. *Hint.* Decompose the determinant into a sum of two summands according to elements of the last row.

b) $x^n + (a_1 + \cdots + a_n) \times x^{n-1}$.

c) $D_n = 0$ if $n > 2$, $D_1 = 1 + x_1y_1$, $D_2 = (x_1 - x_2)(y_1 - y_2)$.

d) 0 if $n > 1$. *Hint.* Decompose the determinant into a sum of determinants according to elements of each column.

e) $1 + \sum_{i=1}^n (a_i + b_i) + \sum_{1 \le i < k \le n} (a_i - a_k) \times (b_k - b_i)$. *Hint.* Decompose the determinant into a sum of two determinants according to elements of the first row.

f) $1 + x_1y_1 + \cdots + x_ny_n$.

1201. $8a + 15b + 12c - 19d$.

1202. $2a - 8b + c + 5d$.

1203. a) $x^n + (-1)^{n+1}y^n$. *Hint.* Decompose the determinant according to elements of the first column.

b) $a_0x_1x_2x_3 \ldots X_n + a_1y_1x_2x_3 \ldots x_n + a_2y_1y_2x_3 \ldots x_n + \cdots + a_ny_1y_2y_3 \ldots y_n$. *Hint.* Decompose the determinant according to elements of the first row and either apply the theorem on determinants with zero block corner or decompose it according to elements of the last column and compose a recurrence relation.

c) $a_0x^n + a_1x^{n-1} + \cdots + a_n$. *Hint.* Decompose according to elements of the first column.

d) $n!(a_0x^n + a_1x^{n-1} + \cdots + a_n)$.

e) $\dfrac{x^{n+1} - 1}{(x - 1)^2} - \dfrac{n + 1}{x - 1}$. f) $\dfrac{nx^n}{x - 1} - \dfrac{x^n - 1}{(x - 1)^2}$.

g) $a_1a_2 \ldots a_n - a_1a_2 \ldots a_{n-1} + a_1a_2 \ldots a_{n-2} - \cdots + (-1)^{n-1}a_1 + (-1)^n$. *Hint.* Either decompose the determinant according to elements of the first column or decompose the determinant according to elements of the last column, then transfer the last row in the first summand to the first place and compose the recurrence relation.

h) $\prod_{i=1}^n (a_i a_{2n+1-i} - b_i b_{2n+1-i})$.

i) $a_1a_2 \ldots a_n \left(a_0 - \dfrac{1}{a_1} - \dfrac{1}{a_2} - \cdots - \dfrac{1}{a_n} \right)$.

1204. *Hint.* Prove that $D_n = D_{n-1} + D_{n-2}$.

1301. a) 301.

b) -153.

c) 1932.

d) −336.

e) −7497. *Hint.* Obtain a zero corner block.

f) 10.

g) −18016.

h) 1.

i) −2639.

j) $\frac{28}{81}$.

k) 1.

l) −21.

m) 60.

n) 78.

o) −924.

p) 800.

q) 301.

1302. a) $n!$.

b) $(-1)^{n-1}n!$. *Hint.* Subtract the last row (or column) from all the others.

c) $(-1)^{\frac{n(n-1)}{2}} b_1 b_2 \ldots b_n$.

d) $x_1(x_2 - a_{12}) \cdot (x_3 - a_{23}) \ldots (x_n - a_{n-1,n})$. *Hint.* Starting from the last row, subtract from each row the previous one.

e) $(-1)^{n(n-1)} 2n$. *Hint.* Starting from the last column, subtract from each column the previous one.

f) $\prod_{k=1}^{n}(1 - a_{kk}x)$.

g) $(-1)^{\frac{n(n+1)}{2}} (n+1)^{n-1}$. *Hint.* Add all columns to the first one.

h) $[(a + (n-1)b](a - b)^{n-1}$.

i) $b_1 \ldots b_n$.

1303. $(-nh)^{n-1}\left[a + \frac{(n-1)}{2}h\right]$. *Hint.* Subtract from each row of the first $(n-1)$ rows the next one and add together all these $n-1$ rows.

1401. a) $n + 1$.

b) $2^{n+1} - 1$.

c) $3^n - 2^{n+1}$.

d) $5 \cdot 2^{n-1} - 4 \cdot 3^{n-1}$.

e) $2^{n+1} - 1$.

f) $\dfrac{\alpha^{n+1} - \beta^{n+1}}{\alpha - \beta}$ if $\alpha \neq \beta$; and $(n+1)\alpha^n$ if $\alpha = \beta$.

g), h) $\prod_{k=1}^{n} k!$.

i) $\prod_{n \geq i > k \geq 1}(x_i - x_k)$.

j) $\prod_{1 \leq i < k \leq n+1}(a_i b_k - a_k b_i)$.

k) $\left(\sum x_{\alpha_1} x_{\alpha_2} \ldots x_{\alpha_{n-s}}\right) \prod_{n \geq i > k \geq 1}(x_i - x_k)$, where the sum is taken over all combinations $\alpha_1, \ldots, \alpha_{n-s}$ of n things $n - s$ at a time. *Hint.* Adjoint the row $1, z, z^2, \ldots, z^{s-1}, z^s, z^{s+1}, \ldots, z^n$ and the column ${}^t(z^s, x_1^s, \ldots, x_n^s)$. Calculate the determinant obtained in two ways: by decomposing according to elements of the adjoint row and as a Vandermonde determinant; compare the coefficients at z^s.

l) $[2x_1 x_2 \ldots x_n - (x_1 - 1)(x_2 - 1) \ldots (x_n - 1)] \prod_{n \geq i > k \geq 1}(x_i - x_k)$. *Hint.* Adjoint the first row $1, 0, 0, \ldots, 0$ and the first column with unit entries, subtract the first column from the others, then represent the unit in the left-hand corner in the form $2 - 1$ and afterwards represent the determinant as a difference of two determinants according to elements of the first rows.

m) $(-1)^{n-1}(n-1)x^{n-2}$.

n) $\dfrac{x(a-y)^n - y(a-x)^n}{x - y}$.

1501. $(a^2 + b^2 + c^2 + d^2)^2$. *Hint.* Multiply the matrix by its transpose. Find the coefficient at a^4 in the expansion of the determinant.

1502. a) 0, if $n > 2$ and $\sin(\alpha_1 - \alpha_2) \sin(\beta_1 - \beta_2)$ if $n = 2$.

b) $\prod_{n \geq i > k \geq l}(a_i - a_k)(b_i - b_k)$.

c) $\dbinom{n}{1}\dbinom{n}{2} \ldots \dbinom{n}{n} \prod_{n \geq i > k \geq 0}(a_k - a_i)(b_i - b_k)(b_i - b_k)$.

d) $\prod_{n \geq i > k \geq 1}(x_i - x_k)^2$.

1503. *Hint.* Multiply by the Vandermonde determinant.

1504. a) $(a + b + c + d)(a - b + c - d)(a + bi - c - di)(a - bi - c + di) = a^4 - b^4 + c^4 - d^4 - 2a^2c^2 + 2b^2d^2 - 4a^2bd + 4b^2ac - 4c^2bd + 4d^2ac$. *Hint.* Make use of Exercise 1503. b) $(1 - \alpha^n)^{n-1}$. *Hint.* Make use of Exercise 1503 and the equality $(1 - \alpha\varepsilon)(1 - \alpha\varepsilon_2) \ldots (1 - \alpha\varepsilon_n) = 1 - \alpha^n$.

1601. a) 2. *Hint.* Show that all three members occuring with plus sign in the expansion of the determinant cannot be equal to 1; consider the determinant with zero entries at the principal diagonal and with units as the other entries.

b) 4. *Hint.* Consider in the expansion of the determinant the product of members with a plus sign and members with a minus sign; calculate the determinant whose entries at the principal diagonal are equal to -1 with all other entries being units.

1602. *Hint.* Make use of the expansion of the determinant.

1604. *Hint.* Apply to the product $A\hat{A}$ the theorem concerning the product of determinants.

1605. *Hint.* Decompose $\det C$ according to elements of columns into the sum of n^m determinants. In each summand factor out b_{jk_j} from the jth column. Show that $\det C = \sum_{k_1,\ldots,k_m=1}^{n} b_{1k_1} \ldots b_{mk_m} A_{k_1\ldots k_m}$. Notice that if $m > n$ then there exist equal numbers among $k_1, \ldots, k_m$ and therefore $A_{k_1\ldots k_m} = 0$. A second way: if $m > n$ then complete A and B to square matrices with the help of $m - n$ zero columns and apply the theorem on products of determinants.

1606. and **1607.** *Hint.* Make use of Exercise 1605.

1608. *Hint.* Decompose according to elements of the last row.

1609. *Hint.* Prove firstly that

$$
\begin{vmatrix} a_{11} + x & \cdots & a_{1n} + x \\ \vdots & \ddots & \vdots \\ a_{n1} + x & \cdots & a_{nn} + x \end{vmatrix} = \begin{vmatrix} a_{11} & \cdots & a_{1n} \\ \vdots & \ddots & \vdots \\ a_{n1} & \cdots & a_{nn} \end{vmatrix} + x \sum_{i,j} A_{ij},
$$

then in the left-hand side of the equality and in the first summand of the right-hand side subtract the first row from all the others and put $x = 1$.

1611. *Hint.* Fulfil over k groups of n rows of D, the transformations which reduce the determinants A to the triangular form; decompose the obtained determinant according to elements of rows with indices $n, 2n, \ldots, kn$ using the Laplace theorem.

1612. a) The sum of all possible products $a_1, \ldots, a_n$, one of which contains all elements, the others being obtained from it by eliminating one or several factors with adjacent indices (if all factors are eliminated then the member is supposed to be equal to 1). *Hint.* Apply the recurrence relation $(a_1 \ldots a_n) = a_n(a_1 \ldots a_{n-1}) + (a_1 \ldots a_{n-2})$.

b) $(a_1 a_2 \ldots a_n) = (a_1 a_2 \ldots a_k)(a_{k+1} a_{k+2} \ldots a_n)$
$+ (a_1 a_2 \ldots a_{k-1})(a_{k+2} a_{k+3} \ldots a_n)$.

c) *Hint.* Induction.

1613. *Hint.* In the case of linear dependent rows of the matrix $(C|D)$, reduce it by elementary row transformations to a matrix with zero row. Apply these elementary transformations to columns of ^tD and ^tC; the obtained matrix differs from $A^tD - B^tC$ by a nonsingular factor. In the case when the minor

$$
\begin{vmatrix} c_{1i_1} & \cdots & c_{1i_k} & d_{1j_1} & \cdots & d_{1j_l} \\ \vdots & \ddots & \vdots & \vdots & \ddots & \vdots \\ c_{ni_1} & \cdots & c_{ni_k} & d_{nj_1} & \cdots & d_{nj_l} \end{vmatrix} \neq 0,
$$

$(k+l=n, i_s \neq j_t)$, consider the product $\begin{pmatrix} A & B \\ C & D \end{pmatrix} \cdot \begin{pmatrix} D & K \\ -{}^t C & L \end{pmatrix}$, where

$$(K|L) = \begin{pmatrix} 0 & \cdots & c'_{1i_1} & \cdots & c'_{1i_k} & \cdots & 0 & \cdots & d'_{1j_1} & \cdots & 0 & \cdots & d'_{1j_l} & \cdots & 0 \\ \vdots & \ddots & \vdots & \ddots & \vdots & \ddots & \vdots & \ddots & \vdots & \ddots & \vdots & \ddots & \vdots & \ddots & \vdots \\ 0 & \cdots & c'_{ni_1} & \cdots & c'_{ni_k} & \cdots & 0 & \cdots & d'_{nj_1} & \cdots & 0 & \cdots & d'_{nj_l} & \cdots & 0 \end{pmatrix},$$

$\begin{pmatrix} c'_{1i_1} & \cdots & c'_{ni_1} \\ \vdots & \ddots & \vdots \\ d'_{1j_l} & \cdots & d'_{nj_l} \end{pmatrix}$ is the inverse of $\begin{pmatrix} c_{1i_1} & \cdots & d_{1j_l} \\ \vdots & \ddots & \vdots \\ c_{ni_l} & \cdots & d_{nj_l} \end{pmatrix}$.

1614. *Hint.* Consider either the product $\begin{pmatrix} A & B \\ C & D \end{pmatrix} \cdot \begin{pmatrix} C^{-1} & D \\ 0 & -C \end{pmatrix}$ or the product $\begin{pmatrix} A & B \\ C & D \end{pmatrix} \cdot \begin{pmatrix} D & 0 \\ -C & D^{-1} \end{pmatrix}$.

1615.

$$\left[(c-a)^n - \binom{n-1}{1}(c-a)^{n-2} + \binom{n-2}{2}(c-a)^{n-4} + \cdots \right]$$
$$\times \left[(c+a)^n - \binom{n-1}{1}(c+a)^{n-2} + \binom{n-2}{2}(c+a)^{n-4} - \cdots \right].$$

Hint. Use the equality

$$\begin{vmatrix} cE & A \\ A & cE \end{vmatrix} = |c^2 E - A^2| = |cE - A||cE + A|.$$

1616. *Hint.* In the determinant D_{n+2} of a matrix obtained from the initial one by adjoining the row $1, x, \ldots, x^{n+1}$ from below, subtract from each column the previous one. Show that $D_{n+2} = (x-1)D_{n-1}$. Decompose D_{n+2} according to elements of the last row.

1617. *Hint.* Decompose D_{2k+1} according to elements of the last column and show that $-D_1, D_2, -D_3, D_4, \ldots$ are solutions of the same system of equations as coefficients of the decomposition $x/(e^x-1) = 1+b_1x+b_2x^2+b_3x^3+\ldots$ (apply the identity $1 = (1+b_1x+b_2x^2+b_3x^3+\ldots)(1+(x/2!)+(x^2/3!)+(x^3/4!)+\ldots))$. Notice that $b_1 = -\frac{1}{2}$ and that $x/(e^x-1) - 1 + \frac{1}{2}x$ is an even function.

1618. *Hint.* Square each determinant.

1619. a),b) $P_n = Q_n = 1$. *Hint.* Show that $Q_n = P_n^2$.

1620. *Hint.* Apply the Gauss formula $n = \sum_{d|n} \phi(d)$ and show that

$$d_{ij} = \sum_{k=1}^{n} p_{ki} p_{kj} \phi(k)$$

where $p_{ij} = 1$ if i divides j and $p_{ij} = 0$ if i does not divide j; decompose the determinant into a sum of n^n summands.

1621. *Hint.* Check that

$$A = \det \left(\frac{1}{1 - x_i y_j} \right)_{i,j=1,\ldots,n} \prod_{i,j=1}^{n} (1 - x_i y_j)$$

is an integer polynomial in $x_1, \ldots, x_n, y_1, \ldots, y_n$ which is antisymmetric in $x_1, \ldots, x_n$ and in $y_1, \ldots, y_n$. Therefore $A = b\Delta(x_1, \ldots, x_n)\Delta(y_1, \ldots, y_n)$ where b is a polynomial in $x_1, \ldots, y_n$. Compare the degrees of A and $\Delta(x_1, \ldots, x_n)\Delta(y_1, \ldots, y_n)$ and show that $b = 1$.

1701.

a) $\begin{pmatrix} 1 & n+m \\ 0 & 1 \end{pmatrix}$.
b) $\begin{pmatrix} \cos(\alpha+\beta) & -\sin(\alpha+\beta) \\ \sin(\alpha+\beta) & \cos(\alpha+\beta) \end{pmatrix}$.
c) $\begin{pmatrix} 1 & 0 \\ 0 & 1 \\ -1 & 0 \end{pmatrix}$.

d) $\begin{pmatrix} 6 & 14 & -2 \\ 10 & -19 & 17 \end{pmatrix}$.
e) $\begin{pmatrix} 6 & 8 & 6 \\ 8 & 19 & 8 \\ 6 & 8 & 6 \end{pmatrix}$.
f) $\begin{pmatrix} 7 & 5 & 0 \\ -7 & -5 & 0 \\ 14 & 10 & 0 \end{pmatrix}$.

g) $\begin{pmatrix} 3 & 3 & 0 & 0 \\ 3 & 3 & 0 & 0 \\ 0 & 0 & -2 & 2 \\ 0 & 0 & 2 & -2 \end{pmatrix}$.
h) $\begin{pmatrix} 2 & 4 & 0 & 0 \\ 3 & 7 & 0 & 0 \\ 0 & 0 & 0 & 7 \\ 0 & 0 & -2 & 3 \end{pmatrix}$.

1702. a) $\begin{pmatrix} -1 & -4 & -1 \\ 2 & 9 & -7 \\ 13 & -9 & 15 \end{pmatrix}$.
b) $\begin{pmatrix} 7 & 4 & 5 & 11 \\ 6 & 4 & -1 & 6 \\ 0 & 0 & 6 & 12 \\ -1 & 2 & 1 & 3 \end{pmatrix}$.

1703. a) $\begin{pmatrix} 9 & 0 & 0 \\ 0 & 9 & 0 \\ 0 & 0 & 9 \end{pmatrix}$.
b) $\begin{pmatrix} 0 & 0 & 1 & 0 \\ 0 & 0 & 0 & 1 \\ 0 & 0 & 0 & 0 \\ 0 & 0 & 0 & 0 \end{pmatrix}$.
c) $\begin{pmatrix} 4 & 0 & 0 & 0 \\ 0 & 4 & 0 & 0 \\ 0 & 0 & 4 & 0 \\ 0 & 0 & 0 & 4 \end{pmatrix}$.

d) $\begin{pmatrix} 0 & 0 & 2 & 9 \\ 0 & 0 & 0 & 6 \\ 0 & 0 & 0 & 0 \\ 0 & 0 & 0 & 0 \end{pmatrix}$.

1704. a) $\begin{pmatrix} \cos n\alpha & \sin n\alpha \\ -\sin n\alpha & \cos n\alpha \end{pmatrix}$. *Hint.* Apply induction.

b) $\begin{pmatrix} \lambda^n & n\lambda^{n-1} \\ 0 & \lambda^n \end{pmatrix}$.

c) $\begin{pmatrix} 3n+1 & -n \\ 9 & -3n+1 \end{pmatrix}$. *Hint*. Notice that the first and the third matrices are inverse to each other and therefore powers are products of n factors.

1705. a) $\begin{pmatrix} 1 & 4 & 0 \\ 0 & 1 & 0 \\ 1 & 4 & 0 \\ 1 & 4 & 0 \end{pmatrix}$. **b)** $\begin{pmatrix} 18 & 18 & 18 \\ 18 & 18 & 18 \\ 18 & 18 & 18 \end{pmatrix}$.

1707. $H^k = \begin{pmatrix} 0 & E \\ 0 & 0 \end{pmatrix}$, where E is the unit matrix of size $n - k$ if $k \le n - 1$ and $H^k = 0$ if $k \ge n$.

1708. *Hint*. Represent $f(x)$ in the form

$$ f(x) = \sum_{k=0}^{n} \frac{f^{(k)}(\lambda)}{k!} (x - \lambda)^k $$

and I in the form $I = \lambda E + H$ where H is from 1707. Make use of Exercise 1707.

1710. a) $\begin{pmatrix} 3 & 1 \\ -4 & -1 \end{pmatrix}$. **b)** $\begin{pmatrix} 1 & 1 & 5 \\ 0 & 1 & 6 \\ 0 & 0 & 1 \end{pmatrix}$.

1711. a) $\begin{pmatrix} 2 & 1 \\ -4 & -2 \end{pmatrix}$. **b)** $\begin{pmatrix} 0 & 1 & -\frac{1}{2} & \frac{1}{3} & -\frac{1}{4} & \cdots & \frac{(-1)^n}{n-1} \\ 0 & 0 & 1 & -\frac{1}{2} & \frac{1}{3} & \cdots & \frac{(-1)^n}{n-2} \\ 0 & 0 & 0 & 1 & -\frac{1}{2} & \cdots & \frac{(-1)^n}{n-3} \\ \cdots & \cdots & \cdots & \cdots & \cdots & \cdots & \cdots \\ 0 & 0 & 0 & 0 & 0 & \cdots & 1 \\ 0 & 0 & 0 & 0 & 0 & \cdots & 0 \end{pmatrix}$.

1714. $\sum_k a_{jk} E_{ik}$.

1705. $\sum_k a_{ki} E_{kj}$.

1716. and 1717. *Hint*. Make use of Exercises 1714 and 1715.

1718. *Hint*. Show that A commutes with $E + E_{ij}$, $i \ne j$ if and only if A commutes with E_{ij}. Apply Exercise 1716.

1719. $A = 0$. *Hint*. After multiplication of A by E_{ji} we obtain a matrix in which the only non-zero entry of the principal diagonal is equal to a_{ij}.

1721. *Hint*. Exploit Exercise 1720.

1722. If $\lambda = 0$. *Hint*. Make use of Exercise 1720.

1724. $\text{tr}[A, B] = 0$. *Hint*. Calculate the square of a matrix with zero trace.

1725. $\begin{pmatrix} AA_1 + BC_1 & AB_1 + BD_1 \\ CA_1 + DC_1 & CB_1 + DD_1 \end{pmatrix}$.

1726. *Hint*. Find entries of principal diagonals in $A^t A$ and $^t A A$.

1727. *Hint.* Let $B = (b_{ij})$ where $b_{ij} = 0$ if $i > j$. By the assumption $b_{ij} = 0$ if $i > j$, $b_{ii} \neq 0$ for all i and $b_{1i}b_{1j} + \cdots + b_{ii}b_{ij} = 0$ if $j \geq i + k$. Show by induction on i that $b_{ij} = 0$.

1728. *Hint.* Notice that $E_{ij} = [E_{ij}, E_{jj}]$ if $i \neq j$ and a matrix $\mathrm{diag}(\alpha_1, \ldots, \alpha_n)$ with zero trace is equal to $\sum_{i=2}^{n} \alpha_i (E_{ii} - E_{11}) = \sum_{i=2}^{n} \alpha_i [E_{i1}, E_{1i}]$.

1729. $A = \mathrm{diag}(h_1, \ldots, h_n)$,

$$B = \begin{pmatrix} 0 & 0 & 0 & \cdots & 0 & 0 \\ h_1 & 0 & 0 & \cdots & 0 & 0 \\ 0 & h_1 + h_2 & 0 & \cdots & 0 & 0 \\ 0 & 0 & h_1 + h_2 + h_3 & \cdots & 0 & 0 \\ \cdots & & & & & \\ 0 & 0 & 0 & \cdots & 0 & 0 \\ 0 & 0 & 0 & \cdots & \sum_{k=1}^{n-1} h_k & 0 \end{pmatrix}$$

where $h_k = (n - 2k + 1)/2$.

1801. a) $X = \begin{pmatrix} 2 & 3 \\ 0 & 2 \end{pmatrix}$, $Y = \begin{pmatrix} -1 & -2 \\ 0 & -1 \end{pmatrix}$.

b) $Y = 2X + \begin{pmatrix} 0 & -1 \\ 1 & 0 \end{pmatrix}$, where X is an arbitrary matrix of size 2.

1803. a) $\begin{pmatrix} 1 & 1 \\ 0 & 0 \end{pmatrix}$. b) $\begin{pmatrix} 11 & 3 \\ -24 & -7 \end{pmatrix}$. c) $\begin{pmatrix} a & b \\ 2a-1 & 2b-3 \end{pmatrix}$ $(a, b \in \mathbb{R})$.

d) $\emptyset$.

e) $\begin{pmatrix} -1 & 2 \\ 0 & 0 \end{pmatrix}$. f) $\begin{pmatrix} 6 & 4 & 5 \\ 2 & 1 & 2 \\ 3 & 3 & 3 \end{pmatrix}$. g) $\begin{pmatrix} 1 & 2 & 3 \\ 4 & 5 & 6 \\ 7 & 8 & 9 \end{pmatrix}$. h) $\begin{pmatrix} 1 & 1 & \cdots & 1 \\ 0 & 1 & \cdots & 1 \\ \cdots & & & \\ 0 & 0 & \cdots & 1 \end{pmatrix}$.

i) $\begin{pmatrix} 1 & 1 & -2 \\ 4 & 2 & -4 \\ -6 & 3 & 3 \end{pmatrix}$. j) $-\frac{1}{6}\begin{pmatrix} -9 & 1 & -5 \\ -6 & -2 & 4 \\ 9 & -3 & -3 \end{pmatrix}$. k) $\frac{1}{3}\begin{pmatrix} 0 & 2 & -1 \\ 0 & -1 & 2 \\ 0 & 0 & 0 \end{pmatrix}$.

l) $\begin{pmatrix} 0 & 1 & 1 \\ 0 & 2 & 0 \\ 0 & 3 & 0 \end{pmatrix}$. m) $\frac{1}{8}\begin{pmatrix} 1 & 7 & -5 \\ 7 & -5 & 1 \\ -5 & 1 & 7 \end{pmatrix}$. n) $\begin{pmatrix} 1 & 1 & -2 \\ 4 & -2 & 4 \\ -6 & 2 & 3 \end{pmatrix}$.

1804. *Hint.* Apply the Kroneker–Capelli theorem.

1805. *Hint.* Reduce A by elementary row transformations of the augmented matrix $(A|B)$ to the row-echelon form.

1806. *Hint.* Indicate a matrix B assuming that the matrix A is in row-echelon form.

1808. a) $\begin{pmatrix} 1 & -3 \\ 0 & 1 \end{pmatrix}$. b) $\begin{pmatrix} 1 & 0 \\ -\frac{3}{2} & \frac{1}{2} \end{pmatrix}$. c) $\begin{pmatrix} -5 & 2 \\ 3 & -1 \end{pmatrix}$. d) $\begin{pmatrix} 7 & -3 \\ -2 & 1 \end{pmatrix}$.

e) $\begin{pmatrix} \frac{1}{5} & 0 & 0 \\ 0 & \frac{1}{3} & 0 \\ 0 & 0 & -\frac{1}{2} \end{pmatrix}$. f) $\begin{pmatrix} 1 & 0 & 0 \\ 0 & 1 & 0 \\ -3 & 0 & 1 \end{pmatrix}$. g) $\begin{pmatrix} \frac{1}{6} & 0 & 0 \\ 0 & -5 & 2 \\ 0 & 3 & -1 \end{pmatrix}$.

h) $\begin{pmatrix} 7 & -3 & 0 \\ -2 & 1 & 0 \\ 0 & 0 & \frac{1}{7} \end{pmatrix}$. i) $\begin{pmatrix} 1 & -1 & 0 \\ 0 & 1 & 0 \\ 0 & -1 & \frac{1}{3} \end{pmatrix}$. j) $\begin{pmatrix} \frac{1}{2} & 0 & 0 \\ -\frac{3}{2} & 1 & -\frac{1}{2} \\ 0 & 0 & \frac{1}{2} \end{pmatrix}$.

k) $\begin{pmatrix} \cos\alpha & -\sin\alpha \\ \sin\alpha & \cos\alpha \end{pmatrix}$.

1809. a) $\begin{pmatrix} 1 & 0 & 0 & 0 \\ 0 & 0 & 0 & 1 \\ 0 & 1 & 0 & 0 \\ 0 & 0 & 1 & 0 \end{pmatrix}$. b) $\begin{pmatrix} 0 & 1 & 0 & 0 \\ 0 & 0 & 0 & 1 \\ 1 & 0 & 0 & 0 \\ 0 & 0 & 1 & 0 \end{pmatrix}$. c) $\begin{pmatrix} \frac{1}{2} & 0 & 0 & 0 \\ 0 & 0 & \frac{1}{2} & 0 \\ 0 & 0 & 0 & 1 \\ 0 & 1 & 0 & 0 \end{pmatrix}$.

d) $\begin{pmatrix} 0 & 0 & 1 & 0 \\ 0 & 0 & 0 & \frac{1}{3} \\ 0 & \frac{1}{2} & 0 & 0 \\ -1 & 0 & 0 & 0 \end{pmatrix}$. e) $\begin{pmatrix} 1 & -1 & 0 & \ldots & 0 & 0 \\ 0 & 1 & -1 & \ldots & 0 & 0 \\ \hdotsfor{6} \\ 0 & 0 & 0 & \ldots & 1 & -1 \\ 0 & 0 & 0 & \ldots & 0 & 1 \end{pmatrix}$.

f) $\begin{pmatrix} 1 & 0 & 0 & 0 & \ldots & 0 & 0 & 0 \\ -1 & 1 & 0 & 0 & \ldots & 0 & 0 & 0 \\ 1 & -1 & 1 & 0 & \ldots & 0 & 0 & 0 \\ -1 & 1 & -1 & 1 & \ldots & 0 & 0 & 0 \\ \hdotsfor{8} \\ \ldots & \ldots & \ldots & \ldots & \ldots & 1 & 0 & 0 \\ \ldots & \ldots & \ldots & \ldots & \ldots & -1 & 1 & 0 \\ \ldots & \ldots & \ldots & \ldots & \ldots & 1 & -1 & 1 \end{pmatrix}$.

g) $\begin{pmatrix} 1 & -1 & 1 \\ -38 & 41 & 34 \\ 27 & -29 & 24 \end{pmatrix}$. h) $\begin{pmatrix} -8 & 29 & -11 \\ -5 & 18 & -7 \\ 1 & -3 & 1 \end{pmatrix}$.

i) $\begin{pmatrix} \frac{7}{3} & 2 & -\frac{1}{3} \\ \frac{5}{3} & -1 & -\frac{1}{3} \\ -2 & 1 & 1 \end{pmatrix}$. j) $\frac{1}{9}\begin{pmatrix} 1 & 2 & 2 \\ 2 & 1 & -2 \\ 2 & -2 & 1 \end{pmatrix}$.

k) $\begin{pmatrix} \frac{2}{2} & -6 & -26 & 17 \\ -17 & 5 & 20 & -13 \\ -1 & 0 & 2 & -1 \\ 4 & -1 & -5 & 3 \end{pmatrix}$.

1810. a) $\begin{pmatrix} A^{-1} & 0 \\ -C^{-1}BA^{-1} & C^{-1} \end{pmatrix}$. b) $\begin{pmatrix} A^{-1} & -A^{-1}BC^{-1} \\ 0 & C^{-1} \end{pmatrix}$.

1811. a) $\begin{pmatrix} -3 & 2 & 0 & 0 \\ 2 & -1 & 0 & 0 \\ 8 & -\frac{9}{2} & 1 & -\frac{3}{2} \\ -1 & \frac{1}{2} & 0 & \frac{1}{2} \end{pmatrix}$. **b)** $\begin{pmatrix} -1 & 3 & -8 & 3 \\ 1 & -2 & 4 & -1 \\ 0 & 0 & 2 & -1 \\ 0 & 0 & 1 & -1 \end{pmatrix}$.

1813. a), b) ± 1.

1814. ± 1.

1815. *Hint.* Utilize the adjoint matrix $\hat{A}$

1816. b) *Hint.* Note that if $\det A = 0$ then the system of equations $\sum_{j=1}^{n} a_{ij} x_j = 0$ has a nonzero solution.

1817. *Hint.* Put $C = (E + AB)^{-1}$ and prove that $(E - BCA)(E + BA) = E$.

1818. *Hint.* Compare ranks of AB and BA with ranks of A and B.

1819. *Hint.* Apply Exercise 1804.

1820. *Hint.* Exploit the correspondence between multiplication by elementary matrices and elementary transformations.

1822. *Hint.* Let $A = (a_{ij})$, $B = (b_{ij})$ be matrices of size n whose entries are polynomials in $2n^2$ variables a_{ij}, b_{ij}, $1 \le i, j \le n$. Then $A^{\wedge} = (\det A) A^{-1}$. Make use of Exercise 1821 for the proof of the first equality. Any values can be substituted instead of a_{ij}, b_{ij}. Similarly one can prove the other equalities.

1823. *Hint.* Apply Exercise 1110f.

1824. *Hint.* Let B_i be a row of length $n-1$ which is obtained from B by excluding ith coordinates. Prove that $C_i{}^t B_i \ne -1$ for some i. Applying Exercise 1822, indicate a nonzero minor of size $n - 1$.

1902. a) $\begin{pmatrix} 1 & 0 \\ 4 & 1 \end{pmatrix} \cdot \begin{pmatrix} 1 & 0 \\ 0 & -3 \end{pmatrix} \cdot \begin{pmatrix} 1 & 2 \\ 0 & 1 \end{pmatrix}$.

b) $(E - E_{12})(E + E_{21}) \times (E - 2E_{22})(E + E_{12})(E + E_{31})(E + E_{32})$
$(E - 3E_{33})(E + E_{13})(E + 2E_{23})$.

Hint. Make use of Exercise 1713.

1903. a) $\begin{pmatrix} 1 & 4 & 9 & 16 \\ 1 & 6 & 15 & 28 \\ 1 & 4 & 12 & 32 \\ 1 & 2 & 3 & 4 \end{pmatrix}$. **b)** $\begin{pmatrix} 1 & 2 & 3 & 4 \\ 2 & 6 & 10 & 14 \\ 3 & 6 & 12 & 24 \\ 4 & 4 & 4 & 4 \end{pmatrix}$.

c) $\begin{pmatrix} -5 & 2 & 3 & 4 \\ -10 & 3 & 5 & 7 \\ -15 & 2 & 4 & 8 \\ 0 & 1 & 1 & 1 \end{pmatrix}$. **d)** $\begin{pmatrix} 1 & 2 & 3 & 4 \\ 2 & 5 & 8 & 11 \\ 3 & 6 & 10 & 16 \\ -2 & -5 & -8 & 11 \end{pmatrix}$.

1906. *Hint.* Make use of Exercise 1904.

1908. In the case of commuting matrices.

1909. *Hint.* For construction of Y utilize matrices U, V such that $UXV = E_{11} + \cdots + E_{rr}$.

1912. Yes, if $n \geq 3$.

1914. $\{\alpha E_{1,n-1} \mid \alpha \in K\}$.

1915. *Hint.* Apply the binomial theorem from Exercise 1706. The statement is not correct.

1917. If $A^n = 0$ then $\det A = 0$. *Hint.* Make use of Exercise 1802.

1919. *Hint.* Apply the formula for the sum of an infinite geometrical progression.

1920. *Hint.* Make use of Exercise 1919.

1922. *Hint.* See Exercises 1917 and 1919.

1923. The statement is not valid.

1927. *Hint.* Exploit calculations of the inverse matrix with the help of elementary transformations.

2001. a) $1 + 18i$.

b) $4i$.

c) $7 + 17i$.

d) $10 - 11i$.

e) $14 - 5i$.

f) $5 + i$.

g) $\frac{13}{2} - \frac{1}{2}i$.

h) $\frac{11}{5} - \frac{27}{5}i$.

i) 4.

j) $52i$.

k) 2.

l) 1.

m) -1.

2002. $i^n = 1$ if $n = 4k$; $i^n = i$ if $n = 4k + 1$; $i^n = -1$ if $n = 4k + 2$; $i^n = -i$ if $n = 4k + 3$ where k is an integer, $i^{77} = i$; $i^{98} = -1$; $i^{-57} = -i$.

2004. a) $z_1 = i$, $z_2 = 1 + i$.

b) $z_1 = 2$, $z_2 = 1 - i$.

c) $\emptyset$.

d) $z_1 = \dfrac{(2 + i)z_2 - i}{2}$.

e) $x = 3 - 11i$, $y = -3 - 9i$, $z = 1 - 7i$.

2005. a) $x = 2$, $y = -3$.

b) $x = 3$, $y = -5$.

2008. a) 0, 1, $-\frac{1}{2} \pm i\frac{\sqrt{3}}{2}$.

b) $0, \pm1, \pm i$.

2009. *Hint.* Induction on the number of operations.

2010. *Hint.* Apply the previous exercise.

2011. a) $\pm\frac{\sqrt[4]{2}}{2}(1 + i)$.

b) $\pm(2 - i)$.

c) $\pm(3 - 2i)$.

d) $z_1 = 1 - 2i$, $z_2 = 3i$.

e) $z_1 = 5 - 2i$, $z_2 = 2i$.

f) $z_1 = 5 - 3i$, $z_2 = 2 + i$.

2101. a) $5(\cos 0 + i \sin 0)$.

b) $\cos\dfrac{\pi}{2} + i \sin\dfrac{\pi}{2}$.

c) $2(\cos \pi + i \sin \pi)$.

d) $3\left(\cos\left(-\dfrac{\pi}{2}\right) + i \sin\left(-\dfrac{\pi}{2}\right)\right)$.

e) $\sqrt{2}\left(\cos\dfrac{\pi}{4} + i \sin\dfrac{\pi}{4}\right)$.

f) $\sqrt{2}\left(\cos\left(-\dfrac{\pi}{4}\right) + i \sin\left(-\dfrac{\pi}{4}\right)\right)$.

g) $2\left(\cos\dfrac{\pi}{3} + i \sin\dfrac{\pi}{3}\right)$.

h) $2\left(\cos\dfrac{2\pi}{3} + i \sin\dfrac{2\pi}{3}\right)$.

i) $2\left(\cos\left(-\dfrac{2\pi}{3}\right) + i \sin\left(-\dfrac{2\pi}{3}\right)\right)$.

j) $2\left(\cos\left(-\dfrac{\pi}{3}\right) + i \sin\left(-\dfrac{\pi}{3}\right)\right)$.

k) $2\left(\cos\dfrac{\pi}{6} + i \sin\dfrac{\pi}{6}\right)$.

l) $2\left(\cos\dfrac{5\pi}{6} + i \sin\dfrac{5\pi}{6}\right)$.

m) $2\left(\cos\left(-\dfrac{5\pi}{6}\right) + i \sin\left(-\dfrac{5\pi}{6}\right)\right)$.

n) $2\left(\cos\left(-\dfrac{\pi}{6}\right) + i \sin\left(-\dfrac{\pi}{6}\right)\right)$.

o) $\dfrac{2}{\sqrt{3}}\left(\cos\dfrac{\pi}{6} + i \sin\dfrac{\pi}{6}\right)$.

p) $2\sqrt{2+\sqrt{3}}\left(\cos\dfrac{\pi}{12}+i\sin\dfrac{\pi}{12}\right)$ or $(\sqrt{6}+\sqrt{2})\left(\cos\dfrac{\pi}{12}+i\sin\dfrac{\pi}{12}\right)$. *Hint.* In order to obtain the second expression for the absolute values apply the formula

$$\sqrt{a\pm\sqrt{b}}=\sqrt{\dfrac{a+\sqrt{a^2-b}}{2}}\pm\sqrt{\dfrac{a-\sqrt{a^2-b}}{2}}.$$

q) $2(\sqrt{2+\sqrt{3}})\left(\cos\left(-\dfrac{5\pi}{12}\right)+i\sin\left(-\dfrac{5\pi}{12}\right)\right).$

r) $\cos(-\alpha)+i\sin(-\alpha).$

s) $\cos\left(\dfrac{\pi}{2}-\alpha\right)+i\sin\left(\dfrac{\pi}{2}-\alpha\right).$

t) $\cos 2\alpha+i\sin 2\alpha.$

u) $2\cos\dfrac{\varphi}{2}\left(\cos\dfrac{\varphi}{2}+i\sin\dfrac{\varphi}{2}\right).$

v) $\cos(\varphi-\psi)+i\sin(\varphi-\psi).$

2102. a) 2^{50}.

b) 2^{150}.

c) -2^{30}.

d) $(2+\sqrt{3})^{12}$.

e) $-2^{12}(2-\sqrt{3})^6$.

f) -2^6.

g) $2^{15}i$.

h) -64.

2103. a) $3+4i$.

b) $5-12i$.

2105. *Hint.* Make use of Exercise 2104.

2106. The equality is obtained only if either $\arg z_1=\arg z_2$ or at least one of the given numbers is equal to zero; utilize the geometrical meaning of a number $\min(|z_1|,|z_2|)|\arg z_1-\arg z_1|.$

2107. *Hint.* Reduce the exercise to the theorem on the sum of squares of lengths of diagonals in a parallelogram.

2110.*Hint.* Prove that $z=\cos\varphi\pm i\sin\varphi.$

2111. a) $4\cos^3 x\sin x-4\cos x\sin^3 x.$ *Hint.* Calculate $(\cos x+i\sin x)^4$ by de Moivre formulae and by the binomial formula.

b) $\cos^4 x-6\cos^2 x\sin^2 x+\sin^4 x.$

c) $5\cos^4 x\sin x-10\cos^2 x\sin^3 x+\sin^5 x.$

d) $\cos^5 x-10\cos^3 x\sin^2 x+5\cos x\sin^4 x.$

2113. a) $\frac{1}{8}(\cos 4x - 4\cos 2x + 3)$; if $z = \cos x + i\sin x$, $\sin x = (z - z^{-1})/2i$, $z^k + z^{-k} = 2\cos kx$.

b) $\frac{1}{8}(\cos 4x + 4\cos 2x + 3)$.

c) $\frac{1}{16}(\sin 5x - 5\sin 3x + 10\sin x)$.

d) $\frac{1}{16}(\cos 5x + 5\cos 3x + 10\cos x)$.

2114. a) *Hint.* Apply the hint of Exercise 2113.

2206. It is not valid: these sets have different cardinalities.

2207. a) $\cos \dfrac{(4k+1)\pi}{12} + i\sin \dfrac{(4k+1)\pi}{12}$ $(0 \le k \le 5)$.

b) $\left[\cos \dfrac{(6k-1)\pi}{30} + i\sin \dfrac{(6k-1)\pi}{30}\right]$ $(0 \le k \le 9)$.

c) $\sqrt{2}\left[\cos \dfrac{(8k-1)\pi}{32} + i\sin \dfrac{(8k-1)\pi}{32}\right]$ $(0 \le k \le 7)$.

d) $\left\{1, -\dfrac{1}{2} \pm i\dfrac{\sqrt{3}}{2}\right\}$.

e) $\{\pm 1, \pm 1\}$.

f) $\left\{\pm 1, \pm\dfrac{1+i\sqrt{3}}{2}, \pm\dfrac{1-i\sqrt{3}}{2}\right\}$.

g) $\left\{\dfrac{\sqrt{3}}{2} + \dfrac{1}{2}i, -\dfrac{\sqrt{3}}{2} + \dfrac{1}{2}i, -i\right\}$.

h) $\{1 \pm i, -1 \pm i\}$.

i) $2\sqrt{1}$.

j) $\left\{\pm\sqrt{2}, \pm\sqrt{2}i, \pm(1+i), \pm(1-i)\right\}$.

k) $\left\{\pm i\sqrt{3}, \pm\dfrac{\sqrt{3}}{2}(\sqrt{3}+i), \pm\dfrac{\sqrt{3}}{2}(\sqrt{3}-i)\}\right\}$.

l) $\left\{\sqrt{3}+i, -1+i\sqrt{3}, -\sqrt{3}-i, 1-i\sqrt{3}\right\}$.

m) $\left\{3+i\sqrt{3}, \sqrt{3}-3i, -3-i\sqrt{3}, -\sqrt{3}+3i\right\}$.

n) $\left\{\dfrac{1}{2}\sqrt[3]{4}(i-1), \dfrac{\sqrt[3]{4}}{4}(1-\sqrt{3}-i(\sqrt{3}+1)), \dfrac{\sqrt[3]{4}}{4}(1+\sqrt{3}-i(\sqrt{3}-1))\right\}$.

o) $\left\{\dfrac{1}{2}\sqrt{2}(\sqrt{2+\sqrt{3}}-i\sqrt{2-\sqrt{3}}), -\dfrac{1}{2}\sqrt{2}(\sqrt{2-\sqrt{3}}-i\sqrt{2+\sqrt{3}}), 1-i\right\}$.

p) $\left\{\pm\left(\dfrac{3}{2}+i\dfrac{\sqrt{3}}{2}\right),\pm\left(\dfrac{\sqrt{3}}{2}-\dfrac{3}{2}i\right)\right\}.$

q) $\left\{\pm\dfrac{\sqrt{2}}{2}\pm i\dfrac{\sqrt{2}}{2}\right\}.$

r) $\left\{+2i,\ -\sqrt{3}-i,\ \sqrt{3}-i\right\}.$

s) $\sqrt[4]{2}\left(\cos\dfrac{\pi+6k\pi}{12}+i\sin\dfrac{\pi+6k\pi}{12}\right),\ k=0,1,2,3.$

2208. a) $\frac{1}{4}(\sqrt{5}-1)$.   b) $\frac{1}{4}\sqrt{10+2\sqrt{5}}$.

2210. $\{\pm1\};\ \left\{1,\ \frac{1}{2}\pm i\frac{\sqrt{3}}{2}\right\};\ \{\pm1,\ \pm i\};\ \left\{\pm1,\ \pm\frac{1}{2}(1+i\sqrt{3}),\ \pm\frac{1}{2}(1-i\sqrt{3})\right\};$
$\left\{\pm1,\ \pm i,\ \pm\frac{\sqrt{2}}{2}(1+i),\ \pm\frac{\sqrt{2}}{2}(1-i)\right\},$
$\left\{\pm i;\ \pm\frac{1}{2}(1+i\sqrt{3}),\ \pm\frac{1}{2}(\sqrt{3}+i),\ \pm\frac{1}{2}(\sqrt{3}-i)\right\}.$

2211. $(-1)^{n-1}$. *Hint.* Multiply each root except 1 and -1 by its inverse.

2213. *Hint.* The greatest common divisor of numbers r and s can be written in the form $ru+sv$.

c) If $\alpha\in U_r$, $\beta\in U_s$, then $(\alpha\beta)^{rs}=1$, i.e. $\alpha\beta\in U_{rs}$ and $U_rU_s\subseteq U_{rs}$; if $\alpha_1\neq\alpha_2$ are elements of U_r, $\beta_1\neq\beta_2$ are elements of U_s then $\alpha_1\beta_1\neq\alpha_2\beta_2$, otherwise $\alpha_1\alpha_2^{-1}=\beta_1\beta_2^{-1}\in U_r\cap U_s$ though $\alpha_1\alpha_2^{-1}\neq1$, $U_r\cap U_s=\{1\}$ (see b)); therefore $|U_rU_s|=rs=|U_{rs}|$, and hence $U_{rs}=U_rU_s$.

2215. ε and $\bar{\varepsilon}$ have the same orders.

2216. See Exercise 2212.

2217. a) $-n(1-z)$ if $z\neq1$ and $\dfrac{n(n+1)}{2}$ if $z=1$.

b) $2(1-z)^{-1}$.

c) either $n=2$, $m=1$ or $n=1$, $m=2$.

2218. *Hint.* a) See Exercise 2214.

b) See Exercise 2216.

2219. *Hint.* See Exercise 2216.

2220. *Hint.* b) Each root is primitive only for one order, so the given sum is equal to the sum of all roots of order n.

c), d) follow from b).

e) See Exercise 2216.

f) Consider primary decomposition of n.

338   A.I. Kostrikin

2222. *Hint.* Represent z in the trigonometrical form.

2223. a) $x - 1$.

b) $x + 1$.

c) $x^2 + x + 1$.

d) $x^2 + 1$.

e) $x^2 - x + 1$.

f) $x^4 - x^2 + 1$.

g) $x^{p-1} + x^{p-2} + \cdots + 1$.

h) $(x^{p^k} - 1)/(x^{p^{k-1}} - 1)$.

2224. *Hint.* a) See the hint to Exercise 2220b.

b)   See Exercises 2219a, b; ε is a primitive root of order n if and only if $-\varepsilon$ is a primitive root of order $2n$ (n is odd).

c)   follows from a) and the inversion formula (see Exercise 505b).

d)   if $\{\varepsilon_i\}$ are all primitive roots of order d of 1 and $\{\varepsilon_{ik} \mid 1 \le k \le d\}$ are all values of roots of order d of ε_i then $\{\varepsilon_{ik} \mid i = 1, \ldots, \phi(k); \ k = 1, \ldots, d\}$ are primitive roots of order n.

e)   See Exercise 2219; for any divisor d of $m = n/p$ we have:

$$\mu\left(\frac{n}{d}\right) = \mu\left(\frac{m}{d}p\right) = \mu\left(\frac{m}{d}\right)\mu(p) = -\mu\left(\frac{m}{d}\right)$$

and all divisors of n are obtained if we add to all divisors of m their products by p; therefore

$$\Phi_m(x) = \prod_{d\mid n}(x^d - 1)^{\mu\left(\frac{n}{d}\right)} = \prod_{d\mid n}(x^d - 1)^{\mu\left(\frac{n}{d}\right)} \prod_{d\mid n}(x^{pd} - 1)^{\mu\left(\frac{n}{d}\right)}$$

$$= \prod_{d\mid m}(x^d - 1)^{-\mu\left(\frac{m}{d}\right)} \cdot \prod_{d\mid n}(x^{pd} - 1)^{\mu\left(\frac{m}{d}\right)} = \frac{\Phi_m(x^p)}{\Phi_m(x)}.$$

2225. a) $\Phi_{10}(x) = \Phi_5(-x) = x^4 - x^3 + x^2 - x + 1$.

b) $\Phi_{14}(x) = x^6 - x^5 + x^4 - x^3 + x^2 - x + 1$.

c) $\Phi_{15}(x) = \Phi_3(x^5)/\Phi_3(x) = \dfrac{x^{10} + x^5 + 1}{x^2 + x + 1} = x^8 - x^7 + x^5 - x^4 + x^3 - x + 1$.

d) $\Phi_{30}(x) = \Phi_{15}(-x) = x^8 + x^7 - x^5 - x^4 - x^3 - x + 1$.

e) $\Phi_{36}(x) = \Phi_6(x^6) = x^{12} - x^6 + 1$.

f) $\Phi_{100}(x) = \Phi_{10}(x^{10}) = x^{40} - x^{30} + x^{20} - x^{10} + 1$.

g) $\Phi_{216}(x) = \Phi_6(x^{36}) = x^{72} - x^{36} + 1$.

h) $\Phi_{288}(x) = \Phi_6(x^{48}) = x^{96} - x^{48} + 1.$

i) $\Phi_{1000}(x) = \Phi_{10}(x^{100}) = x^{400} - x^{300} + x^{200} - x^{100} + 1.$

2226. *Hint.* a), b) follow from 2224c; $\Phi_n(0)$ is equal to the product of all primitive roots of order n by -1.

2227. $\Phi_1(1) = 0$, $\Phi_{p^k}(1) = p$ and $\Phi_n(1) = 1$ for all other n (p is a prime). *Hint.* By Exercise 2224d,e we have $\Phi_{p_1^{k_1}\ldots p_s^{k_s}}(1) = \Phi_{p_1\ldots p_s}(1) = \Phi_{p_2\ldots p_s}(1)/\Phi_{p_2\ldots p_s}(1) = 1$; finally see Exercise 2223h.

2301. a) $2^{n/2} \cos \dfrac{n\pi}{4}$. *Hint.* Calculate $(1+i)^n$ by the binomial formula and by the de Moivre formula.

b) $2^{n/2} \sin \dfrac{n\pi}{4}$.

c) $\dfrac{1}{2}\left(2^{n-1} + 2^{n/2}\cos\dfrac{n\pi}{4}\right)$. *Hint.* Make use of a) and the equalities $\sum_{k=0}^n \binom{n}{k} = 2^n$; $\sum_{k=0}^n (-1)^k \binom{n}{k} = 0.$

d) $\dfrac{1}{2}\left(2^{n-1} + 2^{n/2}\sin\dfrac{n\pi}{4}\right)$.

e) $-\dfrac{n}{1-\varepsilon}$ if $\varepsilon \neq 1$; $\dfrac{n(n+1)}{2}$ if $\varepsilon = 1$. *Hint.* Multiply the given sum by $1 - \varepsilon$ if $\varepsilon \neq 1$.

2302. *Hint.* The left- and right-hand sides of the equalities a) and b) are equal to the real and imaginary parts of the sum $z + \cdots + z^n = z\dfrac{z^n-1}{z-1}$, where $z = \cos x + i \sin x$.

c), d) are similar to a) and b).

f) Factorize the left-hand side into the product $(x-\varepsilon_1)\ldots(x-\varepsilon_{2n})$ and combine factors $x - \varepsilon_i$ and $x - \varepsilon_{n-i} = x - \bar\varepsilon_i$. h) Divide equalities in exercises f), g) by $x^2 - 1$ and by $x - 1$, respectively; put $x = 1$ in the obtained equality.

2303. $x_k = -\left[\sin\dfrac{(2k+1)\pi - 2\varphi}{2n}\right]\left[\sin\dfrac{(2k+1)\pi - 2\varphi - 2n\alpha}{2n}\right]^{-1}$,
$k = 0, 1, \ldots, n - 1$.

Hint. If $z = \cos\varphi + i\sin\varphi$, $t = \cos\alpha + i\sin\alpha$ then $2\cos\varphi = z + z^{-1}$; $2\cos(\varphi + k\alpha) = zt^k + z^{-1}t^{-k}$ and therefore $t(1 + zx)^n + t^{-1}(1 + z^{-1}x)^n = 0$.

2304. d) *Hint.* Make use of Exercise 412.

2305. a) $2^n \cos^n \dfrac{x}{2} \cos\dfrac{n+2}{2}x$.

b) $2^n \cos^n \dfrac{x}{2} \sin\dfrac{n+2}{2}x$.

c) $\dfrac{n}{2} - \dfrac{\sin 4nx}{4 \sin 2x}$.

d) $\dfrac{(n+1) \sin nx - n \cos(n+1)x - 1}{4 \sin^2 \dfrac{x}{2}}$.

e) $\dfrac{(n+1) \sin nx - n \sin(m+1)x}{4 \sin^2 \dfrac{x}{2}}$.

2307. *Hint.* As in Exercises 412 and 2304d $\sin mx / \sin x$ is a polynomial of degree $(m-1)/2$ in $\sin^2 x$ with the leading coefficient $(-4)^{\frac{m-1}{2}}$ whose roots are $\sin^2 \left(\dfrac{2\pi j}{m} \right)$ where $j = 1, 2, \ldots, \dfrac{m-1}{2}$.

2402. a) $\pm \dfrac{1}{2} \pm \dfrac{1}{2} i$.

b) $-1, \dfrac{1}{2} \pm i \dfrac{\sqrt{3}}{2}$.

c) $4 + i\sqrt{3}, 3 + 2i\sqrt{3}, 1 + 2i\sqrt{3}, i\sqrt{3}, 1, 3$.

2403. The distance between the points associated with the given numbers.

2404. a) The vertices of a regular triangle with the center at the origin.

b)   The vertices of a diamond with the center at the origin.

2406. a) The circle of radius 1 with the center at the origin.

b)   A ray outgoing from the origin and forming the angle $\pi/3$ with the positive real half-axis.

c)   The circle of radius 2 with the center at the origin including the boundary.

d)   The interior unit disc with the center at the point $1 + i$.

e)   The circle of radius 5 with the center at the point $-3 - 4i$ including the boundary.

f)   The interior ring located between the circles of radii 2 and 3 with the center at the origin.

g)   The ring located between the circles of radii 1 and 2 with the center at the point $2i$, the circle of radius 1 is included and of radius 2 is not included.

h)   The interior angle containing the positive real semiaxis which is formed by rays outgoing from the origin and forming the angles $-\pi/6$ and $\pi/6$ with the real semiaxis.

i)   The interior angle with the vertex z_0 whose rays form angles α and β with the positive real semiaxis.

j)   The band between the lines $x = \pm 1$ including boundaries.

k)   The interior band between the line $y = 1$ and the real axis.

l) Two lines $y = \pm 1$.

m) The interior band between the lines $x + y = \pm 1$.

n) The ellipse $\dfrac{4x^2}{9} + \dfrac{4y^2}{5} = 1$.

o) The hyperbola $\dfrac{4x^2}{9} - \dfrac{4y^2}{7} = 1$.

p) The parabola $y^2 = 8x$.

2407. *Hint.* The sum of squares of diagonals of a parallelogram is equal to the sum of squares of its sides. Put $z_1 = x_1 + y_1 i$, $z_2 = x_2 + y_2 i$ and interpret the square of the absolute value of a complex number as the scalar square of the vector associated this number.

2408. $z_4 = z_1 - z_2 + z_3$.

2409. $\dfrac{z + w}{2} \pm i \dfrac{z - w}{2}$.

2410. $z_k = c + (z_0 - c)\left(\cos \dfrac{2\pi k}{n} + i \sin \dfrac{2\pi k}{n}\right)$ $(k = 0, 1, 2, \ldots, n - 1)$,

where $c = \dfrac{1}{2}(z_0 + z_1) + \dfrac{1}{2} i \cot \dfrac{\pi}{n}(z_1 - z_0)$ is the center of the polygon.

2411. The circle of radius 1 with the center at the origin except the point $z = -1$.
Hint. Put $t = \tan \dfrac{\varphi}{2}$, $-\pi < \varphi < \pi$.

2412. *Hint.* a) In the proof of necessity check that the vectors $z_3 - z$ and $z_3 - z_2$ are collinear; in the proof of sufficiency subtract $(\lambda_1 + \lambda_2 + \lambda_3)z_1 = 0$ from the given equality.

b)   Make use of the previous exercise.

2413. The circle with ends of a diameter at the points $\dfrac{z_1 + \lambda z_2}{1 + \lambda}$, $\dfrac{z_1 - \lambda z_2}{1 - \lambda}$ if $\lambda \neq 1$; the line passing through the middle of the segment with ends z_1, z_2 and perpendicular to this segment if $\lambda = 1$.

2414. $\sqrt{13} - 1$.

2415. $1 + 3\sqrt{5}$.

2416. The required curve consists of points such that the product of distances between this point and the points $z = \pm 1$ is equal to λ. This curve is a *lemniscate*. If $\lambda = 1$ we obtain the *Bernoulli lemniscate* which is given in the pole coordinates by the equation $r^2 = 2\cos 2\varphi$. *Hint.* Show that the curve has no points on the imaginary axis if $\lambda < 1$.

2424. and **2425.** *Hint.* See H. Cartan. Théorie élémentaire des fonctions analytiques d'une ou plusieurs variables complexes, Hermann, Paris, 1961, ch. 6, § 3, items 5 and 6.

2426. $a = 0$. *Hint.* Consider the image of U under the mapping $z \to 1 + az$.

2501. a) $2x^2 + 3x + 11, 25x - 5$.

b) $\frac{1}{9}(3x - 7), -\frac{1}{9}(26x + 2)$.

2502. a) $x + 1$.

b) $x^3 - x + 1$.

c) $x^3 + x^2 + 2$.

d) 1.

e) $x^2 + 1$.

f) $x^3 + 1$.

g) $x^2 - 2x + 2$.

h) $x + 3$.

i) $x^2 + x + 1$.

j) $x^2 - 2\sqrt{2}x - 1$.

k) 1.

2503. a) $d = x^2 - 2 = -(x + 1)f + (x + 2)g$.

b) $d = 1 = xf - (3x^2 + x - 1)g$.

2504. *Hint.* a) Passing from f and g to fd^{-1} and gd^{-1} one can assume that $d = 1$. If $1 = fw + gh$ then take instead of w the remainder from division of w by g.

b) Compare degrees of gv and $d - fu$.

c) Make use of the fact that u and v are coprime.

2505. a) $u(x) = \frac{1}{3}(-16x^2 + 37x + 26), v(x) = \frac{1}{3}(16x^3 - 53x^2 - 37x + 23)$.

b) $u(x) = 4 - 3x, v(x = 1 + 2x + 3x^2$.

c) $u(x) = 35 - 84x + 70x^2 - 20x^3, v(x) = 1 + 4x + 10x^2 + 20x^3$.

2506. Let

$$P_{r,s}(x) = 1 + \frac{r}{1!}x + \frac{r(r + 1)}{2!}x^2 + \cdots + \frac{r(r + 1)\ldots(r + s - 1)}{s!}x^s.$$

Then $u(x) = P_{m,n-1}(1 - x); v(x) = P_{n,m-1}(x)$.

2508. a) $(x - 1)^3(x + 3)^3(x - 3)$.

b) $(x - 2)(x^2 - 2x + 2)^2$.

c) $(x + 1)^4(x - 4)$.

d) $(x + 1)^4(x - 2)^2$.

e) $(x^3 - x^2 - x - 2)^2$.

f) $(x^2 + 1)^2(x - 1)^3$.

g) $(x^4 + x^3 + 2x^2 + x + 1)^2$.

2601. a) $f(x) = (x - 1)(x^3 - x^2 + 3x - 3) + 5$, $f(x_0) = 5$.

b) $f(x) = (x + 3)(2x^4 - 6x^3 + 13x^2 - 39x + 109) - 327$, $f(x_0) = -327$.

c) $f(x) = (x - 2)(3x^4 + 7x^3 + 14x^2 + 9x + 5)$, $f(x_0) = 0$.

d) $f(x) = (x + 2)(x^3 - 5x^2 + 2) + 1$, $f(x_0) = 1$.

e) $f(x) = (x-1)^5 + 5(x-1)^4 + 10(x-1)^3 + 10(x-1)^2 + 5(x-1) + 1$, $f(x_0) = 1$.

f) $f(x) = (x+1)^4 - 2(x+1)^3 - 3(x+1)^3 - 3(x+1)^2 + 4(x+1) + 1$, $f(x_0) = 1$.

g) $f(x) = (x - 2)^4 - 18(x - 2) + 38$, $f(x_0) = 38$.

h) $f(x) = (x+i)^4 - 2i(x+i)^3 - (1+i)(x+i)^2 - 5(x+i) + 7 + 5i$, $f(x_0) = 7 + 5i$.

i) $f(x) = (x + 1 - 2i)^4 - (x + 1 - 2i)^3 + 2(x + 1 - 2i) + 1$, $f(x_0) = 1$.

2602. a) $f(2) = 18$; $f'(2) = 48$; $f''(2) = 124$; $f'''(2) = 216$; $f^{IV}(2) = 240$; $f^{V}(2) = 120$.

b) $f(1 + 2i) = -12 - 2i$; $f'(1 + 2i) = -16 + 8i$; $f''(1 + 2i) = -8 + 30i$; $f'''(1 + 2i) = 24 + 30i$; $f^{IV}(1 + 2i) = 24$.

c) $f(-2) = 8$; $f'(-2) = 2$; $f''(-2) = 12$; $f'''(-2) = -24$; $f^{IV}(-2) = 24$;

2603. a) 3. b) 4. c) 2. d) 3.

2604. -5.

2605. $a = n$, $b = -(n + 1)$.

2606. $3125b^2 + 108a^5 = 0$, $a \neq 0$.

2608. *Hint.* Calculate the derivative.

2609. *Hint.* Induction on k.

2610. If k is the multiplicity of a as a root of $f'''(x)$ then this multiplicity is equal to $k + 3$.

2611. *Hint.* Induction on the degree of the polynomial.

2612. *Hint.* Show that if x_0 is a root of multiplicity k then $f(x_0) \neq 0$ and x_0 is a root of multiplicity $k + 1$ of the polynomial $f(x)f'(x_0) - f(x_0)f'(x)$ of degree at most n.

2613. *Hint.* Induction on r. Consider the polynomial $xf'(x)$.

2614. *Hint.* a) Apply Exercise 2613.

b) Prove that for any numbers $b_0, \ldots, b_{k-1}$ there exist polynomials $g_i(n)$ of degrees at most $s_i - 1$ such that

$$b_n = \sum_{i=1}^{m} g_i(n)a_i^n, \quad n = 0, \ldots, k - 1.$$

2701. a) $(x-1)(x-2)(x-3)$.

b) $(x-1-i)(x-1+i)(x+1-i)(x+1+i)$.

c) $(x-i\sqrt{3})(x+i\sqrt{3})\left(x-\dfrac{3}{2}-\dfrac{\sqrt{3}}{2}i\right)\left(x-\dfrac{3}{2}+\dfrac{\sqrt{3}}{2}i\right)\left(x+\dfrac{3}{2}-\dfrac{\sqrt{3}}{2}i\right)$
$\left(x+\dfrac{3}{2}+\dfrac{\sqrt{3}}{2}i\right)$.

d) $\prod_{\substack{k=1\\(k,3)=1}}^{3n-1}\left(x-\cos\dfrac{2\pi k}{3n}-i\sin\dfrac{2\pi k}{3n}\right)$.

e) $2^{n-1}\prod_{k=1}^{n}\left(x-\cos\dfrac{(2k-1)\pi}{2n}\right)$.

2702. a) $(x^2+3)(x^2+x+3)(x^2-3x+3)$.

b) $\left(x^2+2x+1+\sqrt{2}+2(x+1)\sqrt{\dfrac{\sqrt{2}+1}{2}}\right)$
$\left(x^2+2x+1+\sqrt{2}-2(x+1)\sqrt{\dfrac{\sqrt{2}+1}{2}}\right)$.

c) $(x^2-x\sqrt{a+2}+1)(x^2+x\sqrt{a+2}+1)$.

d) $\prod_{k=0}^{n-1}\left(x^2-2x\cos\dfrac{(3k+1)2\pi}{3n}+1\right)$.

e) $\left(x^2-2x\cos\dfrac{\pi}{9}+1\right)\left(x^2+2x\cos\dfrac{2\pi}{9}+1\right)\left(x^2+2x\cos\dfrac{4\pi}{9}+1\right)$.

f) $\left(x^2+x\sqrt{2}+1\right)\left(x^2-x\sqrt{2}+1\right)\left(x^2+x\sqrt{2+\sqrt{2}}+1\right)$
$\left(x^2-x\sqrt{2+\sqrt{2}}+1\right)\left(x^2+x\sqrt{2-\sqrt{2}}+1\right)\left(x^2-x\sqrt{2-\sqrt{2}}+1\right)$.

2703. a) $(x-1)^2(x-2)(x-3)(x-1-i)$.

b) $(x-i)^2(x+1+i)$.

2704. a) $(x-1)^2(x-2)(x-3)(x^2-2x+2)$.

b) $(x^2+1)^2(x^2+2x+2)$.

2705. *Hint.* The roots of x^2+x+1, i.e. roots of 1 of order 3 which are different from 1, are roots of $x^{3m}+x^{3n+1}+x^{3p+2}$.

2706. Numbers m, n, p should have the same parity.

2707. $m=6k+1$. *Hint.* Write down the condition under which the roots of x^2+x+1 are multiple roots of $(x+1)^m-x^m-1$.

2708. a) $(x-1)^2(x+2)$.

b) $(x+1)^2(x^2+1)$.

c) $x^{(m,n)} - 1$.

d) $x^{(m,n)} + 1$, if $\dfrac{m}{(m,n)}$ and $\dfrac{n}{(m,n)}$ are odd numbers, and 1 otherwise.

2709. *Hint.* Prove that $f(1) = 0$.

2710. *Hint.* Induction on the degree of $f(x)$. Consider $\dfrac{d}{dx} f(x^n)$.

2711. *Hint.* Divide $f_1(x^3)$ and $f_2(x')$ with remainder by $x^2 + x + 1$.

2712. *Hint.* Notice that $f(x) = g(x)^2 h(x)$ where $h(x)$ has no real roots. Show that $[u(x)^2 + v(x)^2](x^2 + px + q)$ is a sum of squares if $x^2 + px + q$ has no real roots.

2713. and **2714.** *Hint.* See Lang. S. Bull. Amer. Math. Soc. 1990; 23(1): 38–39.

2801. c) *Hint.* Change the variable x by $y - m$ and reduce the statement to b).

2802. a) 2.

b) -3.

c) $-3, 1/2$.

d) $5/2, -3/4$.

e) $\frac{1}{2}, -\frac{2}{3}, \frac{3}{4}$.

f) The polynomial has no rational roots.

g) $-\frac{1}{2}$ of multiplicity two.

h) $\frac{1}{2}$.

2803. *Hint.* Let m be an integer root of $f(x)$. Then $f(x) = (x-m)g(x)$. Hence $f(0) = -mg(0)$, i.e. m is odd. Similarly $f(1) = (1-m)g(1)$, i.e. $1-m$ is odd, a contradiction.

2804. *Hint.* Let r be a rational root of $f(x)$. Then $f(x) = (x-r)g(x)$ where $g(x) \in \mathbb{Z}[x]$ and $f(x_i) = (x_i - r)g(x_i) = \pm 1$. It follows that $x_i - r = \pm 1$.

2805. *Hint.* If the polynomial $f \in \mathbb{Q}[x]$ is irreducible over $\mathbb{Q}$ then $(f, f') = 1$.

2808. *Hint.* Suppose that the coefficients of the product are divisible by a prime p. Reduce modulo p.

2809. *Hint.* a), b) Apply Exercise 2808.

c) Let $x^{105} - 9 = f(x)g(x)$, where $f(x), g(x) \in \mathbb{Q}[x]$ and $a = \sqrt[105]{9}$; then

$$f(x) = (x - \alpha_1 a) \ldots (x - \alpha_k a)(\alpha_i^{105} = 1)$$

and

$$|f(0)| = a^k |\alpha_1 \ldots \alpha_k| = a^k \in \mathbb{Q},$$

which is impossible if $k < n$.

d) Change the variable y by $x - 1$.

e) If $f = gh$ where $g, h \in \mathbb{Z}[x]$ then for any $i = 1, \ldots, n$ we have $g(a_i)h(a_i) = -1$; it follows that $g(a_i) + h(a_i) = 0$, and if degrees of g and h are less than n then $g + h = 0$ and $f = -g^2$.

f) Let $f(x) = g(x)h(x)$, where $g(x), h(x) \in \mathbb{Z}[x]$. It is possible to assume that $g(x), h(x)$ take only positive values. Then $g(a_i) = h(a_i) = 1$ for all i. Therefore it can be assumed that degrees of $g(x)$ and $h(x)$ are equal to n, i.e. $g(x) = 1 + b(x - a_1) \ldots (x - a_n), h(x) = 1 + c(x - a_1) \ldots (x - a_n)$, where $b = c = \pm 1$. In this case $g(x)h(x) \neq f(x)$.

2810. and **2811.** *Hint.* See Selmer E.S., Math. Scand. 1956; 4: 287–302.

2812. *Hint.* See Tverberg H. Math. Scand. 1960; 8: 121–126.

2813. *Hint.* See Ljunggren. Math. Scand. 1960; 8: 65–70.

2814. *Hint.* If the set of these numbers p is finite then $a_0 \neq 0$. Let c be a number which is divisible by all these primes. Then $f(a_0 c) = a_0 r$, where $r \equiv 1 \pmod{c}$ and (for appropriate choice of c) $r \neq \pm 1$; thus $f(x)$ has a root in the residue field modulo any prime divisor of r, a contradiction to the choice of c.

2815. *Hint.* Notice that all elements of $\mathbf{F}_q$ are roots of $x^q - x$.

2816. *Hint.* Consider first the case of a mapping h which takes value 1 at one point in F^n and vanishes at all other points.

2817. *Hint.* See the references in Exercises 2812 and 2813.

2818. – **2820.** *Hint.* See the reference in Exercise 2810.

2821. *Hint.* See Perron O. L. J. Reine angew Math. 1907; 132: 288–307.

2822. a) $x, x + 1, x^2 + x + 1, x^3 + x^2 + 1, x^3 + x + 1, x^4 + x^3 + 1, x^4 + x + 1, x^4 + x^3 + x^2 + x + 1$.

b) $x^2 + 1, x^2 + x + 2, x^2 + 2x + 2$.

Hint. A polynomial of degree 4 is irreducible if and only if it has no roots in the given field and is not a product of two irreducible polynomials of degree two.

c) 14.

d) 8 and 18.

2823. $\dfrac{q(q - 1)}{q}$ and $\dfrac{q(q - 1)(q - 2)}{3}$.

2824. *Hint.* The group $\mathbf{Z}_p^*$ is cyclic of order $p - 1$. Therefore $\mathbf{Z}_p^*$ has a subgroup of order d. All generators of this group are roots of $\Phi_d(x)$.

2825. *Hint.* Let $f(x) = f(x + k)$ for some $1 \leq k \leq p - 1$. Then $f(x) = f(x + kl)$ for all $l \in \mathbb{Z}$. But elements kl run over the whole field $\mathbf{Z}_p$.

2826. *Hint.* Let $H(x) = x^p - x - a = f(x)g(x)$ where $f(x) \in \mathbf{Z}_p[x]$ is irreducible. Notice that $H(x) = H(x + k)$ for all $k \in \mathbf{Z}_p$. Therefore

$f(x)g(x) = f(x+k)g(x+k)$. Notice that $\mathbf{Z}_p[x]$ is a unique factorization domain. Apply Exercise 2825.

2827. *Hint.* See Lang S. Algebra, Addison-Wesley, Reading, 1965, p. 245.

2828. $x = b(a - 1)^{-1}$.

2829. and **2830.** *Hint.* See R. Lidl, H. Niederreiter, Finite Fields, Addison-Wesley, Reading, 1983, ch. 3, § 5.

2831. $a = 0$ and 36. *Hint.* Expand in powers of $x - a$.

2832. – 2834. *Hint.* See E.R. Berlekamp, Algebraic Coding Theory, McGraw-Hill, New York, 1968, ch. 3, § 3.

2901. a) $\dfrac{1}{12(x - 1)} - \dfrac{4}{3(x + 2)} + \dfrac{9}{4(x + 3)}$.

b) $-\dfrac{1}{16}\left(\dfrac{1+i}{x - 1 - i} + \dfrac{1-i}{x - 1 + i} + \dfrac{-1+i}{x + 1 - i} + \dfrac{-1+i}{x + 1 + i}\right)$.

c) $\dfrac{1}{4(x - 1)^2} - \dfrac{1}{4(x + 1)^2}$.

d) $\dfrac{3}{(x - 1)^3} - \dfrac{4}{(x - 1)^2} + \dfrac{1}{x - 1} - \dfrac{1}{(x + 1)^2} - \dfrac{2}{x + 1} + \dfrac{1}{x - 2}$.

e) $-\dfrac{1}{6(x - 1)} + \dfrac{1}{2(x - 2)} - \dfrac{1}{2(x - 3)} + \dfrac{1}{6(x - 4)}$.

f) $-\dfrac{2}{(x - 1)} + \dfrac{-2+i}{2(x - i)} - \dfrac{2+i}{2(x + i)}$.

g) $-\dfrac{1}{4(x - 1)} - \dfrac{1}{4(x + 1)} - \dfrac{i}{4(x - i)} + \dfrac{i}{4(x + i)}$.

h) $\dfrac{1}{3}\left(-\dfrac{1}{(x - 1)} + \dfrac{\varepsilon}{x - \varepsilon} + \dfrac{\varepsilon^2}{x - \varepsilon^2}\right)$, $\varepsilon = -\dfrac{1}{2} + \dfrac{i\sqrt{3}}{2}$.

i) $\sum_{k=-n}^{n} \dfrac{(-1)^{n-k}\binom{n}{k}}{x - k}$.

j) $\dfrac{1}{4(x + 1)} - \dfrac{1}{4(x - 1)} + \dfrac{1}{4(x - 1)^2} + \dfrac{1}{4(x + 1)^2}$.

k) $\dfrac{1}{n}\sum_{k=0}^{n-1} \dfrac{\varepsilon_k}{x - \varepsilon_k}$, $\varepsilon_k = \cos\dfrac{2\pi k}{n} + i\sin\dfrac{2\pi k}{n}$.

2902. a) $-\dfrac{1}{8(x - 2)} - \dfrac{1}{8(x + 2)} + \dfrac{1}{2(x^2 + 4)}$.

b) $\dfrac{1}{8}\left(-\dfrac{x + 2}{x^2 + 2x + 2} - \dfrac{x - 2}{x^2 - 2x + 2}\right)$.

c) $-\dfrac{1}{4(x + 1)} + \dfrac{x - 1}{4(x^2 + 1)} + \dfrac{x + 1}{2(x^2 + 1)^2}$.

d) $\dfrac{3}{16(x-1)^2} - \dfrac{3}{16(x-1)} + \dfrac{1}{16(x+1)^2} + \dfrac{3}{16(x+1)} + \dfrac{1}{4(x^2+1)} + \dfrac{1}{4(x^2+1)^2}.$

e) $\dfrac{1}{n}\sum_{k=1}^{n}(-1)^{k-1}\dfrac{\sin\dfrac{2k-1}{2nz}\pi}{x-\cos\dfrac{2k-1}{2n}\pi}.$

f) $\sum_{i=1}^{n}\dfrac{1}{f'(x_i)(x-x_i)}$ ($x_1,\ldots,x_n$ are roots of $f(x)$).

g) $\dfrac{1}{3(x-1)} - \dfrac{x+2}{3(x^2+x+1)}.$

h) $\dfrac{1}{18}\left(\dfrac{1}{x^2+3x+3} + \dfrac{1}{x^2-3x+3} - \dfrac{2}{x^2+3}\right).$

i) $-\dfrac{1}{x} + \dfrac{7}{x+1} + \dfrac{3}{(x+1)^2} - \dfrac{6x+2}{x^2+x+1} - \dfrac{3x+2}{(x^2+x+1)^2}.$

j) $\dfrac{1}{16(x-1)^2} - \dfrac{3}{16(x-1)} + \dfrac{3}{16(x+1)} + \dfrac{1}{4(x^2+1)} + \dfrac{1}{4(x^2+1)^2} + \dfrac{1}{16(x+1)^2}.$

k) $\dfrac{1}{n}\sum_{k=1}^{n}\dfrac{\cos\dfrac{(2k-1)m\pi}{n} - x\cos\dfrac{(2k-1)(2m+1)}{2n}\pi}{x^2-2x\cos\dfrac{(2k-1)}{2n}\pi+1}.$

2903. $-\sum_{a=0}^{p-1}\dfrac{1}{x-a}.$

2905. *Hint.* Make use of Exercise 2904.

3001. a) $-x^4+4x^3-x^2-7x+5.$

b) $x^3-9x^2+21x-8.$

3005. $f(0) = \dfrac{1}{n}(y_1+\cdots+y_n).$

3006. *Hint.* Reduce the exercise by a change of the variable to the case when $x_1,\ldots,x_n$ are roots of order n of 1 and $x_0=0$; then make use of Exercise 3005.

3007. *Hint.* a) Reduce to Exercise 3005 for the polynomial x^{s+1}.

b) Reduce to Exercise 3005 for the polynomial $x^n - f(x)$.

3008. $f(x) = 1 - \dfrac{2x}{1} + \dfrac{2x(2x-2)}{12} + \cdots + \dfrac{2x(2x-2)\ldots(2x-4n+2)}{(2n)!}(-1)^n.$

3009. $f(x) = x^{p-2}.$

3011. and **3012.** *Hint.* R. Lidl, H. Niederreiter, Finite Fields, Addison-Wesley, Reading, 1983, ch. 7, § 3.

3101. a) $x^4+4x^3-7x^2-22x+24.$

b) $x^4+(3-i)x^3+(3-3i)x^2+(1-3i)x-i.$

c) $x^4 - 3x^3 + 2x^2 + 2x - 4$.

d) $x^4 - 19x^2 - 6x + 72$.

3102. a) $\dfrac{2}{3}$ and $-\dfrac{2}{3}$.

b) a^2 and $(-1)^n b$.

3103. a) 0. b) -1.

3104. $\sigma_i = 0$ if $i < n$ and $\sigma_n = (-1)^{n+1}$.

3105. $\lambda = \pm 6$.

3106. $\lambda = -3$.

3107. $q^3 + pq + q = 0$.

3108. *Hint.* Calculate the product of the roots of $x^{p-1} - 1$ over the residue field modulo p by the Vieta formula.

3109. a) $\sigma_1 \sigma_2 - 3\sigma_3$.

b) $\sigma_1^4 - 4\sigma_1^2 \sigma_2 + 8\sigma_1 \sigma_3$.

c) $\sigma_1^2 \sigma_4 + \sigma_3^2 - 4\sigma_2 \sigma_4$.

d) $\sigma_1^3 - 4\sigma_1 \sigma_2 + 8\sigma_3$.

e) $\sigma_1 \sigma_2 - \sigma_3 + \sigma_1^2 + \sigma_2 + 2\sigma_1 + 1$.

f) $\sigma_3^2 + \sigma_1^2 \sigma_3 - 2\sigma_2 \sigma_3 + \sigma_2^2 - 2\sigma_1 \sigma_3 + \sigma_3$.

g) $3\sigma_1^3 - 9\sigma_1 \sigma_2 + 27\sigma_3$.

h) $\sigma_1 \sigma_2 \sigma_3 - \sigma_1^2 \sigma_4 - \sigma_3^2$.

i) $\sigma_1^3 \sigma_2^2 - 2\sigma_1^4 \sigma_3 - 3\sigma_1 \sigma_2^3 + 6\sigma_1^2 \sigma_2 \sigma_3 + 3\sigma_2^2 \sigma_3 - 7\sigma_1 \sigma_3^2$.

j) $\sigma_1^3 - 4\sigma_1 \sigma_2 + 8\sigma_3$.

k) $\sigma_1^2 - 2\sigma_2$.

l) $\sigma_1^3 - 3\sigma_1 \sigma_2 + 3\sigma_3$.

m) $\sigma_1 \sigma_3 - 4\sigma_4$.

n) $\sigma_2^2 - 2\sigma_1 \sigma_3 + 2\sigma_4$.

o) $\sigma_1^2 \sigma_3 - 2\sigma_2 \sigma_3 - \sigma_1 \sigma_4 + 5\sigma_5$.

p) $\sigma_1 \sigma_2^2 - 2\sigma_1^2 \sigma_3 - \sigma_2 \sigma_3 + 5\sigma_1 \sigma_4 - 5\sigma_5$.

3110. a) -35.

b) 16.

c) $a_1^2 a_2^2 - 4a_1^3 - 4a_2^3 + 18a_1 a_2 a_3 - 27a_3^2$.

d) $\dfrac{25}{27}$.

e) $\dfrac{35}{27}$.

f) $-\dfrac{1679}{625}$.

3112. *Hint.* Make use of the equality $\sigma_{ki} = \sigma_k - x_i\sigma_{k-1,i}$.

3114. $\dfrac{d}{dt}(\ln \lambda_t) = \sum_i \dfrac{x_i}{1 + x_i t} = \sum_i (1 - x_i t + x_i^2 t^2 + \dots) = s_1 - s_2 + \dots$

3115. $\dfrac{d}{dt}(\ln \lambda_t) = \dfrac{\lambda_t'}{\lambda_t} = \dfrac{\sigma_1 + 2\sigma_2 t + \dots + n\sigma_n t^{n-1}}{1 + \sigma_1 t + \dots + \sigma_n t^n}$. *Hint.* From Exercise 3114 it follows that $(1 + \sigma_1 t + \dots + \sigma_n t^n)(s_1 - s_2 t + \dots) = \sigma_1 + 2\sigma_2 t + \dots + n\sigma_n t^{n-1}$. Compare coefficients in the same powers of t.

3116. *Hint.* Make use of Exercise 3115.

3118. $\dfrac{\phi(n)}{\phi\left(\frac{n}{d}\right)}\mu\left(\dfrac{n}{d}\right)$ where $d = (m, n)$.

3119. $s_1 = -1, s_2 = \dots = s_n = 0$.

3120. $s_1 = \dots = s_{n-1} = 0, s_n = n$.

3121. a) $x = 2, x_2 = -1 + i\sqrt{3}, x_3 = -1 - i\sqrt{3}$ up to a permutation.

b) $x_1 = 1, x_2 = 1, x_3 = -2$ up to a permutation.

3124. *Hint.* Factorize $f(x)$ into linear factors and apply Exercise 3123.

3125. $x^3 - 3x^2 + 2x - 1$.

3126. $x^4 - 4x^3 + 10x^2 - x + 9$.

3127. *Hint.* a) Check that $f(x_1, \dots, x_n)$ is divisible by $x_i - x_j$ for all $1 \le i < j \le n$.

b) follows from a).

3128. *Hint.* b) Consider the product

$$\left[\sum_{r \ge 0} h_r t^r (-1)^r\right](1 + x_1 t)\dots(1 + x_n t)$$

and utilize a).

c) Utilize b).

3131. *Hint.* See I. G. Macdonald. Symmetric Functions and Hall Polynomials. Clarendon Press, Oxford, 1979, ch. 1, § 3.

3132. *Hint.* Ibid. Ch. 1, § 4.

3133. *Hint.* Ibid. Ch. 1, § 5.

3201. a) -7.

b) 243.

c) 0.

d) -59.

e) 4854.

3202. a) 3 and -1.

b) $\pm i\sqrt{2}$ and $\pm 2i\sqrt{3}$.

c) $\mathfrak{z} = 1, \mathfrak{z} = \frac{1}{2}\left[-2+\sqrt{2}\pm\sqrt{4\sqrt{2}-2}\right]$.

3203. a) $y^6 - 4y^4 + 3y^2 - 12y + 12 = 0$.

b) $5y^5 - 7y^4 + 6y^3 - 2y^2 - y - 1 = 0$.

c) $x_1 = 1, x_2 = 2, x_3 = 0, x_4 = -2; y_1 = 2, y_2 = 3, y_3 = -1, y_4 = 1$.

d) $x_1 = 0, x_2 = 3, x_3 = 2, x_4 = 2; y_1 = 1, y_2 = 0, y_3 = 2, y_4 = -1$.

e) $x_1 = x_2 = 1, x_3 = -1, x_4 = 2; y_1 = y_2 = -1, y_3 = 1, y_4 = 2$.

3204. *Hint.* If $f = a_0(x - x_1)\ldots(x - x_n)$ and g_1, g_2 have degrees m and k, then $R(f, g_1g_2) = a_0^{m-k}g_1(x_1)g_2(x_1)\ldots g_1(x_n)g_2(x_n) = R(f, g_1)R(f, g_2)$.

3205. *Hint.* Consider the case when $n > 2$ and m is not divisible by n. Then $R(\Phi_n, x^m - 1) = P^{\phi(n)/\phi(n_1)}$ if $n_1 = \frac{n}{d} = p^\lambda$. Otherwise the result is equal to 1.

3206. $R(\Phi_m, \Phi_n) = 0$ if $m = n$; $R(\Phi_m, \Phi_n) = p^{\phi(n)}$ if $m = np^\lambda$ and $R(\Phi_m, \Phi_n) = 1$ otherwise if $m \geq n$.

3207. a) $a(b^2 - 4ac)$.

b) $-27q^2 - 4p^3$.

c) $-27a_3^2 + 18a_1a_2a_3 - 4a_1^3a_3 - 4a_2^3 + a_1^2a_2^2$.

d) 2777.

e) 725.

3208. a) ± 2.

b) $\left\{3, 3\left(-\frac{1}{2}\pm i\frac{\sqrt{3}}{2}\right)\right\}$.

c) $\lambda_1 = 0, \lambda_2 = -3, \lambda_3 = 125$.

d) $\lambda_1 = -1, \lambda_2 = -\frac{3}{2}, \lambda_{3,4} = \frac{7}{2}\pm\frac{9}{2}i\sqrt{3}$.

3209. *Hint.* Make use of the factorization $f = a_0(x - x_1)\ldots(x - x_n)$.

3210. $(-1)^{(n-1)(n-2)/2}$.

3211. $(-1)^{\phi(n)/2}n^{\phi(n)}\left[\prod_{p|n}p^{\phi(n)/(1-p)}\right]$.

3212. $(-1)^{\frac{n(n-1)}{2}}(n!)^{-n}$.

3213. *Hint.* Make use of Exercise 3209.

3214. *Hint.* Make use of Exercises 3201, 3204 and 3123.

3215. *Hint.* See E.R. Berlekamp, Algebraic Coding Theory, McGraw-Hill, New York, 1968, p. 143.

3216. $(-1)^{n(n-1)/2} n^n a^{n-1}$.

3217. a) $1 \cdot 2^2 \cdot 3^3 \cdot \ldots \cdot (n-1)^{n-1} n^n$.

b) $1 \cdot 2^3 \cdot 3^5 \cdot \ldots \cdot n^{2n-1}$;   c) $2^{n-1} n^n$.

3301. a) There are three real roots in the intervals $(-2, -1)$, $(-1, 0)$, $(1, 2)$.

b)   There are three real roots in the intervals $(-2, -1)$, $(-1, 0)$, $(1, 2)$.

c)   There are three real roots in the intervals $(-4, -3)$, $(1, 3/2)$, $(3/2, 2)$.

d)   There is one real root in the interval $(-2, -1)$.

e)   There is one real root in the interval $(1, 2)$.

f)   There are four real roots in the intervals $(-3, -2)$, $(-2, -1)$, $(-1, 0)$, $(4, 5)$.

g)   There are two real roots in the intervals $(-1, 0)$, $(1, 2)$.

h)   There are four real roots in the intervals $(-1, 0)$, $(0, 1)$, $(1, 2)$, $(2, 3)$.

i)   There are two real roots in the intervals $(-1, 0)$, $(0, 1)$.

j)   No real roots.

3302. If $a^5 - b^2 > 0$ then all roots are real. If $a^5 - b^2 < 0$ then the polynomial has only one real root.

3303. If n is odd and $d > 0$ then there are three real roots. If n is odd and $d < 0$ then there is only one real root. If n is even and $d > 0$ then there are two real roots. If n is even and $d < 0$ then there are no roots.

3304. If n is even then $E_n(x)$ has no real roots. If n is odd then E_n has one real root.

3305. *Hint.* Check that the derivative has no roots in the interval $(0,1)$.

3306. *Hint.* a) Numbers $a_1, \ldots, a_m$ are roots of $f'(x)$ with multiplicities $k_1 - 1, \ldots, k_m - 1$, respectively. Moreover in each interval (a_i, a_{i+1}) the derivative $f'(x)$ has a root.

b)   follows from a).

c)   If $c_k = c_{k+1}$ then $x = 0$ is a multiple root of the kth derivative $f^{(k)}(x)$, i.e. by b) $x = 0$ is a root of multiplicity at least $k + 1$ of $f(x)$.

3307. *Hint.* Follows from Exercise 3306c.

3308. *Hint.* Multiply by $x - 1$ and apply Exercise 3307.

3309. *Hint.* Multiply by x^{-n}.

3310. *Hint.* Prove that $f(x) > 0$ if $x \geq (1/n!)$.

3312. One root is in the first and one root is in the fourth quadrant. There are two roots each in the second and in the third quadrants.

3314. *Hint.* Show that the root z satisfies the equation $z^n = (1 - az)/(a - z)$. Check that $|1 - az| = |a - z|$ for the real number a.

3315. *Hint.* Show that $1 - |a_1 z + \cdots + a_n z^n| > 1/(k+1)^n$ if $|z| < 1/(k+1)$.

3316. *Hint.* According to Exercise 2905 $f'/f = (x - a_1)^{-1} + \cdots + (x - a_n)^{-1}$ where a_i are roots of f. Let $x = a - bi$ where $b > 0$. Then

$$\Im \left(\frac{f'(a - bi)}{f(a - bi)} \right) = \sum_{j=1}^{n} \frac{b + \Im a_j}{|a - bi - a_j|^2} > 0.$$

Thus $f'(a - b_i) \neq 0$.

3317. *Hint.* Any convex domain is an intersection of half-planes. Make use of Exercise 3316.

3318. *Hint.* Apply the Sturm theorem. S. Lang, Algebra, Addison-Wesley, Reading, 1965, ch. IX, § 2.

3319. *Hint.* Make use of Exercise 3318.

3320. $x^3 + x^2 - x - 1$, $x^2 \pm x - 1$, $x \pm 1$.

3402. a) $\lambda = \pm 1$.   b) $\lambda \neq (-1)^n$.

3403. *Hint.* Differentiate twice and apply induction in the cases c), d), e). In the cases f), g) use Vandermonde determinant.

3404. *Hint.* a), b) Utilize Vandermonde determinant.

c)   Differentiate and utilize Vandermonde determinant.

3405. *Hint.* If $f_1, \ldots, f_n$ are linearly independent then there exists a point a_1 such that $f_1(a_1) \neq 0$; check that the system $f_i - \dfrac{f_i(a_1)}{f_1(a_1)} f_1, i = 2, \ldots, n$ is linearly independent and finish the proof by induction on n.

3407. *Hint.* a) If char $P \neq 2$ then $1 + 1 = 2$ is an invertible element in P. Therefore for any vector space L over P and any $x \in L$ there exists a vector $y \in L$ such that $y = \frac{1}{2}x$. Then $y + y = x$. If the characteristic of P is equal to 2 then $x + x = 2x = 0$ for any $x \in L$ but in the additive group of integers $1 + 1 \neq 0$ and $2y \neq 1$ for any integer y.

b)   In a vector space over a field of characteristic p we have $px = 0$ for any vector x.

c)   For proof of necessity see the hint to Exercise 3407b; in order to prove sufficiency put

$$[k]a = \underbrace{a + a \cdots + a}_{k \text{ times}}.$$

d) In order to prove the sufficiency put $\dfrac{p}{q}a = b$ for any rational number $\dfrac{p}{q}$
   $(p, q \in \mathbb{Z})$ if b is a solution of the equation $qx = pa$; check that solutions of
   equations $qx = pa$ and $nx = ma$ coincide, if $\dfrac{p}{q} = \dfrac{m}{n}$.

3408. *Hint.* e), f) Induction on n.

3409. a) A system of all one-element subsets of set M form a basis. The
dimension is equal to n.

b) *Hint.* Induction on k.

3410. a) $(1, 2, 3)$.

b) $(1, 1, 1)$.

c) $(0, 2, 1, 2)$.

3411. a) $x_1 = -27x'_1 - 71x'_2 - 41x'_3$; $x_2 = 9x'_1 + 20x'_2 + 9x'_3$; $x_3 = 4x'_1 + 12x'_2 + 8x'_3$; b) $x_1 = 2x'_1 + x'_3 - x'_4$; $x_2 = -3x'_1 + x'_2 + x'_4$; $x_3 = x'_1 - 2x'_2 + 2x'_3 - x'_4$; $x_4 = x'_1 - x'_2 + x'_3 - x'_4$.

3412. $a_0, a_1, \ldots, a_n$; $f(\alpha), f'(\alpha), f''(\alpha)/2!, \ldots, f^{(n)}(\alpha)/n!$;

$$\begin{pmatrix} 1-\alpha & \alpha^2 & -\alpha^3 & \ldots & (-1)^n\alpha^n \\ 0 & 1-2\alpha & 3\alpha^2 & \ldots & (-1)^{n-1}n\alpha^{n-1} \\ \cdots\cdots\cdots\cdots\cdots\cdots\cdots\cdots\cdots\cdots\cdots\cdots \\ 0 & 0 & 0 & \ldots & 1 \end{pmatrix}.$$

3413. a) Two rows will interchange.

b) Two columns will interchange.

c) The matrix will be replaced by one symmetric with respect to its center.

3501. The sets are subspaces in the following cases: b) if a line passes through
the origin; f), g), h), i), j) if $a = 0$; k) only if f is zero sequence; and also l), m).

3502.

a) $((1, 0, 0, \ldots, 0, 1), (0, 1, 0, \ldots, 0, 0), (0, 0, 1, \ldots, 0, 0), \ldots, (0, 0, 0, \ldots, 1, 0))$; $n - 1$;

b) $((1, 0, 0, 0, 0, \ldots, 0), (0, 0, 1, 0, 0, \ldots, 0), (0, 0, 0, 0, 1, \ldots, 0), \ldots)$; $\left[\dfrac{n+1}{2}\right]$;

c) add the vector $(1, 1, 1, \ldots, 1, 1)$ to the vectors from b); $1 + \left[\dfrac{n+1}{2}\right]$ if $n > 1$;

d) $((1, 0, 1, 0, 1, \ldots), (0, 1, 0, 1, 0, \ldots))$; 2 (if $n > 1$);

e) the fundamental system of solutions forms a basis.

3503. a) $\{E_{ij} \mid i, j = 1, 2, \ldots, n\}$; n^2.

b) Matrices $\{E_{ij} + E_{ji} \mid 1 \leq i \leq j \leq n\}$ form a basis; $n(n + 1)/2$.

c) If $\mathrm{char}\,K \neq 2$ then $\{E_{ij} - E_{ji} \mid 1 \le i < j \le n\}$; $n(n-1)/2$; if $\mathrm{char}\,K = 2$ the answer is the same as in b).

f) $\{E_{11} - E_{ii} \mid i = 2, 3, \ldots, n\} \cup \{E_{ij} \mid i, j = 1, 2, \ldots, n; i \neq j\}$; $n^2 - 1$.

3504. The sets form a subspace in the following cases: a) and b) if $a = 0$; c) if $|S| = 1$; e).

3505. The sets form a subspace in the following cases: a), b), d).

3507. a) $\{f(x)(x - \alpha) \mid f(x) \in R[x]_{n-1}\}$.

b) $\{f(x)(x - \alpha)(x - \bar{\alpha}) \mid f(x) \in R[x]_{n-2}\}$.

c) The dimension is equal to $n - k + 1$.

3509. a) The dimension is equal to $\binom{n+k-1}{n-1}$. *Hint.* Take a base consisting of monomials and apply Exercise 3501.

b) $\binom{k+n}{n}$. *Hint.* Put $x_i = \dfrac{y_i}{y_{n+1}}$ and apply a).

3510. a) q^n.

b), c) $(q^n - 1)(q^n - q) \ldots (q^n - q^{n-1})$.

d) $q^{n^2} - (q^n - 1)(q^n - q) \ldots (q^n - q^{n-1})$.

e) $\dfrac{(q^n - 1)(q^n - q) \ldots (q^n - q^{n-k+1})}{(q^k - 1)(q^k - q) \ldots (q^k - q^{k-1})}$. *Hint.* The denominator is equal to the number of distinct bases in k-dimensional subspace.

f) q^{n-r}.

3511. a) (a_1, a_2, a_4); 3.

b) (a_1, a_2, a_5); 3.

3512. b), c) *Hint.* Make use of the formula: $\dim L_1 + \dim L_2 = \dim (L_1 + L_2) + \dim(L_1 \cap L_2)$.

3513. a) No, it is not possible. *Hint.* Take $U = \langle a + b \rangle$, $V = \langle a \rangle$, $W = \langle b \rangle$, where a and b are linearly independent vectors.

b) *Hint.* If $x \in U \cap (VW)$ then $x = v + w$, $w = x - v \in U$ (since $v, x \in U$), i.e. $w \in U \cap W$, and therefore $x \in (U \cap V) + (U \cap W)$. The converse inclusion follows from the fact that $U \cap V$ and $U \cap W$ are contained in U and in $V + W$.

3514. a) 3, 1.

b) 3, 2.

c) 4, 2.

3515. a) (a_1, a_2, b_1); $(3, 5, 1)$.

b) (a_1, a_2, a_3, b_1); $(1, 1, 1, 1, 1)$; $(0, 2, 3, 1, -1)$.

c) $(a_1, a_2, , a_3, b_1); (1, 1, 1, 1, 0); (1, 0, 0, 1, -1).$

d) $(a_1, a_2, b_1), (5, -2, -3, -4).$

e) $(a_1, a_2, a_3, b_1); b_2$ is a basis of the intersection.

3516. a) $x_1 - x_2 - x_4 = 0, x_2 + x_3 - x_4 = 0.$

b) $x_1 - x_2 - 2x_3 = 0, x_1 - x_2 + 2x_4 = 0, 2x_1 + x_2 - x_5 = 0.$

3517. *Hint.* b) Consider $\langle x \rangle, \langle y \rangle, \langle z \rangle$ where the vectors are pairwise linearly independent.

3518. *Hint.* The projection of the vector e_i into L_1 in parallel with L_2 has ith coordinate $(n-1)/n$ and all others are equal to $(-1/n)$; the projection into L_2 in parallel with L_1 has all coordinates equal to $1/n$.

3519. $(-1, -3, 1, 3).$

3521. $A = \frac{1}{2}(A + {}^tA) + \frac{1}{2}(A - {}^tA).$

3522. b) 0 and E_{ij} if $i \leqslant j$, $E_{ij} - E_{ji}$ and E_{ji} if $i > j$.

3523. b) 0 and E_{ij} if $i < j$; E_{ii} and 0 if $i = j$; $E_{ij} + E_{ji}$ and $-E_{ji}$ if $i > j$.

3524. $(q^n - q^m)(q^n - q^{m+1}) \ldots (q^n - q^{n-1})/(q^m - 1) \ldots (q^m - q^{m-1}).$

3601. *Hint.* Induction on m.

3602. a) $q^{nk}.$

b) $(q^k - 1)(q^k - q) \ldots (q^k - q^{n-1})$, where $n \leq k$.

3605. *Hint.* Choose a basis $e_1, \ldots, e_n$ in V such that $A(e_1), \ldots, A(e_k)$ is a basis of Im A and $e_{k+1}, \ldots, e_n \in \text{Ker}A$. Define the action of C and D on $e_{k+1}, \ldots, e_n$ and on $A(e_1), \ldots, A(e_k).$

3606. *Hint.* Choose a basis $e_1, \ldots, e_n$ as in Exercise 3605. Define an action of C on $A(e_1), \ldots, A(e_k).$

3607. *Hint.* Make use of Exercise 3605.

3608. *Hint.* Choose a basis $e_1, \ldots, e_n$ as in Exercise 3605.

3609. *Hint.* a) Apply Lagrange polynomials f_i such that $f_i(i) = 1$, $f_i(j) = 0$, $(i, j = 0, \ldots, n; i \neq j).$

b) Consider polynomials $1, x, \ldots, x^n.$

c) Consider the matrix $(\gamma^i(x^j)) (i, j = 1, \ldots, n+1).$

3610. a) *Hint.* Choose in V an arbitrary system of coordinates and write down the assertion of the exercise in the form of systems of equations.

b) f_i is a Lagrange polynomial:

$$f_i(x) = \frac{(x - 0)(x - 1) \ldots (x - i + 1)(x - i - 1) \ldots (x - n)}{i(i - 1) \ldots 1 \cdot (-1) \ldots (i - n)}.$$

c) $f_i(x) = x^i/i!$ $(i = 0, 1, \ldots, n)$.

3611. *Hint.* Find a basis $(e_1, e_2, \ldots, e_k)$ for which $f(e_1) = 1$, $f(e_2) = \cdots = f(e_n) = 0$.

3612. *Hint.* Make use of systems of linear equations.

3613. *Hint.* Prove that

$$y = x - \frac{f(x)}{f(a)} a \in U.$$

3614. *Hint.* Apply Exercise 3613b.

3615. *Hint.* Define the intersection of kernels in some basis by a homogeneous linear system as in Exercise 3612.

3616. *Hint.* If the system $e_1, \ldots, e_k$ is linearly independent then complete it to a basis and consider the dual basis in V^*.

3617. *Hint.* a) Complete a basis $(e_1, \ldots, e_k)$ of U to a basis $(e_1, \ldots, e_n)$ of V. If $(e^1, \ldots, e^n)$ is the dual basis then prove that $U^\perp = \langle e^{k+1}, \ldots, e^n \rangle$.

b) Make use of a).

c) Make use of b).

3618. *Hint.* Prove that $\mathbb{Q}[x]$ is a countable set and indicate in $\mathbb{Q}[x]^*$ an uncountable set of independent linear functions. For example, for each subset I of natural numbers define the function f_I by the formula $f_I(u) = \sum_{i \in I} u_i$, where $u = \sum_i u_j x^j$.

3619. *Hint.* Make use of Exercise 3526 where $U = \mathrm{Ker} l_1$, $W = \mathrm{Ker} l_2$.

3620. *Hint.* Make use of Exercise 3525.

3621. See S. Lang, Algebra, Addison-Wesley, Reading, 1965, ch. VIII, § 4–6.

3701. a), b), c), f), g), h), i), j), l), m), n), p), q), s).

3702. a) E in the standard basis.

c) In a basis of matrix units E_{ij} the matrix entries $a_{ij,kl}$ of the function f have the form $a_{ij,ji} = 1$, and 0 in the other cases.

d) 0.

g) $a_{ij,ij} = 1$ and 0 in the other cases (see c)).

h) In the basis $(1, i)$ the matrix $\begin{pmatrix} 1 & 0 \\ 0 & -1 \end{pmatrix}$. i) E (see g)). k) $\begin{pmatrix} 0 & 1 \\ -1 & 0 \end{pmatrix}$.

l), m), o) The space has an infinite dimension.

q) $2E$ in the standard basis. r) $\begin{pmatrix} 0 & 1 & -1 \\ -1 & 0 & 1 \\ 1 & -1 & 0 \end{pmatrix}$ in an orthonormal basis of $\mathbb{R}^3$.

3705. a), b), d), e), f).

3706. a) $\begin{pmatrix} 0 & -6 & -9 \\ -2 & 20 & 30 \\ -3 & 30 & 45 \end{pmatrix}$.   b) $\begin{pmatrix} 11 & 8 & 15 \\ 6 & 5 & 12 \\ 11 & 10 & 29 \end{pmatrix}$.

3707. a) $\begin{pmatrix} 1+i & 1-i \\ -3+i & -1-i \end{pmatrix}$.   b) $\begin{pmatrix} 4-2i & -2-i \\ 1+i & -i \end{pmatrix}$.

3708. a) -43.

b) $1-19i$.

3709. a) $-3+7i$.

b) $22+40i$.

3710. a) $\begin{pmatrix} 2 & 5 & -1 \\ -4 & 6 & 8 \\ -10 & -23 & -4 \end{pmatrix}$.   b) $\begin{pmatrix} -2 & 3 & 0 \\ -5 & -10 & 15 \\ 29 & -26 & 3 \end{pmatrix}$.

3711. a) $\begin{pmatrix} 5+5i & 2i \\ 7+2i & -1+4i \end{pmatrix}$.   b) $\begin{pmatrix} 13-i & 7-5i \\ 4 & 3-i \end{pmatrix}$.

3712. a) $\langle (-1, -1, 1) \rangle$, $\langle (10, 7, 1) \rangle$.

b) $\langle (-1, -5, 3) \rangle$, $\langle (1, -2, 1) \rangle$.

3714. a) $\langle (-1, 1, 1) \rangle$, $\langle (-17, -13, 7) \rangle$.

b) $\langle (2, -3, 1) \rangle$, $\langle (-4, -5, 1) \rangle$.

3716. a) $\langle (1, -2, 1) \rangle$, $\langle (-1, -5, 3) \rangle$.

b) $\langle (-1, 1, 1) \rangle$, $\langle (4, 0, -9) \rangle$.

3718. c) *Hint.* Make use of Exercise 3622 and show that the left and the right kernels are ideals of K.

d) $(a - a^q \mid a \in K)$.

3719. c) $\begin{pmatrix} 1 & 0 \\ 0 & 2 \end{pmatrix}$.

d)   *Hint.* Follows from c).

e) $F\sqrt{2}$.

3721. a) if $\lambda = \pm 1$, c).

3722. $F = {}^t A G \bar{B}$.

3723. If F is symmetric then it can be reduced to the form $a E_{11}$, if it is not symmetric then it can be reduced to the form E_{12}.

3724. $F' = {}^t C F \bar{C}$, $\tilde{F} = F \bar{C} = {}^t \bar{C}^{-1} F'$.

3726. *Hint.* An example of discrepancy: the function on $\mathbb{R}^2$ with the matrix $\begin{pmatrix} 0 & 1 \\ 0 & 0 \end{pmatrix}$ in some basis.

3728. b) *Hint.* Example: the function on $\mathbb{R}^2$ with matrix $\begin{pmatrix} 1 & 0 \\ 0 & -1 \end{pmatrix}$ in some basis.

3729. d) *Hint.* If $\varepsilon = 1$ then matrices for f_1 and f_2 in suitable bases have the form

$$\begin{pmatrix} 0 & A \\ {}^t A & 0 \end{pmatrix}, \quad \begin{pmatrix} 0 & B \\ {}^t B & 0 \end{pmatrix},$$

where A and B are nonsingular matrices. Find directly the matrix of a change of basis.

3730. *Hint.* Make use of Exercise 2729.

3731. *Hint.* The proof is similar to that of the theorem on reduction of a symmetric bilinear form to the normal form. It is possible to apply a method similar to Lagrange algorithm: firstly group together monomials with factors y_1 and x_1, then apply induction.

3732. a), b) Functions are not equivalent.

3733. a) $x_1' y_2' - x_2' y_1'$, where $x_1' = x_1 - 2x_3$, $x_2' = x_2 - x_3$, $x_3' = x_3$.

b) $x_1' y_2' - x_2' y_1'$, where $x_1' = x_1 - \dfrac{3}{2} x_3$, $x_2' = 2x_2 + x_3$, $x_3' = x_3$.

c) $x_1' y_2' - x_2' y_1' + x_3' y_4' - x_4' y_3'$, where $x_1' = x_1 - 2x_3$, $x_2' = x_2 - x_4$, $x_3' = x_3$, $x_4' = x_4$.

d) $x_1' y_2' - x_2' y_1'$, where $x_1' = x_1 + x_3$, $x_2' = x_2 + x_3 + x_4$, $x_3' = x_3$, $x_4' = x_4$.

3735. $60t(1-t)$, $t^2(1-t)$, $t(1-t)(t^2 - t - 1)$.

3736. *Hint.* Make use of Exercise 3731b.

3737. *Hint.* The determinant of a skew-symmetric matrix of odd size is equal to zero.

3745. a) $1, 2$.

b) $2, 3, 4$.

c)   The maximal dimension is $\binom{n}{2}$.

3801. a), c), g), i), l), o), q), r).

3803. The functions are not equivalent.

3804. A basis exists only for f_1.

3806. a) $\langle (2, 1, 0) \rangle$.

b) $\langle (-21, 13, 0) \rangle, (-79, 0, 13) \rangle$.

3808. a) $2x_1' y_1' - \frac{1}{2} x_2' y_2' + 3x_3' y_3'$.

b) $x_1' y_1' - x_2' y_2' + 16 x_3' y_3'$.

3809. a) The functions are equivalent.

b)   The functions are not equivalent.

3810. a), c), i), l), o), p).

3811. a) $\lambda > 2$.

b) $|\lambda| < \sqrt{5/3}$.

c) $-0,8 < \lambda < 0$.

d)   For no λ.

3813. *Hint.* Consider the function $x_1^2 + 4x_1x_2 + x_2^2$.

3814. a) $\lambda < -20$.

b) $\lambda < -0,6$.

3815. a) $x_1y_1 + x_1y_2 + x_2y_1 + 2x_2y_2 - 3x_2y_3 - 3x_3y_1 + 2x_1y_3 + 2x_3y_1 - x_3y_3$.

b) $\frac{1}{2}(x_1y_2 + x_2y_1 + x_1y_3 + x_3y_1 + x_2y_3 + x_3y_2)$.

3816. a) $2x_1y_1 - \frac{3}{2}x_1y_2 - \frac{3}{2}x_2y_3 - 2x_3y_1 + \frac{1}{2}x_1y_2 + \frac{1}{2}x_2y_1 - \frac{5}{2}x_2y_3 - \frac{5}{2}x_3y_2 + x_3y_3$.

b) $-2x_2y_2 + \frac{3}{2}x_2y_3 + \frac{3}{2}x_3y_2 - \frac{1}{2}x_1y_3 - \frac{1}{2}x_3y_1 + 2x_3y_3$.

3817. a) The functions are not equivalent.

b)   The functions are equivalent.

3818. a) $y_1^2 + y_2^2 - y_3^2$.

b) $y_1^2 + y_2^2 - y_3^2$.

c) $y_1^2 - y_2^2$.

d) $y_1^2 - y_2^2 - y_3^2 - y_4^2$.

3819. a) The functions are not equivalent.

b)   The functions are equivalent.

3822. $n(n + 1)/2, n(n - 1)/2$.

3823. *Hint.* If $U = \langle e_1, e_2 \rangle$ and $f(e_1, e_2) = f(e_2, e_2)$ choose in $U^{\perp}$ a vector e_3 such that $f(e_2, e_3) \neq 0$.

3826. *Hint.* Reduce the function to the normal form and apply an argument similar to that in the proof of the law of inertia.

3827. n, n.

3829. *Hint.* Consider the appropriate quadratic function.

3831. $n(n + 1)/2$.

3832. *Hint.* Consider values at the points of the form $\lambda x + y$ where $\lambda \in F$.

3901. The following mappings are linear operators: a) if $a = 0$; b) if $a = 0$; c), d), e) if $a = 0$; f), g), h), k).

3905. a) $\{0\}$, V.

b) V, $\{0\}$.

c) V, $\{0\}$ if $\alpha \neq 0$; V if $\alpha = 0$.

d) $\langle b \rangle$, $\langle a \rangle^{\perp}$ if $a, b \neq 0$; $\{0\}$, V if either $a = 0$ or $b = 0$.

e) V, $\{0\}$.

f) $\mathbb{R}[x]_n$, $\{0\}$.

g) $\mathbb{R}[x]_{n-1}$, $\mathbb{R}$.

h) $\mathbb{R}[x]_{n-k}$, $\mathbb{R}[x]_{k-1}$.

k) $\mathbb{R}$, $\{0\}$.

3907. *Hint.* Complete the basis $(e_1, \ldots, e_k)$ of the subspace to the basis $(e_1, \ldots, e_n)$ of the whole space and consider projections onto $\langle e_1, \ldots, e \rangle$ and onto $\langle e_{k+1}, \ldots, e_n \rangle$ (see Exercise 3917).

3909. *Hint.* Complete the basis of the subspace to a basis of the space.

3912. *Hint.* Prove firstly that $\mathrm{rk}\,A = \mathrm{rk}\,BA + \dim(\mathrm{Im}\,A \cap \mathrm{Ker}\,B)$.

3914. $n(n - r)$ where $r = \mathrm{rk}\,A$.

3915. a) $\begin{pmatrix} 1 & 0 & 0 \\ 1 & 2 & 0 \\ 0 & 1 & 3 \end{pmatrix}$.

b) $\begin{pmatrix} \cos\alpha & -\sin\alpha \\ \sin\alpha & \cos\alpha \end{pmatrix}$, if the positive direction of reading of angles coincides with the direction of the shortest rotation which maps the first basic angle into the second one.

c) $\begin{pmatrix} 0 & 0 & 1 \\ 1 & 0 & 0 \\ 0 & 1 & 0 \end{pmatrix}$. d) $\begin{pmatrix} 0 & 0 & 0 \\ 0 & 1 & 0 \\ 0 & 0 & 0 \end{pmatrix}$. e) $\begin{pmatrix} 1 & 0 & -2 \\ 0 & 0 & 0 \\ -2 & 0 & 4 \end{pmatrix}$.

f) $\begin{pmatrix} a & 0 & b & 0 \\ 0 & a & 0 & b \\ c & 0 & d & 0 \\ 0 & c & 0 & d \end{pmatrix}$. g) $\begin{pmatrix} a & c & 0 & 0 \\ b & d & 0 & 0 \\ 0 & 0 & a & c \\ 0 & 0 & b & d \end{pmatrix}$. h) $\begin{pmatrix} 1 & 0 & 0 & 0 \\ 0 & 0 & 1 & 0 \\ 0 & 1 & 0 & 0 \\ 0 & 0 & 0 & 1 \end{pmatrix}$.

i) $\begin{pmatrix} a_1b_1 & a_1b_3 & a_2b_1 & a_2b_3 \\ a_1b_2 & a_1b_4 & a_2b_2 & a_2b_4 \\ a_3b_1 & a_3b_3 & a_4b_1 & a_4b_3 \\ a_3b_2 & a_3b_4 & a_4b_2 & a_4b_4 \end{pmatrix}$, where $A = \begin{pmatrix} a_1 & a_2 \\ a_3 & a_4 \end{pmatrix}$, $B = \begin{pmatrix} b_1 & b_2 \\ b_3 & b_4 \end{pmatrix}$.

j)
$$\begin{pmatrix} a_1+b_1 & b_3 & a_2 & 0 \\ b_2 & a_1+b_4 & 0 & a_2 \\ a_3 & 0 & a_4+b_1 & b_3 \\ 0 & a_3 & b_2 & a_4+b_4 \end{pmatrix}.$$

k)
$$\begin{pmatrix} 0 & 1 & 0 & 0 & \dots & 0 \\ 0 & 0 & 2 & 0 & \dots & 0 \\ 0 & 0 & 0 & 3 & \dots & 0 \\ \multicolumn{6}{c}{\dotfill} \\ 0 & 0 & 0 & 0 & \dots & n \\ 0 & 0 & 0 & 0 & \dots & 0 \end{pmatrix}.$$

l)
$$\begin{pmatrix} 0 & 1 & 0 & 0 & \dots & 0 \\ 0 & 0 & 1 & 0 & \dots & 0 \\ 0 & 0 & 0 & 1 & \dots & 0 \\ \multicolumn{6}{c}{\dotfill} \\ 0 & 0 & 0 & 0 & \dots & 1 \\ 0 & 0 & 0 & 0 & \dots & 0 \end{pmatrix}.$$

3917. *Hint.* The first k entries of the principal diagonal of the matrix of the operator are equal to 1 while all other entries are equal to 0.

3918. The first k columns of matrices consist of the coefficients of expressions of vectors $b_1, \dots, b_k$ via $a_1, \dots, a_n$; other columns are arbitrary.

3919. a)
$$\begin{pmatrix} 4 & 5 & 0 & -1 \\ 1 & 0 & 2 & 3 \\ 2 & 3 & 0 & 3 \\ 1 & 6 & -1 & 7 \end{pmatrix}.$$
b)
$$\begin{pmatrix} -5 & -8 & -6 & -2 \\ 2 & 4 & 4 & 0 \\ -3 & -2 & -1 & -5 \\ 6 & 7 & 6 & 13 \end{pmatrix}.$$

3920.
$$\begin{pmatrix} -\frac{1}{3} & \frac{2}{3} & \frac{2}{3} \\ \frac{2}{3} & \frac{2}{3} & -\frac{1}{3} \\ \frac{2}{3} & -\frac{1}{3} & \frac{2}{3} \end{pmatrix}.$$

3921.
$$\begin{pmatrix} 1 & 2 & 2 \\ 3 & -1 & -2 \\ 2 & -3 & 1 \end{pmatrix}.$$

3923. a) The first k columns of the matrix are zero ones; other k columns are linearly independent.

3924. *Hint.* Consider the subspaces V_i ($i = 1, 2$), consisting of all vectors x for which $(f_i(\mathcal{A}))(x) = 0$.

4001. a) Polynomials of zero degree; $\{0\}$.

b)   Nonzero symmetric and skew-symmetric matrices; $\{1, -1\}$.

c)   Monomials; $\{0, 1, 2, \dots, n\}$.

d)   Monomials; $\{0, \frac{1}{2}, \dots, 1/(n+1)\}$.

4002. *Hint.* The equality $f(ax + b) = \lambda f(x)$ implies $\lambda = a^k$ where k is the degree of $f(x)$.

4004. *Hint.* If $\mathcal{A}(x) = \lambda x$, $\lambda \neq 0$ then $\mathcal{A}^{-1}(x) = (1/\lambda)x$.

4006. *Hint.* Make use of matrices of operators.

4008. *Hint.* a) Make use of Exercise 4007. b) Consider the factor-space with respect to the subspace $\langle a \rangle$ where a is a common eigenvector of all given operators.

4009. *Hint.* Consider $\mathcal{A}^2 - \lambda \mathcal{E}^2$.

4011. *Hint.* Consider the matrix

$$
A = \begin{pmatrix}
-a_{n-1} & -a_{n-2} & \cdots & -a_1 & -a_0 \\
1 & 0 & \cdots & 0 & 0 \\
0 & 1 & \cdots & 0 & 0 \\
\cdots & \cdots & \cdots & \cdots & \cdots \\
0 & 0 & \cdots & 1 & 0
\end{pmatrix}.
$$

4015. a) $\lambda_1 = \lambda_2 = \lambda_3 = -1$; $c(1, 1, -1)$ $(c \neq 0)$.

b) $\lambda_1 = \lambda_2 = \lambda_3 = 2$; $c_1(1, 2, 0) + c_2(0, 0, 1)$ (c_1 and c_2 are not equal to zero simultaneously).

c) $\lambda_1 = 1, \lambda_2 = \lambda_3 = 0$; for $\lambda_1 = 1$ the eigenvectors have the form $c(1, 1, 1)$, for $\lambda_{2,3} = 0$ they have the form $c(1, 2, 3)$ $(c \neq 0)$.

d) $\lambda_1 = \lambda_2 = 1$; for $\lambda_{1,2} = 1$ they have the form $c_1(2, 1, 0) + c_2(1, 0, -1)$ (c_1 and c_2 are not equal to zero simultaneously), for $\lambda_3 = -1$ they have the form $c(3, 5, 6)$ $(c \neq 0)$.

e) $\lambda_1 = 1, \lambda_2 = 2 + 3i, \lambda_3 = 2 - 3i$ (over $\mathbb{C}$); for $\lambda_1 = 1$ $c(1, 2, 1)$, for $\lambda_2 = 2 + 3i$ $c(3 - 3i, 5 - 3i, 4)$, for $\lambda_3 = 2 - 3i$ $c(3 + 3i, 5 + 3i, 4)$, everywhere $c \neq 0$.

f) $\lambda = 2$; $c_1(1, 1, 0, 1) + c_2(0, 0, 1, 1)$ (c_1 and c_2 are not equal to zero simultaneously).

4016. a) $((1, 1, 1), (1, 1, 0), (1, 0, -3))$, $\begin{pmatrix} 1 & 0 & 0 \\ 0 & 2 & 0 \\ 0 & 0 & 2 \end{pmatrix}$.

b) It is not diagonalizable over $\mathbb{R}$ nor over $\mathbb{C}$.

c) $((1, 1, 2), (3 - 3i, 4, 5 - 3i), (3 + 3i, 4, 5 + 3i))$, $\begin{pmatrix} 1 & 0 & 0 \\ 0 & 2 + 3i & 0 \\ 0 & 0 & 2 - 3i \end{pmatrix}$.

d) $((1, 1, 0, 0), (1, 0, 1, 0), (1, 0, 0, 1), (1, -1, -1, -1))$, $\begin{pmatrix} 2 & 0 & 0 & 0 \\ 0 & 2 & 0 & 0 \\ 0 & 0 & 2 & 0 \\ 0 & 0 & 0 & -2 \end{pmatrix}$.

4017. Elements α_k and α_{n-k+1} are either both non-zero or are both zero $(k = 1, \ldots, n)$.

4018. *Hint.* It is possible to take as T the matrix with 1 at the principal diagonal and at the adjacent diagonal below, with -1 at the adjacent diagonal above the principal diagonal and with 0 at the other places. Then B has from the beginning

$n/2$ units at the principal diagonal if n is even and $(n + 1)/2$ units if n is odd; the other entries of the principal diagonal are equal to -1.

4019. *Hint.* Consider the matrix of $\mathcal{A}$ in a basis whose first vectors are eigenvectors with the eigenvalue λ_0.

4020. $\lambda_1, \ldots, \lambda_n, \bar{\lambda}_1, \ldots, \bar{\lambda}_n$.

4021. a) $\lambda_i \lambda_j$ $(i, j = 1, \ldots, n)$.

b) λ_i / λ_j $(i, j = 1, \ldots, n)$.

4022. $\{0\}$ and $\mathbb{R}[x]_k$ $(k = 0, 1, 2, \ldots, n)$.

4027. $\{0\}$ and the linear spans of subsets of the basis.

4028. $V_i = \langle e_1, \ldots, e_i \rangle$ $(i = 1, \ldots, n)$.

4029. $\{0\}$, V, $\langle (2, 2, -1) \rangle$, $U = \langle (1, 1, 0), (1, 0, -1) \rangle$, $\langle (2, 2, -1), a \rangle$, $\langle a \rangle$ where $a \in U$.

4030. V, $\{0\}$, $\langle (1, -2, 1) \rangle$, $\langle (1, 1, 1), (1, 2, 3) \rangle$.

4031. The linear span of any set of monomials of degree at most n.

4032. $\langle \cos kx, \sin kx \rangle$.

4033. *Hint.* Consider eigenspaces U_1, U_{-1} of $\mathcal{A}$ and V_1, V_{-1} of $\mathcal{B}$. In the case, when all intersections $U_i \cap V_j$ are trivial find nonzero vectors $a \in V_1, a + \lambda b \in V_{-1}$ such that $a + b \in V_1, a + \lambda b \in V_{-1}$ for some λ.

4035. a) $\lambda_1 = 1$, $\lambda_{2,3} = 0$; $\langle (1, 1, 1) \rangle$ for $\lambda_1 = 1$, $\langle (1, 1, 0), (1, 0, -3) \rangle$ for $\lambda_{2,3} = 0$.

b) $\lambda_1 = 3$, $\lambda_{2,3} = -1$; $\langle (1, 2, 2) \rangle$ for $\lambda_1 = 3$, $\langle (1, 1, 0), (1, 0, -1) \rangle$ for $\lambda_{2,3} = -1$.

c) $\lambda_{1,2,3} = -1$; V.

d) $\lambda_{1,2} = 2, \lambda_{3,4} = 0$; $\langle (1, 0, 1, 0), (1, 0, 0, 1) \rangle$ for $\lambda_{1,2} = 2$,
    $\langle (1, 0, 0, 0), (0, 1, 0, 1) \rangle$ for $\lambda_{3,4} = 0$.

4037. *Hint.* Make use of Exercise 4036.

4040. Eigenvalues of L_A are eigenvalues of A.

4041. *Hint.* a) Reduce X to the row-echelon form.

b) Utilize a).

4042. *Hint.* Induction on the dimension of the space.

4043. *Hint.* Induction on the degree of minimal annihilator polynomial.

4101. a) $\begin{pmatrix} 3 & 0 & 0 \\ 0 & -1 & 1 \\ 0 & 0 & -1 \end{pmatrix}$.

b) $\mathrm{diag}(1, 2 + 3i, 2 - 3i)$.

c) $\begin{pmatrix} -2 & 0 & 0 \\ 0 & 1 & 0 \\ 0 & 0 & 1 \end{pmatrix}$.

d) $\begin{pmatrix} -1 & 0 & 0 \\ 1 & -1 & 0 \\ 0 & 0 & -1 \end{pmatrix}$. e) $\begin{pmatrix} 0 & 0 & 0 \\ 0 & 2 & 0 \\ 0 & 0 & 2 \end{pmatrix}$. f) $\begin{pmatrix} 1 & 0 & 0 & 0 \\ 0 & 1 & 1 & 0 \\ 0 & 0 & 1 & 1 \\ 0 & 0 & 0 & 1 \end{pmatrix}$.

g) Two boxes of size 2 with 0 at the principal diagonal.

h) A box with 1 at the principal diagonal.

i) A box with 1 at the principal diagonal.

j) A box with n at the principal diagonal.

k) $\mathrm{diag}(1, 2, \ldots, n)$.

l) $\mathrm{diag}(\varepsilon_0, \varepsilon_1, \ldots, \varepsilon_{n-1})$, where ε_i is a root of order n of 1 $(i = 1, \ldots, n)$.

m) A box with α at the principal diagonal. *Hint.* A nonzero minor of size $n - 1$ is located in the right top corner of the matrix $A - \lambda E$; $A - \lambda E$; find elementary divisors of $A - \lambda E$.

4103. a)
$$\begin{pmatrix} f(\alpha) & \dfrac{f'(\alpha)}{1!} & \dfrac{f''(\alpha)}{2!} & \cdots & \dfrac{f^{(n)}(\alpha)}{n!} \\ 0 & f(\alpha) & \dfrac{f'(\alpha)}{1!} & \cdots & \dfrac{f^{(n-1)}(\alpha)}{(n-1)!} \\ 0 & 0 & f(\alpha) & \cdots & \dfrac{f^{(n-2)}(\alpha)}{(n-2)!} \\ \cdots & \cdots & \cdots & \cdots & \cdots \\ 0 & 0 & 0 & \cdots & f(\alpha) \end{pmatrix}.$$

b) A Jordan box with α^2 at the principal diagonal if $\alpha \neq 0$; if $\alpha = 0$ then two Jordan boxes with zero at the principal diagonals whose sizes are equal to $n/2$ if n is even and to $(n - 1)/2$, $(n + 1)/2$ if n is odd.

4104. Two boxes with α at the principal diagonal. Their sizes are equal to $n/2$ if n is even and to $(n - 1)/2$, $(n + 1)/2$ if n is odd. *Hint.* Make use of Exercises 4102 and 4103.

4105. a) In each box of the Jordan form of A replace λ ($\lambda \neq 0$) by λ^2; if the diagonal entries of a box of size k are equal to 0 and $k = 2l$ then the box is replaced by two boxes of sizes l, if $k = 2l + 1$ then replace the box by two boxes of sizes $l + 1$ and l.

b) In the Jordan form of A the diagonal entries are replaced by their inverses.

4106. a) The diagonal matrix with entries ± 1 at the principal diagonal. A is a reflection of the space V with respect to some subspace L_1 and in parallel with some complementary subspace L_2.

b) The diagonal matrix where entries of the principal diagonal are equal either to
1 or to 0. $\mathcal{A}$ is a projection of the space V onto some subspace L_1 in parallel
with some complementary subspace L_2.

4107. The entries of the principal diagonal are roots of 1.

4110. a) $\begin{pmatrix} 2 & 1 & 0 \\ 0 & 2 & 0 \\ 0 & 0 & 2 \end{pmatrix}$, $((1, 4, 3), (1, 0, 0), (3, 0, 1))$.

b) $\begin{pmatrix} 0 & 1 & 0 \\ 0 & 0 & 0 \\ 0 & 0 & 0 \end{pmatrix}$, $((1, -3, -2), (1, 0, 0), (1, 0, 1))$.

c) $\begin{pmatrix} 1 & 1 & 0 & 0 \\ 0 & 1 & 0 & 0 \\ 0 & 0 & 1 & 1 \\ 0 & 0 & 0 & 1 \end{pmatrix}$, $((1, 1, 1, 1), (-1, 0, 0, 0), (1, 1, 0, 0), (0, 0, -1, 0))$.

d) $\begin{pmatrix} 2 & 1 & 0 & 0 \\ 0 & 2 & 0 & 0 \\ 0 & 0 & 2 & 0 \\ 0 & 0 & 0 & 1 \end{pmatrix}$, $((-1, -1, -1, 0), (2, 1, 0, 0), (1, 0, 0, -1), (3, 6, 7, 1))$.

4111. One Jordan box.

4112. *Hint.* Make use of the Jordan form of the matrix of $\mathcal{A}$.

4113. *Hint.* Make use of the Jordan form of matrices of operators.

4114. *Hint.* Make use of the Jordan form of B.

4116. The eigenvalue is equal to 1, boxes have sizes 1, 3, 5.

4117. The eigenvalue is equal to 0, boxes have sizes $n + 1, n, \ldots, 2, 1$.

4118. *Hint.* Make use of reduction to the Jordan form.

4120. *Hint.* Make use of the Jordan form of the kth power of the Jordan box
(see Exercise 4103).

4121. a) $\pm \frac{1}{4} \begin{pmatrix} 7 & 1 \\ -1 & 9 \end{pmatrix}$.

b) $\pm \frac{1}{5} \begin{pmatrix} 12 & 2 \\ 3 & 13 \end{pmatrix}$, $\pm \begin{pmatrix} 0 & 2 \\ 3 & 1 \end{pmatrix}$.

4122. a) $2^{50} \begin{pmatrix} -24 & 25 \\ -25 & 26 \end{pmatrix}$.   b) $\begin{pmatrix} -7 & 4 \\ -14 & 8 \end{pmatrix}$.

4123. $(t - \lambda_1) \ldots (t - \lambda_n)$.

4124. $(t - \alpha)^n$.

4126. a) $t - 1$.

b) t.

c) $t^2 - t$.

d) $t^2 - 1$.

e) t^k.

f) $t(t-1)\ldots(t-n)$.

g) $(t-1)\left(t-\dfrac{1}{2}\right)\ldots\left(1-\dfrac{1}{n+1}\right)$.

h) $(t^2+1)\ldots(t^2+n^2)$.

i) $(t^2+1)\ldots\left(t^2+\dfrac{1}{n^2}\right)$.

j) Coincides with the minimal polynomial of A_3.

k) $(t-1)^2$.

4127. a) $(t-2)^3$.

b) $t^2 - 5t + 6$.

4128. $(t-1)^2(t-2)$, $V = L_1 \oplus L_2$, where L_1 has a basis $(e_1, e_2 - e_3)$ and L_2 has a basis (e_2).

4130. It contains boxes of size one with entry 1 and of sizes one and two with 1 and 0 at the principal diagonal.

4133. *Hint.* Compare the dimensions of the space of polynomials in A and the space of matrices commuting with A.

4134. c) *Hint.* Make use of b).

4138. *Hint.* Make use of Exercise 4134 and of decomposition of a space into a direct sum of cyclic subspaces.

4142. c) *Hint.* Prove by induction on l that there exists $B_i \in I$ such that $p(A+B_i)$ is divisible by $p^l(A)$ in the ring $K[A]$.

4143. *Hint.* Deduce from Exercise 4136 and the previous exercise, proving firstly that all elements of I are nilpotent.

4144. *Hint.* Make use of Exercises 4139, 4142.

4145. *Hint.* a) Make use of Jordan form of A.

b) Utilize Exercise 3117.

4147. *Hint.* Follows from Exercise 4116.

4202. *Hint.* See A.I. Kostrikin, Yu.I. Manin. Linear Algebra and Geometry, Gordon and Breach, Reading, 1989, part 1, § 10.

4203. *Hint.* Make use of Exercises 4202 and 4201.

4204. *Hint.* Make use of Exercise 4203.

4213. *Hint.* See N. Bourbaki, Théorie Spectrale, Hermann, Paris, 1976, ch. 1, § 2, 5.

4214. *Hint.* See R. Horn, C.R. Johnson, Matrix Analysis, Cambridge University Press, 1986, chs. 5, 6.

4219. a) $\begin{pmatrix} 2e^2 & -e^2 \\ e^2 & 0 \end{pmatrix}$.   b) $\begin{pmatrix} 4e-3 & 2-2e \\ 6e-6 & 4-3e \end{pmatrix}$.

c) $\begin{pmatrix} 3e-1 & e & -3e+1 \\ 3e & e+3 & -3e-3 \\ 3e-1 & e+1 & -3e \end{pmatrix}$.

d) $\begin{pmatrix} 3 & -15 & 6 \\ 1 & -5 & 2 \\ 1 & -5 & 2 \end{pmatrix} + 2\pi \in E$, where $n \in \mathbb{Z}$.

e) $\begin{pmatrix} 1 & -1 \\ 1 & -1 \end{pmatrix}$.

4220. $\det e^A = e^{\operatorname{tr} A}$.

4222. – 4233. *Hint.* See R. Horn, C.R. Johnson, Matrix Analysis, Cambridge University Press, 1986, ch. 8, § 81–84.

4234. a) $x = (1, 1)$, $\rho(A) = 3$.   b) $x = (1, 1)$, $\rho(A) = 7$.

c) $x = (5, 3, 1)$, $\rho(A) = 5$.   d) $x = (1, 0, 1, 0)$, $\rho(A) = 6$.

4304. a) Subspace of scalar matrices.

b) Subspace of skew-symmetric matrices.

c) Subspace of symmetric matrices.

d) Subspace of lower niltriangular matrices.

4313. Both matrices are equal to G^{-1}.

4314. a) ${}^t S^{-1}$.

b) ${}^t \bar{S}^{-1}$.

4315. a) $((1, 2, 2, -1), (2, 3, -3, 2), (2, -1, -1, -2))$.

b) $((1, 1-1, -2), (2, 5, 1, 3))$.

c) $((2, 1, 3, -1), (3, 2, -3, -1))$.

4316. For example, a) $((2, -2, -1, 0), (1, 1, 0, -1))$.

b) $((0, 1, 0, -1), (1, 0, -1, 0))$.

4318. a) $x_2 + x_4 = 0$.

b) $x_1 + x_2 + x_3 + x_4 = 0,$
$-18x_1 + x_2 + 18x_3 + 11x_4 = 0.$

4319. a) $(1, -1, -1, 5), (3, 0, -2, -1)$.

b) $(3, 1, -1, -2), (2, 1, -1, 4)$.

c) $\left(0, -\frac{3}{2}, \frac{3}{2}, 0\right), \left(7, -\frac{5}{2}, -\frac{5}{2}, 2\right).$

4321. a) $\sqrt{14}$.   b) 2.   c) $1/5$.   d) $\sqrt{3/5}$.   e) $\sqrt{5/7}$.

4326. *Hint.* See Exercise 4325d.

4328. a) 6, 6, 6; $60°$.

4332. 0, if n is odd; $\dfrac{1}{2}\dbinom{n}{k} = \dbinom{2k-1}{k-1}$, if $n = 2k$.

4333. $a\sqrt{n}$; $\arccos \dfrac{1}{\sqrt{n}}$.

4334. $R = \dfrac{a\sqrt{n}}{2}$; $R < a$ if $n < 4$, $R = a$ if $n = 4$ and $R > a$ if $n > 4$.

4336. a) 8. b) 4. c) 12714. d) 0.

4338. a) $60°$.   b) $30°$.

4341. $\arccos \sqrt{k/n}$.

4342. $\arccos(2/3)$. *Hint.* Let $a_i = \overline{A_0 A_i}$ $(i = 1, 2, 3, 4)$; show that the square of the cosine of the angle between the vectors $a_1 t_1 + a_2 t_2$ and $a_3 t_3 + a_4 t_4$ is equal to

$$\frac{(t_1 + t_2)^2 (t_3 + t_4)^2}{4(t_1^2 + t_1 t_2 + t_2^2)(t_3^2 + t_3 t_4 + t_4^2)};$$

find the maximum of the function $(t_1 + t_2)^2$ under the restriction $t_1^2 + t_1 t_2 + t_2^2 = 1$.

4343. $45°$. *Hint.* Find the minimum of angles of vectors of the second plane with their orthogonal projections to the first plane.

4344. b) $P_0(x) = 1$, $P_1(x) = x$, $P_2(x) = \frac{1}{2}(3x^2 - 1)$, $P_3(x) = \frac{1}{2}(5x^2 - 3x)$, $P_4(x) = \frac{1}{8}(35x^4 - 30x^2 + 3)$.

c)
$$P_k(x) = \sum_{j=0,\ j \geq k/2} (-1)^{k-j} \frac{1 \cdot 3 \cdot 5 \dots (2j-1)}{(k-j)!(2j-k)!2^{k-j}} x^{2j-k}$$
$$= \frac{1}{2^k k!} \sum_{j=0,\ j \geq k/2} (-1)^{k-j} \binom{k}{j} \frac{(2j)!}{(2j-k)!} x^{2j-k}.$$

d) $\sqrt{2/2k+1}$.

e) 1.

4345. a) $\sqrt{\Delta}$, where

$$\Delta = \begin{vmatrix} 1 & \dfrac{1}{2} & \cdots & \dfrac{1}{n+1} \\ \dfrac{1}{2} & \dfrac{1}{3} & \cdots & \dfrac{1}{n+2} \\ \multicolumn{4}{c}{\cdots\cdots\cdots\cdots} \\ \dfrac{1}{n+1} & \dfrac{1}{n+2} & \cdots & \dfrac{1}{2n+1} \end{vmatrix} = \dfrac{(1!2!\ldots n!)^3}{(n+1)!(n+2)!\ldots(2n+1)!}.$$

b) $\dfrac{1}{\binom{2n}{n}\sqrt{4n+1}}.$

4402. $\bar{G}^{-1t}\bar{A}\bar{G}.$

4403. $\begin{pmatrix} 3 & 6 \\ -1 & -3 \end{pmatrix}.$

4404. Projection onto the bisector of the second and the fourth quadrants in parallel with the coordinate axis.

4409. a) $D^* = -D.$

b) *Hint.* Integrate by parts.

4410. *Hint.* See the hint to Exercise 4409.

4413. *Hint.* Make use of Exercise 4412 and of connections between Hermitian and quadratic functions.

4414. *Hint.* The assumption of the exercise is equivalent to the equality $(\mathcal{A}\mathcal{A}^*x, x) = (\mathcal{A}^*\mathcal{A}x, x)$ for all $x \in V$. Make use of Exercises 4401e and 4413.

4415. *Hint.* If $\mathcal{A} = \mathcal{B} - \lambda\mathcal{E}$ then $\mathcal{A}^* = \mathcal{B} - \bar{\lambda}\mathcal{E}$, apply Exercise 4414 where x is an eigenvector of the operator $\mathcal{B}$ with the eigenvalue λ.

4416. *Hint.* Make use of Exercise 4415.

4417. *Hint.* a) Make use of Exercises 4415, 4406 and 4401a.

b) Make use of a) and Exercise 4402.

c) If $\mathcal{A}$ is normal then the statement follows from Exercise 4415, in order to prove the converse statement show as in b) that $\mathcal{A}$ has an orthonormal eigenbasis.

4419. *Hint.* Make use of diagonalizability of the matrix of an operator in some orthonormal basis.

4420. *Hint.* Make use of an orthonormal eigenbasis of $\mathcal{A}$ and $\mathcal{B}$.

4421. *Hint.* Make use of an orthonormal eigenbasis.

4422. *Hint.* Make use of an orthonormal eigenbasis and interpolating polynomial.

4423. *Hint.* Apply Exercise 4401c,d,e.

4424. *Hint.* a), b) Make use of Exercise 4423 for the subspace $\operatorname{Ker} f(\mathcal{A})$.

c) There exist polynomials $a(x)$, $c(x)$ such that $a(x)f_1(x) + c(x)f_2(x) = 1$, deduce from here that $\operatorname{Ker} f(\mathcal{A}) = \operatorname{Ker} f_1(\mathcal{A}) \oplus \operatorname{Ker} f_2(\mathcal{A})$; if $x \in \operatorname{Ker} h(\mathcal{A})$, $y \in \operatorname{Ker} f(\mathcal{A})$ then by Exercises 4423 and 4424a,b we have $(x, y) = (c(\mathcal{A})f_2(\mathcal{A})x, y) = (x, \bar{c}(\mathcal{A}^*)\bar{f}_2(\mathcal{A}^*)y) = 0$.

d) By Exercises 4407, 4423 and 4421a–c we have $\operatorname{Ker} f(\mathcal{A})^{\perp} = \operatorname{Im} \bar{f}(\mathcal{A}^*) \subseteq \operatorname{Ker} \bar{f}(\mathcal{A}^*)^{n-1} = \operatorname{Ker} f(\mathcal{A})^{n-1}$, hence $V = \operatorname{Ker} f(\mathcal{A}) + \operatorname{Ker} f(\mathcal{A})^{n-1}$, i.e. $f(x)^{n-1}$ annihilates $\mathcal{A}$ if $n \geq 2$.

4425. *Hint.* Make use of Exercise 4423.

4426. *Hint.* Make use of Exercise 4425.

4427. *Hint.* a) Make use of Exercise 4426. b) follows from Exercises 4406 and 4429. Induction on dimension with the help of Exercises 4427 and 4428.

4504. a) $\begin{pmatrix} 1 & 0 \\ 0 & 3 \end{pmatrix}$, $\left(\frac{1}{\sqrt{2}}(1, -1), \frac{1}{\sqrt{2}}(1, 1) \right)$.

b) $\begin{pmatrix} 9 & 0 & 0 \\ 0 & 18 & 0 \\ 0 & 0 & -9 \end{pmatrix}$, $\left(\frac{1}{3}(2, 2, 1), \frac{1}{3}(2, -1, -2), \frac{1}{3}(1, -2, 0) \right)$.

c) $\begin{pmatrix} 9 & 0 & 0 \\ 0 & 9 & 0 \\ 0 & 0 & 27 \end{pmatrix}$, $\left(\frac{1}{\sqrt{2}}(1, 1, 0), \frac{1}{\sqrt{18}}(1, -1, -4), \frac{1}{3}(2, -1, 1) \right)$.

d) $\begin{pmatrix} 6 & 0 & 0 \\ 0 & 6 & 0 \\ 0 & 0 & 3 \end{pmatrix}$, $\left(\frac{1}{\sqrt{6}}(1, -2, 1), \frac{1}{\sqrt{2}}(-1, 0, 1), \frac{1}{\sqrt{3}}(1, 1, 1) \right)$.

e) $\begin{pmatrix} 1 & 0 & 0 \\ 0 & 1 & 0 \\ 0 & 0 & -1 \end{pmatrix}$, $\left(\frac{1}{\sqrt{2}}(1, 0, 1), (0, 1, 0), \frac{1}{\sqrt{2}}(1, 0, -1) \right)$.

f) $\begin{pmatrix} 1 & 0 & 0 & 0 \\ 0 & 1 & 0 & 0 \\ 0 & 0 & -1 & 0 \\ 0 & 0 & 0 & -1 \end{pmatrix}$,

$\left(\frac{1}{\sqrt{2}}(1, 0, 0, 1), \frac{1}{\sqrt{2}}(0, 1, 1, 0), \frac{1}{\sqrt{2}}(1, 0, 0, -1), \frac{1}{\sqrt{2}}(0, 1, -1, 0) \right)$.

g) $\begin{pmatrix} 2 & 0 & 0 & 0 \\ 0 & 2 & 0 & 0 \\ 0 & 0 & -2 & 0 \\ 0 & 0 & 0 & -2 \end{pmatrix}$,

$\left(\frac{1}{\sqrt{2}}(1, 1, 0, 0), \frac{1}{2}(1, -1, 1, 1), \frac{1}{\sqrt{2}}(0, 0, 1, -1), \frac{1}{2}(1, -1, -1, -1) \right)$.

4507. a) $\begin{pmatrix} 5 & 0 \\ 0 & -1 \end{pmatrix}$, $\left(\dfrac{1}{\sqrt{3}}(1+i,1), \dfrac{1}{\sqrt{6}}(1+i,-2) \right)$.

b) $\begin{pmatrix} 2 & 0 \\ 0 & 4 \end{pmatrix}$, $\left(\dfrac{1}{\sqrt{2}}(1,-i), \dfrac{1}{\sqrt{2}}(1,i) \right)$.

c) $\begin{pmatrix} 2 & 0 \\ 0 & 8 \end{pmatrix}$, $\left(\dfrac{1}{\sqrt{6}}(2-i,-1), \dfrac{1}{\sqrt{6}}(1,2+i) \right)$.

4509. *Hint.* Commuting operators have a common eigenvector x; consider orthocomplement to $\langle x \rangle$.

4510. *Hint.* Apply Vieta formulae and Descartes' theorem.

4511. *Hint.* Apply Exercise 4401e.

4513. *Hint.* Make use of the diagonalizability of A in some orthogonal basis and apply Exercise 4510.

4514. $\begin{pmatrix} 3 & 2 & 0 \\ 2 & 4 & 2 \\ 0 & 2 & 5 \end{pmatrix}$.

4515. *Hint.* By Exercise 4513 we have $A = A_1^2$, $B = B_1^2$ where A_1, B_1 are non-negative self-conjugate operators; if A is positive then $AB = A_1[A_1 B_1)(A_1 B_1)^*]A_1^{-1}$; apply Exercise 4511.

4518. *Hint.* Prove that the rank of $A - \lambda E$ is not less than $n - 1$ for any λ.

4519. The functions in principal axes and matrices of changes of basis A ($^t x = A^t y$) are given:

a) $3y_1^2 + 6y_2^2 + 9y_3^2$, $\dfrac{1}{3}\begin{pmatrix} 2 & -1 & 2 \\ 2 & 2 & -1 \\ -1 & 2 & 2 \end{pmatrix}$.

b) $9y_1^2 + 18y_2^2 - 9y_3^2$, $\dfrac{1}{3}\begin{pmatrix} 2 & 2 & -1 \\ -1 & 2 & 2 \\ 2 & -1 & 2 \end{pmatrix}$.

c) $3y_1^2 + 6y_2^2 - 2y_3^2$, $\dfrac{1}{\sqrt{6}}\begin{pmatrix} \sqrt{2} & 1 & \sqrt{3} \\ -\sqrt{2} & -1 & \sqrt{3} \\ \sqrt{2} & -2 & 0 \end{pmatrix}$.

d) $5y_1^2 - y_2^2 - y_3^2$, $\dfrac{1}{\sqrt{6}}\begin{pmatrix} \sqrt{2} & 1 & \sqrt{3} \\ \sqrt{2} & 1 & -\sqrt{3} \\ \sqrt{2} & -2 & 0 \end{pmatrix}$.

e) $3y_1^2 - 6y_2^2$, $\dfrac{1}{6}\begin{pmatrix} 4 & \sqrt{2} & 3\sqrt{2} \\ 2 & -4\sqrt{2} & 0 \\ 4 & \sqrt{2} & -3\sqrt{2} \end{pmatrix}$.

f) $2y_1^2 + 4y_2^2 - 2y_3^2 - 4y_4^2$, $\dfrac{1}{2}\begin{pmatrix} 1 & 1 & 1 & 1 \\ -1 & 1 & 1 & -1 \\ -1 & -1 & 1 & 1 \\ 1 & -1 & 1 & -1 \end{pmatrix}$.

g) $5y_1^2 - 5y_2^2 + 5y_3^2,$ $\dfrac{1}{\sqrt{5}} \begin{pmatrix} 2 & 1 & 0 & 0 \\ 1 & -2 & 0 & 0 \\ 0 & 0 & 2 & 1 \\ 0 & 0 & -1 & 2 \end{pmatrix}.$

h) $2y_1^2 - 4y_2^2,$ $\dfrac{1}{\sqrt{2}} \begin{pmatrix} 1 & 0 & 1 & 0 \\ 1 & 0 & -1 & 0 \\ 0 & 1 & 0 & 1 \\ 0 & 1 & 0 & -1 \end{pmatrix}.$

i) $9y_1^2 + 9y_2^2 + 9y_3^2,$ $\dfrac{1}{3} \begin{pmatrix} 3 & 0 & 0 & 0 \\ 0 & 1 & 2 & 2 \\ 0 & 2 & 1 & -2 \\ 0 & 2 & -2 & 1 \end{pmatrix}.$

j) $4y_1^2 + 4y_2^2 + 4y_3^2 - 6y_4^2 - 6y_5^2,$ $\dfrac{1}{\sqrt{10}} \begin{pmatrix} \sqrt{10} & 0 & 0 & 0 & 0 \\ 0 & \sqrt{2} & 0 & 2\sqrt{2} & 0 \\ 0 & -2\sqrt{2} & 0 & \sqrt{2} & 0 \\ 0 & 0 & 1 & 0 & 3 \\ 0 & 0 & 3 & 0 & -1 \end{pmatrix}.$

4604. *Hint.* Apply Exercise 4603 and orthogonalization.

4605. *Hint.* b) Put $w = x - \dfrac{\|x\|}{\|y\|} y.$

4606.

a) $\begin{pmatrix} 1 & 0 & 0 \\ 0 & 1 & 0 \\ 0 & 0 & -1 \end{pmatrix};$ $\left(\dfrac{1}{\sqrt{3}}(1,1,1), \dfrac{1}{\sqrt{2}}(1,0,-1), \dfrac{1}{\sqrt{6}}(1,-2,1) \right).$

b) $\begin{pmatrix} 1 & 0 & 0 \\ 0 & 0 & 1 \\ 0 & -1 & 0 \end{pmatrix};$ $\left(\dfrac{1}{\sqrt{2}}(1,1,0), (0,0,1), \dfrac{1}{\sqrt{2}}(1,-1,0) \right).$

c) $\begin{pmatrix} 1 & 0 & 0 \\ 0 & \dfrac{1}{2} & -\dfrac{\sqrt{3}}{2} \\ 0 & \dfrac{\sqrt{3}}{2} & \dfrac{1}{2} \end{pmatrix};$ $\left(\dfrac{1}{\sqrt{3}}(1,1,1), \dfrac{1}{\sqrt{6}}(2,-1,-1), \dfrac{1}{\sqrt{2}}(0,1,-1) \right).$

d) $\begin{pmatrix} 1 & 0 & 0 \\ 0 & \dfrac{1}{2} & -\dfrac{\sqrt{3}}{2} \\ 0 & \dfrac{\sqrt{3}}{2} & \dfrac{1}{2} \end{pmatrix};$ $\left(\dfrac{1}{\sqrt{2}}(1,1,0), \dfrac{1}{\sqrt{2}}(1,-1,0), (0,0,1) \right).$

e) $\begin{pmatrix} 1 & 0 & 0 \\ 0 & \dfrac{2\sqrt{2}-1}{4} & -\dfrac{\sqrt{7+4\sqrt{2}}}{4} \\ 0 & \dfrac{\sqrt{7+4\sqrt{2}}}{4} & \dfrac{2\sqrt{2}-1}{4} \end{pmatrix}$;

$\left(\dfrac{1}{\sqrt{5-2\sqrt{2}}}(1-\sqrt{2},1,-1), \dfrac{1}{\sqrt{2}}(0,1,1), \dfrac{1}{\sqrt{10-4\sqrt{2}}}(-2,1,-\sqrt{2}-1) \right).$

f) $\begin{pmatrix} 1 & 0 & 0 & 0 \\ 0 & 1 & 0 & 0 \\ 0 & 0 & 1 & 0 \\ 0 & 0 & 0 & -1 \end{pmatrix}$,

$\left(\dfrac{1}{2}(1,1,1,-1), \dfrac{1}{2}(1,1,-1,1), \dfrac{1}{2}(1,-1,1,1), \dfrac{1}{2}(-1,1,1,1,1) \right).$

g) $\begin{pmatrix} 1 & 0 & 0 & 0 \\ 0 & -1 & 0 & 0 \\ 0 & 0 & 0 & 1 \\ 0 & 0 & -1 & 0 \end{pmatrix}$,

$\left(\dfrac{1}{\sqrt{2}}2(1,1,0,0), \dfrac{1}{\sqrt{2}}(0,0,1,-1), \dfrac{1}{\sqrt{2}}(1,-1,0,0), \dfrac{1}{\sqrt{2}}(0,0,1,1) \right).$

h) $\begin{pmatrix} 1 & 0 & 0 \\ 0 & \dfrac{1}{2} & \dfrac{\sqrt{3}}{2} \\ 0 & -\dfrac{\sqrt{3}}{2} & \dfrac{1}{2} \end{pmatrix}$, $\left(\dfrac{1}{\sqrt{3}}(1,1,1), \dfrac{1}{\sqrt{6}}(2,-1,-1), \dfrac{1}{\sqrt{2}}(2,-1,-1) \right).$

i) $\begin{pmatrix} 1 & 0 & 0 \\ 0 & 0 & -1 \\ 0 & 1 & 0 \end{pmatrix}$, $\left(\dfrac{1}{3}(-1,2,2), \dfrac{1}{3}(2,2,-1), \dfrac{1}{3}(-2,1,-2) \right).$

j) $\begin{pmatrix} 1 & 0 & 0 \\ 0 & \dfrac{2}{7} & -\dfrac{3\sqrt{5}}{7} \\ 0 & \dfrac{3\sqrt{5}}{7} & \dfrac{2}{7} \end{pmatrix}$, $\left(\dfrac{1}{\sqrt{3}}(1,1,1), \dfrac{1}{\sqrt{2}}(1,-1,0), \dfrac{1}{3\sqrt{10}}(3,5,-8) \right).$

k)
$$\begin{pmatrix} 1 & 0 & 0 \\ 0 & \frac{1}{12}(-2+7\sqrt{2}) & -\frac{1}{12}\sqrt{42+28\sqrt{2}} \\ 0 & \frac{1}{12}\sqrt{42+28\sqrt{2}} & \frac{1}{12}(-2+7\sqrt{2}) \end{pmatrix},$$

$$\left(\frac{1}{\sqrt{42+28\sqrt{2}}}(-2-\sqrt{2},-4-3\sqrt{2},\sqrt{2}), \frac{1}{84}(6\sqrt{2},-2-\sqrt{2},2-\sqrt{2}), \right.$$
$$\left. \frac{1}{84}\left(0,\sqrt{42-28\sqrt{2}},\sqrt{42+28\sqrt{2}}\right) \right).$$

l)
$$\begin{pmatrix} 1 & 0 & 0 \\ 0 & \frac{1}{2} & -\frac{\sqrt{3}}{2} \\ 0 & -\frac{1}{2}\sqrt{3} & \frac{1}{2} \end{pmatrix}, \left(\left(\frac{\sqrt{2}}{2},\frac{\sqrt{2}}{2},0 \right), \left(\frac{\sqrt{2}}{2},-\frac{\sqrt{2}}{2},0 \right), (0,0,-1) \right).$$

4607. a) $\left(\begin{matrix} e^{i\alpha} & 0 \\ 0 & e^{-i\alpha} \end{matrix} \right), \left(\frac{1}{\sqrt{2}}(1,i),\frac{1}{\sqrt{2}}(1,-i) \right).$

b) $\frac{1}{\sqrt{3}}\left(\begin{matrix} 1+i\sqrt{2} & 0 \\ 0 & 1-i\sqrt{2} \end{matrix} \right), \left(\frac{1}{4-2\sqrt{2}}(1,-i(1-\sqrt{2})), \right.$
$$\left. \frac{1}{4-2\sqrt{2}}(i(1-\sqrt{2}),-1) \right).$$

c) $\left(\begin{matrix} 1 & 0 & 0 \\ 0 & i & 0 \\ 0 & 0 & -i \end{matrix} \right), \left(\frac{1}{3}(2,-2i,i),\frac{1}{3}(2,i,-2i),\frac{1}{3}(-i,2,2) \right).$

d) $\left(\begin{matrix} \frac{1+i\sqrt{3}}{2} & 0 \\ 0 & \frac{1-i\sqrt{3}}{2} \end{matrix} \right), \left(\frac{1}{\sqrt{23-4\sqrt{3}}}(4,(\sqrt{3}-2)i), \right.$
$$\left. \frac{1}{\sqrt{23-4\sqrt{3}}}((\sqrt{3}-2)i,4) \right).$$

e) $\left(\begin{matrix} i & 0 \\ 0 & -i \end{matrix} \right), \left(\frac{1}{\sqrt{2}}(1,-i),\frac{1}{\sqrt{2}}(-i,1) \right).$

4608. *Hint.* The matrix is similar to the diagonal matrix $\left(\begin{matrix} e^{i\alpha} & 0 \\ 0 & e^{-i\alpha} \end{matrix} \right)$, which is in turn similar to the matrix $\left(\begin{matrix} \cos\alpha & -\sin\alpha \\ \sin\alpha & \cos\alpha \end{matrix} \right)$ by Exercise 4607a.

4611. b) *Hint.* Make use of diagonal form of matrices of unitary and Hermitian operators.

4612. *Hint.* Any orthonormal basis in V with the same orientation as (e_1, e_2, e_3) by an operator of the form $\mathcal{A}_\varphi \mathcal{B}_\theta \mathcal{A}_\psi$ can be transformed to (e_1, e_2, e_3).

4613. a) *Hint.* If e_1, e_2, e_3 is a basis in V then operators of rotations of the planes $\langle e_1, e_2 \rangle$ and $\langle e_2, e_3 \rangle$ have the required representation; apply Exercise 4607a.

4614. *Hint.* A rotation of two-dimensional plane is a product of two reflections; for the proof of the second statement notice that if $A = A_1 \ldots A_m$ then $\mathrm{Ker}(A - \mathcal{E}) \supseteq \cap_{i=1}^m \mathrm{Ker}(A_i - \mathcal{E})$.

4615. *Hint.* Use orthonormal eigenbasis.

4616. *Hint.* Apply Exercise 4515.

4617. a) $\dfrac{1}{\sqrt{2}} \begin{pmatrix} 3 & 1 \\ 1 & 3 \end{pmatrix} \cdot \dfrac{1}{\sqrt{2}} \begin{pmatrix} 1 & -1 \\ 1 & 1 \end{pmatrix}$.

b) $\dfrac{1}{\sqrt{2}} \begin{pmatrix} 5 & -3 \\ -3 & 5 \end{pmatrix} \cdot \dfrac{1}{\sqrt{2}} \begin{pmatrix} 1 & -1 \\ 1 & 1 \end{pmatrix}$.

c) $\dfrac{1}{3} \begin{pmatrix} 14 & 2 & -4 \\ 2 & 17 & 2 \\ -4 & 2 & 14 \end{pmatrix} \cdot \dfrac{1}{3} \begin{pmatrix} 2 & -1 & 2 \\ 2 & 2 & -1 \\ -1 & 2 & 2 \end{pmatrix}$.

4618. *Hint.* Prove that $\mathcal{B}^2 = A A^*$.

4619. *Hint.* Let $e^{i\alpha_1}, \ldots, e^{i\alpha_n}$ be all distinct eigenvalues of A, find a polynomial $f(t)$ of degree at most n such that $f(e^{i\alpha_j}) = e^{i\alpha_j/k}$ for all $1 \le j \le n$. Check that $f(A)^k = A$.

4620. *Hint.* Make use of diagonalizability of the operator.

4621. *Hint.* Represent A as a square of a positive self-conjugate operator C. Show that the operator $C^{-1}ABC$ is self-conjugate.

4622. *Hint.* Apply Exercise 4621 and represent A and B as squares of positive (non-negative) self-conjugate operators.

4624. *Hint.* Apply Exercise 4623.

4629. *Hint.* Make use of pole decomposition of A.

4630. *Hint.* a) Apply Exercises 4625–4627.

b) Follows from a).

c) Utilize the Vandermonde determinant $W(1, \varepsilon, \varepsilon^2, \ldots, \varepsilon^{n-1})$, where $\varepsilon = \cos\dfrac{2\pi}{n} + i\sin\dfrac{2\pi}{n}$.

4631. *Hint.* Apply Exercises 4604, 4605.

4632. *Hint.* Apply Exercise 4332.

4701. a), d).

4702. 21.

4703. a) $B(v_3, v_4, v_5) = 0$, $-B \otimes A(e_1, e_1 + e_2, e_2 + e_3, e_2, e_2) = 1$, $(A \otimes B - B \otimes A)(v_1, v_2, v_3, v_4, v_5) = 1$.

b) $A(e_1 + e_2, e_2 + e_3) = 2,$ $B(e_3 + e_1, e_2, e_2) = 2,$ $A(e_2, e_2) = 0,$
   $B(e_1 + e_2, e_2 + e_3, e_2 + e_1) = 8,$ $(A \otimes B - B \otimes A)(v_1, v_2, v_3, v_4, v_5) = 4.$

4704. 0.

4705. 0.

4706. $(A \otimes B)(e_1, e_2, e_3, e_3, e_3) = 0,$ $(B \otimes A)(e_1, e_2, e_3, e_3, e_3) = 1.$

4707. a) 4. b) −9. c) 3.

4709. a) $T(v, f) = f(Av)$, where

$$A = \begin{pmatrix} 0 & 0 & 0 & 0 \\ 1 & 0 & 0 & 0 \\ 0 & 1 & 0 & 0 \\ 0 & 0 & 1 & 0 \end{pmatrix} = (\text{Im } A)^{\perp} = \langle e^{\perp} \rangle;$$

therefore $\{ f \in V^* \| T(v, f) = 0 \text{ for any } v \in V \}.$

b) $\langle e_4 \rangle$.

4710. $p^2(4p - 3)$.

4711. a) 2. b) 1. c) 2.

4713. a) 5. b) 1. c) 3.

4714. a) e_3.

b) $5e_3 + 5e_4$.

4715. a) $(2e^1 - e^3) \otimes (2e_1 + 2e_2)$.

b) $2e^1 \otimes e_3$.

4716. a) $(2e_1 + e_2) \otimes e_3 + (e_1 + e_2) \otimes e_4; e^1 \otimes (2e^3 - e^4) - e^2 \otimes (e_3 - e_4)$.

b) $(3e_1 + 2e_2) \otimes (e_3 + e_4) - (2e_1 + e_2 + e_3 + e_4) \otimes e_2;$
   $(e^1 + e^2) \otimes e^3 + (e^1 + e^3) \otimes (e^4 - 2e^3)$.

c) $(e_1 + e_2) \otimes (e_1 + e_2 + e_3 + e_4) + (3e_1 + 2e_2 + 2e_3 + 3e_4) \otimes e_4;$
   $e^2 \otimes e^2 - (e^1 + e^3 + e^4) \otimes e^3 + (e^1 + e^2 + e^3 + e^4) \otimes e_4$.

d) $(2e_1 + e_2) \otimes e_1 + 2(e_1 + e_2) \otimes e_2 + 3(e_3 + e_4) \otimes e_3 + 4(e_3 + 2e_4) \otimes e_4;$
   $e^1 \otimes (e^1 - e^2) - 2e^2 \otimes (e^1 - 2e^2) + 3e^3 \otimes (2e^3 - e^4) - 4e^4 \otimes (e^3 - e^4)$.

4717. *Hint.* Consider the eigenbasis.

4718. b) $(\text{tr} A)^k$. c) d^{2n}.

4719. a) Three boxes of size 2 with 1, 2, 3 at the principal diagonal.

b) One box of size 1 and one box of size 3 with 2 at the principal diagonal.

c) Two boxes of size 3 with 0 at the principal diagonal.

4802. $\frac{2}{3}n(n + 1)(n - 1)$, where $n = \dim V$.

4805. *Hint.* Calculate dimensions.

4807. *Hint.* Prove that the trace of $\Lambda^q \mathcal{A}$ coincides up to a sign with qth coefficient of the characteristic polynomial.

4808. a) Two Jordan boxes of sizes 5 and 1 with 1 at the principal diagonal.
b) Jordan box of size 3 with 6 at the principal diagonal and three boxes of size 1 with 4, 6 and 9 at the principal diagonal.

c)   Two Jordan boxes of size 2 with 2 at the principal diagonal; one box of size 2 with -2 at the principal diagonal and four boxes of size 1 with 1, 4, -4 and -4 at the principal diagonal.

4809. *Hint.* Make use of hint for Exercise 4807.

4814. *Hint.* Apply Exercise 4812.

4815. *Hint.* Consider a basis containing x.

4909. $^t x = B x' + {}^t a, \; x' = B^{-1} x - B^{-1 t} a.$

4910. a) $x_1 - 2x_2 - x_3 + x_4 = -2,$ 　　b) $3x_1 - 2x_2 - x_3 - x_4 = 1,$
$\qquad\quad 2x_1 + 7x_2 + 3x_3 + x_4 = 6, \qquad\qquad 6x_3 + 5x_4 = 1,$
$\qquad\quad x_1 = t_1 + 3t_2, \qquad\qquad\qquad\quad x_1 = t_1 + 2t_2,$
$\qquad\quad x_2 = -t_1 + t_2, \qquad\qquad\qquad\; x_2 = t_1 + 3t_2,$
$\qquad\quad x_3 = 2 + 2t_1 - 3t_2, \qquad\qquad\; x_3 = 6 - 5t_1,$
$\qquad\quad x_4 = -t_1 - 4t_2. \qquad\qquad\qquad x_4 = -7 + 6t_1.$

4911. The first equality takes place if $\langle a_1, \dots, a_s \rangle$ contains the origin, otherwise the second equality takes place.

4914. *Hint.* If $P_i = a_i + L_i \; (i = 1, \dots, s)$, then

$$\langle P_1 \cup \dots P_s \rangle = a_1 + (L_1 + \cdots + L_s + \langle \overline{a_1 a_2}, \dots, \overline{a_1 a_s} \rangle).$$

4916. a) $\dim \langle P_1 \cup P_2 \rangle = 3$, $\dim P_1 \cap P_2 = 1$.

b) $\dim \langle P_1 \cap P_2 \rangle = 4$, $P_1 \cap P_2 = \emptyset$ the degree of parallelelism is equal to 1;

c) $\dim \langle P_1 \cap P_2 \rangle = 4$, the planes P_1 and P_2 are crossing.

4918. The hyperplane parallel to P_1, and to P_2. It passes through one of the points of the form mentioned.

4920. a) $x_1 = 3t, x_2 = -1 + 3t, x_3 = 3 - t, x_4 = 1 - t.$

b) $x_1 = 1 + 2t, x_2 = -3 + 6t, x_2 = -2 + 2t, x_4 = -2 + 4t, x_5 = 5t.$

c)   There is no line.

4922. $1 - x_1 - \cdots - x_n, x_1, \dots, x_n.$

4923. The line contains at least three points if $|K| \geq 3$; the statement is not valid if $|K| = 2$.

4924. For any point a there exists a vector v such that $f(a + v) = a + v.$

4928. If $\operatorname{char}K \nmid n$ then $\frac{1}{n}(a + f(a) + f^2(a) + \cdots + f^{n-1}(a))$ is a fixed point for any point a.

4929. *Hint.* See Exercise 4928.

4931. a) $a + \langle e_1, e_2 \rangle$, where $a = (-1, 0, -1)$, $e_1 = (1, 2, 3)$, $e_2 = (1, 1, 1)$.

b) $a + \lambda e_1$, $a + \langle e_1 \rangle$, $a = a + \lambda e_1 + \langle e_2 \rangle$, $a + \langle e_1, e_2 \rangle$, $a + \lambda e_1 + \langle e_2, e_3 \rangle$, where $a = (3, 3, 4)$, $e_1 = (1, 2, 2)$, $e_2 = (-1, -2, -1)$, $e_2 = (1, 1, 0)$, λ is arbitrary.

c) $a, a + \langle \lambda e_1 + \mu e_3 \rangle, a + \langle e_1, e_2 \rangle, a = a + \langle e_1, e_2 + \lambda e_3 \rangle$, where $a = (0, -1, -4)$, $e_1 = (1, 4, 3)$, $e_2 = (1, 0, 0)$, $e_3 = (3, 0, 1)$, λ, μ are arbitrary.

d) $a + \lambda e_1 + \langle e_1, e_2 \rangle, a + \langle e_1, e_2 \rangle, a + \lambda e_1 + \langle e_1 + e_2, e_3 \rangle$, where $a = \left(\frac{7}{2}, \frac{15}{2}, 7 \right)$, $e_1 = (2, 1, 0)$, $e_2 = (-1, 0, 1)$, $e_3 = (3, 5, 6)$, λ is arbitrary.

4933. a) A transformation exists.

b) A transformation does not exist.

c) A transformation exists.

4934. a) A transformation exists.

b) A transformation does not exist.

4935. P_1; P_2; $\prod_i \setminus P_i$ $(i = 1, 2)$, where $\prod_i$ is a hyperplane containing P_i and parallel to P_j $(j \neq i)$; arbitrary hyperplanes which are parallel simultaneously to P_1 and P_2.

4936. *Hint.* Apply Exercise 4923.

4937. *Hint.* b) Show with the help of Exercise 4936 that f preserves parallelism of lines; define the mapping $Df : V \to V$ by the formula $Df\,\overline{(ab)}\,\overline{f(a)f(b)}$. Show that $f(x + y) = f(x) + f(y)$. Let v be some vector from V. Making use of the relations $Df(\alpha v) = \sigma(\alpha)Df(v)$ define a mapping $\sigma : K \to K$. Show that it does not depend on v and it is an automorphism of the field K.

5003. *Hint.* If $M' \neq \emptyset$ then $H = \{x \mid \sum_{i \in J} f_i(x) = 0\}$ is a hyperplane of support of M and $M \cap H = M^J$. Conversely, let Γ be a side of M which is an intersection with H. Let a be an internal point of Γ and let $J = \{i \mid f_i(a) = 0\}$. Then M^J is a side of M containing Γ and a is its internal point. It follows that $M^J \subseteq H$ and therefore $M^J = \Gamma$.

5004. *Hint.* Consider the system of barycentric coordinates associated with the points $a_0, a_1, \ldots, a_n$.

5007. The vertices are $A = (1, 1, 1)$, $B = (1, 1, -2)$, $C = (1, -2, 1)$, $D = (-2, 1, 1)$ and $E = \left(-\frac{1}{2}, -\frac{1}{2}, -\frac{1}{2} \right)$. The polyhedron is the union of triangular pyramids with common base BCD.

5008. a) Tetrahedron with vertices

$$(1, 0, 0, 0), \quad (0, 1, 0, 0), \quad (0, 0, 1, 0), \quad (0, 0, 0, 1).$$

b)   Octahedron with vertices

$$(1, 1, 0, 0), \ (1, 0, 0, 1, 0), \ (1, 0, 0, 1), \ (0, 1, 1, 0), \ (0, 1, 0, 0, 1), \ (0, 0, 1, 1).$$

c)   Triangular prism whose vertices of one base are

$$(1, 0, 0, 0), \ (0, 1, 0, 0), \ (0, 0, 1, 0)$$

and the vertices of the other base are

$$(1, 0, 0, 1), \ (0, 1, 0, 1), \ (0, 0, 0, 1, 1).$$

d)   Parallelogram with vertices

$$(1, 0, 1, 0), \ (1, 0, 0, 1), \ (0, 1, 1, 0), \ (0, 1, 0, 1).$$

5015. *Hint.* Reduce the proof with the help of Exercise 5014 to the case when $S = M \cup \{a\}$, where M is n-dimensional simplex and $a \notin M$. Apply Exercise 5014 and notice that any segment $\overline{ab}$, where $b \in M$, intersects some $(n-1)$-dimensional side Γ of the simplex M which does not contain the point b; hence it is contained in the union of simplices M and $\mathrm{conv}(\Gamma \cup \{a\})$.

5016. *Hint.* Make use of Exercise 5015.

5017. Rays starting at the point a and intersecting M^0 are sweeping out the angle whose size does not surpass π.

5018. *Hint.* Induction on n. Consider the arbitrary hyperplane H passing through a. Prove that if some neighborhood of a in H is contained in M then M lies in one side of H. Otherwise, by inductive hypothesis, there exists in H a hyperplane $a+W$ (where W is $(n-2)$-dimensional subspace of the vector space V corresponding to A) such that $M \cap H$ lies in one side of H. Let U be a two-dimensional complementary subspace to W. Consider the projection N of the set M into two-dimensional plane $P = a+U$ in parallel W (see Exercise 4938). Prove that $a \notin N^0$ and apply Exercises 5017, 5012.

5020. *Hint.* Choose a point $b \in M$ and for any point $a \notin M$ conduct a hyperplane of support through a point of the segment $\overline{ab}$, which is the nearest to a.

5021. *Hint.* Show that any hyperplane of support of a closed convex cone passes through the origin.

5022. *Hint.* The set M of all non-negative linear combinations of functions $f_1, \ldots, f_m$ is a closed convex cone in vector space L of all affine linear functions in A. If M does not contain positive constants then by Exercise 5021 there exists a linear function φ on L such that $\varphi(1) = 1$ and $\varphi(f) \leq 0$ if $f \in M$. Show that any linear function φ on L, for which $\varphi(1) = 1$, is of the form $\varphi(f) = f(a)$ where a is some point of A (not depending on f).

5023. *Hint.* b) Prove with the help of Exercise 5020 that for any point $a \neq M$ there exists a linear function $f \in M^*$ such that $f(a) > 1$.

5024. *Hint.* Induction on dimension of the space. Firstly prove that any non-extreme point belongs to an interval connecting boundary points. Then by inductive hypothesis deduce that any boundary point belongs to a convex hull of the set of extreme points lying in a hyperplane of support passing through this point.

5027. *Hint.* Follows from Exercises 5024 and 5026.

5028. *Hint.* It suffices to consider the case when the affine hull of the given points coincides with the whole space. Then identify the affine space with a vector space, taking as a zero some internal point of the convex hull M of the given points. Prove that the convex set M^* defined as in Exercise 5023 is a convex polyhedron. Apply Exercises 5027 and 5023b.

5029. a) $x_1 \geq 0, x_2 \geq 0, x_3 \geq 0, x_4 \geq 0, x_1 + x_3 \leq 1, x_1 + x_3 \leq 1, x_1 + x_4 \leq 1, x_2 + x_3 \leq 1, x_2 + x_4 \leq 1$; three-dimensional sides are four quadrangular pyramids $Oabcd, Odefa, Odefb, Oabce$ with the vertices d, e, a, b and four tetrahedrons $acdf, acef, bcdf, bcef$.

b) $x_1 \geq 0, x_2 \geq 0, x_3 \geq 0, x_4 \geq 0, x_1 + x_4 \leq 1, x_2 + x_4 \leq 1; x_3 + x_4 \leq 1$; three-dimensional sides are the parallelepiped $Oabcdefg$ and six quadrangular pyramids with common vertex h, the bases of which are the two-dimensional sides of the parallelepiped mentioned.

5031. *Hint.* Consider the convex set $N - M$ in the space V, which consists of the vectors connecting points of M with points of N. Prove that it is closed and deduce from Exercise 5020 the existence of a linear function φ on V such that $\varphi(\overline{xy}) \geq 1$ for all $x \in M, y \in N$. Take as f a suitable affine linear function whose linear part coincides with φ.

5032. *Hint.* In the space L consider the closed convex cone K consisting of all affine linear functions which are non-negative on M. Assume that $K \cap N = \emptyset$ and deduce from Exercise 5031 the existence of a linear function on L which (i) is non-negative on K, (ii) is negative on N (iii) satisfies the condition $\varphi(1) = 1$. Show that $\varphi(f) = f(a)$ where $a \in M$, a contradiction with the assumption.

5033. *Hint.* Obviously $\max_{x \in M} \min_{y \in N} F(x, y) \leq \min_{y \in N} \max_{x \in M} F(x, y)$. Let $\max_{x \in M} \min_{y \in N} F(x, y) = c$. Then for any point $x \in M$ there exists a point $y \in N$ such that $F(x, y) \leq c$. Prove with the help of Exercise 5032 that there exists a point $y_0 \in N$ such that $F(x, y_0) \leq c$ for all $x \in M$. Deduce from here that $\min_{y \in N} \max_{x \in M} F(x, y) = c$. Similarly prove that there exists a point $x_0 \in M$ such that $F(x_0, y) \geq c$ for all $y \in N$.

5034. *Hint.* Prove all statements by induction on $n + m$.

5035. a) It is bounded.

b), c) It is not bounded.

d), e), f) It is bounded.

5036. a) $(0, 0, 3, 0, 2)$, $(0, 1, 3, 0, 0)$, $(0, 0, \frac{19}{5}, \frac{2}{5}, 0)$, $(\frac{3}{2}, 0, 0, 0, \frac{5}{2})$, $(\frac{17}{8}, 0, 0, \frac{5}{8}, 0)$, $(\frac{3}{2}, \frac{5}{4}, 0, 0, 0)$.

b) $(6, 4, 0)$, $(0, 12, 2)$.

c) $(0, 0, 9, 20)$, $(0, 3, 0, 14)$, $(4, 0, 13, 0)$.

5037. a) $z_{\max} = 9$.

b) $z_{\min} = -4$;

c) $z_{\max} = 90$;

d) $z_{\max} = 35/4$.

5101. If $a_0, a_1, \ldots, a_n$ are the required points then scalar products of vectors $\overline{a_0 a_1}, \ldots, \overline{a_0 a_m}$ can be expressed via distances between the given points; the Gram matrix made from them should be either positively definite (in the case a)) or non-negative definite (in the case b)) (see Exercise 4311).

5102. a) 4.

b)   3.

c)   2.

d)   A set does not exist.

5106. a) $(5, -4, 4, 0) + \langle(3, -4, 3, -1)\rangle$.

b) $(5, 0, 2, 11) + \langle(3, -1, 2, 5)\rangle$.

5107. a) 5.   b) 6.   c) 7.   d) $\sqrt{(581/27)}$.

5108. $\left|c - \sum_{i=1}^{n} a_i b_i\right| / \sqrt{\sum_{i=1}^{n} a_i^2}$.

5109. $2^{n+\frac{1}{2}} / \sqrt{2n+1} \binom{2n}{n}$. *Hint.* Make use of an orthogonal basis consisting of Legendre polynomials (see Exercise 4344.)

5110. $\pi/2^n$. *Hint.* Utilize the representation of $\cos^{n+1} x$ as a trigonometrical polynomial.

5111. $P \cap Q$ consists of one point.

5112. a) $-x_1 + 3x_2 + 2x_3 + x_4 = 6$, $x_1 + 2x_2 + 3x_3 - x_4 = 4$.

b) $(3, -2, 1, 4) + \langle(2, 3, -1, -2), (3, 2, -5, 1)\rangle$.

5114. a) 22/3.   b) 5.   c) 7.   d) 6.

5115. $d\sqrt{\dfrac{n+1}{2(k+1)(n-k)}}$.

5117. Pairs $\{P_1, P_2\}$ and $\{Q_1, Q_2\}$ are metric congruent, but they are not congruent to the pair $\{R_1, R_2\}$. All distances are equal to 36; cosines of angles for the first pairs are equal to $-3/5$ and $4/5$, for the third pair to $-1/\sqrt{5}$ and $2/\sqrt{5}$.

5118. a) $(2, -3, -4, 1, 0) + \langle (18, 0, -13, -1, 5) \rangle$.

b) $(5, 2, 2, -5, -6) + \langle (0, 3, -2, -2, 1), (1, 0, 1, -1, 0) \rangle$. *Hint.* Make use of Exercise 5113d.

5120. *Hint.* Make use of Exercise 4604.

5122. *Hint.* Prove that there exists an orthogonal operator which maps the unit basic vectors which are orthogonal to the sides of the first tetrahedron to unit basic vectors which are orthogonal to corresponding sides of the second tetrahedron.

5123. a) The rotation on $-\pi/2$ around the point $(1,3)$.

b) The rotation on $\pi/4$ around the point $(-1/\sqrt{2}, 1 + 1/\sqrt{2})$.

5124. a) The composition of the reflection with respect to the line with directing vector $a = (1, 1)$ passing through the point $(1/2,0)$ and of the parallel transfer on the vector $\frac{1}{2}a$.

b) The reflection with respect to the line with directing vector $(\sqrt{3}, 1)$ and passing through the point $(2,0)$.

5125. a) The rotation on $\pi/3$ around the axis with directing vector $a = (-2, -2, 1)$ passing through the point $(1, 2, 0)$.

b) The composition of the rotation on $\pi/2$ around the axis with directing vector $a = (-2, -2, 1)$ and passing through the point $(2, -1, -\frac{2}{3})$, and of the parallel transfer on the vector $2a$.

c) The composition of the rotation on $\pi - \arcsin(5/14)$ around the axis with directing vector $a = (1, 1, 1)$ passing through the point $(-1, 2, 1)$, and of the parallel transfer on the vector a.

5126. a) The composition of the rotation on $\pi/2$ around the axis with directing vector $(2, 2, -1)$ passing through the point $P = (0, 1, -1)$ and of the reflection with respect to the orthogonal plane passing through the point P.

b) The composition of the reflection with respect to the plane $x - 2y + z = 3$ and of parallel transfer on the vector $(3, 2, 1)$.

c) The composition of the rotation on $\arccos(1/3)$ around the axis with directing vector $(1, 0, -1)$ passing through the point $P = (1, -1, 0)$, and of the reflection with respect to the orthogonal plane passing through the point P.

d) The reflection with respect to the plane $3x - y - 2z + 7 = 0$.

5202. *Hint.* Transfer the origin to the point b utilizing the formula from Exercise 5201.

5203. *Hint.* Make use of Exercise 5202.

5204. *Hint.* If we introduce the extended column of coordinates

$$\tilde{X} = {}^t(x_1, \ldots, x_n, 1),$$

then $Q(a_0 + x) = {}^t\tilde{X} A_Q \tilde{X}$ and $\tilde{X} = \tilde{T}\tilde{X}'$.

5205. *Hint.* Utilize the Taylor expansion of the polynomial $Q(x_1, \ldots, x_n)$ at the point $(x_1^0, \ldots, x_n^0)$:

$$Q(x_1, \ldots, x_n) = Q(x_1^0, \ldots, x_n^0) + \sum_{i=1}^{n} \frac{\partial Q}{\partial x_i}(x_1^0, \ldots, x_n^0)(x_i - x_i^0)$$

$$+ \frac{1}{2} \sum_{i,j=1}^{n} \frac{\partial^2 Q}{\partial x_i \partial x_j}(x_1^0, \ldots, x_n^0)(x_i - x_i^0)(x_j - x_j^0).$$

5206. a) The point $-1/(n-1), \ldots, -1/(n-1)$.

b)   The hyperplane $x_1 + \cdots + x_n + 1 = 0$.

c)   If n is even then the center is the point $(x_1^0, \ldots, x_n^0)$ where

$$x_i^0 = \begin{cases} (-1)^{i/2} & \text{if } i \text{ is even,} \\ (-1)^{n+1-i/2} & \text{if } i \text{ is odd.} \end{cases}$$

If $n = 4k + 3$ then the center is the line

$$(0, -1, 0, 1, \ldots, -1, 0) + t(1, 0, -1, 0, \ldots, 0, -1);$$

if $n = 4k + 1$ then the center is empty.

d)   The center is empty.

5207. a) 9.   b) 17.

5208. a) $3n - 1$.

b) $n^2 + 3n - 1$.

5209. *Hint.* Apply Exercise 5205.

5210. *Hint.* Make use of Exercises 5201 and 5205.

5211. a) $x_1 + 2x_2 + 2x_3 + x_4 = 1$.

b) $x_1 + 2\sum_{i=2}^{n-1} x_i + x_n + 2 = 0$.

5212. *Hint.* Make use of Exercises 5203 and 5210.

5213. *Hint.* Make use of Exercise 5203.

5214. *Hint.* Make use of Exercise 5213.

5215. a) $(-1, 2, 3)$ and $(-2, -1, -4)$.

b)   The lines entirely lies in the quadric.

c)   The line is tangent to the quadric at the point $(-3, 0.0)$.

5216. $(x_1 + \sqrt{12})/2 = x_2 = -x_3$ and $(x_1 - \sqrt{12})/2 = x_2 = -x_3$. *Hint.* The required line can be defined either by equations $(x - a)/2 = y - b = -z$ on or by equations $x = a - 2z$, $y = b - z$. After substitution of these values of x and

y into the equation of the quadric we obtain an identity. Since all coefficients of the obtained equality are vanishing we can determine the values of parameters a and b.

5217. Two complex conjugate lines are:

$$t\left(1, i\sqrt{\frac{3}{2}}, -\frac{3}{2}\right) \quad \text{and} \quad t\left(1, -i\sqrt{\frac{3}{2}}, -\frac{3}{2}\right).$$

5218. a) $x_1^2 + 5x_2^2 + 4x_3^2 + 4x_1x_2 - 2x_2x_3 - 4x_1x_3 = 1$.

b) $x_1^2 + 2x_2^2 + x_3^2 - 4x_1x_2 + 6x_2x_3 - 2x_1x_3 + 20x_2 + 12x_3 + 12 = 0$.

5219. a) An ellipse.

b) A hyperbola.

c) A pair of intersecting lines.

d) An empty set.

5220. a) The affine type of the quadric is determined by the canonical equation $y_1^2 + y_2^2 + \cdots + y_n^2 + 2y_{n+1} = 0$, the metric type is determined by the equation $y_1^2 y_2^2 + \cdots + y_{n-1}^2 + (n+1)y_n^2 + 2y_{n+1} = 0$.

b) The affine type of the quadric is determined by the equation $y_1^2 - y_2^2 - y_3^2 - \cdots - y_n^2 = -1$, the metric type is determined by the equation $(n-1)y_1^2 - y_2^2 - y_3^2 - \cdots - y_n^2 = 1$.

5221. a) $\left(-1, \frac{3}{2}, 0\right)$, a hyperboloid of one sheet.

b) The line of centers is defined by the equations $x_1/3 = x_2/2 = (x_3 - 2)/1$, an elliptic cylinder.

c) The center is empty, an elliptic paraboloid.

d) $\left(\frac{14}{3}, 3, \frac{1}{3}\right)$, a hyperboloid of one sheet.

e) The pair of intersecting planes $(x_1 + x_2 + x_3 - 1)(x_1 + x_2 - x_3 + 1) = 0$.

f) The sphere $(x_1 - 1)^2 + \left(x_2 + \frac{2}{3}\right)^2 + x_3^2 = \frac{16}{9}$.

g) The circular cylinder $(x_1 - 1)^2 + \left(x_2 + \frac{2}{3}\right)^2 = \frac{16}{9}$.

h) The circular cone $(x_1 - 1)^2 + \left(x_2 + \frac{2}{3}\right)^2 - \left(x_3 - \frac{2}{3}\right)^2 = 0$.

i) The pair of parallel planes $(2x_1 - x_2 + 6)(2x_1 - x_2 - 6) = 0$.

j) The ellipsoid $y_1^2/49 + 4y_2^2/49 + 9y_3^2/49 = 1$, the center is $(3, -1, 2)$; the large, the average and the small axis are parallel to the axes Ox_1, Ox_2 and Ox_3, respectively.

k) The rotation hyperboloid of one sheet $y_1^2/4 - y_2^2/16 - y_3^2/16 = -1$; the center is $(-4, 0, -6)$, the axis of rotation is parallel to Ox_1.

l) The circular cone $y_1^2 - y_2^2/3 + y_3^2 = 0$; the top is $(3, 5, -2)$ and the axis of rotation is parallel to Ox_2.

m) The paraboloid of rotation, the top is $\left(10, -\frac{1}{2}, -\frac{5}{2}\right)$; the axis of rotation is parallel to Ox_1.

5222. a) The circular cone $-y_1^2 + y_2^2 + y_3^2 = 0$, directing vector of its axis is $\left(1/\sqrt{2}, 1/\sqrt{2}, 0\right)$.

b) The hyperbolic paraboloid $y_1^2 - y_2^2 = 2y_3$; the top is $(0,0,0)$, directing vectors in the canonical system of coordinates are: $e_1' = \left(1/\sqrt{2}, 1/\sqrt{2}, 0\right)$, $e_2' = \left(-1/\sqrt{2}, 1/\sqrt{2}, 0\right)$, $e_3' = (0, 0, 1)$.

c) The parabolic cylinder $x_3^2 = 5x_1$, directing vectors in the canonical system of coordinates are $e_1' = \left(\frac{3}{5}, \frac{4}{5}, 0\right)$, $e_2' = \left(-\frac{4}{5}, \frac{3}{5}, 0\right)$, $e_3' = (0, 0, 1)$.

d) The circular cone $-4y_1^2 + y_2^2 + y_3^2 = 0$, the directing vector of its axis is $-\left(2/\sqrt{5}, 1/\sqrt{5}, 0\right)$.

e) The hyperbolic cylinder $y_3^2 - 2y_1^2 = 1$, the directing vector of the axis of the hyperbola is $\left(1/\sqrt{2}, 1/\sqrt{2}, 0\right)$; the directing vector of elements of the cylinder are $\left(-1/\sqrt{2}, 1/\sqrt{2}, 0\right)$.

f) The circular cylinder $y_1^2 + y_3^2 = 4/25$; its axis passes through the point $(0, 0, -215)$ and has the directing vector $\left(-2/\sqrt{5}, 1/\sqrt{5}, 0\right)$.

g) The parabolic cylinder $y_1^2 = 5y_2$, the top of the parabola is $O' = \left(-1, -\frac{12}{25}, -\frac{16}{25}\right)$, directing vectors in the canonical system of coordinates are: $e_1' = \left(0, -\frac{3}{5}, -\frac{4}{5}\right)$ (the directing vector of the axis of the parabola is in the side of concavity), $e_2' = (1, 0, 0)$, $e_3' = \left(0, \frac{4}{5}, -\frac{3}{5}\right)$ (directing vector of elements of the cylinder).

h) The parabolic cylinder $y_3 = 2y_1^2$, the top of the parabola is $O' = (0, 0, 1)$, the directing vector of its axis in the side concavity is $(0,0,1)$, the directing vector of elements of the cylinder is $\left(-1/\sqrt{2}, 1/\sqrt{2}, 0\right)$.

i) The rotation hyperbola of one sheet $3y_1^2/2 + 3y_2^2/2 - 3y_3^2 = \frac{85}{4}$, the center is $O' = \left(\frac{14}{9}, -\frac{7}{18}, -\frac{14}{9}\right)$, the directing vector of the axis of rotation is $\left(\frac{2}{3}, \frac{1}{3}, -\frac{2}{3}\right)$.

j) The paraboloid of rotation $y_1^2 + y_2^2 = \frac{2}{3}y_3$, the top is $O' = (1, 0, -1)$, the directing vector of the axis of rotation is $\left(\frac{2}{3}, \frac{1}{3}, -\frac{2}{3}\right)$.

k) The rotation hyperbola of two sheets $2y_1^2 + 2y_2^2 - 4y_3^2 = -1$, the center is $O' = \left(-\frac{1}{2}, -\frac{1}{2}, -\frac{1}{2}\right)$, the directing vector of the axis of rotation is $\left(1/\sqrt{3}, 1/\sqrt{3}, 1/\sqrt{3}\right)$.

l) The ellipsoid of rotation $y_1^2 + y_2^2 + (y_3^2/4) = 1$, the center is $O' = (1, 1, 1)$, the directing vector of the axis of rotation is $\left(1/\sqrt{3}, 1/\sqrt{3}, 1/\sqrt{3}\right)$.

m) The rotation hyperboloid of two sheets $6y_1^2 + 6y_2^2 - 2y_3^2 = -1$, the center is $O' = \left(-\frac{1}{3}, \frac{2}{3}, \frac{2}{3}\right)$, the directing vector of the axis of rotation is $\left(1/\sqrt{2}, 0, 1/\sqrt{2}\right)$.

n) The parabolic cylinder $y_2^2 = \frac{4}{3}y_1$, $O' = (2, 1, -1)$, $e_1' = \left(\frac{2}{3}, \frac{2}{3}, \frac{1}{3}\right)$, $e_2' = \left(\frac{2}{3}, -\frac{1}{3}, -\frac{2}{3}\right)$, $e_3' = \left(\frac{1}{3}, -\frac{2}{3}, \frac{2}{3}\right)$.

o) The elliptic cylinder $(y_1^2/2) + y_2^2 = 1$, $O' = (0, 1, 0)$, $e_1' = \left(1/\sqrt{3}, 1/\sqrt{3}, -1/\sqrt{3}\right)$, $e_2' = \left(1/\sqrt{6}, -2/\sqrt{6}, -1/\sqrt{6}\right)$, $e_3' = \left(1/\sqrt{2}, 0, -1/\sqrt{2}\right)$.

p) The elliptic paraboloid $y_1^2 + (3y_2^2/2) = 2y_3$, $O' = (2, 2, 1)$, $e_1' = \left(1/\sqrt{2}, -1/\sqrt{2}, 0\right)$, $e_2' = \left(1/3\sqrt{2}, 1/3\sqrt{2}, -4/3\sqrt{2}\right)$, $e_3' = \left(\frac{2}{3}, \frac{2}{3}, \frac{1}{3}\right)$.

q) The hyperbolic paraboloid $y_1^2 - y_2^2 = 2y_3$, $O' = (0, 0, 1)$, $e_1' = \left(1/\sqrt{2}, -1/\sqrt{2}, 0\right)$, $e_2' = \left(1/3\sqrt{2}, 1/3\sqrt{2}, -4/3\sqrt{2}\right)$, $e_3' = \left(\frac{2}{3}, \frac{2}{3}, \frac{1}{3}\right)$.

r) The hyperbolic paraboloid $(y_1^2/2) + y_2^2 = 2y_3$, $O' = (1, 2, 3)$, $e_1' = \left(-\frac{2}{3}, \frac{1}{3}, \frac{2}{3}\right)$, $e_2' = \left(\frac{1}{3}, -\frac{2}{3}, \frac{2}{3}\right)$, $e_3' = \left(-\frac{2}{3}, -\frac{2}{3}, -\frac{1}{3}\right)$.

s) $-(y_1^2/9) - (y_2^2/9) - (y_3^2/9) + (y_4^2/3) = 1$; $O' = (0, 1, 2, 3)$, $e_1' = \left(1/\sqrt{2}, 1/\sqrt{2}, 0, 0\right)$, $e_2' = \left(1/\sqrt{6}, -1/\sqrt{6}, 2/\sqrt{6}, 0\right)$, $e_3' = \left(1/2\sqrt{3}, -1/2\sqrt{3}, -1/2\sqrt{3}, -3/2\sqrt{3}\right)$; $e_4' = \left(\frac{1}{2}, -\frac{1}{2}, -\frac{1}{2}, \frac{1}{2}\right)$.

t) $y_1^2 + y_2^2 + y_3^2 = 9$; $O' = (0, 0, 0, 0)$, $e_1' = \left(\frac{1}{2}, \frac{1}{2}, -\frac{1}{2}, -\frac{1}{2}\right)$, $e_2' = \left(\frac{1}{2}, -\frac{1}{2}, \frac{1}{2}, -\frac{1}{2}\right)$, $e_3' = \left(\frac{1}{2}, -\frac{1}{2}, -\frac{1}{2}, \frac{1}{2}\right)$; $e_4' = \left(\frac{1}{2}, \frac{1}{2}, \frac{1}{2}, \frac{1}{2}\right)$.

5223. If $-\frac{1}{2} < a < 1$.

5224. All appropriate coefficients of their equations, except possibly constant terms, are proportional.

5225. The canonical equation of the quadric in the space is $(a - b)y_1^2 + \cdots + (a - b)y_{n-1}^2 + (a + (n - 1)b)y_n^2 + 2cy_{n+1} = 0$.

5227. If $a = b$ (the plane of dimension $n - 1$).

5228. *Hint.* a) The required quadratic equation is of the form $\omega \wedge \omega = 0$ where $\omega = x_0 e_1 \wedge e_2 + x_1 e_1 \wedge e_3 + x_2 e_1 \wedge e_4 + x_3 e_2 \wedge e_3 + x_4 e_2 \wedge e_4 + x_5 e_3 \wedge e_4$.

b) Let $U \in \Lambda^r V$, $W \subset V$ be a minimal subspace such that U is contained in the image of the embedding $\Lambda^r W \to \Lambda^r V$. Consider the subspace $W' = \{\omega \in W : \omega \wedge U = 0\}$. It is clear that decomposability of U is equivalent to the equality $W = W'$. Since pairings $\Lambda^k V \otimes \Lambda^{n-k} V \to \Lambda^n V \cong K$ are nondegenerate, it is possible to define a pairing $\Lambda^{r-1} V^* \otimes \Lambda^r V \to V$. Let the image of $\Theta \otimes U$ in V be denoted by $i(\Theta)U$. Then W can be characterized as the image of the mapping $\Lambda^{r-1} V^* \to V$ defined by $\Theta \to i(\Theta)U$. The condition $W = W'$ is now equivalent to the condition $(i(\Theta)U) \wedge U = 0$ for all $\Theta \in \Lambda^{r-1} V^*$. This is the required system of quadratic equations. In particular, if $r = 2$ then $(i(v^*)U) \wedge U = \frac{1}{2} i(v^*)(U \wedge U)$ for all $v^* \in V^*$ and therefore the decomposability of U is equivalent to the validity of the equality $U \wedge U = 0$. If $n = 4$ then the condition $U \wedge U = 0$ gives a unique quadratic equation.

5301. a) $(x, y) \to \left(\dfrac{x}{x - y}, \dfrac{1}{x - y} \right)$.

b) $(x, y) \to \left(\dfrac{1 - y}{x}, \dfrac{x + y}{x} \right)$.

5302. a) $(x, y) \to \left(\dfrac{x}{1 - y}, \dfrac{1 + y}{1 - y} \right)$.

b) $(x, y) \to \left(\dfrac{1}{x}, \dfrac{y}{x} \right)$.

5303. a) $(x, y) \to \left(\dfrac{2x + 1}{x + 2}, \dfrac{y\sqrt{3}}{x + 2} \right)$.

b) $(x, y) \to \left(\dfrac{2x - 1}{x - 2}, \dfrac{y\sqrt{3}}{x - 2} \right)$.

5304. a) $(x, y, z) \to \left(\dfrac{1 + z}{y}, \dfrac{1 - z}{y}, \dfrac{x}{y} \right)$.

b) $(x, y, z) \to \left(\dfrac{x + y}{z + 1}, \dfrac{z - 1}{z + 1}, \dfrac{x - y}{z + 1} \right)$.

c) $(x, y, z) \rightarrow \left(\dfrac{z}{x}, \dfrac{y}{x}, \dfrac{1}{x} \right)$.

5305. a) $\min(k - 1, n - k)$.

b) $\min(k, n - k - 1)$.

5310. *Hint.* Consider complementations to affine charts.

5311. $q^n + q^{n+1} + \cdots + 1$.

5312. $\dfrac{(q^{n+1} - 1)(q^{n+1} - q) \ldots (q^{n+1} - q^k)}{(q^{k+1} - 1)(q^{k+1} - q) \ldots (q^{k+1} - q^k)}$.

5313. $\dfrac{(q^{n+1} - 1)(q^{n+1} - q) \ldots (q^{n+1} - q^n)}{q - 1}$.

5317. *Hint.* Consider both $P(V)$ and $P(V^*)$.

5320. *Hint.* Make use of the the previous exercise.

5322. a) $a_3 = \dfrac{(a_1 + a_2)a_2}{3a_1 - a_2}$.

b) $a_2 = \dfrac{a_1(l - a_1)}{l + a_1}$.

5325. *Hint.* Choose an affine chart in which the line is infinitely far.

5326. *Hint.* Choose an affine chart in which two pairs of opposite sides of hexagon are pairs of parallel lines.

5334. *Hint.* This line is obtained by application of a correlation corresponding to the given circle to the given point.

5401. a) It is not associative.

b) It is associative.

c) It is not associative.

d) It is not associative.

e) It is associative.

f) It is not associative.

g) It is associative.

5402. All elements of the form $e_a = \begin{pmatrix} 1 & a \\ 0 & 0 \end{pmatrix}$ are left neutral; there are no two-sided and right neutral elements. The only right invertible elements with respect to e_a are $\begin{pmatrix} x & y \\ 0 & 0 \end{pmatrix}$ if $x \neq 0$; the only left invertible elements are $\begin{pmatrix} x & ax \\ 0 & 0 \end{pmatrix}$ if $x \neq 0$.

5403. Each element is right neutral; each element is left invertible with respect to any neutral element x; the only right invertible element with respect to x is x if $|M| = 1$.

5404. It is a semigroup. There is no neutral elements if $|M| > 1$.

5405. a) 3.

b) It is not a group.

5406. *Hint.* Consider the mapping $A \to \bar{A}$.

5501. In a) all sets except $\mathbb{N}$ are groups. In c) all sets except $\mathbb{N}_0$ are groups. The sets d), e), f), g), h), i) if either $r = 1$ or $r = 0$, and k) if $\varphi_k = 2k\pi/n$ (assuming that $\varphi_1 < \varphi_2 < \cdots < \varphi_n$), are groups.

5502. It is isomorphic to the group i) if $r = 1$.

5504. The following mappings form a group: d), e), h), i), j), k).

5505. The following sets form a group: a), d), e) if $d = 1$, f), h), i), k), l), m), n), o), p), q) if $\lambda < 0$, r).

5510. a) and c) hold.

5513. *Hint.* Consider the element $(xy)^2$.

5514. a), e), f) are homomorphisms.

5515. For commutative groups.

5516. It is a homomorphism.

5517. $\{\mathbb{Z}, n\mathbb{Z}, \mathbf{UT}_2(\mathbb{Z})\}, \{\mathbb{Q}, \mathbf{UT}_2(\mathbb{Q})\}, \{\mathbb{R}, \mathbf{UT}_2(\mathbb{R})\}, \{\mathbb{C}, \mathbf{UT}_2(\mathbb{C})\}, \{\mathbb{Q}^*\}, \{\mathbb{R}^*\}, \{\mathbb{C}^*\}$.

5518. $[k] \to [2^k]$ and $[k] \to [3^k]$.

5519. *Hint.* See Exercise 5513 if $x^2 = e$ holds identically in the group; otherwise find non-commuting elements x and y for which $x^2 = y^3 = 1$.

5520. There is no other automorphisms.

5522. a) Isosceles nonregular triangle or a pair of points.

b) $[KB]\cap[LC]\cap[MA]$ where K, L, M are middles of sides of a regular triangle ABC.

c) A regular triangle.

d) A parallelogram or a rectangle.

5523. $\mathbf{D}_4$ is isomorphic to the group from Exercise 5504k; $\mathbf{Q}_8$ is isomorphic to the group from Exercise 5505d.

5526. a) $\mathbf{Z}_2$.   b) $\mathbf{Z}_{p-1}$.   c) $\mathbf{S}_3$.   d) $\mathbf{S}_3$.   e) $\mathbf{D}_4$.   f) $\mathbf{S}_4$.

5528. a) $\{e, (123), (132)\}$.

b), f) *Hint.* See Exercise 5502.

5530. *Hint.* Make use of Exercise 5529.

5531. *Hint.* Make use of Exercises 5523 and 5529.

5533. These groups are not pairwise isomorphic. *Hint.* Consider centers of the groups.

5601. *Hint.* b) If $A \cup B$ is a subgroup and $x \in A \setminus B$, $y \in B \setminus A$ then consider the element xy.

c) Consider $x \in (C \setminus A) \cap (C \setminus B)$.

5602. *Hint.* For any element a of a subsemigroup there exist distinct integers k and l such that $a^k = a^l$, hence $a \cdot a^{k-l-1} = a^{k-l} = e$ and therefore the element a is invertible in the subsemigroup; the statement is not valid for $\mathbb{N} \subset \mathbb{Z}$.

5603. a) 6.   b) 5.   c) 12.   d) 8.   e) 4.   f) 8.   g) 2.

5604. *Hint.* Consider the case when $E + pX$ has prime order.

5606. a) 2.   b) 4.   c) 20.   d) 0.

5607. *Hint.* b) Apply a).

c) Consider the permutations (123), (12) and (13).

5608. *Hint.* a) For coprime integers p and q there exist integers u and v such that $pu + qv = 1$.

b) follows from a).

c) Consider (12) and (123).

5609. *Hint.* Make use of the fact that the order of a cycle is equal to its length.

5611. $n/\mathrm{GCD}(n, k)$.

5613. $p^m = p^{m-1}$.

5614. *Hint.* a) See Exercise 5611.

b) See the hint to Exercise 5608.

c) Consider the least natural number s for which $a^s \in H$.

d) Make use of c). If d_1 and d_2 are distinct divisors of n then the corresponding subgroups have different orders.

5615. *Hint.* If $x^k = e$ and $x = a^l$ then $a^{kl} = e$, hence $kl : n$ and $l : \mathrm{GCD}(n, k)$; the element a^k has the order $n/\mathrm{GCD}(n, k)$ (see Exercise 5610) and therefore it satisfies the condition if $\mathrm{GCD}(n, l) = n/k$.

5617. *Hint.* Let $n = |G|$, $d = d(G)$ and m be the least common multiple of orders of elements G.

a) By the Lagrange theorem $d|n$ whence $x^d = 1$ and d is divisible by the order of any element of the group, i.e. $m|d$.

b) Let $d = p_1^{k_1} \ldots p_s^{k_s}$ be the prime decomposition; by a) G contains an element x whose order is equal to $p_1^{k_1}l$, where l and p_1 are coprime; then the order of x^l is equal to $p_1^{k_1}$; similarly we obtain elements $x_2, \ldots, x_s$; then the product

$x_1, \ldots, x_s$ (see Exercise 5608a) has order d. Statements b) and c) are invalid for S_3.

5618. $U_{p\infty}$.

5619. b) It is not valid: in the group G of bijections of a plane onto itself the composition of symmetries with respect to two parallel lines is a parallel transfer.

c)   The set of roots of all orders of 1; the set of diagonal matrices with roots of 1 at the principal diagonal.

5620. It is not valid: in $GL_2(\mathbb{R})$ the elements of order 2 do not form a subgroup (see the answer to Exercise 5619b).

5621. Z_{p^k} (p is a prime).

5622. a) *Hint.* Write out explicitly all the subgroups (see Exercise 5614d).

b) Z_{p^k} (p is a prime). *Hint.* Notice that the group is the union of its cyclic subgroups; if they form a chain then the group is cyclic, next make use of Exercise 5614d.

c) $Z_{p^n} U_{p\infty}$. *Hint.* Let p be the least order of elements of the group; then p is a prime since $p = kl$ implies that the subgroup $\langle x \rangle$ has an element of order k; $\langle x \rangle_p$ is the least nonidentical subgroup contained in all other subgroups, so orders of all elements are divisible by p and are in fact powers of p.

5623. $\bigcup_{n \in N} \langle \frac{1}{n!} \rangle$.

5624. $\cos \dfrac{2k\pi}{n} + i \sin \dfrac{2k\pi}{n} \to [k]$.

5625. a) $\cong$ b);   c) $\cong$ f);   d) $\cong$ e) $\cong$ g).

5626. *Hint.* a) If the group G has no elements of order 2 then $G = \{(x, x^{-1})$ $(x \neq e)\} \cup \{e\}$ and $|G|$ is odd.

b)   See Exercise 5513.

5627. *Hint.* If the order of a subgroup H is equal to n then $x^n = e$ for any $x \in H$, whence $H \subset U_n$, but U_n is cyclic, apply Exercise 5614c.

5628. *Hint.* See Exercise 5622c.

5629. *Hint.* b) Show that if finite abelian group contains at most one subgroup of any given order then it is cyclic; apply a).

5630. a) $E, S_3, \langle (ij) \rangle, \langle (123) \rangle$.

b) $E, D_4, \langle (13) \rangle, \langle (24) \rangle, \langle (12)(34) \rangle, \langle (13)(24) \rangle, \langle (14)(23) \rangle, \langle (1234) \rangle, V_4$.

c) $E, Q_8, \langle i \rangle, \langle j \rangle, \langle k \rangle$.

d) $E, A_4, \langle (12)(34) \rangle, \langle (13)(24) \rangle, \langle (14)(24) \rangle, V_4, \langle (123) \rangle, \langle (124) \rangle, \langle (134) \rangle,$ $\langle (234) \rangle$.

5631. a) $(ij) = (1i)(1j)(1i)$.

5632. a) $\mathbf{D}_4$.

b) $\mathbf{D}_2(\mathbb{R})$ if $a \neq b$; $\mathbf{SL}_2(\mathbb{R})$ if $a = b$.

c) $\langle g \rangle$.

5633. a) $\mathbf{D}_4$.

b) $\mathbf{S}_3$ as a subgroup of $\mathbf{S}_4$ consisting of all permutations fixing the element 4.

c) $\{e, (12), (34), (12)(34)\}$.

d) $\mathbf{S}_4$.

e) $\mathbf{A}_4$.

5638. *Hint.* Make use of Exercise 5637.

5701. a) Two orbits; the first consists only of one zero vector, the other consists of all nonzero vectors.

b) Each orbit consists of all vectors of the same length.

c) For any $I \subseteq \{1, 2, \ldots, n\}$ the corresponding orbit O_I consists of the vectors x in which the coordinate x_i is equal to 0 if and only if $i \in I$. There are 2^n different orbits.

d) There are $n + 1$ different orbits $O, O_1, \ldots, O_n$ where O consists only of zero vector and O_i, $i \geq 1$ consists of all vectors $x = \sum_{i=1}^{n} x_i e_i$ such that $x_i \neq 0$ and $x_j = 0$ for all $j > i$.

5702. a) G_a contains only the identical operator.

b) G_a consists of operators with matrices $A = (a_{ij})$ such that $\sum_{j=1}^{n} a_{ij} = 1$ for all $i = 1, 2, \ldots, n$.

5703. a) The group of orthogonal operators on the plane $\langle x \rangle^{\perp}$.

b) The group of rotations of the plane $\langle x \rangle^{\perp}$.

5704. a) The orbit of G is equal to X.

b) G_U consists of all matrices of the form

$$\begin{pmatrix} \overbrace{\begin{matrix} 1 & \cdots & 0 \\ \vdots & \ddots & \vdots \\ 0 & \cdots & 1 \\ 0 & \cdots & 0 \\ \vdots & \ddots & \vdots \\ 0 & \cdots & 0 \end{matrix}}^{k} & \begin{matrix} * & \cdots & * \\ \vdots & \ddots & \vdots \\ * & \cdots & * \\ * & \cdots & * \\ \vdots & \ddots & \vdots \\ * & \cdots & * \end{matrix} \end{pmatrix}$$

5705. c) G_f consists of all upper-triangular matrices in the base $e_1, \ldots, e_n$.

5709. The orbits are a) $\{1, 5, 4, 9\}$, $\{2, 8\}$, $\{3\}$, $\{6, 10, 7\}$. b) $\{1, 7, 2, 4\}$, $\{3, 6\}$, $\{5, 8, 9\}$, $\{10\}$.

5710. a) $\begin{pmatrix} \pm 1 & 0 \\ 0 & \pm 1 \end{pmatrix}$.

b)   *Hint.* Consider, for example, the mapping

$$\begin{pmatrix} 1 & 0 \\ 0 & 1 \end{pmatrix} \mapsto e, \qquad\qquad \begin{pmatrix} 1 & 0 \\ 0 & 1 \end{pmatrix} \mapsto (12)(34),$$

$$\begin{pmatrix} -1 & 0 \\ 0 & 1 \end{pmatrix} \mapsto (13)(24), \qquad\qquad \begin{pmatrix} -1 & 0 \\ 0 & 1 \end{pmatrix} \mapsto (14)(23),$$

and establish an isomorphism by enumerating the sides of a diamond.

c)   Two orbits: $\{A, C\}$ and $\{B, D\}$,

$$G_A = G_C = \left\{ \begin{pmatrix} 1 & 0 \\ 0 & 1 \end{pmatrix}, \begin{pmatrix} -1 & 0 \\ 0 & 1 \end{pmatrix} \right\};$$

$$G_B = G_D = \left\{ \begin{pmatrix} 1 & 0 \\ 0 & 1 \end{pmatrix}, \begin{pmatrix} 1 & 0 \\ 0 & -1 \end{pmatrix} \right\}.$$

5711. The group contains n different rotations of an n-gon around its center and n axial symmetries; $|\mathbf{D}_n| = 2n$.

5712. a) 24.

b)   12.

c)   60. *Hint.* All vertices of a regular polyhedron form one orbit with respect to the action of the group of rotations of the polyhedron. Thus the order of the stationary subgroup is equal to the number of edges outgoing from one vertex.

5713. *Hint.* a) Associate each rotation of a cube with a permutation of the set of its diagonals.

b)   Associate each rotation of a tetrahedron with a permutation of the set of its vertices.

c)   Associate each isometry of a tetrahedron with a permutation of the set of its vertices; the obtained mapping into $\mathbf{S}_4$ is injective because each affine transformation is uniquely determined by images of four points in the general position; deduce surjectivity from the fact that the image contains the subgroup $\mathbf{A}_4$ and some odd permutation.

5714. a) 4.   b) 5.

5715. a) The orbit of G is equal to Y.

b) $G_a = 1$.

5717. a) $\{az \mid |a| = 1\}$.

b) The orbit of the origin is the whole disc.

c) 1.

5719. By the assumption $m = hm_0$ for some $h \in G$. It follows that $gm = g(hm_0) = (gh)m_0 = (hg)m_0 = h(gm_0) = hm_0 = m$.

5720. *Hint.* a) Notice that $ag_1 H = ag_2 H \to g_1 H = g_2 H$; and $xH = a(a^{-1}xH)$ for each $x \in G$.

b) Check that $\sigma_{ab} = \sigma_a\sigma_b$.

c) Prove that conditions $gH = agH$ and $a \in gHg^{-1}$ are equivalent.

5721. a) Cosets $\{e\}, \{x\}, \{x^2\}, \{x^3\}$ are singletons, enumerate them by numbers 1, 2, 3, 4. Then $\sigma_x = (1234)$, $\sigma_{x^2} = (13)(24)$, $\sigma_{x^3} = (1432)$, σ_e is the identical permutation.

b) Let x be the given symmetry and y be a rotation of the square on $90°$. Then $G = H \cup yH \cup y^2H \cup y^3H$. Enumerate cosets in this order. Then σ_e is the identical permutation, $\sigma_y = (1234)$, $\sigma_{y^2} = (13)(24)$, $\sigma_{y^3} = (14)(23)$, $\sigma_x = (24)$, $\sigma_{xy} = (12)(34)$, $\sigma_{y^2x} = (13)$, $\sigma_{y^3x} = (14)(23)$. (*Hint.* Apply the relation $xy = y^{-1}x$ for calculations).

5723. a) The subgroup generated by the Klein group and by the cycle (12).
b) The set of all powers of the given permutation.

5724. a) The subgroup of diagonal matrices.

b) The whole group.

c) The set of matrices of the form $\begin{pmatrix} a+b & 2a \\ 3a & 4a+b \end{pmatrix}$ where $a, b \in \mathbb{R}$ and $b^2 + 5ab - 2a^2 \neq 0$.

d) The set of matrices of the form $\begin{pmatrix} a & b \\ 0 & a \end{pmatrix}$ where $a, b \in \mathbb{R}$, $a \neq 0$.

5725. a) The subgroup of all diagonal matrices.

b) The subgroup of all matrices of the form $\begin{pmatrix} A & 0 \\ 0 & B \end{pmatrix}$, where A and B are nonsingular matrices of sizes k and $n - k$ respectively.

5726. A_1 and A_3 are conjugate since they have the same Jordan canonical form, A_1 and A_2 are not conjugate since they have different Jordan canonical forms.

5727. a) C_{ij} as a group is generated by matrices $E + \lambda E_{pq}$ where $j \neq p \neq q \neq i$.

b) λE, $\lambda^n = 1$.

c) $E + {}^t ab$ where a, b are rows such that $b^t a = 0$. *Hint.* The last statement follows from c).

5728. a) $SO_2(\mathbb{R})$. b) $\pm E$, symmetries with respect to OX and to OY.

5730. a) $S_3 = \{e\} \cup \{(12), (13), (23)\} \cup \{(123), (132)\}$.

b) $A_4 = \{e\} \cup \{(12)(34), (13)(24), (14)(23)\} \cup \{(123), (134), (142), (243)\} \cup \{(132)(143), (124), (234)\}$.

c)   Symmetries with respect to lines joining midpoints of opposite sides of the square, rotations of the square on angles $\pm\pi/2$, central symmetry of the square, the identical mapping.

5731. a) The identical group.

b)   A group of order 2. *Hint.* Since all nonidentical elements of the group are conjugate the order n of the group is divisible by $n - 1$.

c)   The group is either isomorphic to the group of permutations S_3 or is the group of order 3. *Hint.* Each group has a class containing only the unit. Let n be the order of G and k, l be the numbers of conjugate elements in each of the other classes, $k \leq l$. Then n is divisible by k and l, $1 + k + l = n$. The only possible solutions are: 1) $n = 3, k = l = 1$, 2) $n = 4, k = 1, l = 2$ (this solution can be rejected since the groups of order 4 are abelian and they have 4 classes), 3) $n = 6, k = 2, l = 3$; in order to establish an isomorphism $G \cong S_3$ consider the action of G by conjugations (see Exercise 5722) on the class containing 3 elements.

5732. a) $\{(12)(34), (13)(24), (14)(23)\}$.

b) $\{(123), (132), (124), (142), (134), (143), (234), (243)\}$.

5733. *Hint.* Let $a = (i_1 \ldots i_k)(i_{k+1} \ldots i_l) \ldots$ be a decomposition of the permutation a into disjoint cycles. Write down the permutation $c = bab^{-1}$ in the form

$$\begin{pmatrix} i_1 & \cdots & i_k & i_{k+1} & \cdots & i_l & \cdots \\ j_1 & \cdots & j_k & j_{k+1} & \cdots & j_l & \cdots \end{pmatrix}.$$

Then $c = (j_1 \ldots j_k)(j_{k+1} \ldots j_l) \ldots$.

5734. a) 5.

b)   7.

c)   11.

d) $(n + 6)/2$ if n is even, and $(n + 3)/2$ if n is odd. *Hint.* In order to find the number of elements conjugate to the given one it suffices to find the order of its centralizer; notice that the rotation around the center on angle π maps an n-gon onto itself if n is even.

5735. *Hint.* Necessity follows from the equality of traces of conjugate matrices. In order to prove the sufficiency of the equality $\varphi_1 + \varphi_2 = 2\pi k$ take $\mathrm{diag}(-1, -1, 1)$ as a conjugating matrix for canonical forms.

5736. *Hint.* a) Conjugate subgroups have the same order.

b) $K = gHg^{-1}$ where $g = \mathrm{diag}(2, 1)$.

5737. a) $N(H) = \left\langle H \begin{pmatrix} 0 & 1 \\ 1 & 0 \end{pmatrix} \right\rangle.$

b) $N(H)$ consists of all non-singular matrices of size 2 such that $a_{21} = 0$;

c) $N(H)$ consists of 8 permutations written in answers to Exercise 5701b.

5738. a) Aut$G$ is the cyclic group of order 4, consisting of automorphisms of raising to powers $k = 1, 2, 3, 4$.

b) Aut$G$ is a group of order 2 in which the only nonidentical automorphism is the automorphism of raising to the fifth power.

5739. *Hint.* a) Each automorphism of the group S_3 is determined by its action on three elements of order 2.

b) Any permutation of nonunit elements of the group V_4 determines an automorphism.

5740. a) AutZ_9 is the cyclic group of order 6 generated by the automorphism of squaring.

b) The group is not cyclic because $|\mathrm{Aut}Z_9| = 4$ but the square of each automorphism is the identical mapping.

5741. $|\mathrm{AutAutAut}Z_9| = 1$. *Hint.* Make use of Exercises 5738 and 5740.

5742. and 5743. *Hint.* M.I. Kargapolov, Y.I. Merzlyakov. Basic Group Theory. Nauka, Moscow, 1982, ch. 2, § 5.3.

5744. Let $D_4 = \langle a, b \mid a^4 = b^2 = (ab)^2 = 1 \rangle$. Then $\mathrm{Aut}D_4 = \langle \varphi, \psi \rangle$ where $\varphi(a) = a, \varphi(b) = ba, \psi(a) = a^{-1}, \varphi(b) = b$. Thus $\varphi^4 = \psi^2 = (\varphi\psi) = 1$, i.e. $\mathrm{Aut}D_4 \simeq D_4$; $\mathrm{Int}D_4 = \langle \varphi^2, \psi \rangle$.

5745. Let $D_n = \langle a, b \mid a^n = b^2 = (ab)^2 = 1 \rangle$. Then $\mathrm{Aut}D_n = \langle \varphi, \psi_k, (k, n) = 1 \rangle$, where $\varphi(a) = a, \varphi(b) = ba, \varphi(a) = a^k, \psi(b) = b$, where $(k, n) = 1, 1 \leq k \leq n - 1$.

5801. *Hint.* b) Apply the theorem concerning the determinant of a product of matrices.

c) Apply the theorem concerning the parity of a product of permutations.

5803. a) A_3.

b) V_4.

c) V_4 and A_4. *Hint.* Note that the order of a subgroup divides the order of a group and that normal subgroup together with any element contains all its conjugates. Apply Exercises 5727 and 5730.

5804. *Hint.* For example, $K = \{(12)(34)\}$, $H = V_4$.

5805. *Hint.* $xyx^{-1}y^{-1} = x(yx^{-1}y^{-1}) = (xyx^{-1})y^{-1} \in A \cap B$.

5806. *Hint.* Let $c \in C$ and $G = H \cup Hx$ be coset partition of G. Then any element of C can be written either in the form hch^{-1} or in the form $hxcx^{-1}h^{-1}$ where $h \in H$.

5807. Five classes of conjugate elements consisting of 1, 15, 20, 12 and 12 elements. *Hint.* Make use of Exercises 5733 and 5806. The group A_5 consists of four classes of elements which are conjugate in S_5 and whose representatives are e, (12)(34), (124), (12345). The first and the second classes contain 1 and 15 elements respectively and therefore they are classes of conjugate elements in A_5. The third class does not split into two classes in A_5, for we can take $x = (45)$ (see the hint to Exercise 5806), but then $(45)(123)(45)^{-1} = (123)$. Finally the fourth class splits into two classes in A_5, for the number, 24, of its elements does not divide the order of A_5.

5808. *Hint.* According to Exercise 5807 if the order of a normal subgroup divides 60 then this order is a sum of numbers 1, 15, 20, 12, 12 with the coefficients 0 or 1, and one of the summands is equal to 1 since e belongs to every subgroup.

5809. *Hint.* Prove c) first. The center consists of $\pm E$. There are no other subgroups of order 2 and therefore all of them are normal (see Exercise 5802). Classes of conjugate elements are $\{E\}$, $\{-E\}$, $\{\pm I\}$, $\{\pm J\}$, $\{\pm K\}$.

5810. Subgroups D_k in D_n where k divides n and subgroups of rotations in D_n.

5812. λE.

5813. *Hint.* c) follows from Exercise 5604.

d) By c) the group G maps injectively under the natural homomorphism $SL_n(2)$ onto $SL_n(Z_3)$.

5814. *Hint.* If α_g is the automorphism $x \to gxg^{-1}$ then α_e is the identical automorphism, $(\alpha_g)^{-1} = \alpha_{g^{-1}}$, $\alpha_g\alpha_h = \alpha_{gh}$, and $(\varphi\alpha_g\varphi^{-1})(x) = \varphi(g\varphi^{-1}(x)g^{-1}) = \varphi(g)x\varphi(g^{-1}) = \alpha_{\varphi(g)}(x)$ for any $\varphi \in \text{Aut}G$.

5816. a) S_2 if $n = 2$ and $\{e\}$ if $n \neq 2$;

b) A_3 if $n = 3$ and $\{e\}$ if $n \neq 3$.

c) The center is identical if n is odd and contains the rotation on angle π if n is even.

5817. *Hint.* The element belongs to the center if and only if it is equal to all its conjugates. Therefore if the center is trivial then $p^n = 1 + p^{k_1} + \cdots + p^{k_i}$ $(k_i \geq 1)$

(the number of elements in any class of conjugate elements divides the order of the group). But then 1 is divisible by p.

5818. b) The center consists of the matrices of $E + bE_{13}$.

c) A class of conjugate elements containing a noncentral element $E + aE_{12} + bE_{13} + cE_{23}$ consists of the matrices $E + aE_{12} + xE_{13} + cE_{23}$ $(x \in \mathbf{Z}_p)$.

5819. a) $\{\lambda E\}$.

b) $\{\pm E\}$.

c) The whole group.

d) $\{E\}$.

e) $\{\pm E\}$.

f) $\{\alpha E \mid \alpha^n = 1\}$.

g) $\{E + \lambda E_{1n}\}$.

5822. *Hint.* The group H is isomorphic to a factor-group of G.

5823. A homomorphism is determined by the image of a generator a. Here are all possible images of this element: a) any element of the group; the number of homomorphisms is equal to n. b) $e, b^3, b^6, b^9, b^{12}, b^{15}$. c) e, b, b^2, b^3, b^4, b^5. d) e, b^5, b^{10}. e) e.

5824. *Hint.* Find the image of $a/2$ if $a \mapsto 1$.

5825. a) $\mathbf{Z}_n$. b) $\mathbf{Z}_4$. c) $\mathbf{Z}_3$. d) $\mathbf{Z}_2$.

5826. *Hint.* Construct a linear mapping F^n onto F^{n-k} with the kernel H.

5827. *Hint.* Consider mappings: a) $x \to \cos 2\pi x + i \sin 2\pi x$; b) $z \to \frac{z}{|z|}$; c) $z \to |z|$; d) $z \to z^n$; e) $z \to z^n$; f) $z \to \left(\frac{z}{|z|}\right)^n$; g) $z \to \frac{z}{|z|}$; h) $z \to |z|$.

5828. *Hint.* In order to prove the existence of an isomorphism $X/Y \cong \mathbf{Z}$ find a homomorphism from X onto $\mathbf{Z}$ with the kernel Y.

5829. *Hint.* Use the fact that each element $g \in G$ has a unique decomposition kh where $k \in K, h \in H$. Prove that the mapping $g \to k$ is a homomorphism $G \to K$.

5830. *Hint.* According to Exercise 5713 the group S_4 acts on a cube. Enumerate three pairs of opposite sides of the cube as 1, 2, 3. Thus we obtain the action of the group on the set $\{1, 2, 3\}$. Check that the kernel of this action is the subgroup V_4.

5831. *Hint.* Check that the intersection N of all subgroups of G which are conjugate in G to H is a normal subgroup of G. Show with the help of Exercise 5720 that the factor-group G/N is isomorphic to some subgroup of the group S_k.

5832. *Hint.* Let N be the normal subgroup of G constructed in the hint to Exercise 5831. Then $p!$ is divisible by $|G/N|$ and by $|G/N| \geq p$, for $N \subseteq H$. By assumption, p is the minimal prime divisor of $|G|$. Hence $|G/N|$ has no prime divisors less than p since $|G|$ is divisible by $|G/N|$. On the other hand in prime decomposition of $p!$ all prime divisors, except p, are less than p. Therefore $|G/N| = p$, i.e. indices and orders of subgroups N and H coincide. The inclusion $N \subseteq H$ implies $N = H$ and therefore the subgroup H is normal.

5833. *Hint.* Any linear operator acts on one-dimensional subspaces. Check that in two-dimensional space over $\mathbf{Z}_3$ there are four one-dimensional subspaces which can be permuted by the appropriate linear operator. Check finally that the kernel of the action coincides with the center of the group $\mathbf{GL}_2(\mathbf{Z}_3)$.

5834. *Hint.* A proper subgroup of order n contains all cosets of the form $(k/n) + \mathbf{Z}$, where k is any integer.

5835. *Hint.* Consider the mapping which associates with each $g \in G$ an automorphism $x \to gxg^{-1}$.

5836. *Hint.* If $G/Z = \langle a\mathbf{Z} \rangle$ then elements $x, y \in G$ can be factorized $x = a^k z_1$, $y = a^l z_2$, and therefore $xy = yx$.

5837. *Hint.* Make use of Exercises 5817 and 5836.

5838. *Hint.* Make use of Exercises 5835 and 5836.

5839. $p^2 + p - 1$ where p classes consist of one element, and the others consist of p elements. *Hint.* Deduce from Exercises 5817 and 5836 that the center $\mathbf{Z}$ has order p. The centralizer of any element $a \notin \mathbf{Z}$ has order p^2 since it contains $\mathbf{Z} \cup \{a\}$ and does not coincide with the whole group. The number of elements conjugate with a is equal to $p^3 : p^2 = p$.

5840. *Hint.* a) Check that the products $a_0 b_1 \ldots a_{n-1} b_{n-1} a_n$ of elements of maximal subgroups A and B form a subgroup C strictly containing A and B (hence, coinciding with G). Elements of $A \cap B$ commute with elements of C since A and B are commutative.

b)   Let H be some maximal subgroup of G: $H \neq \{e\}$ since G is not a cyclic group. Put $|H| = m$ and $|G| = n = lm$. Since H is a maximal subgroup and the group G is simple, it follows that the normalizer N of H in G coincides with H, i.e. there exist l different maximal subgroups conjugate with H. If we admit that pairwise intersections contain only $\bar{e}$ then their union contains $1 + l(m - 1)$ elements of G. Since $lm - l + 1 < n$ there exists an element which does not belong to any of them. Hence there exists a maximal subgroup K containing this element which is not conjugate with H. Let again $|K| = m_1$ and $n = l_1 m_1$. Then, assuming as above that $l + l_1$ maximal subgroups have $\{e\}$ as the pairwise intersection, we obtain $1 + l(m - 1) + l_1(m_2 - 1) \geq 1 + (n/2) + (n/2) > n$ elements of G.

c)   One of the maximal subgroups is noncommutative since otherwise, as is shown in a), b) the group G has a nontrivial center and therefore is not simple.

5841. *Hint.* See D. Gorenstein, Finite Groups, Harper and Row, 1968, ch. 2, § 8.

5842. *Hint.* Let G be a finite subgroup in $\mathbf{SL}(2, \mathbb{Q})$. Introduce in the space $\mathbb{R}^2$ a new scalar product $(x, y)_G = \sum_{g \in G}(gx, gy)$, where $(x, y) = x_1 y_1 + x_2 y_2$ for rows $x = (x_1, x_2)$ and $y = (y_1, y_2)$. Show that with respect to this scalar product each operator g is orthogonal. Therefore G consists of rotations and reflections. Deduce that $G \subseteq \mathbf{D}_n$ for some n. Since $\mathrm{tr}\, g \in \mathbb{Q}$ deduce from Exercise 413 that $n = 3, 4$ or 6.

5843. – 5847. *Hint.* See D.A. Suprunenko. Matrix groups, Nauka, Moscow, 1972, ch. 3.

5848. *Hint.* See O'Meara. A general isomorphism theory for linear groups. J. Algebra. 1977, 44(1): 93–142. See Dieudonné. La Géométrie des Groupes Classiques, Springer-Verlag, Berlin, 1971.

5901. a) $(q^n - 1)(q^n - q) \ldots (q^n - q^{n-1})$. *Hint.* If we have already chosen i first rows then we have $q^n - q^i$ possibilities for selection of the next $(i + 1)$st row: in fact the $(i + 1)$st row is an arbitrary row of length n which does not belong to the linear span of preceding rows. There exist altogether q^n rows of length n over a field with q elements, and q^i of them are linear combinations of i preceding (independent) rows.

b) $1/(1 - q)(q^n - 1)(q^n - q) \ldots (q^n - q^{n-1})$. *Hint.* The subgroup $\mathbf{SL}_n(q)$ is the kernel of a homomorphism $A \to \det A$ from $\mathbf{GL}_n(q)$ onto the multiplicative group of the field $\mathbf{Z}_q$ (containing $q - 1$ elements). By the homomorphism theorem $|\mathbf{GL}_n(q)/\mathbf{SL}(q)| = q - 1$; now apply a) and the Lagrange theorem.

5902. a) The groups are not isomorphic. *Hint.* Find the numbers of elements of second order in these groups.

b) The groups are not isomorphic. *Hint.* Notice that the matrix $2E$ lies in the center of $\mathbf{SL}_2(3)$, apply Exercise 5811a.

5903. a) 2-subgroups $\langle (12) \rangle$, $\langle (13) \rangle$, $\langle (23) \rangle$; 3-subgroup $\langle (123) \rangle$.

b) 2-subgroup $\mathbf{V}_4$; 3-subgroups $\langle (123) \rangle$, $\langle (124) \rangle$, $\langle (134) \rangle$, $\langle (234) \rangle$.

5904. a) The first and the second (see the answer to Exercise 5903a) Sylow 2-subgroups are conjugate by the permutation (23), the first and the third by (13).
b) The first and the second Sylow 3-subgroups are conjugate by the permutation $(12)(34)$, the first and the third by $(13)(24)$, the first and the fourth by $(23)(14)$.

5905. *Hint.* Enumerate vertices of a square and obtain an isomorphic representation of the group $\mathbf{D}_4$ by permutations: $\mathbf{D}_4 \simeq P \subset \mathbf{S}_4$. Since $|\mathbf{D}_4| = 8$ and $|\mathbf{S}_4| = 24 = 8 \cdot 3$, P is a Sylow 2-subgroup of $\mathbf{S}_4$. Other Sylow 2-subgroups of $\mathbf{S}_4$ are conjugate to P and therefore isomorphic.

5906. a) It is contained in the subgroup

$$\{e, (1324), (1423), (12)(34), (13)(24), (14)(23), (12), (34)\}.$$

b)   It is contained in the subgroup

$$\{e, (1234), (1432), (13)(24), (12)(34), (14)(23), (13), (24)\}.$$

c)   It is contained in each of the three Sylow 2-subgroups.

5907. *Hint.* These groups are nonisomorphic by Exercise 5902. If some non-abelian group G of order 8 has a noncentral subgroup of order 2 then by Exercises 5720 and 5905 $G \cong \mathbf{D}_4$. Otherwise let e and $-e$ be central elements of G (by Exercises 5817 and 5818 the center of G consists of two elements). Let $i, j \in G$ and $ij \neq ji$. Put $k = ij, i^{-1} = -i, j^{-1} = -j, k^{-1} = -k$. Check that the natural mapping from G onto the quaternion group is an isomorphism.

5908. *Hint.* Solving, in the group $\mathbf{SL}_2(3)$, the equation $X^2 = E$ we obtain only two solutions: $X = \pm E$. Similarly we find six elements of order 4, solving the equation $X^2 = -E$. There are no square roots of these elements, i.e. $\mathbf{SL}_2(3)$ has no elements of order 8. Thus we have 8 elements whose orders are powers of 2. $\mathbf{SL}_2(3)$ has only one Sylow 2-subgroup, since $|\mathbf{SL}_2(3)| = 24 = 8 \cdot 3$ by Exercise 5902. Hence, this subgroup is normal. It is non-abelian, since, for example, elements

$$\begin{pmatrix} 0 & 1 \\ -1 & 0 \end{pmatrix} \quad \text{and} \quad \begin{pmatrix} -1 & -1 \\ -1 & 1 \end{pmatrix}$$

have order 4 and do not commute. Now make use of Exercise 5907.

5909. a) 5. b) 10. c) 6.

5910. p^m, where $m = \left[\dfrac{n}{p}\right] + \left[\dfrac{n}{p^2}\right] + \left[\dfrac{n}{p^3}\right] + \dots$.

5911. $(p-2)!$. *Hint.* The number $p!$ is divisible by p but it is not divisible by p^2. Hence, each Sylow p-subgroup consists of powers of one cycle $(i_1 i_2 \dots i_p)$. The number of these cycles is equal to $(p-1)!$, and the number of different generators of a cyclic subgroup of order p is equal to $p-1$.

5912. *Hint.* Apply the theorem on conjugacy of Sylow subgroups.

5913. a) *Hint.* $|\mathbf{SL}_2(p)| = p(p-1)(p+1)$ (see Exercise 5902). Therefore, Sylow p-subgroup has order p.

b)   The normalizer consists of all matrices of the form $\begin{pmatrix} x & y \\ 0 & x^{-1} \end{pmatrix}$, where $x \neq 0$.

c)   Since the order of the normalizer is equal to $p(p-1)$ its index and the number of distinct Sylow p-subgroups is equal to $p+1$.

d)   *Hint.* Make use of Exercise 5901.

e)   The set of all matrices of the form $\begin{pmatrix} x & y \\ 0 & z \end{pmatrix}$, where $x, z \neq 0$.

f) $p+1$.

5914. *Hint.* Prove that the order of the subgroup and the maximal power of p dividing $|\mathbf{GL}_n(p)|$ are equal to $p^{n(n-1)/2}$ (see Exercise 5901).

5915. a) If p is odd then the Sylow p-subgroup is unique and consists of rotations of regular n-gon on angles $2\pi k/p^l, 0 \leq k < p^l$, where p^l is the maximal power of p dividing n. Let $n = 2^l \cdot m$, where m is odd. Then $\mathbf{D}_n$ contains m different Sylow 2-subgroups. Each of them can be obtained in the following way. Choose a regular 2^l-gon whose vertices are contained among vertices of the given n-gon (centers are the same). *Hint.* Consider all isometries mapping the 2^l-gon onto itself.

b) In case $p = 2$ conjugating elements are rotations on angles $2\pi k/m, 0 \leq k < m - 1$.

5916. *Hint.* Let $|G| = p^l \cdot m$, where m is not divisible by p, and $|\mathrm{Ker}\varphi| = p^s \cdot t$, where t is not divisible by p. Then $H \simeq G/\mathrm{Ker}\varphi$, and by the Lagrange theorem the order of the Sylow p-subgroup P of H is equal to p^{l-s}. On the other hand, $|P \cap \mathrm{Ker}\varphi| \leq p^s$, for $|\mathrm{Ker}\varphi|$ is divisible by $|P \cap \mathrm{Ker}\varphi|$. Hence $|\varphi(P)| = |P/P \cap \mathrm{Ker}\varphi| \geq p^{i-s}$.

5917. *Hint.* Clearly, $P \subseteq \varphi_A(P) \times \varphi_B(P)$, where φ_A and φ_B are homomorphisms of projections onto A and B, respectively. This inclusion is in fact an equality (compare orders $|P|$, $|\varphi_A(P)|$ and $|\varphi_B(P)|$).

5918. *Hint.* a) Let $|G| = p^l \cdot m$ and $|H| = p^s \cdot t$, where m, t are not divisible by p. Then the order of the p-subgroup PH/H of G/H is at most p^{t-s}. Hence the order of the kernel $P \cap H$ of the natural homomorphism $P \rightarrow PH/H$ is at least p^s.

b) Take as P and H for example, distinct Sylow 2-subgroups of $\mathbf{S}_3$ (see Exercise 5903).

5919. *Hint.* See Exercise 5831.

5920. *Hint.* Apply the theorem on the number of distinct Sylow p-subgroups. This number divides the order of the group and is congruent to 1 modulo p. Apply Exercises 5912 and 5805.

5921. Five Sylow 2-subgroups and one Sylow 5-subgroups (see the hint to Exercise 5920).

5922. *Hint.* a) Apply Exercise 5831 to Sylow 3-subgroup H.

b) If the Sylow 5-subgroup is not normal then, by the theorem on the number of Sylow subgroups, the group has 16 distinct 5-subgroups. Since their pairwise intersections are trivial the group has at most $80 - 16 \cdot 4 = 16$ elements whose orders are powers of 2. These elements form only one Sylow 2-subgroup, and therefore it is normal.

c) The solution is similar to b).

5923. *Hint.* a) See the hint to Exercise 5920.

b)   Consider all matrices of the form $\begin{pmatrix} a & b \\ 0 & 1 \end{pmatrix}$, where $b \in \mathbf{Z}_q$, and a belongs
to a subgroup of order p of the multiplicative group of the field $\mathbf{Z}_q$. (This
subgroup exists since $|q - 1|$ is divisible by p.)

5924. 49.

5925. *Hint.* Induction on order of the group.

5926. *Hint.* Induction on order of the group. Choose in G a normal subgroup
of index p.

6001. *Hint.* If $\mathbf{Z} = A \oplus B$, where $A \neq 0$, $B \neq 0$, and $m \in A$, $n \in B$, then
$mn \in A \cap B = \{0\}$. The similar argument can be applied to the group $\mathbb{Q}$.

6002. *Hint.* The groups S_3, A_4, S_4 have no normal subgroups whose intersection
is the unit subgroup. In $\mathbf{Q}_8$ every nontrivial subgroup contains -1; therefore the
listed groups have no direct decomposition.

6003. *Hint.* If $\langle a \rangle$ is an additive cyclic group of order $n = n_1 \cdot n_2$, where
$(n_1, n_2) = 1$, then $\langle a \rangle = \langle a^{n_1} \rangle + \langle a^{n_2} \rangle$ (these subgroups have orders n_2 and n_1,
respectively, hence their intersection is trivial).

6005. a) $\langle a \rangle_6 = \langle a^3 \rangle \times \langle a^2 \rangle$.

b) $\mathbf{Z}_{12} \simeq \mathbf{Z}_3 \oplus \mathbf{Z}_4$.

c) $\mathbf{Z}_{60} \simeq \mathbf{Z}_3 \oplus \mathbf{Z}_4 \oplus \mathbf{Z}_5$. *Hint.* Indicate generators in summands.

6006. *Hint.* It follows from the representation of complex numbers in the
trigonometrical form.

6007. *Hint.* An element of $\mathbf{Z}_{2^n}$ is invertible if and only if its class contains an
odd number, therefore the order of the multiplicative group of the ring $\mathbf{Z}_{2^n}$ is equal
to 2^{n-1}. The element $3 = 1 + 2 \pmod{2^n}$ has order 2^{n-2} and its cyclic subgroup
has a trivial intersection with the subgroup $\{\pm 1\}$; thus their product has order 2^{n-1},
i.e. it coincides with the whole group $\mathbf{Z}_{2^n}^*$.

6008. a) The product of orders of the factors.

b)   The least common multiple of orders of the components.

6009. *Hint.* Show by the previous exercise that $(A_1 + A_2 + \cdots + A_{i-1}) \cap A_i = \{0\}$
for any i.

6011. *Hint.* If $m = p_1^{k_1} \ldots p_r^{k_r}$ then the group has elements of orders
$p_1^{k_1}, \ldots, p_r^{k_r}$ (see, for example, Exercise 6003). Show by Exercises 6008, 6007
that the sum of these elements has order m. The group S_3 has elements of order 2
and 3, but it has no elements of order 6. Make use of Exercise 5608b.

6012. $\{\pm 1\} \times \langle 2 \rangle = \{\pm 1\} \times \langle -2 \rangle$.

6013. *Hint.* One of the summands coincides with A, while the other is generated
by a sum of a generator of the group $\mathbf{Z}$ and some element of A. Thus there exist
$|A|$ direct decompositions.

6014. Each class of $A \times B$ is a product of a class of A by a class of B.

6016. *Hint.* Take as C a subgroup generated by preimages of basic elements of A/B.

6017. *Hint.* $G = A \oplus \text{Ker} \pi$.

6018. *Hint.* The commutativity of B is essential since images of A_1 and A_2 commute under any homomorphism $\varphi : A_1 \times A_2 \to B$.

6020. a), b), c) $\mathbf{Z}_6$.

d) $\text{Hom}(A_1, B) \oplus \text{Hom}(A_2, B)$.

e) $\text{Hom}(A, B_1) \oplus \text{Hom}(A, B_2)$.

f) $\mathbf{Z}_d$, where $d = (m, n)$.

g) $\mathbf{Z}_n$.

h) $\{0\}$.

i) $\mathbf{Z}$.

6021. *Hint.* Associate $\varphi(1)$ with a homomorphism $\varphi : \mathbf{Z} \to A$.

6024. a) $\mathbf{Z}$.

b) $\mathbf{Z}_n$.

c) $\mathbf{Q}$. *Hint.* Show, that if $\varphi : \mathbf{Q} \to \mathbf{Q}$ is an endomorphism then $\varphi(r) = r\varphi(1)$.

6025. *Hint.* a) A map $x \to nx$ has a trivial kernel if and only if the group has no elements whose orders divide n. If $n = p_1^{k_1} \ldots p_r^{k_r}$ is a prime decomposition then prime components of the groups with respect to primes $p_1, \ldots, p_r$ are equal to 0.

b) Surjectivity of the map means that the equation $nx = g$ is solvable in the group for any g.

6026. *Hint.* Associate with the endomorphism φ a matrix in the same way as was done for linear operators.

6027. a) $\mathbf{Z}_2$.

b) $\mathbf{Q}^*$.

c) The unit group if $n = 1$; the cyclic group of order 2 if $n = 2$; $\mathbf{Z}_2 \times \mathbf{Z}_{2^{n-2}}$ if $n > 2$ (see Exercise 6006).

d) The group of all integer matrices with determinants ± 1. *Hint.* In all cases apply Exercises 6023 and 6024.

6028. *Hint.* a) $\langle a \rangle_{30} = \langle a_1 \rangle_2 \oplus \langle a_2 \rangle_{15}$, where $a_1 = 15a$, $a_2 = 2a$. Under any automorphism we have $\varphi(\langle a_1 \rangle) = \langle a_1 \rangle$, $\varphi(\langle A_2 \rangle) = \langle a_2 \rangle$, since a_1 and a_2 have coprime orders. Note also that $\langle a_1 \rangle$ has only identical automorphism.

b) Let $\mathbf{Z} = \langle a \rangle$, $\mathbf{Z}_2 = \langle b \rangle$; under any automorphism we have $\varphi(\mathbf{Z}_2) = \mathbf{Z}_2$ and $\varphi(b) = b$. Moreover $\varphi(a)$ can be equal to one of the elements

$a, -a, a + b, -a + b$. Check that the squares of all these automorphisms are identical.

6029. *Hint.* With the notation of the previous exercise we have $\varphi(a) = na + \varepsilon b$, $\varphi(b) = \delta b$, where $n \in \mathbb{Z}$, $\varepsilon, \delta = 0, 1$. Endomorphisms φ_1, φ_2, where $\varphi_1(a) = a$, $\varphi_1(b) = 0$, $\varphi_2(b) = 0$, $\varphi_2(a) = b$ do not commute.

6030. *Hint.* Any prime component is invariant under any endomorphism of a group; make use of Exercise 6020.

6031. *Hint.* The induction on the number of generators of a group. If the group is cyclic and it is equal to $\langle a \rangle$ with the operation of addition, U is its nonzero subgroup, and k is the least positive number such that $ka \in U$, then U is generated by ka. Suppose that $ma \in U$. Divide m with remainder by k: $m = qk + r$. Then $ra = ma - q(ka) \in U$, hence, $r = 0$ and $ma = q(ka)$. Assume that the statement is proved for a group with $n - 1$ generators, $G = \langle a_1, \ldots, a_{n-1} \rangle$ and $U \subseteq G$ is a subgroup. Consider elements $u = m_1 a_1 + \cdots + m_n a_n \in U$. If $m_n = 0$ for all $u \in U$ then $U \subseteq \langle a_1, \ldots, a_{n-1} \rangle$ and the proof follows from the inductive hypothesis. Otherwise, let m_n^0 be the least positive number such that there exists $u^0 \in U$ for which $u^0 = m_1^0 a_1 + \cdots + m_n^0 a_n$. Obviously any number m_n which occurs in the decomposition of some $u \in U$ is divisible by m_n^0, $m_n = qm_n^0$. Then $u - qu^0 \in U \cap \langle a_1, \ldots, a_{n-1} \rangle$. This subgroup, by hypothesis, is generated by $n - 1$ elements. Then U is generated by these elements and by u^0.

6032. *Hint.* a) If φ is a homomorphism from a group G onto itself which is not an automorphism, then $\mathrm{Ker}\,\varphi \subset \mathrm{Ker}\,\varphi^2 \subset \ldots$ is a strictly ascending chain of subgroups and its union can not be generated by finitely many elements because each of them belongs to a member of the chain with a finite index. It remains to apply the previous exercise.

b)   Consider differentiation.

6033. *Hint.* If free abelian groups of ranks m and n $(m \neq m)$ are isomorphic then the rank is not an invariant of a free abelian group. But its invariance can be proved in the same way as the main lemma on linear dependence. One can also apply the following argument: if G is a free abelian group of rank n then $|G/2G| = 2^n$.

6034. *Hint.* Make use of the uniqueness of decomposition of finitely generated abelian groups.

6035. *Hint.* Induction on the order of a group and the number m.

6036. *Hint.* Make use of the proof of the theorem on uniqueness of decomposition of finite abelian groups.

6037. *Hint.* Make use of the theorem on uniqueness of decomposition.

6040. a) It does.

b)   It does not.

c) It does.

6041. (3,27). *Hint.* Show that $\langle a \rangle_9 \oplus \langle b \rangle_{27} = \langle a \oplus 3b \rangle \oplus \langle b \rangle$.

6042. a) The groups are not isomorphic. *Hint.* The second group is cyclic and the first one is not.

b) The groups are isomorphic.

c) The groups are not isomorphic.

6043. a) 3. b) 4.

6046. *Hint.* Prove that a finite abelian group is not cyclic if it contains a subgroup of type (p, p) (see Exercise 6040). Notice that the equation $x^p = 1$ has at most p solutions in a field.

6047. *Hint.* Let $a_1, \ldots, a_n$ be a maximal independent system of elements. Consider an element $1 + a_1 \ldots a_n$ and deduce from here that the group F^* is finite.

6048. *Hint.* Apply Exercise 6046.

6050. *Hint.* If y_j $(j = 1, \ldots, n)$ is a basis then x_i $(i = 1, \ldots, n)$ can be expressed via this basis with integer matrix of coefficients B. Then $AB = E$ and $\det A = \pm 1$, where $A = (a_{ij})$.

6051. *Hint.* Make use of the proof of the main theorem on finitely generated abelian groups based on the reduction of integer matrices to the diagonal form by elementary row and column transformations.

6052. a) $\mathbf{Z}_2 \oplus \mathbf{Z}_2 \oplus \mathbf{Z}_3$.

b) $\mathbf{Z}_{31}$.

c) $\mathbf{Z}_2 \oplus \mathbf{Z}_3 \oplus \mathbf{Z}_3$.

d) $\mathbf{Z}_2 \oplus \mathbf{Z}_4$.

e) $\mathbf{Z}_4 \oplus \mathbf{Z}$.

f) $\mathbf{Z}_2 \oplus \mathbf{Z}_2 + \mathbf{Z}$.

g) $\mathbf{Z}_3$.

h) $\mathbf{Z} \oplus \mathbf{Z}$.

i) $\mathbf{Z}$.

j) $\{0\}$.

6053. 3.

6055. *Hint.* Taking into account Exercises 6030 and 6024 it remains to show that the endomorphism ring of a finite prime-power noncyclic group is noncommutative. Without loss of generality it is possible to consider only the group $\langle a \rangle_{p^k} \oplus \langle b \rangle_{p^l}$, $k \geq l$. By Exercise 6020 any endomorphism of this group has the form: $\varphi(a) = s_1 a + t_1 b$, $\varphi(b) = s_2 a + t_2 b$, where s_2 is divisible by p^{k-l}.

Then, for example, automorphisms φ, ψ such that $\varphi(a) = a, \varphi(b) = 0, \psi(a) = b$, $\psi(b) = 0$ do not commute.

6056. *Hint.* Prove that H is finitely generated. Choose a system of elements $e_1, \ldots, e_k$ in H, maximally independent over $\mathbb{R}$. Prove that H is generated by $e_1, \ldots, e_k$ and the finite set $H \cap D$, where $D = \{\sum x_i e_i \mid 0 \le x_i \le 1\}$.

6059. *Hint.* Apply Exercise 6056.

6060. *Hint.* A map $x \to nx$ is an automorphism of the cyclic group $\langle a \rangle$ (it has a trivial kernel), therefore $nx = a$ for some x.

6063. *Hint.* Clearly the group $\mathbb{Q}$ is divisible. If $\varepsilon^{p^k} = 1$ then there exists δ such that $\delta^p = \varepsilon$. If $q \neq p$ is a prime then $(q, p^k) = 1$. Apply Exercises 6060 and 6061.

6065. *Hint.* It follows from the assumption that the sum of subgroups A and B is direct; it is necessary to show that it is equal to G. Assume that there exists $g \notin A \oplus B$. The subgroup $\langle g \rangle$ has nonzero intersection with $A \oplus B$ for otherwise the sum $A \oplus B \oplus \langle g \rangle$ would be direct and instead of B it would be possible to take $B \oplus \langle g \rangle$, a contradiction with the maximality of B. Let $ng \in A \oplus B$. We can suppose that n is a prime (if this is not the case then instead of g we can take $(n/p)g$ for some $p|n$). Thus $ng = a + b, a \in A, b \in B$. Since A is divisible there exists an element a_1 such that $na_1 = a$. Thus $ng_1 = b$, where $g_1 = g - a$ does not belong to $A \oplus B$. By the choice of the subgroup B we have $A \cap \langle g_1, B \rangle \neq 0$. Thus some element $a' \in A$ can be expressed in the form $a' = kg_1 + b', b' \in B$, $0 < k < n$. Since $(k, n) = 1$ there exist u, v such that $ku + nv = 1$ and therefore $g_1 = kug_1 + nvg_1$. Since $ng_1 \in A \oplus B, kg_1 = a' - b' \in A \oplus B$, then $g_1 \in A \oplus B$, which is a contradiction.

6066. *Hint.* Let D be the sum of all divisible subgroups. It is not difficult to check that D is divisible. Let $a \in D$, then $a = a_1 + \cdots + a_k$, where $a_i \in A_i$ $(i = 1, \ldots, k)$ and A_i are divisible summands of D. If $na_i' = a_i, i = 1, \ldots, k$, then $n \left(\sum_{i=1}^k a_i' \right) = a$. By the previous exercise, the whole group has a direct decomposition $D \oplus B$. If B has a divisible subgroup then this subgroup is contained in D, a contradiction. Hence B has no divisible subgroups. The factor-group of the whole group by D is isomorphic to B.

6067. *Hint.* Make use of Exercise 6016.

6068. *Hint.* Make use of Exercise 6067.

6069. *Hint.* Apply Exercise 6067.

6101. *Hint.* a) Consider elements, conjugate with the transposition (12) with the help of a power of the given cycle.

b) Elements from A_n are products of even numbers of transpositions, and $(ij)(jk) = (ijk), (ij)(kl) = (ikj)(ikl)$. (See Exercises 315 and 316).

6102. *Hint.* Make use of reduction of matrices by elementary row transformations to the row-echelon form.

6103. *Hint.* A nonsingular matrix can be reduced to the diagonal form by elementary row transformations, i.e. by multiplication by corresponding elementary matrix from the left.

6105. *Hint.* See D. Gorenstein, Finite Groups, Harper and Row, 1968, § 44.

6107. a) $\{1, a\}$, $\{5, a\}$, $\{2, 3\}$, $\{4, 3\}$, where a is any element from $\mathbf{Z}_6$.

b)   Two different transpositions or a transposition and a threefold cycle.

c)   Any two elements of order 4 which are inverse to each other.

d)   A rotation σ of a square on the angle $\pm(\pi/2)$ and any axial symmetry τ; also τ and $\tau\sigma$.

e) $\{a, b\}$, $\{a, a + b\}$, $\{b, a + b\}$.

6110. *Hint.* If $g_1, \ldots, g_n$ is a finite system of generators, and $f_1, f_2, \ldots, f_k, \ldots$ is another system of generators, then elements $g_1, \ldots, g_n$ can be expressed via the second system. Each expression contains finitely many elements of the second systems, say, $f_1, \ldots, f_m$. Then $f_1, \ldots, f_m$ generate the group.

6111. The normal closure of A is generated as a subgroup by elements

$$B^i A B^{-i} = \begin{pmatrix} 1 & 2^i \\ 0 & 1 \end{pmatrix} \quad (i \in \mathbb{Z}).$$

Hence it is isomorphic to the additive group of rational numbers of the form $m/2^k$. This subgroup is not finitely generated.

6112. *Hint.* a) Induction on a number of all possible cancellations.

b)   The operation is correctly defined by a). The associative law obviously holds. The empty word is the unit. An inverse of a word $u = x_{i_1}^{\varepsilon_1} \ldots x_{i_n}^{\varepsilon_n}$ is equal to the word $x_{i_n}^{-\varepsilon_n} \ldots x_{i_1}^{-\varepsilon_1}$.

6113. *Hint.* Define the homomorphism φ as follows: if $u = x_{i_1}^{\varepsilon_1} \ldots x_{i_n}^{\varepsilon_n}$ then $\varphi(u) = g_{i_1}^{\varepsilon_1} \ldots g_{i_n}^{\varepsilon_n}$. This is the unique possible definition.

6114. *Hint.* Every reduced word can be written in the form $u = vwu^{-1}$, where the first and the last letters of w are not coprime. Then $u^n = vw^n v^{-1}$, where the length of w^n is equal to the length of w multiplied by n. In general $d(u^n) = d(u) + (n-1)d(w)$, and therefore $u^n \neq 1$ (the empty word).

6115. *Hint.* Assume that commuting elements u, v are reduced. Let $d(u) \leq d(v)$. 1) If in uv more than half of the word u can be canceled then we pass to the words u, uv. (The second one is shorter than v, and these words commute as well as u with v.) 2) If in vu more than half of the word u cancels then as above we consider the word u^{-1}, vu. 3) If more than half of the second factor cancels in the word vu^{-1} then we pass to u^{-1}, $u^{-1}v$. 4) If more than half of the first factor cancels in vu^{-1} then we pass to uvu^{-1}. 5) In the remaining case, $u = u_1u_2$, where $d(u_1) = d(u_2)$, $v = u_2^{-1}v'$, and the factors have no cancellations. It follows from $uv = vu$ that u, $v' = u_2^{-1}v'u_1u_2$. Since $v'u_1u_2$ has no more cancellations than u_1 we have $u_1 = u_2^{-1}$ and $u = 1$. 6) Applying transformations 1) – 4), we finally come to case 5). Considering the previous step we find a generating element g such that u and v can be expressed via g.

6116. *Hint.* In any commutator, and in any product of commutators, the sum of exponents of all occurrences of x_i is equal to 0 for any i. In a word u, let the sum of exponents in some x_i be equal to $k \neq 0$. According to Exercise 6113 we construct a homomorphism of a free group into $\mathbb{Z}$ such that $x_i \to 1$, $x_j \to 0$ $(j \neq i)$. Then u passes to $k \neq 0$ and, hence, does not belong to the commutant.

6117. Words with irreducible presentation uw_1u^{-1}, where w_1 is a cyclic permutation w.

6118. *Hint.* Let F be a free group with free generators $x_1, \ldots, x_n$, and let A be a free abelian group with a basis $a_1, \ldots, a_n$. If a homomorphism $F \to A$ extends a mapping $x_i \to a_1, \ldots, x_n \to a_n$ (see Exercise 6113) then its kernel is equal to the commutant.

6119. *Hint.* Apply Exercise 6116.

6120. *Hint.* A subgroup of index 2 is normal in any group. The exercise is reduced to determination of all possible homomorphisms of the free group onto the group $\langle a \rangle_2$. If x_1, x_2 are free generators of the free group, then according to Exercise 6113 it is necessary to find all possibilities for images of x_1 and x_2. The answer: $\varphi_1(x_1) = a$, $\varphi_1(x_2) = 1$, $\varphi_2(x_1) = a$, $\varphi_2(x_2) = a$, $\varphi_3(x_1) = 1$, $\varphi_3(x_2) = a$, i.e. there are three subgroups of index 2.

6122. *Hint.* Obviously, under any homomorphism of the group $F = (x_1, x_2)$ into $\mathbb{Z}_n \times \mathbb{Z}_n$ a commutant and x_1^n, x_2^n go to the unit. The factor-group by the subgroup N generated by the commutant and elements x_1^n, x_2^n is isomorphic to $\mathbb{Z}_n \times \mathbb{Z}_n$. Therefore N is the kernel of any surjective homomorphism $F \to \mathbb{Z}_n \times \mathbb{Z}_n$.

6123. a) 16.

b)   36. *Hint.* Apply Exercise 6113.

6125. *Hint.* According to Exercise 6113 construct a homomorphism φ of a free group F with free generators $x_1, \ldots, x_n$ into H such that $\varphi(x_i) = h_i$, $i = 1, 2, \ldots, n$. Under this homomorphism the least normal subgroup R, containing words $R_i(x_1, \ldots, x_n)$, $i \in I$ maps to the unit. If $N = \text{Ker}\varphi$ then $\text{Im } \varphi \simeq F/N \simeq (F/R)/(N/R)$.

6127. *Hint.* Prove that every element can be expressed in the form a^ib^j, $0 \leq i < 2, 0 \leq j < 7$.

6128. *Hint.* Deduce from the defining relations that the order of the group ≤ 8. Apply Exercise 6125.

6129. *Hint.* Deduce from the defining relations that the order of the group $\leq 2n$. Apply Exercise 6125.

6130. *Hint.* Deduce from the defining relations that the order of the group ≤ 8. Apply Exercise 6125.

6131. *Hint.* According to Exercise 6125, consider the homomorphism from this group onto the group of matrices mentioned for which

$$x_1 \to \begin{pmatrix} -1 & 0 \\ 0 & 1 \end{pmatrix}, \quad x_2 \to \begin{pmatrix} -1 & 1 \\ 0 & 1 \end{pmatrix}.$$

(The square of the second matrix is equal to E.) Utilize the fact that the subgroup generated by $x_1 x_2$ is normal.

6132. *Hint.* See the hint to Exercise 6131.

6133. *Hint.* See J. Milnor. Introduction to Algebraic K-theory, Princeton University Press, 1971, § 5.

6134. *Hint.* Each coset by H has the form $g^i H$, $i \in \mathbb{Z}$, and therefore any element of the group has the form $g^i h$, $h \in H$.

6135. *Hint.* Let $\langle h \rangle$ be the infinite cyclic subgroup generated by h; the factor-group G/H is the infinite cyclic group generated by gH. By the previous exercise $G = \langle g \rangle \langle h \rangle$. Since H is normal $ghg^{-1} \in H$ and the mapping $x \to gxg^{-1}$ ($x \in H$) is an automorphism of H. Therefore ghg^{-1} (and h) is a generator of H. Hence, ghg^{-1} is equal either to h or to h^{-1}. Hence one of the relations $ghg^{-1} = h$, $ghg^{-1} = h^{-1}$ holds in the group. In the first case the group is a free abelian group since it is generated by the elements x_1, x_2 with the defining relation $x_1 x_2 x_1^{-1} = x_2$. Consider the group with generators x_1, x_2 and the defining relation $x_1 x_2 x_1 = x_2^{-1}$. It follows from the defining relation that in this group the cyclic subgroup generated by x_2 is normal, and the factor-group by this subgroup is an infinite cyclic group (considering the homomorphism to $\mathbb{Z}$ such that $x_1 \to 1, x_2 \to 0$). The element x_2 has also infinite order. In fact, consider a homomorphism from the group into the group of matrices of the form $\begin{pmatrix} \pm 1 & n \\ 0 & 1 \end{pmatrix}$, $x_2 \to \begin{pmatrix} 1 & 1 \\ 0 & 1 \end{pmatrix}$ (see Exercise 6125).

6136. The least normal subgroup generated by x is isomorphic to the additive group of all numbers of the form $(m/2^k)$, $m, k \in \mathbb{Z}$. *Hint.* Consider a homomorphism into the group of matrices of size two under which $x_1 \to \begin{pmatrix} 2 & 0 \\ 0 & 1 \end{pmatrix}$, $x_2 \to \begin{pmatrix} 1 & 1 \\ 0 & 1 \end{pmatrix}$. (Compare with Exercise 6111).

6201. a) $\begin{pmatrix} \frac{1}{a} & 0 \\ 0 & a \end{pmatrix}$. b) $\begin{pmatrix} 1 & \frac{b}{c} + \frac{ay - bx}{cz} - \frac{y}{z} \\ 0 & 1 \end{pmatrix}$. c) $\begin{pmatrix} 1 & \lambda(\beta - \beta^{-1}) \\ 0 & 1 \end{pmatrix}$.

6202. *Hint.* a) $g[a, b]g^{-1} = [gag^{-1}, gbg^{-1}]$.

b) $[aG', bG'] = [a, b]G' = G'$.

c) If $[aN, bN] = N$ then $[a, b]N = N$ and $[a, b] \in N$.

6203. *Hint.* $\varphi([a, b]) = [\varphi(a), \varphi(b)]$.

6204. *Hint.* If $\varepsilon : G \to G/G'$ is the natural homomorphism and $\varphi : G/G' \to A$ is a homomorphism into an abelian group A then $\varphi\varepsilon : G \to A$ is also a homomorphism. This correspondence is bijective by Exercise 6202c and the fact ε is surjective.

6205. *Hint.* It follows from the theorem on determinants of products of matrices that $|ABA^{-1}B^{-1}| = 1$.

6206. *Hint.* It follows from the fact that $[(a_1, b_1), (a_2, b_2)] = ([a_1, a_2], [b_1, b_2])$.

6207. a) $\mathbf{A}_3$, 2.

b) $\{e, (12)(34), (13)(24), (14)(23)\}$, 3.

c) $\mathbf{A}_4$, 2. f) $\{\pm 1\}$, 4.

6208. a) $\mathbf{A}_4$. *Hint.* A commutator is an even permutation and by Exercise 6201c the commutant contains all threefold cycles; $\mathbf{A}_n$ is generated by threefold cycles (see Exercise 6101).

b) If an element $a \in \mathbf{D}_n$ is a rotation on angle $2\pi/n$ then $\mathbf{D}'_n = \langle a \rangle$, if n is odd, and $\mathbf{D}'_n = \langle a^2 \rangle$, if n is even.

6210. *Hint.* a) Induction with application of the previous exercise.

b) Induction with application of Exercise 6202.

6211. *Hint.* a) Follows from the fact that the commutant of a subgroup is contained in the commutant of a group.

b) Follows from Exercise 6203.

c) Induction with application of Exercise 6206.

d) Since $B^{(k)} = \langle e \rangle$ then $G^{(k)} \subseteq A$ and $G^{(k+l)} = \langle e \rangle$, where $A^{(l)} = \langle e \rangle$.

6212. *Hint.* See Exercises 6207 and 6208.

6214. *Hint.* Follows from Exercise 6213c, since the commutant of this group is contained in $\mathbf{UT}_n(K)$.

6215. *Hint.* If a group G has a series, mentioned in the exercise, then $G^{(l)} = \langle e \rangle$ by Exercise 6202c. If a group is solvable then factors of its series of commutants $G^{(i)}/G^{(i+1)}$ are abelian, therefore it is possible to insert a chain of subgroups between $G^{(i)}$ and $G^{(i+1)}$ and obtain a series with the required properties.

6216. *Hint.* According to Exercise 5817 the center of a finite p-group G is non-trivial. Let A be a subgroup of order p lying in the center. Then A is normal in G. Complete the proof by induction on the order of the group. Pass to G/A (which is a p-group) and make use of Exercise 6212.

6217. *Hint.* If $q > p$ then Sylow q-subgroup is normal in a group (see the hint to Exercise 5920).

6218. *Hint.* a) Sylow 5-subgroup is normal since the index of its normalizer is a divisor of 4 and it is congruent to 1 modulo 5.

b) If Sylow 3-subgroups in a group of order 12 are not normal then the number of these subgroups is at least 8. But by Sylow's theorem there exists a subgroup of order 4 and it is unique.

c) If $p > q$ then the number m of subgroups of order p^2 is congruent to 1 modulo p only if $m = 1$. If $p < q$ then the number of q-subgroups is congruent to 1 modulo q and divides either p or p^2. This number cannot divide p. Hence, it is equal to p^2. Thus the number of elements of order q is equal to $p^2(q-1)$. However, a subgroup of order p^2 exists, hence it is unique ($p^2q = p^2(q-1) + p^2$).

d) The Sylow 7-subgroup is normal.

e) The Sylow 5-subgroup is normal.

f) Apply the argument of Exercises 6216, 6218c and the fact that if k is an index of the normalizer of a Sylow subgroup then the group can be represented by permutations on sets of Sylow subgroups, i.e. on a set of k symbols.

6220. *Hint.* Make use of Exercise 6219.

6221. *Hint.* Make use of Exercise 6201c.

6226. *Hint.* See J.E. Humphreys Linear Algebraic Groups, Springer-Verlag, New York, 1975, ch. 17, § 17.6.

6227. *Hint.* a) Since the order $q - 1$ of the multiplicative group $\mathbf{Z}_q$ is divisible by p, then there exist $p - 1$ integers r (see Exercise 6046).

b) The group, consisting of matrices $\begin{pmatrix} r^i & x \\ 0 & 1 \end{pmatrix}$, where r is a number from a) considered modulo q, $x \in \mathbf{Z}_q$ $(0 \le i < p)$, is noncommutative. It suffices to consider matrices $\begin{pmatrix} r & 0 \\ 0 & 1 \end{pmatrix}$ and $\begin{pmatrix} 171 \\ 0 & 1 \end{pmatrix}$. This group has order pq. Let G be a non-abelian group of order pq and let $A = \langle a \rangle$ be its Sylow subgroup of order q. Let $B = \langle b \rangle$ be the Sylow subgroup of order p of G. Then by Sylow's theorem (see also Exercise 6217) the subgroup A is normal in G. Therefore $bab^{-1} = a^{s^i}$. In particular, $b^p ab^{-p} = a = a^{s^p}$; therefore $s^p \equiv 1 \pmod{q}$, since G is non-abelian. Replacing the element b by its kth power ($1 < k < p$), we may replace s by any number with similar properties. Therefore if G_1 and G_2 are two non-abelian groups of order pq, then it is possible to choose elements a_i, b_i $(i = 1, 2)$ similar to a and to b and having properties: $a_i^q = e$,

$b_i^q = e$, $b_i a_i b_i^{-1} = a_i^r$, where $r^p \equiv 1 \pmod q$. An isomorphism between these groups is established by the correspondence $\varphi(a_1^s b_1^t) = a_2^s b_2^t$, where $0 \le s < q, 0 \le t < p$.

6228. *Hint.* b) The product of these permutations in the given order is a cycle of length 7. According to a) the factor-group of this group by the commutant is trivial, therefore the group coincides with its commutant.

c) The group has a homomorphic mapping onto the group from b) and by Exercise 6125 it is not solvable.

6229. It is not solvable if a system of free generators contains more than one element, for in this case every nontrivial normal subgroup is not abelian. *Hint.* See also Exercise 6211b.

6301. The following sets form a ring: a), b), d), f), g), h), j), k), l), m), n) if $D \equiv 1 \pmod 4$.

6302. The following sets form a ring: c), d), e), f), g) if $D \equiv 1 \pmod 4$, h), i) *Hint.* Make use the fact that $\sqrt[3]{2}$ is not a root of a quadratic polynomial over $\mathbb{Q}$.

6303. All sets form a ring except h).

6304. It does not form a ring.

6305. *Hint.* See Exercise 102.

6307. 6302c; 6304d if $n > 2$; 6302e if $D = c^2$ ($c \in \mathbb{Z}$); 6302f if $D = c^2$ ($c \in K$); 6303a; 6303b; 6303e if $|R \setminus D| > 1$; 6303i; 6305.

6310. *Hint.* Notice that $(xy)^{-1} = y^{-1}x^{-1}$.

6311. a) $\mathbb{Z}_n^*$ consists of all classes $[k]$ such that k and n are coprime; zero divisors are all classes $[k]$ such that k and n have a nontrivial common divisor; nilpotent elements are all classes $[k]$ such that each prime divisor of n divides k.

b) $\mathbb{Z}_{p^n}^*$ consists of all classes $[k]$ such that k is not divisible by p; zero divisors are all classes of the form $[pm]$; every zero divisor is nilpotent.

c) Similar to a) where instead of n one takes the polynomials f.

d) The set of matrices (α_{ij}) such that $\alpha_{ii} \ne 0$ ($i = 1, \ldots, n$); $\alpha_{ii} = 0$ for at least one i; $\alpha_{ii} = 0$.

e) The set of matrices A such that $\det A \ne 0$, $\det A = 0$, $\operatorname{tr} A = 0$, respectively.

f) The set of functions which do not vanish at any point; the set of functions vanishing at some point; the zero function.

g) Invertible elements are the series with nonzero constant term; there are no nontrivial zero divisors and nilpotent.

6312. *Hint.* a) A mapping $x \to ax$ ($a \in R$, $a \ne 0$) is a bijection and therefore $ax = a$ for some $x \in R$; each element $b \in R$ has a factorization $b = ya$ and therefore $bx = b$, i.e. x is a left unit.

b) An element is not a right zero divisor if it is right invertible, thus $x \to xa$ is a bijection.

c) If $ab = 0$ and a is not a right zero divisor then elements $x_1 a, \ldots, x_n a$ are distinct and one of them is equal to 1. The statement c) is not valid for the algebra over $\mathbf{Z}_2$ with a basis (x, y) and multiplication table $xy = y^2 = 0$, $yx = y$, $x^2 = x$. The statement b) is not valid for the infinite-dimensional algebra over $\mathbf{Z}$ with a basis $(y^k x^l \mid k, l \in \mathbf{N})$ (elements x and y do not commute) and with multiplication

$$y^k x^l \cdot y^r x^s = \begin{cases} y^k x^{l-r+s} & \text{for } l > r, \\ y^k x^s & \text{for } l = r, \\ y^{k+r-l} x^s & \text{for } l < r. \end{cases}$$

6313. *Hint.* If $ab = 1$ then $(ba - 1)b = 0$.

6314. *Hint.* b) See the answer to Exercise 6313. c) See the answer to Exercise 6312.

6316. *Hint.* a) A mapping $[x]_k \to ([x]_k, [x]_l)$ is an isomorphism.

c) A pair $([x], [y])$ is invertible in $\mathbf{Z}_k \times \mathbf{Z}_l$ if and only if $[x]$ is invertible in $\mathbf{Z}_k$ and $[y]$ is invertible in $\mathbf{Z}_l$; $\varphi(n)$ is the number of generators in $\mathbf{Z}_n$.

6318. *Hint.* b), c) Consider the linear mapping $\varphi_a : A \to A$ defined by the formula $\varphi_a(x) = ax$.

6319. *Hint.* There exists an annihilator polynomial for each element of the algebra.

6320. a) $\mathbb{C} \oplus \mathbb{C}$, $\mathbb{C}[x]/\langle x^2 \rangle$.

b) Algebras from a) and the following three algebras: $\mathbb{C}e \oplus \mathbb{C}e$, where $e^2 = 0$; $\mathbb{C}e \oplus \mathbb{C}f$, where $e^2 = ef = fe = 0$, $f^2 = e$; $\mathbb{C}e \oplus \mathbb{C}f$, where $e^2 = 0$, $f^2 = f$.

6321. a) $\mathbb{R} \oplus \mathbb{R}$, $\mathbb{C}$, $\mathbb{R}[x]/\langle x^2 \rangle$.

b) Algebras from a) and the algebras: $\mathbb{R}e \oplus \mathbb{R}e$, where $e^2 = 0$; $\mathbb{R}e \oplus \mathbb{R}f$, where $e^2 = 0$, $f^2 = f$, and the vector space $\mathbb{R}e \oplus \mathbb{R}f$, where $e^2 = ef = fe = 0$, $f^2 = e$.

6322. a) It is not an algebra.

d) All quaternions $x_1 i + x_2 j + x_3 k$, such that $x_1^2 + x_2^2 + x_3^2 = 1$.

6323. *Hint.* Make use of a basis in $T(V)$ constructed with the help of V.

6324. *Hint.* b) Make use of a basis in $\Lambda^k(V)$ constructed via a basis of V.

c) If x is a nilpotent element of a ring then $\alpha + x$ is invertible if $\alpha \neq 0$.

6327. *Hint.* Apply the operators $p_1^{l_1} \ldots p_n^{l_n} q_1^{t_1} \ldots q_n^{t_n}$ to monomials $x_1^{m_1} \ldots x_n^{m_n}$.

6330. *Hint.* b) Zeros of a continuous function form a closed subset. If $fg = 0$ then the union of zeros of f and g is equal to $[0, 1]$.

6401. *Hint.* Apply division with remainder. a) $n\mathbb{Z}$.

b) $f(x)K[x]$.

6402. *Hint.* a) Consider the ideal $(2, x)$.

b) Consider the ideal (x, y).

6403. *Hint.* a) If a nonzero matrix X belongs to an ideal I then $AXB \in I$ has the form $E_{11} + \cdots + E_{rr} \in I$ for some matrices A, B. Hence $AXBE_{11} = E_{11} \in I$, and therefore $E = E_{11} + \cdots + E_{nn} \in I$.

6405. Each ideal consists of all matrices of the form $\begin{pmatrix} a_1 & a_2 \\ 0 & a_3 \end{pmatrix}$, such that entries a_k form ideals in $\mathbb{Z}$, I_k $(k = 1, 2, 3)$, where $I_1 \subseteq I_2$ and $I_3 \subseteq I_2$.

6407. 0; the whole algebra; all matrices with zero first (second) column; all matrices with the same columns.

6408. a) 0, L and a subalgebra $\langle e \rangle$.

b) $0, L\langle 1+e \rangle$ and $\langle 1-e \rangle$. Every ideal, different from 0 and L is a one-dimensional subspace of L.

6412. a) $\langle p \rangle$, where p is a prime.

b) $\langle p(x) \rangle$, where $p(x)$ is a polynomial of degree one.

c) $\langle p(x) \rangle$, where $p(x)$ is either a polynomial of degree one or a polynomial of degree two which has no real roots.

6413. It is not the case.

6414. *Hint.* b) Suppose that for each point $a \in [0, 1]$ there exists a function f_a in the ideal such that $f_a(a) \neq 0$. Since the function $f_a^2(x)$ is continuous then $f_a^2(x)$ is positive on some neighborhood $(a - \varepsilon_a, a + \varepsilon_a)$ of the point a (and it is non-negative at all other points). From every cover of the segment $[0, 1]$ we can choose a finite cover. Hence there exist finitely many functions $f_1, \ldots, f_k$ in the ideal such that $f_1^2(x) + \cdots + f_k^2(x) > 0$ for all x.

6415. *Hint.* Consider the ideal generated by $a \neq 0$. A ring with zero multiplication whose additive group is a cyclic simple group has no nontrivial ideals, but it is not a field.

6416. *Hint.* Prove that total right zero divisors (i.e. elements $a \in R$ for which $Ra = 0$) form left ideal and therefore they are equal to zero. If $ba \neq 0$ then $Ra = R$. Deduce from here that R has no zero divisors and that nonzero elements of the ring form a multiplicative group.

6417. *Hint.* Let $R \ni a \neq 0$. Then $Ra \supseteq Ra^2 \supseteq \ldots$ and $Ra^k = Ra^{k+1}$ for some k. Hence $a^k = ba^{k+1}$ and $1 = ba$.

6418. *Hint.* Put $\delta_1(a) = \min \delta_1(ax)$, where $x \in K \setminus \{0\}$.

6419. *Hint.* a) Consider the norm $\delta(x + iy) = x^2 + y^2$.

b) Elements 2 and $1 \pm \sqrt{3}$ are prime, and $4 = 2 \cdot 2 = (1 + i\sqrt{3})(1 - i\sqrt{3})$ are two nonassociated prime decompositions.

c) Consider the norm $\delta(x + iy) = x^2 + y^2$.

6420. *Hint.* See Exercise 819.

6422. *Hint.* Let $R \subseteq A \subseteq Q$ and let I be an ideal of A. Prove that $I = \langle r_0 \rangle$, where r_0 generates the ideal consisting of all numerators of elements of I.

6423. *Hint.* Let $R[x]$ be principal ideal ring. For $0 \neq a \in R$ consider an ideal $I = \langle x, a \rangle$ of $R[x]$. Since $a \in R$, $I = \langle f_0 \rangle$, where f_0 is a constant, i.e. $I = R[x]$. It follows that $1 = u(x)x + v(x)a$ and $v(0) = 1$, hence R is a field. Notice that $F[x, y] \cong F[x][y]$.

6424. $(x^n), n \geq 0$.

6426. *Hint.* a) Present the unit element in the form $1 = a_1 + a_2$, where $a_1 \in I_1$, $a_2 \in I_2$.

b) By induction reduce to the case $n = 2$. For each $i \geq 2$ it is possible to find elements $a_i \in I_i$ and $b_i \in I_i$ such that $1 = a_i + b_i$. Then $1 = \prod_{i=1}^{n}(a_i + b_i) \in I_1 + \prod_{i=2}^{n} I_k$. Hence, $I_1 + \prod_{i=2}^{n} I_i = A$, and according to a) it is possible to find $y_1 \equiv 1 \pmod{I_1}$ and $y_1 \equiv 0 \pmod{\prod_{i=2}^{n} I_i}$. Similarly find $y_2, \ldots, y_n \in A$ such that $y_j \equiv 1 \pmod{I_j}$ and $y_j \equiv 0 \pmod{I_i}$ if $i \neq j$. Then the element $x = x_1 y_1 + \cdots + x_n y_n$ satisfies the requirements of the exercise.

6427. a) No, it is not the case.

b) Yes, it is valid.

6429. a) *Hint.* Make use of Exercise 6311c.

6435. a) $n \to 0$.

b) $n \to i; n \to 0$.

c) $n \to 0$.

d) Any homomorphism has the form $n \to n e_i$, where e_i is an idempotent of the matrix ring; 8 homomorphisms corresponding to the idempotents O, E, E_{11}, $E_{22}, E_{11} + E_{12}, E_{21} + E_{22}, E_{11} + E_{21}, E_{12} + E_{22}$.

6436. a) $n \to na$, where a is an arbitrary fixed element of $\mathbb{Q}$.

b) $n \to 0, n \to n$.

6437. *Hint.* Prove that the kernel of a homomorphism either is equal to zero or coincides with the field.

6439. *Hint.* Consider the homomorphisms a) $f(x) \to f(\alpha)$; b) $f(x) \to f(l)$; c) $f(x) \to f\left(\dfrac{-1+i\sqrt{3}}{2}\right)$.

6440. A field is obtained if $f_1(x) = x^2 + x + 1$, isomorphic factor-rings if $f_1(x) = x^2$ and $f_2(x) = x^2 + 1$. *Hint.* Consider multiplication tables for the factor-rings.

6441. The factor-rings are not isomorphic. *Hint.* The first factor-ring has a nonzero element whose cube is equal to zero; the second factor-ring has no elements with this property.

6442. They are not isomorphic.

6443. *Hint.* Under multiplication by $x - a \in F[x]$ any element of the first module vanishes; the second module does not have this property. Both factor-rings are isomorphic to F.

6444. *Hint.* Let $\langle (x - a)(x - b) \rangle = I_1$, $\langle (x - c)(x - d) \rangle = I_2$. Write down an arbitrary element of $F[x]/I_1$ in the form $\alpha(x-a)+\beta(x-b)+I_1$ and associate with it the element $k\alpha(x - c) + k\beta(x - d) + I_2 \in F[x]/I_2$, where $k = (a - b)/(c - d)$.

6445. A_1 and A_3, A_2 and A_5.

6446. a) The algebras are isomorphic.

b) The algebras are not isomorphic.

6447. a) The algebras are isomorphic.

b) The algebras are not isomorphic.

6448. *Hint.* Find the inverse of f by the method of indeterminate coefficients.

6450. *Hint.* Similar to Exercise 6316.

6451. *Hint.* See Exercise 6415.

6452. *Hint.* Embed a domain into a field.

6453. *Hint.* a) Find zero divisor.

b) Prove that each nonzero element has an inverse.

c) Prove that the ring has no zero divisors if n is a prime, which is not a sum of two squares, and that a finite nonzero commutative ring without zero divisors is a field.

6454. *Hint.* Consider the mapping $a_0 x^k + \cdots + a_k \to \bar{a}_0 x^k + \cdots + \bar{a}_k$, where $\bar{a}_i = a_i + \langle n \rangle$ $(i = 0, \ldots, k)$.

6455. p^n.

6456. *Hint.* a) Introduce a structure of a ring on the direct sum $S = R \oplus \mathbb{Z}$.

b) If R is an algebra over a field K then decompose the algebra over K into a direct sum $S = R \oplus K$.

c) Associate with each element a of A a linear operator φ_a on the vector space A over K such that $\varphi_a(x) = ax$.

d) Apply b).

6457. *Hint.* Prove that $I_k + \cap_{i \neq k} I_i = A$ for any $k = 1, \ldots, s$. Deduce from here that the mapping f is surjective.

6458. *Hint.* Apply the homomorphism $f(x) \to (f(1), f(-1))$.

6460. *Hint.* Show that $I \cap \mathbb{Z} \neq 0$ and I contains a polynomial which is nontrivial modulo $I \cap \mathbb{Z}$.

6461. – 6463. *Hint.* Apply the homomorphism theorem.

6464. *Hint.* c) The condition $\det(a_{ij}) \neq 0$ follows from surjectivity of the composition $\Lambda(V) \xrightarrow{\varphi} \Lambda(V) \to \Lambda(V)/I_2$, where I_2 is the ideal generated by $\Lambda^2(V)$. In order to prove that φ is an automorphism it is necessary to show that $\varphi(e_i) \wedge \varphi(e_j) + \varphi(e_j) \wedge \varphi(e_i) = 0$ for all i, j, and that φ is surjective. It suffices to show the last statement for the mapping φ with the unit matrix (a_{ij}). The proof can be given by induction on k, starting by inclusion $\Lambda^n V \subset \operatorname{Im} \varphi$.

6465. *Hint.* b) The annihilator is generated by an idempotent $1 - e$, where e is a generator of the given ideal.

6466. *Hint.* If the ideals $I_1, \ldots, I_n$ are generated by orthogonal idempotents $e_1, \ldots, e_n$ then $I_1 + \cdots + I_n$ is generated by the idempotent $e_1 + \cdots + e_n$.

6467. d) *Hint.* For example, $M_2 = \left\{ \begin{pmatrix} a & a \\ b & b \end{pmatrix} \right\} \oplus \left\{ \begin{pmatrix} a & 2a \\ b & 2b \end{pmatrix} \right\}$, where a, b are arbitrary elements of the field.

e) $\varphi\left(\begin{pmatrix} a & 0 \\ c & 0 \end{pmatrix} \right) = \begin{pmatrix} a & a \\ c & c \end{pmatrix}$, $\varphi\left(\begin{pmatrix} 0 & b \\ 0 & d \end{pmatrix} \right) = \begin{pmatrix} b & 2b \\ d & 2d \end{pmatrix}$.

6468. $M_2(K) = I \oplus J$.

6470. *Hint.* Consider the kernel of the homomorphism $\mathbb{Z}_{mn} \to \mathbb{Z}_m \oplus \mathbb{Z}_n$ for which $l + mn\mathbb{Z} \to (l + m\mathbb{Z}, l + n\mathbb{Z})$.

6471. *Hint.* Apply Exercise 6470 if n is square-free.

6472. *Hint.* Prove that the ideal consisting of all matrices $a E_{1n}$ is a nonzero ideal of this algebra with zero multiplication.

6473. *Hint.* If $R = I_1 \oplus \cdots \oplus I_n$ is a decomposition of R into a direct sum of simple rings and e is an idempotent of R then $e = e_1 + \cdots + e_n$, where $e_i \in I_i$ are idempotents. Prove that there are finitely many idempotents in I_i (make use of Exercise 6416). Apply Exercise 6415.

6474. *Hint.* If $A = I_1 \oplus \cdots \oplus I_n$ is a completely reducible algebra (I_k is a simple algebra) then $I_1 \oplus \cdots \oplus I_{k-1} \oplus I_{k+1} \oplus \cdots \oplus I_n$ is its maximal ideal ($k = 1, 2, \ldots, n$).

6475. *Hint.* Make use of Exercise 6415.

6476. Finite cyclic groups of square-free orders. *Hint.* A cyclic group has no proper subgroups if and only if its order is prime; decompose a cyclic group into a direct sum of primary cyclic groups.

6477. *Hint.* Let $R = I_1 \oplus \cdots \oplus I_n$ be a decomposition of R into a direct sum of minimal left ideals. If $I \subset R$ then there exists $I_{k_1} \not\subseteq I$ and then $I_{k_1} \cap I = 0$. If $I_{k_1} \oplus I \neq R$ then there exists $I_{k_2} \not\subseteq I_{k_1} \oplus I$ and $I_{k_2} \cap (I_{k_1} \oplus I) = 0$. Eventually we have $I_{k_1} \oplus \cdots \oplus I_{k_s} \oplus I = R$ (for some $s < n$).

6478. *Hint.* a) If $R = I_1 \oplus \cdots \oplus I_n$ is a decomposition of R into a direct sum of minimal left ideals and I is a left ideal of R then $R = I_1 \oplus \cdots \oplus I_k \oplus I$ for an appropriate enumeration of summands (see the hint to Exercise 6477) and $I \simeq R/(I_1 \oplus \cdots \oplus) \simeq I_{k+1} \oplus \cdots \oplus I_n$. b) $R = I \oplus J$ (Exercise 6477), $1 = e_1 + e_2$, where $e_1 \in I$, $e_2 \in J$; prove that e_1, e_2 are idempotents and $I = Re_1$.

6479. *Hint.* Consider a cyclic group of prime order with zero multiplication. See the hint to Exercises 6477b and 6416.

6480. *Hint.* See Exercise 6477.

6483. *Hint.* See Exercise 6470.

6484. Linear spans of vectors $e_{i_1}, \ldots, e_{i_s}$, where $1 \leq i_1 < \cdots < i_s \leq n$. *Hint.* Prove that if the submodule A contains a vector $\alpha_{i_1} e_{i_1} + \cdots + \alpha_{i_s} e_{i_s}$, where $\alpha_{i_1} \ldots \alpha_{i_s} \neq 0$ then $e_{i_1}, \ldots, e_{i_s} \in A$.

6485. *Hint.* The mapping $k \to kk_0$, where k_0 is fixed, k is an arbitrary element of R, induces an isomorphism of R-module R with the left ideal $I = Rk_0$. Conversely, the existence of an isomorphism of R-module R with a left ideal $I \subseteq R$ means that $I = Rk_0$, where k_0 is the image of 1 under the isomorphism.

6486. *Hint.* $F[x] = F[x] \circ 1 \oplus F[x] \circ x \oplus \cdots \oplus F[x] \circ x^{k_1}$ where $F[x] \circ x^i \simeq F[x]$ (an isomorphism of $F[x]$-modules).

6501. *Hint.* Let I be an ideal of $A[x]$. It is easy to see that the set of coefficients a_i of polynomials $a_0 + a_1 x + \cdots + a_i x^i$ from J is an ideal I of A. The sequence of ideals $I_0 \subseteq I_1 \subseteq I_2 \subseteq \ldots$ stabilizes, say, on I_r. Let a_{ij} $(i = 0, \ldots, r, j = 1, \ldots, n)$ be generators of I_i. For each pair of indices i, j choose a polynomial f_{ij} in J of degree i with the leading coefficient a_{ij}. Then $\{f_{ij}\}$ is a set of generators of J. It is possible to show by induction on the degree that each $f \in J$ belongs to the ideal generated by all polynomials f_{ij}.

6502. *Hint.* Make use of the previous exercise.

6503. *Hint.* d) Write down the formula for the inverse element.

e)   Consider three cases. *Case 1.* One of the elements $-\alpha, -\beta, -\alpha\beta$ is equal to γ^2 for some $\gamma \in F$. Assume for the definition that $-\alpha = \gamma^2$, $\gamma \in K$. Then A has zero divisors and an isomorphism $A \cong F_2$ can be explicitly defined by the formulas

$$1 \to \begin{pmatrix} 1 & 0 \\ 0 & 1 \end{pmatrix}, \quad i \to \begin{pmatrix} \gamma & 0 \\ 0 & -\gamma \end{pmatrix}, \quad j \to \begin{pmatrix} 0 & \beta \\ -1 & 0 \end{pmatrix}, \quad k \to \begin{pmatrix} 0 & \gamma\beta \\ \gamma & 0 \end{pmatrix}.$$

Case 2. A has a zero divisor $u + p$, where $u = \gamma \cdot 1$, $\gamma \in K$, $\gamma \neq 0$, $p = x_1 i + x_2 j + x_3 k$ is a pure quaternion. Then (see d)) $N(u+P) = \gamma^2 - p^2 = 0$. Put $i' = p$ and add i' to the basis i', j', k' of the space of pure quaternions, such that $i'^2 = \gamma^2$, $j'^2 = -\beta$, $i'j' = -j'i' = k'$. Reduce case 2 to case 1.

Case 3. A has a zero divisor $p = x_1 i + x_2 j + x_3 k$. If $x_1 \neq 0$ consider the quaternion,

$$u + x_1 \left(1 + \frac{u^2}{4\alpha x_1^2}\right) i + x_2 \left(1 - \frac{u^2}{4\alpha x_1^2}\right) j + x_3 \left(1 - \frac{u^2}{4\alpha x_1^2}\right) k.$$

and reduce case 3 to case 2. If $x_1 = 0$ then case 1 obtains.

f) Under matrix representation e) the pure imaginary quaternions are defined by the condition $\operatorname{tr} P = 0$. Thus all nilpotent matrices (and only they) represent purely imaginary zero divisors in A.

g) Make use of d), f).

h) Multiplying if necessary by nonzero element $\lambda \in F$ we can assume that the determinant of a matrix Q is a square in F; then Q in some system of coordinates has the form $\alpha x_1^2 + \beta x_2^2 + \alpha\beta x_3^2$. Notice that the vector product depends on the orientation. Check that under a change of orientation of W we replace the algebra A for the dual algebra A° whose multiplication $*$ is connected with multiplication $\cdot$ in A by the rule $a \cdot b = b * a$ (as vector spaces A and $A^\cdot$ coincide). If A is a quaternion algebra then $A \simeq A^\circ$.

6504. *Hint.* c) Consider F-linear mapping $x \to ax$.

e) Reduce the statement to the case of a simple algebra with unity. The minimal ideals are of the form Ae, where $e^2 = e$.

6505. *Hint.* c) The subspace A_0 of pure quaternions of an algebra $A = C_Q^+(F)$ is defined by the condition $x = -\bar{x}$, where the bar designates the natural involution in Clifford algebra: $\bar{1} = 1$, $\bar{e}_i = e_i$, $\overline{e_i e_j} = e_j e_i$. Take elements $e_1 e_2 - \frac{1}{2} Q(e_1, e_2)$, $e_2 e_3 - \frac{1}{2} Q(e_2, e_3)$, $e_1 e_3 - \frac{1}{2} Q(e_1, e_3)$ as a basis of A_0.

6507. *Hint.* a) It is enough to consider the case of an irreducible polynomial f over $\mathbb{Q}$. Then $K = \mathbb{Q}[X]/(f(X))$ is a finite extension of degree n over $\mathbb{Q}$; let x be a class $X \pmod{f(X)}$. Then the mapping $K \to K$ defined by the formula $a \to xa$ is a linear mapping from n-dimensional vector space K over $\mathbb{Q}$ into itself. Its minimal polynomial coincides with the minimal polynomial of x.

6508. The algebra has no zero divisors.

6509. *Hint.* Let $I = \langle f_1, \ldots, f_n \rangle$. For any point $z \in \mathbb{C}$ define a non-negative integer $n(z) == \min_i \gamma_z(f_i)$, where $\gamma_z(f_i)$ denotes the order of zero of the function f_i at the point z (if $f_i(z) \neq 0$ then $\gamma_z(f_i) = 0$). Let (z_k) be a sequence of points in $\mathbb{C}$ such that $n(z_k) \neq 0$. Construct an entire function f with the sequence (z_k) as the sequence of zeros with multiplicities $n(z_k)$ and show that $I = \langle f \rangle$.

6510. a) $D = 0$. *Hint.* Consider $x = y = 1$.

b) $f(x)D$, where $f(x) \in \mathbb{Z}[x]$, D the usual differentiation.

c) $\sum f_i D_i$, where $f_i \in \mathbb{Z}[x_1, \ldots, x_n]$, D_i is a partial differentiations.

6512. *Hint.* See I.N. Herstien. Noncommutative Rings, Mathematical Associa-tion of America, John Wiley, § 4.6.

6513. *Hint.* See Dixmier J. Algèbres Enveloppantes, Gauthier-Villard Éditeurs, Paris, 1974, § 4.3.

6515. and **6516.** *Hint.* See Borevich Z.I., Shafarevich I.R. Number Theory, Academic Press, New York, 1966.

6602. a) If $\sqrt{n} \notin \mathbb{Q}$.

b)   If $n < 0$.

c) $n = 2$ if $p = 3$; $n = 2, 3$ if $p = 5$; $n = 3, 5, 6$ if $p = 7$.

6605. *Hint.* A multiplicative group of the field with 4 elements has order 3. In order to construct this field it is enough to find a matrix of size 2 over $\mathbb{Z}_2$ satisfying the equation $A^2 + A + E = 0$, i.e. $\operatorname{tr} A = \det A = 1$. For example, take the matrix $\begin{pmatrix} 0 & 1 \\ 1 & 1 \end{pmatrix}$. Then the field consists of the elements $O, E, A, A+E$. If $n = 6$ consider the orders of elements in the additive group.

6606. $\{ke \mid k \in \mathbb{Z}\}$. *Hint.* The additive group of a proper subfield has order p and contains this subfield.

6607. *Hint.* Prove for the field $\mathbb{Q}$ that integers are fixed under any automorphism; for the field $\mathbb{R}$ note that non-negative numbers are squares, and therefore their images are also non-negative. It follows from $x > y$ that $\varphi(x) = \varphi(x-y) + \varphi(y) > \varphi(y)$. Use rational approximations of real numbers.

6608. $z \to z$ and $z \to \bar{z}$. *Hint.* Consider the image of i.

6609. $x + y\sqrt{2} \to x - \sqrt{2}$ is the unique automorphism. *Hint.* Consider the image of $\sqrt{2}$.

6610. *Hint.* Notice that if $m = 1$ then the binomial coefficients $\binom{p}{n}$ are divisible by p; induction on m.

b)   A nonzero homomorphism from a field into itself is an automorphism.

6612. If $m/n = r^2$ $(r \in \mathbb{Q} \setminus \{0\})$.

6614. *Hint.* The additive group of the field K with 4 elements cannot be cyclic. Thus all nonzero elements have order 2, $K = \{0, 1, a, a + 1\}$; multiplication table is unique because $a(a + 1) = 1$.

6615. *Hint.* For example, the field of rational functions with complex coefficients.

6617. $F_p(X)$.

6618. a) $\{-1, -3 + 2\sqrt{2}\}$.

b) $\emptyset$; 13 is not a square in $\mathbb{Q}(\sqrt{2})$.

c) $\emptyset$.

d) $\emptyset$.

6619. a) $\emptyset$.

b) $(2, 3, 2)$.

6623. All of them.

6625. *Hint.* The multiplicative group of a field with n elements has order $n - 1$.

6626. $x = a$.

6628. a) 3 and 5.

b) $2, 3, 8, 9$.

6629. *Hint.* Show that if $a \neq 0$ then $(ba^{-1})^3 = 1$ and 3 divides $2^n - 1$, a contradiction.

6630. *Hint.* Let $F^* = \langle x \rangle$. Prove that x is algebraic over the prime subfield. This subfield is different from $\mathbb{Q}$, since $\mathbb{Q}^*$ is not cyclic.

6631. a) $\{\pm 1\}$.

b) $\emptyset$.

6632. *Hint.* a) $\mathbb{Z}_p$ has no elements of order 2. Since $p > 2$, the mapping $k \to k^{-1}$ is a bijection and $\sum_{i=1}^{p-1} k^{-1} = \sum_{i=1}^{p-1} k$.

b)   Similarly to a); $(p^2 - 1)|8$.

6635. and **6636.** *Hint.* See Platonov V.P., Rapinchuk A.S. Algebraic Groups and Number Theory, Nauka, Moscow, 1991. ch. 1, § 1.1.

6637. and **6638.** *Hint.* Solution is similar to Exercises 6635 and 6636.

6642. – **6645.** *Hint.* See Borevich Z.I., Shafarevich I.R. Number Theory, Academic Press, 1966.

6646. *Hint.* Make use of Exercise 6645.

6647. *Hint.* Make use of valuations of the field of p-adic numbers.

6701. *Hint.* Reduce by induction on s to the case $s = 1$; construct a basis of A over K using a basis of A over K_1 and a basis of K_1 over K.

6706. *Hint.* Apply Exercise 6704.

6707. *Hint.* Reduce by induction on s to the case $s = 2$; apply Exercises 6701, 6704, 6705.

6708. *Hint.* If the polynomial $p(x)$ is irreducible then it has a root in $K[x]/\langle p(x)\rangle$.

6709. *Hint.* a) Induction on degree of $f(x)$, apply Exercise 6708.

b) Apply a) to $f_1(x) \ldots f_l(x)$.

6710. *Hint.* Consider degrees of extensions in the tower of fields $K \subset K(\alpha) \subset K(\theta, \eta)$, where η is a root of $h(x) - \alpha$ in some extension of L. Apply Exercises 6701, 6702.

6711. *Hint.* a), b) Compare factorization of $x^n - \alpha$ into linear factors in its splitting field with possible factorization of this polynomial in K.

c) For example, the polynomial $x^4 + 1$ over the field of real numbers.

6712. *Hint.* $f(x) = \prod_{i \in \mathbf{F}_p}(x - x_0 - i)$, where $\mathbf{F}_p$ is the field with p elements contained in K. Prove that if in some extension L of K the polynomial $f(x)$ has a root then $f(x)$ can be factorized over L into a product of linear factors. Deduce that all irreducible factors of $f(x)$ over K have the same degrees.

6713. a) 1.

b) 2.

c) 2.

d) 6.

e) 8.

f) $p - 1$.

g) $\phi(n)$.

h) $p(p - 1)$.

i) 2^r, where r is the rank of the matrix (k_{ij}), $i = 1, \ldots, s, j = 0, \ldots, t$ over the residue field modulo 2 and $\bar{k}_{ij}$ is the residue class modulo 2, of the exponent k_{ij} in factorization $a_i = (-1)^{k_{i0}} \cdot \prod_{j=1}^{t} p_j^{k_{ij}}$ of a_i into a product of powers of distinct primes $p_1, \ldots, p_t$. (It is admitted that some $k_{ij} = 0$.)

g) *Hint.* Show that if ζ is a primitive root of 1 of order n and $\mu_\zeta(x)$ is its minimal polynomial over $\mathbb{Q}$, then ζ_p is also a root of $\mu_\zeta(x)$ for any prime $p|n$; for otherwise if $x^n - 1 = \mu_\zeta(x)h(x)$ then ζ is a root of $h(x^p)$, a contradiction with the fact that $x^n - 1$ has no multiple factors over residue field modulo p.

h) *Hint.* Apply Exercise 6711.

i) *Hint.* If K is the required field then consider $K^{*2} \cap \mathbb{Q}^*$ and apply induction on n.

6714. $F(X, Y)/F(X^p, Y^p)$, where F is a field of characteristic p. *Hint.* If the field K is finite then make use of Exercise 5633. Let K be infinite and $L = K(a_1, \ldots, a_s)$. By induction on s the question on existence of a primitive element is reduced to the case $s = 2$; show that for some $\lambda \in K$ an element $a_1 + \lambda a_2$ is not contained in any proper intermediate subfield. Conversely, show that if $L = K(a)$, then any intermediate subfield is generated over K by coefficients of some divisor in $L[x]$ of the minimal polynomial $\mu_a(x)$ of a over K.

6715. *Hint.* Choose a basis of $L(x)$ over $K(x)$ consisting of elements of L.

6717. *Hint.* Prove, by induction on i ($0 \le i \le m$), that under some enumeration of $b_1, \ldots, b_n$ the system $a_1, \ldots, a_i, b_{i+1}, \ldots, b_n$ is a maximal system of algebraically independent over K elements in L.

6718. *Hint.* a) Show that the number of maximal ideals does not surpass $(A : K)$. Show next that if an element $a \in A$ is not nilpotent then a maximal ideal in a set of ideals having empty intersection with $\{a, a^2, \ldots, \}$ is a maximal ideal of A.
b) Make use of a).

e) Show for the proof of uniqueness that under any representation of $A^{\mathrm{red}} = \prod_{j=1}^{t} L_j$ the fields L_j are isomorphic to factor-algebras by all possible maximal ideals of A.

6720. *Hint.* Induction on n. Linear dependence of f_i leads to a contradiction since f_i are homomorphisms of algebras.

6723. *Hint.* a) Any K-homomorphism $A \to B$ has a unique extension to L-homomorphism $A_L \to B$.
b) Make use of a).

6724. *Hint.* Take any component of F_L as E.

6725. *Hint.* For the proof of b)$\Rightarrow$ a) notice that if L_i is any component of A_L and $\bar{a}_1, \ldots, \bar{a}_s$ are images of $a_1, \ldots, a_s$ in L_i then $L_i = L(\bar{a}_1, \ldots, \bar{a}_s)$. For the proof of implication a)$\Rightarrow$ b) apply Exercises 6722a, 6719 and 6718e to $K[a]$.

6726. *Hint.* Apply Exercise 6725.

6727. *Hint.* b) Notice that each of the fields L_1, L_2 is splitting for the other one; obtain from here K-embeddings $L_1 \to L_2$ and $L_2 \to L_1$.

6728. *Hint.* Make use of Exercises 6727c and 6722.

6729. *Hint.* a) Choose a splitting field of A containing L, and apply Exercise 6722.
b) Apply a) and Exercise 6723b.

6730. *Hint.* Apply Exercises 6725, 6729b.

6731. *Hint.* b) For example, $\mathbb{Q} \subset \mathbb{Q}(\sqrt{2}) \subset \mathbb{Q}(\sqrt[4]{2})$.

6732. *Hint.* In order to prove the two last relations, reduce the general case to two special cases when $a \in K$ and when $L = K(a)$. In the first case exploit a basis of L/F associated with the tower of fields. In the second case choose, in L/K, a basis of powers of a. For the proof of the first relation note that $\chi_{L/K}(a, x) = \mathbf{N}_{L(x)/K(x)}(a - x)$.

6733. *Hint.* Make use of Exercise 6732.

6734. *Hint.* If $\mathrm{tr}_{L/K}(a) \neq 0$ for some $a \in L$ then $(x, ax^{-1}) \to \mathrm{tr}_{L/K}(a) \neq 0$ for all $x \neq 0$ from L.

6735. *Hint.* Each of conditions a) – c) is equivalent to the fact that $A_L \simeq \Pi L$ for any splitting field L. Forms of traces in A and in A_L are nondegenerate simultaneously. Nilpotent elements are always contained in the kernel of the form of the trace.

6737. *Hint.* Make use of Exercises 6722 and 6712.

6738. *Hint.* Make use of the equality $\mathrm{tr}_{A/K}(a) = \mathrm{tr}_{A_L/L}(a)$.

6739. a) *Hint.* Make use of Exercises 6714, 6722.

b) b is a primitive element. a is not a primitive element.

6740. *Hint.* Make use of Exercise 6722c.

6741. *Hint.* Make use of Exercises 6734 and 6735d.

6742. A polynomial $x^p - t$ over the field of rational functions $K(t)$, where K is an arbitrary field of characteristic $p \neq 0$.

6743. *Hint.* Make use of Exercise 6719.

6744. *Hint.* For the proof of the converse statement, apply Exercise 6741.

6745. *Hint.* Let L be a splitting field of $f(x)$. Show that $B_L \simeq \prod_{i=1}^n A_i$, where $A_i \simeq A_L, n = \deg f$.

6746. *Hint.* For the proof of implication c) $\Rightarrow$ a) represent A as a factor-algebra $K[x_1, \ldots, x_s]/\left(\mu_{a_1}(x_1), \ldots, \mu_{a_s}(x_s)\right)$. Apply Exercise 6745.

6747. *Hint.* a) Make use of Exercises 6746, 6745, 6742, 6711.

6748. *Hint.* Consider $\mu_a(x)$ for any element $a \in L$.

6750. *Hint.* Apply Exercises 6748 and 6749.

6751. *Hint.* b) Prove with the help of Exercise 6726 that for any splitting field E of the extension L/K the number of distinct K-embeddings $L \to L$ is equal to $(K_s : K)$.

6752. *Hint.* a) Calculate the number of distinct K-embeddings of F into some splitting field of the extension F/K.

6753. *Hint.* Consider a tower of fields $K \subset K_s \subset L$ and apply Exercises 6731, 6738.

6754. *Hint.* a) Apply Exercises 6730 and 6735.

b)   Apply Exercise 6727.

6755. a) $G(\mathbb{C}/\mathbb{R})$ consists of the identical automorphism and of the complex conjugation.

b), c) $\mathbf{Z}_2$.

d) $\mathbf{Z}_2 \oplus \mathbf{Z}_2$.

6756. a) $\{e\}$.

b) $\mathbf{S}_2$.

c) $\mathbf{S}_2$.

d) $\mathbf{S}_3$.

e) $\mathbf{D}_4$.

f) $\mathbf{Z}_{p-1}$.

g) $\mathbf{Z}_n^*$.

h) The *semi-direct* product of the group $\mathbf{Z}_p$ and its group of automorphisms.

i) Direct product of r copies of the group $\mathbf{Z}_2$. (See the answer to Exercise 6713.)

6757. *Hint.* Each element $a \in L$ is a root of a separable polynomial over K of degree $\leq |G|$, namely, $f(x) = \prod_{\sigma \in G}(x - \sigma(a))$. Notice that any (finite) separable extension has a primitive element. Prove that $(L : K) = |G|$.

6758. *Hint.* Consider the action of $\mathbf{S}_n$ on the field of rational functions $K(a_1, \ldots, a_n)$ and apply Exercise 6757.

6759. *Hint.* Embed the group G into a symmetric group and apply Exercise 6757.

6760. *Hint.* Apply Exercise 6757.

6761. *Hint.* Prove firstly that any Galois extension $L/\mathbb{R}$ distinct from $\mathbb{R}$ has a degree which is a power of 2. Note that a finite 2-group is solvable and $\mathbb{R}$ has no finite extension of degree ≥ 2. Show that $L = \mathbb{C}$.

6763. *Hint.* Consider an action of elements of the Galois group on $\sqrt{D}$.

6764. *Hint.* Apply the linear independence of automorphisms (Exercise 6721) and prove that L is a cyclic module over $K[\varphi]$.

6766. *Hint.* The group $\mathbf{S}_n$ acts by permutations on components of $A = \prod K_i$ ($K_i \simeq K$). Notice that K_i are unique minimal ideals of A.

6767. *Hint.* Notice that $\tau(x) = \sum_\sigma \sigma(\tau x)e_\sigma = \sum_\sigma \sigma(x)\tau(e_\sigma)$ for $x \in L$.

6768. *Hint.* Make use of Exercise 6720 (or interpret $(\varphi_i(y_j))$ as the matrix of a change of basis; A is assumed to be embedded into A_L).

6769. *Hint.* See Exercise 6764 if K is finite. Let K be infinite, $\omega_1, \ldots, \omega_n$ be some basis of L over K and $\omega = a_1\omega_1 + \cdots + a_n\omega_n$ be an arbitrary element either of L (if $a_i \in K$) or of L_L (if $a_i \in L$). The hypothesis of Exercise 6768, which guarantees that elements $\{\sigma(\omega), \sigma \in G\}$ form a basis of L (respectively, of L_L), means that for some polynomial $f(x_1, \ldots, x_n) \in L[x_1, \ldots, x_n]$ its value

$f(a_1, \ldots, a_n) \neq 0$. Use then the existence of a normal basis of L_L (Exercise 6767).

6770. *Hint.* If the characteristic of $K \neq 2$ then

$$K(x_1, \ldots, x_n)^{A_n} = K(\sigma_1, \ldots, \sigma_n, \Delta),$$

where $\sigma_1, \ldots, \sigma_n$ are elementary symmetric polynomials in $x_1, \ldots, x_n$,

$$\Delta = \prod_{j > i}(x_j - x_i).$$

In the case of an arbitrary characteristic we have $K(x_1, \ldots, x_n)^{A_n} = K(\sigma_1, \ldots, \sigma_n, y)$, where $y = \sum_{\sigma \in A_n} \sigma\left(\prod_{i=1}^{n} x_i^{i-1}\right)$.

6771. $\mathbb{C}(x_1^n, x_1^{n-2}x_2, \ldots, x_1 x_{n-1}, x_n)$. *Hint.* Apply Exercise 6760.

6772. $\mathbb{C}(y_1^n, y_1^{n-2}y_2, \ldots, y_1 y_{n-1}, y_n)$, where $y_i = \sum_{k=1}^{n} \varepsilon^{-ik} x_k$, ε is a primitive root of unit of order n. *Hint.* In the space of linear forms in $x_1, \ldots, x_n$, choose a basis consisting of eigenvectors of σ; then apply Exercise 6771.

6773. The group $\mathbb{Z}_n$. *Hint.* The decomposition field L of $x^n - a$ over K has the form $L = K(\theta)$, where θ is some root of $x^n - a$ in L. The group $G(L/K)$ is generated by an automorphism σ such that $\sigma(\theta) = \varepsilon\theta$, where ε is a generator of the (cyclic) group of roots of 1 of order n. Apply Exercise 6711.

6774. *Hint.* Let ε be a generator of the group of roots of 1 of order n in K; let $y \in L$ be an element of L such that $\sum_{i=1}^{n} \varepsilon^{-i}\sigma^i y \neq 0$ (does it exist?); then $a = \left(\sum_{i=1}^{n} \varepsilon^{-i}\sigma^i y\right)^n$. Consider eigenvectors of σ in L.

6775. *Hint.* If $L = K(\theta_1, \ldots, \theta_s)$, then for any $\sigma \in G(L/K)$ we have $\sigma(\theta_i) = \varepsilon_i(\sigma)\theta_i$, where $\varepsilon_i(\sigma)^n = 1$. Conversely, if the group $G(L/K)$ is abelian of period n then use the following fact: a set of commuting diagonalizable linear operators has a common eigenbasis (see Exercise 4007).

6776. *Hint.* Apply a bilinear map $G(L/K) \times A \to U_n$ to $\sigma \in G(L/K)$ and to $\bar{a} \in A$ ($a \in \langle K^{*n}, a_1, \ldots, a_s \rangle$), $(\sigma, \bar{a}) \to (\sigma\theta) \cdot \theta^{-1}$, where $\theta \in L$ and $\theta^n = a$.

6777. $L \to (L^{*n} \cap K^*)/K^{*n}$; if $A = B/K^{*n}$, $B = \langle K^{*n}, a_1, \ldots, a_s \rangle$ is a subgroup of K^* then $A \to L = K(\theta, \ldots, \theta_s)$, where $\theta_i^n = a_i$. *Hint.* Make use of Exercise 6776.

6778. *Hint.* Let $G(L/K) = \langle \sigma \rangle$. In order to find θ use a root vector of height 2 of σ. For the proof of the converse statement, apply Exercise 6712.

6779. *Hint.* If $L = K(\theta_1, \ldots, \theta_s)$, then for any $\sigma \in G(L/K)$ we have $\sigma(\theta_i) = \theta_i + \gamma_i$, $\gamma_i \in \mathbb{F}_p$ (see Exercise 6712). Conversely, let $G = G(L/K)$ be the direct product of s copies of cyclic groups of order p. Choose in G subgroups H_i ($i = 1, \ldots, s$) of index p, such that $\cap_{i=1}^{s} H_i = \{e\}$; then $L^{H_i} = K(\theta_i)$ (see Exercise 6778) and $L = K(\theta_1, \ldots, \theta_s)$.

6780. *Hint.* Consider a bilinear map $G(L/K) \times A \to \mathbf{F}_p$, such that for $\sigma \in G(L/K)$, $\bar{u} \in A$ ($a \in \langle \rho(k), a_1, \ldots, a_s \rangle$), we have $(\sigma, a) \to \sigma(\theta) - \theta$, where $\theta \in L$ and $\rho(\theta) = a$.

6781. $L \to (\rho(L) \cap K)/\rho(K)$; if $A = B/\rho(K)$, $B = \langle \rho(K), a_1, \ldots, a_s \rangle$ then $A \to K(\theta_1, \ldots, \theta_s)$, where $\rho(\theta_i) = a_i$. *Hint.* Apply Exercises 6779, 6780.

6801. *Hint.* Make use of Exercise 5633.

6802. *Hint.* a) If $|L| = q$, then L is the decomposition field of $x^q - x$.

b) Use the hints to a) and to Exercise 6727b.

6803. *Hint.* See the hint to Exercise 2802; in a) make use of the fact that $x^q - x$ has no multiple roots.

6804. *Hint.* Make use of Exercise 5633.

6805. *Hint.* b) Decompose σ into a product of disjoint cycles.

6806. *Hint.* Let $b = \prod_{j=1}^k p_j^{n_j}$, where p_j are distinct primes. Decompose the ring $\mathbf{Z}_8$ into a direct product of residue rings modulo $p_j^{n_j}$. If $b = p^n$, p is a prime then represent the set of residue classes as a union of subsets containing all elements of the same order in the additive group of the residue ring. Next make use of the structure of the group of invertible elements in the residue ring modulo p^n.

6807. *Hint.* Calculate the number of inversions of σ ordering elements of G in the following way: $0, x_1, \ldots, x_n, -x_n, \ldots, -x_1$, where $\{x_1, \ldots, x_n\} = S$.

6808. *Hint.* a) Apply Exercise 6807, taking some subsets S_1 and S_2 of G_1 and G_2 respectively. Put $S = S_1 \cup \varphi^{-1}(S_2)$, where $\varphi : G \to G_2$ is the canonical homomorphism.

6809. *Hint.* Make use of Exercise 6807.

6810. *Hint.* Make use of Exercise 6809.

6811. *Hint.* The set R of pairs (x, y), where $1 \le x \le (a-1)/2$, $1 \le y \le (b-1)/2$, is the union of four subsets:

$$R_1 = \{(x, y) \in R \mid ay - bx < -b/2\},$$
$$R_2 = \{(x, y) \in R \mid -b/2 < ay - bx < 0\},$$
$$R_3 = \{(x, y) \in R \mid 0 < ay - bx < a/2\},$$
$$R_4 = \{(x, y) \in R \mid a/2 < ay - bx\}.$$

Considering the bijection $(x, y) \to ((a+1)/2 - x, (b+1)/2 - y)$ show that $|R_1| = |R_4|$. Applying Exercise 6809 prove that $\left(\dfrac{a}{b}\right) = (-1)^{|R_2|}$, $\left(\dfrac{b}{a}\right) = (-1)^{|R_3|}$.

6812. *Hint.* Represent the matrix of $\mathcal{A}$ as a product of elementary matrices.

6813. – 6815. *Hint.* See Lidl R., Niedereiter H. Finite Fields, Addison-Wesley, 1983, ch. 2, § 3.

6904. a) Yes. b) No. c) Yes. d) Yes. e) No. f) Yes.

6905. All subspaces mentioned except d), e), h) are invariant.

6907. $\begin{pmatrix} 1 & -t & t^2 \\ 0 & 1 & -2t \\ 0 & 0 & 1 \end{pmatrix}$ (in the basis $1, x, x^2$).

6908. $\begin{pmatrix} \cos t & \sin t \\ -\sin t & \cos t \end{pmatrix}$ (in the basis $\sin x, \cos x$).

6910. *Hint.* Decompose the space $\mathbf{M}_n(K)$ into a sum of subspaces consisting of all matrices with at most one (fixed) nonzero column.

6911. *Hint.* Prove first that a subspace of $\mathbf{M}_n(K)$ which is invariant under all operators $\mathrm{Ad}(A)$, where A runs over all diagonal matrices, is a linear span of some set of matrix units E_{ij} ($i \neq j$), and of some subspace of diagonal matrices.

6912. *Hint.* Prove first that any subspace of $\mathbf{M}_n(K)$ which is invariant under all operators $\Phi(A)$, where A runs over all diagonal matrices, is a linear span of some matrices $aE_{ij} + bE_{ji}$ ($i \neq j$), and of some subspace of diagonal matrices.

6913. *Hint.* Find the general form of matrices X such that

$$X \begin{pmatrix} 0 & 1 \\ 1 & 0 \end{pmatrix} = \begin{pmatrix} 0 & 1 \\ 1 & 0 \end{pmatrix} X, \quad X \begin{pmatrix} 0 & -1 \\ 1 & -1 \end{pmatrix} = \begin{pmatrix} 0 & 1 \\ -1 & -1 \end{pmatrix} X,$$

and show that $\det X = 0$ always.

6916. *Hint.* a) Let $H \subseteq W$ be invariant subspace and $x \in H$. Consider a vector $\pi x - x$, where $\pi = \{ij\}$.

6917. *Hint.* First determine the subspaces invariant under the restriction of Θ to the subgroup of diagonal matrices.

6925. *Hint.* Make use of Exercise 6924c and of partition of the group G on left cosets by H.

6926. a) m.

b)   2.

c)   1.

d) $m + 1$.

6927. *Hint.* If A and B are commuting operators, then every eigenspace of A is invariant under B.

7002. *Hint.* Make use of Exercise 6928.

7005. In both cases each irreducible representation of H occurs with multiplicity 2.

7006. Only trivial representations for groups of odd order; groups of even order have a homomorphism onto the subgroup $\{-1, 1\}$ of $\mathbf{GL}_1(\mathbb{R}) \simeq \mathbb{R}^*$.

7007. *Hint.* Apply the theorem asserting that a real operator has a two-dimensional invariant subspace.

7009. a) $[n/2] + 1$. *Hint.* Make use of Exercise 7008.

7015. For $\mathbf{S}_3$: the trivial representation and the representation which associates the sign with a permutation. *Hint.* Apply the theorem on commutants and Exercise 6207a. For $\mathbf{A}_4$: apply the theorem on commutants, and Exercise 6207b.

7016. *Hint.* Apply the theorem on commutants, and Exercise 6208.

7017. *Hint.* Take the representation from Exercise 6913.

7031. *Hint.* Decompose the regular representation into a sum of irreducible subrepresentations.

7032. *Hint.* Prove that the subgroup generated by $\mathcal{A}$ and $\mathcal{B}$ in $\mathbf{GL}(V)$ is isomorphic to $\mathbf{S}_3$.

7034. a) $1, 1, 2$.

b) $1, 1, 1, 3$.

c) $1, 1, 2, 3, 3$.

d) $1, 1, 1, 1, 2$.

e) If $n = 2k$ then there exist four one-dimensional representations and $k - 1$ two-dimensional representations; if $n = 2k + 1$ then there exist two one-dimensional representations and k two-dimensional ones.

f) $1, 3, 3, 4, 5$. *Hint.* Apply the main theorem and Exercise 6916.

7037. a), b), c) It is not possible.

7038. *Hint.* If the subgroup exists, then there exists, an exact two-dimensional representation of the group $\mathbf{S}_4$.

7042. Only for abelian groups.

7043. *Hint.* Induction on order of the group.

7045. *Hint.* Make use of Exercise 6925.

7046. *Hint.* Note that there exist finitely many nonisomorphic groups of fixed order, and each of them has finitely many nonisomorphic representations of a given dimension.

7048. *Hint.* Notice that groups of orders p and p^2 are abelian.

7049. p^2 of one-dimensional representations and $p - 1$ of p-dimensional. *Hint.* Notice that the center of the given group has order p and the number of classes of conjugate elements is equal to $p^2 + p - 1$. Since the factor-group by the center is commutative, the commutant of the given group has order p. These parameters determine the number of one-dimensional representations. Note that the given group has a normal subgroup of index p. Prove that the dimension of an irreducible representation is at most p.

7102. A basis consists of one vector $\sum_{\sigma \in S_3}(\text{sgn}\sigma)\sigma$. The dimension is equal to 5.

7103. $\{\varepsilon - a, \varepsilon^2 - a^2, \ldots, \varepsilon^{n-1} - a^{n-1}\}$.

7109. a) Let $e_1 = \frac{1}{6}\sum_{\sigma \in S_3}\sigma$, $e_2 = \frac{1}{6}\sum_{\sigma \in S_3}(\text{sgn}\sigma)\sigma$. Commutative ideals are: $0, \mathbb{C}e_1, \mathbb{C}e_2, \mathbb{C}e_1 \oplus \mathbb{C}e_2$.

b) Let $Q_8 = \{E, \bar{E}, I, \bar{I}, J, \bar{J}, K, \bar{K}\}$, $e_1 = (E + \bar{E})(E + I + J + K)$, $e_2 = (E + \bar{E})(E + I - J - K)$, $e_3 = (E + \bar{E})(E - I - J - K)$, $e_4 = (E + \bar{E})(E + I - J + K)$. Commutative ideals are: linear span of any subset of the vectors $\{e_1, e_2, e_3, e_4\}$.

c) Let $e_1 = \frac{1}{10}\sum_{A \in D_5} A$, $e_2 = \frac{1}{10}\sum_{A \in D_5}(\det A)A$. Commutative ideals are: $0, \mathbb{C}e_1, \mathbb{C}e_2, \mathbb{C}e_1 \oplus \mathbb{C}e_2$.

7110. If G is infinite then $x = 0$; if it is finite then $x = \alpha\sum_{g \in G} g, \alpha \in F$.

7111. A basis of the center of $F[G]$ is formed by the elements $\sum_{g \in C} g$, where C runs over all classes of conjugate elements in G.

7116. *Hint.* Make use of Schur's lemma.

7119. Only for $G = \{e\}$.

7122. a) 2. b) 1. c) 2. d) 4.

7124. Let ε be a primitive root of 1 of order 3 in $\mathbb{C}$;

$$r_0 = \frac{1}{3}(e + a + a^2) \in \mathbb{R}[\langle a\rangle_3] \subset \mathbb{C}[\langle a\rangle_3],$$

$$r_1 = \frac{1}{3}(e + \varepsilon a + \varepsilon^2 a^2) \in \mathbb{C}[\langle a\rangle_3],$$

$$r_2 = \frac{1}{3}(e + \varepsilon^2 a + \varepsilon a^2) \in \mathbb{C}[\langle a\rangle_3].$$

$\mathbb{R}[\langle a\rangle_3] = F_0 \oplus F_1$, where $F_0 = \mathbb{R}r_0 \simeq \mathbb{R}$ and

$$F_1 = \left\{\alpha_0 e + \alpha_1 a + \alpha_2 a^2 \Big| \sum_{i=0}^{2}\alpha_i = 0, \ \alpha_i \in \mathbb{R}\right\} \simeq \mathbb{C}.$$

Under an isomorphism $\mathbb{C} \to F_1$ we have $1 \to e - r_0$, $\varepsilon \to a(e - r_0)$. $\mathbb{C}[\langle a\rangle_3] = F_0' \oplus F_1' \oplus F_2'$. The fields $F_i' = \mathbb{C}r_i$ are isomorphic to $\mathbb{C}$.

7125. *Hint.* Exploit the irreducibility of $x^{p-1} + x^{p-2} + \cdots + x + 1$ over $\mathbb{Q}$.

7127. a) Idempotents: $2e + 2a, 2e + a$. Ideals: $\mathbb{F}_3 e_1, \mathbb{F}_3 e_2$.

b) The only idempotent is the unit of the group algebra. Ideal: $\mathbb{F}_2(1 + a)$.

c) Idempotents: $\frac{1}{2}(1 + a), \frac{1}{2}(1 - a)$. Ideals: $\mathbb{C}e_1, \mathbb{C}e_2$.

d) Idempotents: $\frac{1}{3}(1 + a + a^2)$, $\frac{1}{3}(2 - a - a^2)$. Ideals: $\mathbb{R}e_1$, $\mathbb{R}[\langle a \rangle_3]e_2$.

7128. *Hint.* Check the similar statement for the matrix algebra $\mathbf{M}_n(\mathbb{C})$, and apply the theorem concerning the structure of the group algebra of a finite group.

7129. a) 8. b) 32.

7130. a) $\{e\}$.

b) $G \simeq \mathbf{Z}_2$.

c) Either $G \simeq \mathbf{Z}_3$ or $\mathbf{S}_3$. *Hint.* Notice that n is equal to the number of classes of conjugate elements in G.

7134. $U = F[G](a - e)^2$ if $p = 2$.

7136. a) *Hint.* Consider the case $G = H$. Induction on the order of H.

b) $n = 2$.

7139. a) $P/H \simeq a/(g - ge)A \oplus A/(g - \varepsilon^2 e)A$, where ε is a primitive root of 1 of order 3 in $\mathbb{C}$.

b) $P/H = 0$.

c) $P/H \simeq A$.

7140. $\mathrm{Ker}\varphi = 0$.

7141. *Hint.* Consider the similar question for the ring of polynomials $A = F[t]$.

7144. a) The element is prime.

b) $(g_1 - g_2)^2(-g_1^{-1}g_2^{-1} - g_1^{-2})$.

7145. a) 0. b) $F[\langle g_1 \rangle]$. c) F.

7201. *Hint.* Make use of Exercise 6921.

7202. *Hint.* Make use of Exercise 6911. Find a possible diagonal form of the matrix of $\Phi(g)$.

7203. and **7204.** *Hint.* Make use of Exercise 7201.

7205. *Hint.* Notice that a sum of n roots of 1 is equal to n if and only if all summands are equal to 1.

7206. *Hint.* Make use of Exercise 7205, and prove that any subgroup of index p in A is a subgroup of elements of some $(n - 1)$-dimensional subspace.

7207. *Hint.* Let χ be a character of a representation Φ. Apply Exercise 7205 and prove that $\Phi(g) = E$ if $g \in H$. Similarly, show that $g \in K$ if and only if the matrix $\Phi(g)$ is scalar.

7208. *Hint.* Make use of *Maschke's theorem* and of properties of the commutant.

7209. *Hint.* Make use of Maschke's theorem and of properties of the commutant.

7220.

	1	-1	i	j	k
χ	2	-2	0	0	0

Hint. Make use of Exercise 7024.

7221. $\chi_\Phi(\sigma)$ is the number of elements of the set $\{1, 2, 3, \ldots, n\}$ fixed by σ.

7222. Let $\mathbf{D}_n = \langle a, b \mid a^2 = b^n = e, aba = b^{-1} \rangle$. Then $\chi(b^k) = 2 \cos 2\pi k/n$, $\chi(ab^k) = 0$.

7223.

	e	(12)	(123)	(12)(34)	(1234)
χ	3	1	1	-1	-1

Hint. Make use of Exercise 7019.

7224.

	e	(12)	(123)	(12)(34)	(1234)
χ	3	-1	0	-1	1

Hint. Make use of Exercise 7019.

7225. *Hint.* Make use of Exercise 7204.

7226. a) Two characters: a trivial one and $\sigma \to \text{sgn}\,\sigma$.

b)

	e	(123)	(132)	(12)(34)
φ_0	1	1	1	1
φ_1	1	ε	ε^2	1
φ_3	1	ε^2	ε	1

where ε is a primitive root of 1 of order 3 in $\mathbb{C}$.

c)

	1	-1	i	j	k
φ_0	1	1	1	1	1
φ_1	1	1	-1	-1	1
φ_2	1	1	-1	1	-1
φ_3	1	1	1	-1	-1

d)   See a).

e)   Let $D_n = \langle a, b \mid a^2 = b^n = e, aba = b^{-1} \rangle$. If n is odd then there exist two one-dimensional characters: a trivial one and $a^i b^j \to (-1)^i$. If n is even then there exist four characters: a trivial one and $a^i b^j \to (-1)^i$, $a^i b^j \to (-1)^j$, $a^i b^j \to (-1)^{i+j}$.

7227. $n^{n/2}$. *Hint.* Calculating the product of the matrix by its adjoint, use the relations of orthogonality for characters.

7228. a)

	e	(12)	(123)
φ_0	1	1	1
φ_1	1	-1	1
φ_2	2	0	-1

Hint. Make use of Exercises 7226 and 7019.

b)

	e	(12)	(123)	(12)(34)	(1234)
φ_0	1	1	1	1	1
φ_1	1	-1	1	1	-1
φ_2	3	1	0	-1	-1
φ_3	3	-1	0	-1	1
φ_4	2	0	-1	2	0

Hint. Make use of Exercises 7226, 7223, 7224, 7220.

c)

	1	-1	i	j	k
φ_0	1	1	1	1	1
φ_1	1	1	1	-1	-1
φ_2	1	1	-1	1	-1
φ_3	1	1	-1	-1	1
φ_4	2	-2	0	0	0

Hint. Make use of Exercises 7226 and 7220.

d)

	e	b	b^2	a	ab^2
φ_0	1	1	1	1	1
φ_1	1	-1	1	-1	1
φ_2	1	-1	1	1	-1
φ_3	1	1	1	-1	-1
φ_4	2	0	-2	0	0

Hint. Make use of Exercises 7226 and 7222.

e)

	e	b	b^2	a
φ_0	1	1	1	1
φ_1	1	1	1	1
φ_2	2	$2\cos\frac{2\pi}{5}$	$2\cos\frac{4\pi}{5}$	0
φ_3	2	$2\cos\frac{4\pi}{5}$	$2\cos\frac{2\pi}{5}$	0

Hint. Make use of Exercises 7226 and 7222.

f)

	e	$(12)(24)$	(123)	(132)
φ_0	1	1	1	1
φ_1	1	1	ε	ε^2
φ_2	1	1	ε^2	ε
φ_3	3	-1	0	0

where ε is a root of 1 of order 3 in $\mathbb{C}$. *Hint.* Make use of Exercise 7226.

7229. It cannot take these values. *Hint.* A scalar square of the given function is not an integer.

7230. With the notation of Exercise 7228c, $F = 2\varphi_4 + 0, 5\varphi_1 + 0, 5\varphi_3$.

7231. With the notation of the answer to Exercise 7228a, we have $f_1 = -\varphi_0 + 3\varphi_1 + 2\varphi_2$, $f_2 = 4\varphi_1 + \varphi_2$. It follows that f_1 is not a character of any representation. f_2 is a character of a direct sum of an irreducible two-dimensional representation of a group S_3 and four copies of a nontrivial one-dimensional representation of this group.

7232. *Hint.* a) Prove that the mapping from A into $\mathbb{C}$, which moves χ to $\chi(a)$ for some $a \in A$, is a character of A. Prove that the induced mapping $A \to \hat{A}$ is an isomorphism.

7233. *Hint.* c) Deduce the equality $f(a) = \sum_{\chi \in \hat{A}} f(\chi) \cdot \chi(a)$ with the help of a). Prove that under the isomorphism from Exercise 7232c $\hat{\hat{f}}$ goes to $(|A|)^{-1} f$.

7234. *Hint.* Make use of the equality $(f, f)_A = \sum_{\chi \in \hat{A}} (f, \chi)_A^2$.

7237. a) $\chi_\Psi = \Psi_0 + \Psi_1 + \Psi_2$.

b) $\chi_\Psi = \Psi_0 + \Psi_1 + \Psi_2 + \Psi_4$.

c) $\chi_\Phi = \Psi_0 + \Psi_1 + \Psi_2 + \Psi_3$.

7238. $\chi_\Psi = n \cdot \chi_\Phi$.

7239. n^{m-1}. *Hint.* Prove that all irreducible representations of G occur in $\rho^{\otimes m}$ with the same multiplicity.

7240. a) $\chi_{\rho^2} = \Psi_0 + \Psi_1 + \Psi_2$.

b) $\chi_{\rho^3} = \Psi_0 + \Psi_1 + 3\Psi_2$.

7241. If $\bar{n} = [(n+1)/2]$ and $\bar{m} = [m/2]$, then the multiplicity is equal to $\binom{\bar{n}-1}{\bar{m}}$.

7242. *Hint.* Consider the representation in a space of skew-symmetric twice contravariant tensors.

7243. In the notation of the answer to Exercise 7228: a) φ_1.

b) $\varphi_0 + \varphi_2$.

c) $\varphi_0 + \varphi_1$.

d) $\varphi_1 + \varphi_2$.

7302. a) $\begin{pmatrix} 0 & -1 \\ 1 & 0 \end{pmatrix}$. b) $\begin{pmatrix} 0 & 1 \\ 1 & 0 \end{pmatrix}$. c) $\begin{pmatrix} 1 & 0 \\ 0 & 0 \end{pmatrix}$. d) $\begin{pmatrix} 1 & 0 \\ 0 & -1 \end{pmatrix}$. e) $\begin{pmatrix} 0 & 1 \\ 0 & 0 \end{pmatrix}$.

f) $\begin{pmatrix} 0 & 1 \\ 0 & 1 \end{pmatrix}$.

7303. b) and d); c) and f).

7304. Representations are equivalent if and only if, for any k and λ, the Jordan canonical forms of A and $-A$ have the same number of Jordan boxes of size k with the eigenvalue λ.

7305. a) An arbitrary representation has the form $R_A(t) = e^{(\ln t)A}$, where $a \in M_n(\mathbb{C})$.

b) An arbitrary representation is equivalent to a representation of the form:

$$R_{A,B}(t) = \begin{pmatrix} e^{\ln |t| \cdot |A|} & 0 \\ 0 & (\text{sgn} t) e^{\ln |t| \cdot B} \end{pmatrix}, \quad A \in M_p(\mathbb{C}), \quad B \in M_q(\mathbb{C}).$$

Hint. Consider the image of $-1 \in \mathbb{R}^*$ under the given representation. Prove that its eigensubspaces are invariant. Apply a).

c)   An arbitrary representation is equivalent to a representation of the form

$$z \to \begin{pmatrix} z^{k_1} & & & 0 \\ & z^{k_2} & & \\ & & \ddots & \\ 0 & & & z^{k_n} \end{pmatrix} \quad (k_1, \ldots, k_n \in \mathbb{Z}).$$

Hint. Prove that a representation of the additive group $\mathbb{C}$ has the form P_A, $e^{2\pi i A} = E$ (see Exercise 7301), if it is the composition of the homomorphism $\mathbb{C} \to \mathbb{C}^*$ $(t \to e^t)$ and a representation of $\mathbb{C}^*$. Prove next that the matrix A is similar to an integer diagonal matrix.

d)   An arbitrary representation is equivalent to a representation of the form

$$z \to \begin{pmatrix} z^{k_1} & & & 0 \\ & z^{k_2} & & \\ & & \ddots & \\ 0 & & & z^{k_n} \end{pmatrix} \quad (k_1, \ldots, k_n \in \mathbb{Z}).$$

Hint. Consider a representation of the additive group of the field $\mathbb{R}$ obtained as the composition of the homomorphism $\mathbb{R} \to \mathbf{U}$ $(t \to e^{it})$ and a representation of the group $\mathbf{U}$. Apply Exercise 7301.

7306. Yes, it can. *Hint.* Prove that any nonsingular complex square matrix has the form e^A. Apply Exercise 7301.

7307. Linear spans of sets of eigenvectors of A.

7309. *Hint.* Consider restriction of Φ_n to the subgroup of diagonal matrices.

7310. *Hint.* e) Prove that the equality takes place on the subset of diagonalizable matrices.

7311. *Hint.* Notice that $\mathbf{SU}_2 = \{A \in \mathbb{H} \mid (A, A) = 1\}$. Prove that if $A \in \mathbf{SU}_2$ has eigenvalues $e^{\pm i\varphi}$, then $P(A)$ is a rotation of the space $\mathbb{H}_0$ on the angle 2φ around an axis passing through $A - \frac{1}{2}(\operatorname{tr}A)E \in \mathbb{H}_0$.

b)   Prove that the group $R(\mathbf{SU}_2 \times \mathbf{SU}_2)$ acts transitively on the unit sphere in $\mathbb{H}$. Make use of a).

c)   A complexification of $\mathbb{H}_0$ is a subspace of matrices of the form $\begin{pmatrix} a & b \\ c & -a \end{pmatrix}$ in $\mathbf{M}_2(\mathbb{C})$. The required isomorphism associates a polynomial $f(x, y) = -bx^2 + 2axy + cy^2$ to the matrix mentioned.

7313. and **7314.** *Hint.* Suprunenko D.A. Group of Matrices, Nauka, Moscow, 1972, ch. 5.

Theoretical material

1   Affine and Euclidean geometry

An *affine space* over a field K, is a pair (A, V), consisting of the vector space V over K and the set A whose elements are called points. It is assumed that the pair (A, V) has an operation of addition of points and vectors

$$(a, v) \rightarrow a + v \in V.$$

This operation satisfies the following conditions:

(1) $(a + v_1) + v_1 = a + (v_1 + v_2)$ for all $a \in A$, $v_1, v_2 \in V$;

(2) $a + 0 = a$ for all $a \in A$;

(3)   for any two points $a, b \in A$ there exists a unique vector $v \in V$ such that $a + v = b$ (this vector is denoted by $\overline{ab}$).

The term 'affine space' is frequently applied only to the first member of the pair (A, V). In this case V is called the *vector space associated* with the given affine space.

The *dimension* of the affine space (A, V) is the dimension of the vector space V.

Any vector space V can be considered as an affine space in the following sense: put $A = V$ and define the operation of addition as addition in the space V.

An *affine subspace* or a *plane* in an affine space (A, V) is a pair (P, U), where U is a subspace of V and P is a non-empty subset of A such that

(1) $p + u \in P$ for all $p \in P, u \in U$;

(2) $\overline{pq} \in U$ for all $p, q \in P$.

The pair (P, U) is itself an affine space.

The term 'affine subspace' or 'plane' is frequently applied only to the first member of the pair (P, U). In this case the subspace U is uniquely determined by P and it is called the *directing subspace* of the given affine subspace.

A one-dimensional affine subspace is called a *line*. An affine subspace whose dimension equals the dimension of the space minus one is called a *hyperplane*.

If S is a non-empty subset of an affine space A, then the least plane in A containing S is called the *affine hull* of S. It is denoted by $\langle S \rangle$. The set of $k+1$ points $a_0, a_1, \ldots, a_k$ in affine space A is *affine independent* if $\dim \langle a_0, a_1, \ldots, a_k \rangle = k$. In this case we say that the points $a_0, a_1, \ldots, a_k$ are in the *general position*.

Two disjoint planes (P_1, L_1), (P_2, L_2) in an affine space are *parallel* if either $L_1 \subset L_2$ or $L_2 \subset L_1$. Planes are *crossed* if $L_1 \cap L_2 = \{0\}$. In the general case, the number $\dim(L_1 \cap L_2)$ is the *degree of parallelism* of (P_1, L_1) and (P_2, L_2).

A system of affine coordinates in an affine space (A, V) is a set $(a_0, e_1, \ldots, e_n)$ consisting of a point a_0 (the origin) and the basis $(e_1, \ldots, e_n)$ of the vector space V. Coordinates of a point $a \in A$ with respect to this system are coordinates of the vector $\overline{a_0 a_1}$ with the basis $(e_1, \ldots, e_n)$.

An *affine mapping* from an affine space (A, V) to an affine space (B, W) is a pair (f, Df) consisting of a mapping $f : A \to B$ and a linear mapping $Df : V \to W$, such that $f(a + v) = f(a) + Df(v)$ for all $a \in A$, $v \in V$. Bijective affine mapping from an affine space into itself is an *affine transformation*. Frequently the term 'affine mapping' (or 'transformation') is applied only to the first member of the pair (f, Df), namely, to f. The linear mapping Df is called the *linear part* or the *differential* of f.

All affine transformations of an affine space form a group which is called an *affine group*. It is denoted by Aff A.

An affine transformation with an identical differential is called a *parallel transfer*. The parallel transfers form a subgroup of Aff A which can be identified with the additive group of V: each vector $v \in V$ corresponds to the parallel transfer $t_0 : a \to a + v$.

A *configuration* in an affine space A is an ordered set of affine subspaces $\{P_1, \ldots, P_s\}$. Two configurations $\{P_1, \ldots, P_s\}$ and $\{Q_1, \ldots, Q_s\}$ in A are affine congruent if there exists an affine transformation f such that $f(P_i) = Q_i$ for $i = 1, \ldots \sigma$.

It is assumed in the following that (A, V) is an affine space over the field of real numbers.

For any points $a, b \in A$, $a \neq b$, the set of all points $\lambda a + (1 - \lambda)b$, where $0 \leq \lambda \leq 1$, is called an interval connecting a with b. A non-empty subset $M \subseteq A$ is *convex* if, for any two of its points, M contains the interval connecting these points. The dimension of a convex set is the dimension of its affine hull. A convex set is *solid* if its affine hull coincides with the whole space.

A point of a convex set M is termed *internal* if it belongs to an open kernel of M in its affine hull. Otherwise a point is termed *boundary*. A point of a convex set M is *extreme* if it is not an internal point of any interval belonging to M.

The open kernel of a convex set M in a space A is denoted by M°. (If M is not solid then $M^\circ = \emptyset$.)

The least convex set containing a given non-empty set $S \subseteq A$ is called the *convex hull* of S and it is denoted by conv S. The convex hull of $n + 1$ points in the general position is called an *n-dimensional simplex*.

For any non-constant affine linear function f on a space A, the set defined by the inequality $f(x) \geq 0$ is convex. It is called the *half-space* bounded by the hyperplane $\{x \mid f(x) = 0\}$. Each hyperplane bounds two half-spaces. A set S lies in one part of a hyperplane H if it is contained in one of two half-spaces bounded by H. In this case, if $S \cap H = \emptyset$, we say that S strictly lies in one part of H.

 A hyperplane H having a common point with a closed convex set M is called a *basic hyperplane* of M if M lies in one part of H. A nonempty intersection of finitely many half-spaces is called a *convex polyhedron*. Note that a convex polyhedron is a set of all points whose coordinates satisfy some compatible system of non-strong linear inequalities. A subset of n-dimensional affine space, defined in some system of affine coordinates by inequalities $0 \leq x_i \leq 1$ $(i = 1, \ldots, n)$, is called an *n-dimensional parallelepiped*.

The intersection of a convex polyhedron M with a basic hyperplane is called a *side* of M. Zero-dimensional sides are called *vertices*, one-dimensional sides are called *edges*. Any side of a convex polygon is also a convex polygon.

A subset K of some space V is a *convex cone* if $x + y \in K$, $\quad \lambda x \in K$ for all $x, y \in K$ and all $\lambda > 0$. Any convex cone is a convex set in the space V considered as an affine space.

An affine space (E, V) over the field of real numbers is a *Euclidean space* if the vector space V has a Euclidean vector space structure. One can define in Euclidean space E the distance between two points: if $a, b \in E$ then $\rho(a, b) = |\overline{ab}|$, where $|v| = \sqrt{(v, v)}$ is the length of $v \in V$. Similarly one can define the distance between two planes $P, Q \subseteq E$:

$$\rho(P, Q) = \inf\{\rho(a, b) \mid a \in P, \quad b \in Q\}.$$

The *angle* between two planes is defined as the angle between their directing subspaces. Planes are *perpendicular* if the angle between them is equal to $\pi/2$.

An *isometry* of Euclidean space is an affine transformation whose differential is an orthogonal operator. An isometry f is *proper* if $\det Df = 1$. Isometries preserve the distances between points. All isometries of an Euclidean space E form a group, denoted by $\mathrm{Isom}\, E$. The group of proper isometries is denoted by $\mathrm{Isom}_+ E$.

Two configurations $\{P_1, \ldots, P_s\}$ and $\{Q_1, \ldots, Q_s\}$ in an Euclidean space E are *metric congruent* if there exists an isometry f of E such that $f(P_i) = Q$, $(i = 1, \ldots, s)$.

2 Hypersurfaces of the second order

Let (A, V) be an affine space over a field K. A *quadratic function* $Q : A \to K$ on A is a function of the form

$$Q(a_0 + x) = q(x) + l(x) + c,$$

where a_0 is a point of A; $q : V \to K$ is a quadratic function, and $l : V \to K$ is a linear function, $c \in K$. The quadratic function q, not depending on a_0, is called the *quadratic part* of Q. The linear function l is called the *linear part* of Q with respect to the point a_0. A *hypersurface of the second order* in A or a *quadric* is a set of the form

$$X = X_Q = \{a \in A \mid Q(a) = 0\}.$$

A quadratic function Q can be written in an affine system of coordinates $(a_0; e_1, \ldots, e_n\}$ in the expanded form

$$Q(a_0 + x) = \sum_{i,j=1}^{n} a_{ij}x_i x_j + 2\sum_{i=1}^{n} b_i x_i + c,$$

where $a_{ij} = a_{ji}$. The symbol A_0 denotes the matrix (a_{ij}) of the quadratic function q with the basis $(e_1, \ldots, e_n)$. The symbol A_Q denotes the matrix

$$A_Q = \left(\begin{array}{c|c} A_q & \begin{array}{c} b_1 \\ \vdots \\ b_n \end{array} \\ \hline b_1 \ldots b_n & c \end{array} \right).$$

The determinants of A_Q and A_q are denoted by Δ and δ. Thus A_Q, A_q, Δ, and δ depend on the choice of the system of coordinates $(a_0; e_1, \ldots, e_n)$, however the ranks of matrices A_Q and A_q do not depend on this choice. A quadric is *nonsingular* if $\Delta \neq 0$. It is *singular* if $\Delta = 0$.

A hyperplane

$$\sum_{i,j=1}^{n} (a_{ij}x_j^0 + b_i)(x_i - x_i^0) = 0$$

is *tangent* to X_Q at a point $a_0 = (x_1^0, \ldots, x_n^0) \in X_Q$, if $\sum_{j=1}^{n} a_{ij}x_j^0 + b_i \neq 0$ for some i. If this condition is not satisfied then the point $a_0 \in X_Q$ is *singular*.

A vector $r = (r_1, \ldots, r_n)$ is a *vector of asymptotic direction* if $q(r) = 0$.

A point $a \in Q$ is a *central point* of a quadratic function Q (or of a quadric X_Q) if the linear part of Q with respect to a_0 vanishes. Thus $Q(a_0 + x) = Q(a_0 - x)$ for all $x \in L$. The *center of a function* Q (or a quadric X_Q) is the set of its central points.

A cone $\{a \in A \mid q(a - a_0) = 0\}$, where a_0 is a central point, does not depend on the choice of the point. It is called the *asymptotic cone* of X_Q.

For any quadratic function Q there exists an affine system of coordinates in A such that Q is of one of the following forms:

a)   if q is nonsingular then

$$Q(x_1, \ldots, x_n) = \sum_{i=1}^{n} \lambda_i x_i^2 + c \quad (\lambda_i, c \in K, \ \lambda_i \neq 0);$$

b) if q is singular of rank r and the center of Q is non-empty then

$$Q(x_1, \ldots, x_n) = \sum_{i=1}^{r} \lambda_i x_i^2 + c \quad (\lambda_i, c \in K, \ \lambda_i \neq 0);$$

c) if q is singular of rank r and the center of Q is empty then

$$Q(x_1, \ldots, x_n) = \sum_{i=1}^{r} \lambda_i x_i^2 + x_{r+1}.$$

The equation of a quadric in some affine system of coordinates has, over the field of real numbers, one of the following forms:

$$(I_{r,s}): \ x_1^2 + \cdots + x_s^2 - x_{s+1}^2 - \cdots - x_r^2 = 1 \quad (0 \leq s \leq r \leq n);$$

$$(I'_{r,s}): \ x_1^2 + \cdots + x_s^2 - x_{s+1}^2 - \cdots - x_r^2 = 0 \quad (0 \leq s \leq r \leq n, \ s \geq r/2);$$

$$(II_{r,s}): \ x_1^2 + \cdots + x_s^2 - x_{s+1}^2 - \cdots - x_r^2 = 2x_{r+1}(0 \leq s \leq r \leq n-1, \ s \geq r/2).$$

3 Projective space

The *projective space* associated with a vector space V over a field K is the set $P(V)$ of one-dimensional subspaces of V. Elements of $P(V)$ are called *points* of the projective space. Projective spaces have the following structures.

a) $P(V)$ has distinguished subsets, called projective subspaces, or planes. A subset of $P(V)$ is a *projective subspace* if it is the subset of all one-dimensional subspaces of some subspace of V.

b) There exists a family of injective mappings from affine spaces into $P(V)$, called *affine charts*. An affine chart $\varphi : A \to P(V)$ is defined by an affine hyperplane A in V (considered as an affine space which does not contain the origin). Each point of A is associated with a unique one-dimensional subspace of V containing this point.

c) A family of bijective mappings from $P(V)$ into itself is given, called the *projective transformations*. A projective transformation α is a mapping $\alpha : P(V) \to P(V)$ induced by a nonsingular linear operator $\mathcal{A} : V \to V$, i.e. the image under α of an element of $P(V)$ is equal to the image under $\mathcal{A}$ of the one-dimensional subspace representing this element.

A projective transformation maps subspaces into subspaces. The composition of an affine chart with a projective transformation is again an affine chart. The *dimension* of a space $P(V)$ is equal to dim $V - 1$. A projective space of dimension

n is frequently denoted by $\mathbf{P}^n$. An affine chart is often identified with the affine space A and with its image in $P(V)$. Coordinates given in some affine charts are called *nonhomogeneous coordinates* in projective space. These coordinates are not defined on the whole space.

Given a system of coordinates in a space V (with respect to some basis) and a point in $P(V)$, we can consider coordinates of every vector of appropriate one-dimensional subspace. These coordinates are determined up to proportionality. They are called the *homogeneous coordinates* of a point of the projective space.

Let V have a system of coordinates $x_0, x_1, \ldots, x_r$. The affine chart defined by the hyperplane $x_i = 1$ is called the ith *coordinate affine chart*.

If q is a quadratic function on V then the set of one-dimensional subspaces contained in the cones $q(x) = 0$ is a *quadric* in $P(V)$.

For four different points, p_1, p_2, p_3, p_4, in the projective line $\mathbf{P}^1$, we can define the *cross-ratio* $\delta(p_1, p_2, p_3, p_4)$, which is an element of the basic field. In order to calculate the cross-ratio take an arbitrary system of homogeneous coordinates in $\mathbf{P}^1$, and denote by $\Delta(p_i, p_j)$ the determinant of the matrix of size 2 composed by coordinates of points p_i and p_j. Then

$$\delta(p_1, p_2, p_3, p_4) = \frac{\Delta(p_1, p_4)}{\Delta(p_3, p_4)} \cdot \frac{\Delta(p_3, p_2)}{\Delta(p_1, p_2)}.$$

The right-hand side of this formula does not depend on the system of coordinates. In nonhomogeneous coordinates

$$\delta(p_1, p_2, p_3, p_4) = \frac{(x_4 - x_1)(x_2 - x_3)}{(x_4 - x_3)(x_2 - x_1)},$$

where x_i is the coordinate of p_i.

Consider an affine chart and an affine system of coordinates in this chart. We can define subsets of the projective space by equations with respect to the non-homogeneous coordinates. These subsets are contained inside the chart. In some cases it is possible to complement them to a projective space: a subset of a chart defined by linear equations (i.e. an affine subspace), has a unique complementation inside the projective subspace. A subset of a chart which is an affine quadric has a unique complementation to a projective quadric. These complementations are considered in some exercises.

Points of projective spaces which do not belong to the given affine chart are called *points at infinity* with respect to this chart.

4   Tensors

A *tensor of type* (p, q) or the *p times covariant and q times contravariant tensor*, in a vector space V, is a function on $\underbrace{V \times \cdots \times V}_{p \text{ times}} \times \underbrace{V^* \times \cdots \times V^*}_{q \text{ times}}$, which is linear as a function on each of its $p + q$ arguments. Tensors of type (p, q) form a vector space $\mathbf{T}_p^q(V)$. It is naturally identified with *the tensor product*

$$\underbrace{V \otimes \cdots \otimes V}_{q \text{ times}} \otimes \underbrace{V^* \otimes \cdots \otimes V^*}_{p \text{ times}}.$$

The *coordinates of the tensor* $T \in \mathbf{T}_p^q(V)$ (in some basis of V) are denoted by $t_{i_1 \dots i_p}^{j_1 \dots j_q}$.

If $\operatorname{char} K = 0$, then consider in $\mathbf{T}_0^q(V)$ the linear operators Sym and Alt:

$$(\operatorname{Sym} T)(f_1, \dots, f_q) = \frac{1}{q!} \sum_{\sigma \in S_q} T(f_{\sigma(1)}, \dots, f_{\sigma(q)}),$$

$$(\operatorname{Alt} T)(f_1, \dots, f_q) = \frac{1}{q!} \sum_{\sigma \in S_q} (\operatorname{sgn} \sigma) T(f_{\sigma(1)}, \dots, f_{\sigma(q)}).$$

These operators are projections, onto the subspaces $S^q(V)$, and $\Lambda^q(V)$ of symmetric and skew-symmetric tensors respectively.

Elements of $\Lambda^p(V)$ are often called *p-vectors*. (In the case $p = 2$, *p*-vectors are also called *bivectors*.)

The space $\mathbf{S}(V) = \oplus_{q=0}^{\infty} S^q(V)$ with the operation of multiplication $xy = \operatorname{Sym}(x \otimes y)$ is an algebra called the *symmetric algebra of the space V*. The space $\Lambda(V) = \oplus_{q=0}^{\infty} \Lambda^q(V)$, with the operation $x \wedge y = \operatorname{Alt}(x \otimes y)$ is an algebra called the *external algebra of the space V* or the *Grassman algebra*.

5 Elements of representation theory

Different authors traditionally use different approaches in presenting the main definitions and basic results of the representation theory of groups. The development of the theory clarifies connections between various definitions and shows the ways 'of translation from one language to another'. We are not adopting a single approach for exposition of the material. It seems to us that students should become acquainted with the different approaches accepted in the literature, and that teachers should have the opportunity of finding exercises which use a convenient language.

The main features of the basic approaches to the theory of representation of groups, and some variants of terminology, are presented here.

A. Terminology of linear representations. A *linear representation* of a group G in a space V is a homomorphism $\Phi : G \to \mathbf{GL}(V)$ from the group G into the group of nonsingular linear operators on V. The dimension of V is the *dimension* or the *degree of the representation*. A *homomorphism of a representation* Φ of G in a space V into a representation Ψ of G in a space W is a linear mapping $\alpha : V \to W$, such that $\alpha(\Phi(g)v) = \Psi(\alpha(v))$ for all $g \in G$, $v \in V$. If a homomorphism α is an isomorphism of spaces then the representations Φ and Ψ are *isomorphic*.

A subspace $U \subset V$ of a representation Φ of a group G is *invariant* if $\Phi(g)U = U$ for all $g \in G$. A representation of a nonzero degree is *irreducible* if the zero subspace and the whole space are the only invariant subspaces.

B. Terminology of matrix representations. A *matrix representation* of a group G of degree n over a field F is a homomorphism $\rho : G \to \mathbf{GL}_n(F)$ from G into the group of invertible matrices of size n over F. Two matrix representations ρ and σ of a group G of the same size n over F are *equivalent (isomorphic)* if there exists a nonsingular matrix $C \in \mathbf{M}_n(F)$ such that $\rho(g) = C^{-1}\sigma(g)C$ for all $g \in G$.

A matrix representation is *reducible* if it is equivalent to a representation in which all matrices have the same 'zero corner', i.e. they have the form $\begin{pmatrix} A & C \\ 0 & B \end{pmatrix}$, where A and B are matrices of sizes r and s fixed for all $g \in G$.

C. Terminology of linear G-spaces. Let G be a group and let V be a linear space. V is said to have a linear G-space structure if there exists a pairing $G \times V :\to V, (g, v) \to g * v$, such that $v \to g * v$ is a linear operator on V and $g_1 * (g_2 * v) = (g_1 g_2) * v$ for all $g_1, g_2 \in G$, $v \in V$. Two G-spaces V and W are isomorphic if there exists an isomorphism of spaces $\alpha : V \to W$ such that $\alpha(g * v) = g \cdot \alpha(v)$ for all $g \in G$, $v \in V$.

A subspace U of G-space V is *invariant* if $g * u \in U$ for all $g \in G$, $u \in U$. A nonzero G-space V is *irreducible* if it has only trivial invariant subspaces.

D. Terminology of modules over group algebras. A space V is a *module* over the group algebra $F[G]$, or $F[G]$-module if there exists a pairing $F[G] \times V \to V, (a, v) \to a \cdot v$, such that $a_1(a_2 \cdot v) = (a_1 a_2) \cdot v, a \cdot (v_1 + v_2) = a \cdot v_1 + a \cdot v_2$. Two $F[G]$-modules V and W are isomorphic if there exists a linear isomorphism $\alpha : V \to W$ such that $\alpha(a \cdot v) = a \cdot \alpha(v)$ for all $a \in F[G]$, $v \in V$.

A subspace U of $F[G]$-module V is a *submodule* if $a \cdot u \in U$ for all $a \in F[G]$, $u \in U$. A nonzero module V is *simple* or *irreducible* if it has only trivial submodules.

Note that the group G is contained in $F[G]$. (We identify elements of G with sums having one nonzero coefficient which is equal to 1.) If V is a $F[G]$-module then V has a natural structure of G-space, namely $(g, v) \to g \cdot v$.

Conversely, if V is a G-space then we can put

$$\left(\left(\sum \alpha_g \cdot g\right), v\right) \to \sum \alpha_g(g * v).$$

It follows that V is a $F[G]$-module.

If Φ is a linear representation of G in V then the operation $(g, v) \to \Phi(g)v$ induces a structure of G-space on V.

If V is a G-space and $\Phi(g) : v \to g * v$ then $\Phi(g)$ is a linear operator on V. It is easy to show that $g \to \Phi(g)$ is a linear representation of G in V.

Given a linear representation of G in n-dimensional space V we choose a basis of V, and associate with each element $g \in G$ the matrix of the operator $\Phi(g)$ in this basis. We obtain a mapping from G to $\mathbf{GL}_n(F)$ which is a matrix representation of G. If we choose another basis then we obtain an equivalent matrix representation.

Given an n-dimensional matrix representation ρ of G let there be associated with each element $g \in G$ an operator of multiplication by the matrix $\rho(g)$ in the space F^n. Then we obtain a linear representation of G in F^n.

It is clear that these transformations from $F[G]$-modules to G-spaces, linear and matrix representations, and their converses, transform irreducible representations into other irreducible representations and similarly, isomorphic objects remain isomorphic after transformation.

The operation of multiplication in $F[G]$ produces an $F[G]$-module structure in the space $V = F[G]$; the corresponding linear representation of G in V is called *regular*. We can also define a regular representation by considering the space V with the basis $(e_g/g \in G)$ and defining the mapping $R : G \to \mathbf{GL}(V)$ by the rule $R(h)e_g = e_{hg}$ for all $g, h \in G$. The basis (e_g) is called a *canonical basis* of the space of the regular representation.

The main theorems on representations of groups are as follows.

Theorem. Let G' be the commutant of a group G and $\varphi : G \to G/G'$ be the canonical homomorphism. Then the formula $\psi \to \psi \circ \varphi$ establishes a one-to-one correspondence between one-dimensional representations of groups G and G/G'.

Theorem (Maschke). Let a group G be finite and let $\operatorname{char} F$ not divide $|G|$. Then any finite-dimensional representation of G over F is isomorphic to the direct sum of the irreducible representations.

Theorem. Let G be a finite group and F be an algebraically closed field. Let $\operatorname{char} F$ not divide $|G|$. Then the number of distinct irreducible representations of G over F is equal to the number of classes of conjugate elements in G. The sum of the squares of the dimensions of these representations is equal to the order of G.

List of definitions

The definitions for the notation in Chapters 11, 12 and 15 are given in the Theoretical Material section.

Absolute value on a field: see Exercise 6635.

Action of a group on a set: a group G acts on a set M if each element $g \in G$ induces a bijection $M \to M, m \to gm$ and $g_1 g_2(m) = (g_1 g_2)(m), 1m = m$, for all $g_1, g_2 \in G, m \in M$.

Algebra

—   **$\sim$ of formal power series** (in a variable x over a field K): a set of formal expressions $\sum_{k=0}^{\infty} a_k x^k (a_k \in K)$ with natural addition and multiplication by elements of K and with the operation of multiplication

$$\sum_{k=0}^{\infty} a_k x^k \cdot \sum_{k=0}^{\infty} b_k x^k = \sum_{k=0}^{\infty} c_k x^k, \quad \text{where} \quad c_k = \sum_{\substack{i+j=k, \\ i \geq 0, j \geq 0}} a_i b_j.$$

—   **$\sim$ of generalized quaternions**: see Exercise 6503.

—   **$\sim$ of differential operators**: see Weyl algebra in Exercise 6327.

—   **Banach $\sim$**: complete normed algebra.

—   **External $\sim$ of a vector space**: see Theoretical Material IV and see Exercise 6324

—   **Grassman $\sim$ of a vector space**: the external algebra of the space.

—   **Group $\sim$** (of a group G over a field F): the set of finite formal linear combinations $\sum \alpha_g g$ ($g \in G, a_g \in F$) with natural addition and multiplication by elements of F and with the operation of multiplication

$$\alpha_g g \cdot \alpha_h h = \alpha_g \alpha_h gh,$$

which is extended to linear combinations by the law of distributivity.

—   **Noetherian $\sim$** (commutative): a commutative algebra in which any ascending sequence of ideals is finite.

—   **Normed $\sim$** (over a normed field K): an algebra having a function $\|x\|$, $x \in A$, taking the non-negative real values such that

       a) $\|x\| \geq 0$ and $\|x\| = 0$ if and only if $x = 0$;

       b) $\|x + y\| \leq \|x\| + \|y\|$;

c) $\|\lambda x\| = |\lambda| \cdot \|x\|$, where $\lambda \in K, x \in A$;

d) $\|xy\| \le \|x\| \cdot \|y\|$.

— **Semisimple** $\sim$: an algebra in which the zero ideal is the only two-sided ideal consisting of nilpotent elements; in the commutative case, an algebra in which zero is the only nilpotent element.

— **Separable** $\sim$: see Exercise 6735.

— **Simple** $\sim$: an algebra in which the zero ideal and the whole algebra are the only two-sided ideals.

— **Symmetric** $\sim$ **of a vector space**: see Theoretical Material IV and Exercise 6325.

— **Tensor** $\sim$ **of a vector space**: see Exercise 6323.

— **Weil** $\sim$: see Exercise 6327.

Annihilator: see Exercise 6465.

Axis of a homology: see Exercise 5323.

Center of a homology: see Exercise 5323.

— $\sim$ **of a group** (ring): the set of all elements commuting with all elements of the group (ring).

Centralizer of an element of a group: the set of all elements of the group, commuting with the given element.

Circulant: see Exercise 1503.

Codimension of a subspace: the difference between the dimension of the space and the dimension of the subspace.

Commutant of a group: the subgroup generated by all commutators of elements of the group.

Commutator of elements x and y

— $\sim \sim \sim \sim \sim \sim$ **of a group**: the element $xyx^{-1}y^{-1}$.

— $\sim \sim \sim \sim \sim \sim$ **of a ring**: the element $xy - yx$.

Completion of a metric space: the completion with respect to Cauchy sequences.

Component of semisimple (reduced) algebra: see Exercise 6718.

Continuant: see Exercise 1612.

Coordinates

— **Barycentric** $\sim$: the coordinates $\lambda_0, \lambda_1, \ldots, \lambda_n$ of a point x in affine space with respect to a system of points $x_0, x_1, \ldots, x_n$, in a general position which are defined by the equality

$$x = \sum_{i=0}^{n} \lambda_i x_i, \qquad \text{where} \quad \sum_{i=0}^{n} \lambda_i = 1.$$

Correlation: see Exercise 5329.

Cross-ratio: see Exercise 5328.

Decrement of a permutation: the difference between the degree of the permutation and the number of cycles in its decomposition into a product of disjoint cycles (cycles of length 1 are included).

Degree
 — $\sim$ **of a separable algebra**: see Exercise 6728.
 —   **Transcendence** $\sim$ **of an extension**: see Exercise 6717.

Elementary row (column) transformations of a matrix: multiplication of a row (column) by an invertible element (type I); addition to a row (column) of another row (column) multiplied by an element (type II); sometimes permutation rows (columns).

K**-embedding**: an injective K-homomorphism of algebras over some extension of the field K.

Extension
 —   **Galois** $\sim$: see Exercise 6754.
 —   **Normal** $\sim$: see Exercise 6730.
 —   **Separable** $\sim$: see Exercise 6739.
 —   **Purely nonseparable** $\sim$: see Exercise 6748.

Field
 — $\sim$ **of decomposition of a polynomial**: the least splitting field of the polynomial.
 —   **Normed** $\sim$: a field with an absolute value, see Exercise 6635.
 —   **Splitting** $\sim$ **of an algebra**: see Exercises 6727, 6725.
 —   **Splitting** $\sim$ **of a polynomial**: an extension of the field of coefficients of the polynomial over which the polynomial can be factorized into a product of linear factors.

Flag of subspaces: see Exercise 5705.

Function
 — $\sim$ **of Möbius**: the function of natural number n which is defined by the equality

$$\mu(n) = \begin{cases} 1, & \text{if } n = 1, \\ (-1)^r, & \text{if } n \text{ is a product of } r \text{ distinct primes}, \\ 0 & \text{otherwise.} \end{cases}$$

 —   **Euler** $\sim$: if $n = 1$ then it is equal to 1, if $n > 1$ then it is equal to the number of natural numbers less than n and coprime with n.

Gram determinant: the determinant of a Gram matrix.

Group
 —   **Dihedral** $\sim$, $\mathbf{D}_n$: the group of isometries of the plane which map a regular n-gon onto itself.

— $\sim$ **of quaternions Q_8**: the set of the elements $\pm 1, \pm i, \pm j, \pm k$ with the
         same multiplication of elements as in the skew-field of quaternions.
—   **Divisible** $\sim$: an abelian group in which the equation $nx = a$ has a
         solution for any element a and any integer n.
—   **Dual abelian** $\sim$: see Exercise 7214.
—   **Galois** $\sim$ **of an extension**: see Exercise 6754.
—   **Galois** $\sim$ **of a polynomial**: see Exercise 6756.
—   **Kleinian** $\sim$, V_4: the group of the permutations

$$\{e, (12)(34), (13), (24), (14)(23)\}$$

and any group isomorphic to it.
—   **Periodic** $\sim$: a group in which all elements have finite orders.
— p-**group**: a group in which all elements have orders p^n ($n \in \mathbb{N}$).

Homology: see Exercise 5323.

Homomorphism unitary: a homomorphism of rings (algebras) such that a unit
goes to a unit.

Isometry: a mapping from an Euclidean space into itself preserving distances
between points.

Ideal
—   **Maximal** $\sim$: an ideal of a ring (of an algebra) which is not contained
         strictly in any proper ideal of the ring (the algebra).
—   **Prime** $\sim$ (of a commutative ring): an ideal such that the factor-ring
         (factor-algebra) by this ideal has no zero divisors.

Idempotent: an element of a ring whose square is equal to this element.

Idempotents orthogonal: a set of idempotents such that any product of distinct
idempotents is equal to zero.

Involution: see Exercise 6503.

Lie superalgebra: see Exercise 6506.

Matrix
— $\sim$ **units**: the square matrices E_{ij} ($i, j = 1, \dots, n$) in which the only
         nonzero entry is equal to 1 and it is located at (i, j).
—   **Adjoint** $\sim$: transpose to the matrix composed by cofactors of elements
         of the given matrix.
—   **Elementary** $\sim$: a matrix of one of the forms: $E + (\gamma - 1)E_{ii}$ ($\gamma \neq 0$)
         (type I), $E + \alpha E_{ij}$ ($\alpha \neq 0$, $i \neq j$) (type II); sometimes permutation
         matrices are also called elementary.
—   **Gram** $\sim$ (of a system of vectors $e_1, \dots, e_n$ in an Euclidean space): the
         matrix $((e_i, e_j))$ of size n.
—   **Hermitian** $\sim$: a complex matrix A such that $^t A = \bar{A}$ where $\bar{A}$ is a
         matrix obtained from $^t A$ by replacing its elements with the complex
         conjugates.

— **Nilpotent** $\sim$: a matrix some power of which is equal to a zero matrix (a nilpotent element of the ring of matrices).
— **Niltriangular** $\sim$: an upper-triangular matrix with zero entries at the principal diagonal.
— **Orthogonal** $\sim$: a matrix A such that $^t A = A^{-1}$.
— **Periodic** $\sim$: a matrix some power of which is equal to the identity matrix.
— **Permutation** $\sim$: a matrix obtained from the identity matrix by a permutation of its rows.
— **Skew-Hermitian** $\sim$: a complex matrix A such that $^t A = -\bar{A}$, where $\bar{A}$ is the matrix obtained from A by replacing its entries with the complex conjugates.
— **Skew-symmetric** $\sim$: a matrix A such that $^t A = -A$.
— **Symmetric** $\sim$: a matrix A such that $^t A = A$.
— **Triangular** $\sim$: an upper or lower triangular matrix.
— **Unimodular** $\sim$: a matrix with determinant 1.
— **Unitriangular** $\sim$: a triangular matrix with units at the principal diagonal.
— **Unitary** $\sim$: a complex matrix A such that $^t \bar{A} = A^{-1}$, where $^t \bar{A}$ is the matrix obtained from $^t A$ by replacing its elements with the complex conjugates.
— **Upper-triangular** $\sim$: a matrix whose entries below the principal diagonal are equal to 0.

Module
— **Completely reducible** $\sim$: see Exercise 6476.
— **Cyclic** $\sim$: a module with a fixed element m_0 such that any element m of the module can be represented in the form $m = am_0$ where a is an element of the ring.
— **Irreducible** $\sim$: a nonzero module in which the zero submodule and the whole module are the only submodules.
— **Reducible** $\sim$: a nonzero module having proper submodules.
— **Unitary** $\sim$: a module in which the unit of the ring acts identically.

Nilpotent element of a ring: an element some power of which is equal to 0.

Nilradical of a ring: the greatest (in the sense of the set-theoretical inclusion) two-sided ideal of the ring consisting of nilpotent elements.

Norm of an element of an algebra: see Exercise 6732.

Normalizer of a subgroup: the greatest subgroup in which the given subgroup is normal.

Normal closure of an element of a group: the greatest normal subgroup containing the given element.

Operator

— **Adjoint** $\sim$ (to an operator $\mathcal{A}$): the linear operator $\mathcal{A}^*$ such that $(\mathcal{A}x, y) = (x, \mathcal{A}^*y)$.

— **Hermitian** $\sim$: a linear operator $\mathcal{A}$ on an Hermitian space such that $(\mathcal{A}x, y) = (x, \mathcal{A}y)$ for all vectors x and y (i.e. $\mathcal{A}^* = \mathcal{A}$).

— **Normal** $\sim$: a linear operator on a metric space commuting with its adjoint operator.

— **Orthogonal** $\sim$: a linear operator $\mathcal{A}$ preserving the scalar product of vectors i.e. $(\mathcal{A}x, \mathcal{A}y) = (x, y)$ for all vectors x and y (in other words, $\mathcal{A}^* = \mathcal{A}^{-1}$).

— **Self-adjoint or symmetric** $\sim$: a linear operator on an Euclidean or Hermitian space, such that $(\mathcal{A}x, y) = (x, \mathcal{A}y)$ for all vectors x and y (i.e. $\mathcal{A}^* = \mathcal{A}$).

— **Semisimple** $\sim$: a linear operator such that any invariant subspace has an invariant complement subspace.

— **Skew-Hermitian** $\sim$: a linear operator A on an Hermitian space such that $(\mathcal{A}x, y) = -(x, A^*y)$ for all vectors x and y (i.e. $\mathcal{A}^* = -\mathcal{A}$).

— **Skew-symmetric** $\sim$: a linear operator $\mathcal{A}$ such that $(\mathcal{A}x, y) = -(y, \mathcal{A}x)$ for all vectors x and y (i.e. $\mathcal{A}^* = -\mathcal{A}$).

— **Unitary** $\sim$: a linear operator $\mathcal{A}$ on an Hermitian space preserving the scalar product of vectors, $(\mathcal{A}x, \mathcal{A}y) = (x, y)$ for all vectors x, y (i.e. $\mathcal{A}^* = \mathcal{A}^{-1}$).

Orbit of an element: the set of the images of an element under multiplication by all elements of the group.

Orientation of a vector space: see Exercise 5228.

Parallelepiped (with edges $a_1, \ldots, a_k$): the set of linear combinations $\sum_{i=1}^{k} \lambda_i a_i$ $(0 \leq \lambda_i \leq 1, i = 1, \ldots, k)$.

Partition of a number n: see Exercise 3129.

Permutation: a bijection from a finite set onto itself.

Period of a group: the least natural number n such that $x^n = e$ for all elements x of the group.

Periodic part of a group: the set of elements of group of finite orders.

Polynomial

— **Cyclotomic** $\sim \Phi_n(x)$: a polynomial $\prod_{k=1}^{\phi(n)}(x-\varepsilon_k)$, where $\varepsilon_1, \dots, \varepsilon_{\phi(n)}$ are primitive roots of 1 of order n.

— **Legendre** $\sim$: see Exercise 4344.

— **Characteristic** $\sim$ **of an element of an algebra**: see Exercise 6732.

— **Minimal** $\sim$ **of a linear operator**: a polynomial of the least degree annihilating the given operator; the minimal polynomial of the matrix of the operator.

— **— $\sim$ $\sim$ of a matrix**: a polynomial of the least degree annihilating the given matrix.

— **Separable** $\sim$: see Exercise 6742.

Projection (on a subspace U in parallel with a complement subspace V): a linear operator which maps a vector $x = u + v$ ($u \in U$, $v \in V$) to the vector u.

Quadratic reciprocity law: see Exercise 6811.

Quadric

— **k-planar** $\sim$: see Exercise 5331.

— **Plücker** $\sim$: see Exercise 5233.

Pseudoreflection: see Exercise 3913.

Quaternion: an element of the skew-field of quaternions.

— **Pure** $\sim$: a quaternion with a zero real part.

Ring

— **Completely reducible** $\sim$ (left, right): see Exercises 6471 and 6477.

— **$\sim$ of Gaussian integers**: the ring consisting of complex numbers $x + yi$ ($x, y \in \mathbb{Z}$):

— **$\sim$ of polynomials in noncommuting variables** $x_1, \dots, x_n$ (over a ring A): the set of all formal expressions

$$\sum a_{k_1 \dots k_m} x_{k_1} \cdots x_{k_m} \quad (a_{k_1 \dots k_m} \in A, \ m \geq 0, \ 1 \leq k_1, \dots, k_m \leq n)$$

with natural operations of addition and multiplication of monomials

$$a_{k_1 \dots k_m} x_{k_1} \cdots x_{k_m} \cdot b_{i_1 \dots i_s} x_{i_1} \cdots x_{i_s} = a_{k_1 \dots k_m} b_{i_1 \dots i_s} x_{k_1} \cdots x_{k_m} x_{i_1} \cdots x_{i_s},$$

which are extended to the sums by distributivity.

— **$\sim$ without zero divisors** (or a domain): a ring which has no nontrivial zero divisors.

— **Lie** $\sim$: see Exercise 6511.

— **Noetherian** $\sim$ (commutative): a commutative ring in which any strict ascending sequence of ideals is finite.

— **Simple** $\sim$: a ring with nonzero multiplication in which zero and the whole ring are the only two-sided ideals.

Reflection (in a space U in parallel with a complement subspace V): the linear operator which maps each vector $x = u + v$ ($u \in U$, $v \in V$) to the vector $u - v$.

Root (complex) of 1: a complex number some power of which is equal to 1.

   — $\sim$ **of order** n: a complex number, the nth power of which is equal to 1.

   —   **Primitive** $\sim$ **of order** n: a root of 1 of order n which is not a root of 1 of order less than n.

Semi-direct product of groups G **and** H: the set $G \times H$ with the operation

$$(x, y)(z, t) = (x \cdot \varphi(y)(z), yt),$$

where $\varphi : H \rightarrow \text{Aut}G$ is some homomorphism.

Skew-field (division ring) of quaternions: the vector space over the field $\mathbb{R}$ with a basis $1, i, j, k$, where 1 is a unit with respect to multiplication, $i^2 = j^2 = k^2 = -1$, $ij = -ji = k$, $jk = -kj = i$, $ki = -ik = j$; the algebra of generalized quaternions in which $\alpha = \beta = 1$.

Subgroup maximal: a subgroup which is not strictly contained in any proper subgroup.

Subspace

   —   **Complement** $\sim$ (to a subspace U): a subspace V such that the whole space is equal to $U \oplus V$.

   —   **Totally isotropic** $\sim$ (with respect to a symmetric or sesquilinear function $f(x, y)$): a subspace on which $f(x, y)$ vanishes.

Sylow p**-subgroup**: a maximal p-subgroup.

Symbol

   — $\sim$ **of Jacobi**: see Exercise 6806.

   — $\sim$ **of Kronecker**: $\delta_{ij} = 1$, $\delta_{ij} = 0$ if $i \neq j$ ($i, j = 1, \ldots, n$).

   — $\sim$ **of Legendre**: see Exercise 6806.

Symmetric difference: see Exercise 102.

Trace

   — $\sim$ **of a matrix**: the sum of the entries of the principal diagonal.

   — $\sim$ **of an element of an algebra**: see Exercise 6732.

   — $\sim$ **of an operator**: the trace of the matrix of the given operator.

Zero divisor in a ring: a nonzero element a for which there exists an element $b \neq 0$ such that $ab = 0$ (left zero divisor).

List of symbols

$^t A$	transpose of the matrix A
$\hat{A}$	adjoint matrix of the matrix A
$\mathcal{A}^*$	adjoint operator of the linear operator $\mathcal{A}$
$\mathbf{A}_n$	alternating group of degree n (the group of even permutations of the set $\{1, 2, \ldots, n\}$)
A_L	the algebra over an extension L of a field K which is obtained from an algebra A over K, $A_L = L \otimes_K A$
$\lvert A \rvert$	the number of elements in a set A
$[A, B]$	the commutator $AB - BA$ of matrices A and B
$\mathrm{Aut}\, G$	the group of automorphisms of a group G
Alt	an alternating operator in a space $\mathbf{T}_0^q(V)$
(a)	the ideal of the ring generated by the element a
$\langle a \rangle$	the subgroup (subring, subalgebra, subspace) generated by a
$\langle a \rangle_n$	the cyclic group of order n generated by a
$\arg z$	the argument of the complex number z; it is assumed that $0 \le \arg z < 2\pi$
$\mathbb{C}$	the set (field, additive group) of complex numbers
$\mathbf{D}_n$	the dihedral group (group of symmetries of regular n-gon)
$\mathbf{D}_n(A)$	the set of diagonal matrices of size n over the ring A
$\mathcal{D}$	the operator of differentiation in a functional space
$\mathrm{diag}(\lambda_1, \ldots, \lambda_n)$	the diagonal matrix with entries $\lambda_1, \ldots, \lambda_n$ at the principal diagonal
$\mathrm{End}\, A$	the endomorphism ring of the abelian group A (of the ring A)

e^A	the sum of the Taylor series of the function e^x at $x = A$ (A is a matrix)
E_{ij}	the matrix unit, i.e. a matrix in which the only nonzero entry is located at (i, j) and is equal to 1
$\mathbf{F}_q$	a field with q elements
G_a	the stabilizer of an element $a \in M$ under the action of a group G in a set M
G'	the commutant of group G
$\mathbf{GL}(V)$	the group of nonsingular linear operators on the vector space V
$\mathbf{GL}_n(F)$	the group of nonsingular linear operators on an n-dimensional vector space over a field F; the group of nonsingular matrices of size n over F
$\mathbf{GL}_n(q)$	the same as $\mathbf{GL}_n(\mathbf{F}_q)$
$\mathbb{H}$	the skew-field (division ring) of quaternions
$\mathrm{Hom}(A, B)$	the group of homomorphisms from the group A into the abelian group B
K^*	the group of invertible elements of the ring K
$K(a)$	the extension of the field K obtained by addition of an element a
$F[G]$	the group algebra of the group G over the field K
$K[x]$	the ring of polynomials in a variable x with coefficients in the ring K
$K[x]_n$	the set of polynomials from the ring $K[x]$ of degree at most n
$K(x)$	the field of rational functions in a variable x with coefficients in the field K
$K[[x]]$	the power series ring in a variable x with coefficients in the ring K
$K[x_1, \ldots, x_n]$	the ring of polynomials in variables $x_1, \ldots, x_n$ with coefficients in the ring K
$K\{x_1, \ldots, x_n\}$	the ring of polynomials in non-commuting variables $x_1, \ldots, x_n$ with coefficients in the ring K
$L(V)$	the set of linear operators in the vector space V
$\ln A$	the sum of the Taylor series of the function $\ln(1-x)$ at $x = E - A$ (A is a matrix)

$\mathbf{M}_n(K)$	the ring (the algebra) of matrices of size n over the ring K
$\mathbb{N}$	the set of natural numbers
$N(A)$	the nilradical of the algebra A
$N(H)$	the normalizer of the subgroup H
$\mathbf{N}_{A/K}(a)$	the norm of an element a in the algebra A over the field K
$n\mathbf{Z}$	the set of integers divisible by the number n
$\mathbf{O}_n(K)$	the group of orthogonal matrices of size n over the field K
$\mathbb{Q}$	the set (the field, additive group) of rational numbers
$\mathbb{Q}_p$	the field of p-adic numbers
$\mathbb{R}_+$	the set (multiplicative group) of positive real numbers
$\mathrm{rk}\,A$	the rank of the matrix A
$\mathrm{rk}\,\mathcal{A}$	the rank of the linear operator $\mathcal{A}$
$\langle S \rangle$	the subgroup (the subring, subalgebra, subspace) with a set of generators S; the affine hull of the set S
$\mathbf{S}_n$	the symmetric group of degree n (the group of permutations of the set $\{1, \ldots, n\}$)
$\mathbf{S}_X$	the group of one-to-one transformations of the set X
$\mathbf{SL}_n(K)$	the group of matrices with determinants 1 over the field K
$\mathbf{SL}_n(q)$	the same as $\mathbf{SL}_n(\mathbf{F}_q)$
$\mathbf{SO}_n(K)$	the group of orthogonal matrices with determinants 1 over the field K
$\mathbf{SU}_n(\mathbb{C})$	the group of unitary complex matrices with determinants 1
$\mathbf{SU}_n$	the same as $\mathbf{SU}_n(\mathbb{C})$
$S(V)$	the symmetric algebra of the vector space V
$S^q(V)$	the qth symmetric power of the vector space V
Sym	the operator of symmetrization in the space $\mathbf{T}_0^q(V)$
$\mathbf{T}(V)$	the tensor algebra of the vector space V
$\mathbf{T}_p^q(V)$	the vector space of tensors of type (p, q) in the vector space V
$\mathrm{tr}\,A$	the trace of the matrix A
$\mathrm{tr}\,\mathcal{A}$	the trace of the linear operator $\mathcal{A}$

$\mathrm{tr}_{A|K}(a)$	the trace of an element a of the algebra A over the field K
$\mathbf{U}$	the group of complex numbers with absolute value 1
$\mathbf{U}_n$	the group of complex roots of order n of 1
$\mathbf{U}_{p^\infty}$	the group of complex roots of order p^n of 1 $(n \in \mathbb{N})$ (p is a prime)
U°	the orthocomplement of the subset U in a dual vector space
$U^\perp$	the orthocomplement of the subset U of a vector space with respect to a given bilinear function
$\mathbf{UT}_n(K)$	the group of unitriangular matrices of size n over the field K
$\mathbf{V}_4$	the Klein group
V^*	the vector space dual to the space V
$V(a_1, \ldots, a_k)$	the volume of the parallelepiped with edges $a_1, \ldots, a_k$
$x \wedge y$	the product of elements x, y in the Grassman algebra of a vector space
$\mathbb{Z}$	the set (the ring, additive group) of integers; infinite cyclic group
$\mathbf{Z}_n$	the cyclic group of order n, the residue ring modulo n
$\mathbb{Z}_p$	the ring of p-adic integers
$\mathbb{Z}[i]$	the ring of Gaussian integers
$\sqrt[n]{z}$	the set of complex roots of order n of the number $z \in \mathbb{C}$
$\mu(n)$	the Möbius function
$\mu(a)$	the minimal polynomial of the algebraic element a
$\Lambda(V)$	the external algebra (Grassman algebra) of the vector space V
$\Phi_n(x)$	the cyclotomic polynomial $\prod_{k=1}^{\phi(n)}(x-\varepsilon_k)$, where ε_k is a primitive root of order n of 1 $(k = 1, \ldots, \phi(n))$
$\phi(n)$	the Euler function
$\chi_{A|K}(a, x)$	the characteristic polynomial of an element a of the algebra A over the field K
1_X	the identity mapping on the set X
2^X	the set of all subsets of the set X

Index

461